Texts in Applied Mathematics 27



**Springer**
*New York*
*Berlin*
*Heidelberg*
*Barcelona*
*Budapest*
*Hong Kong*
*London*
*Milan*
*Paris*
*Santa Clara*
*Singapore*
*Tokyo*

## Texts in Applied Mathematics

1. *Sirovich:* Introduction to Applied Mathematics.
2. *Wiggins:* Introduction to Applied Nonlinear Dynamical Systems and Chaos.
3. *Hale/Koçak:* Dynamics and Bifurcations.
4. *Chorin/Marsden:* A Mathematical Introduction to Fluid Mechanics, 3rd ed.
5. *Hubbard/West:* Differential Equations: A Dynamical Systems Approach: Ordinary Differential Equations.
6. *Sontag:* Mathematical Control Theory: Deterministic Finite Dimensional Systems.
7. *Perko:* Differential Equations and Dynamical Systems, 2nd ed.
8. *Seaborn:* Hypergeometric Functions and Their Applications.
9. *Pipkin:* A Course on Integral Equations.
10. *Hoppensteadt/Peskin:* Mathematics in Medicine and the Life Sciences.
11. *Braun:* Differential Equations and Their Applications, 4th ed.
12. *Stoer/Bulirsch:* Introduction to Numerical Analysis, 2nd ed.
13. *Renardy/Rogers:* A First Graduate Course in Partial Differential Equations.
14. *Banks:* Growth and Diffusion Phenomena: Mathematical Frameworks and Applications.
15. *Brenner/Scott:* The Mathematical Theory of Finite Element Methods.
16. *Van de Velde:* Concurrent Scientific Computing.
17. *Marsden/Ratiu:* Introduction to Mechanics and Symmetry.
18. *Hubbard/West:* Differential Equations: A Dynamical Systems Approach: Higher-Dimensional Systems.
19. *Kaplan/Glass:* Understanding Nonlinear Dynamics.
20. *Holmes:* Introduction to Perturbation Methods.
21. *Curtain/Zwart:* An Introduction to Infinite-Dimensional Linear Systems Theory.
22. *Thomas:* Numerical Partial Differential Equations: Finite Difference Methods.
23. *Taylor:* Partial Differential Equations: Basic Theory.
24. *Merkin:* Introduction to the Theory of Stability.
25. *Naber:* Topology, Geometry, and Gauge Fields: Foundations.
26. *Polderman/Willems:* Introduction to Mathematical Systems Theory: A Behavioral Approach.
27. *Reddy:* Introductory Functional Analysis: with Applications to Boundary Value Problems and Finite Elements.

B. Daya Reddy

# Introductory Functional Analysis

## With Applications to Boundary Value Problems and Finite Elements

With 145 Illustrations

Springer

B. Daya Reddy
Department of Mathematics and
Applied Mathematics
University of Cape Town
7700 Rondebosch
South Africa

*Series Editors*



Mathematics Subject Classification (1991): 46-01, 65N30

Library of Congress Cataloging-in-Publication Data
Reddy, B. Dayanand, 1953–
Introductory functional analysis : with applications to boundary value problems and finite elements / B. Daya Reddy.
p. cm. — (Texts in applied mathematics ; 27)
Includes bibliographical references and index.
ISBN 0-387-98307-4 (alk. paper)
1. Functional analysis. I. Title. II. Series.
QA320.R433 1997
515′.7—dc21 97-24052

Printed on acid-free paper.



Production managed by Anthony K. Guardiola; manufacturing supervised by Joe Quatela.
Camera-ready copy prepared from the author's LaTeX files.
Printed and bound by Maple-Vail Book Manufacturing Group, York, PA.
Printed in the United States of America.

9 8 7 6 5 4 3 2 1

ISBN 0-387-98307-4 Springer-Verlag New York Berlin Heidelberg SPIN 10557902

# Series Preface

Mathematics is playing an ever more important role in the physical and biological sciences, provoking a blurring of boundaries between scientific disciplines and a resurgence of interest in the modern as well as the classical techniques of applied mathematics. This renewal of interest, both in research and teaching, has led to the establishment of the series: *Texts in Applied Mathematics* (*TAM*).

The development of new courses is a natural consequence of a high level of excitement on the research frontier as newer techniques, such as numerical and symbolic computer systems, dynamical systems, and chaos, mix with and reinforce the traditional methods of applied mathematics. Thus, the purpose of this textbook series is to meet the current and future needs of these advances and encourage the teaching of new courses.

*TAM* will publish textbooks suitable for use in advanced undergraduate and beginning graduate courses, and will complement the *Applied Mathematical Sciences* (*AMS*) series, which will focus on advanced textbooks and research level monographs.

# Preface

A proper understanding of the theory of boundary value problems, as opposed to a knowledge of techniques for solving specific problems or classes of problems, requires some background in functional analysis. The same is true of the finite element method: there is much that can be learned and practised – for example, the basic theory of the method, computational aspects, and so on – without knowledge of even the most basic notions of functional analysis. But for anyone wishing to gain a proper understanding of qualitative aspects of boundary value problems, or of aspects of the finite element method such as those that lead to the development of error estimates, some background in functional analysis is an essential prerequisite.

The issue of an adequate mathematical background is somewhat more straightforward in the case of students of mathematics who have taken courses in real and complex analysis, followed by a course in functional analysis. Such students are ideally equipped to follow courses that deal with existence theory for boundary value problems, and with qualitative aspects of the finite element method. This text has arisen out of a recognition, though, that there are many students, researchers, and practitioners who have not been exposed to the kind of mathematical background just referred to, but who nevertheless wish to become acquainted with the basic notions of functional analysis and its application to the kinds of problems that arise typically in physics and engineering.

Up to the mid-1970s the availability of source material to which such individuals could refer, at least in the English language, was limited almost entirely to the standard texts on real and functional analysis, written by mathematicians for mathematicians having the standard background. The task facing the engineer or applied scientist was thus quite daunting. Fortunately the situation has progressed markedly since then. There is now available a wide range of texts that present functional analysis, often with one or more applications taken from engineering and physics, in a manner accessible to readers not having the standard prerequisites. The styles differ, sometimes quite considerably, from one text to another, although this is not a bad thing given the diversity of interests and backgrounds of the potential readership.

This text is a further addition to the set of books that present functional analysis and its applications to nonspecialists. The approach taken is, first, to assume that readers have no more by way of relevant background than elementary courses in linear algebra, vector analysis, and differential equations, and wish to learn the elements of linear functional analysis.

The book begins with an introductory chapter, which is somewhat in the nature of a prologue, and which presents in mostly descriptive form a motivation for studying functional analysis from the viewpoint of those involved in the study of problems from physics and engineering. The remainder of the book is then divided into parts: Part I is devoted to linear functional analysis, Part II to an introduction to elliptic boundary value problems, and Part III comprises a study of the finite element method.

Two applications are treated in detail in this text: elliptic boundary value problems and the finite element method. In both cases any prior exposure to these areas will represent an advantage to those using this book; indeed, it is expected that such prior exposure will in many cases have provided the motivation to study the material presented here. The presentation of these applications starts more or less at the beginning, so that those having no background in these areas could use this text to acquire such background. On the other hand, it may be the case that the motivation to learn functional analysis arises from an interest in an area of application other than those treated in this text. Such readers might well prefer to focus on Part I of the book.

The incorporation of applications and other illustrative material is approached in two distinct ways. In Part I of the book new concepts, often of an abstract nature, are rendered more accessible by the copious use of concrete worked examples. There is little reference in this part of the book to applications in physics and engineering, for the simple reason that such examples are less well suited to laying bare the essential features of the many new concepts that accompany any introduction to functional analysis. In Parts II and III, it is appropriate and desirable to illustrate abstract concepts by recourse to concrete problems and examples taken from physics and engineering, and this is the approach taken here. I have used as examples problems such as heat conduction, as well as problems in solid and structural mechanics – elasticity, beams, and plates – and return regularly in Parts II and III to these examples in order to motivate and illustrate aspects of the theory of elliptic boundary value problems, and finite elements.

The style adopted in this text differs from that to be found in most texts on analysis, in that it is adapted to the goal of making the subject matter accessible. Thus proofs are sometimes omitted when these are felt to shed little additional light on the relevant topic. It will also be found in places that the presentation of detailed mathematical argument is eschewed in favor of a more descriptive approach, again for the purpose of rendering the material more accessible.

In addition to the many examples, each chapter ends with a collection of exercises for the reader. Some of these consolidate material presented in the chapters, and many exercises serve the purpose of amplification and supplementation. In both cases the exercises are to be regarded as an essential component of the text. Solutions to most of the exercises are presented at the end of the text.

Many individuals have assisted, in various ways, in the completion of this book. I am particularly grateful to Christiaan le Roux, Jean Lubuma, and Sizwe Mabizela, all of whom gave most generously of their time in reading and criticizing a preliminary version of the text. They offered detailed criticism on aspects of style and substance, and pointed out a number of errors. David Davidson organized a study group which worked through most of the book; I found the comments of this group of engineering scientists very helpful indeed. Weimin Han deserves special thanks for his constructive suggestions; so too does Brendt Wohlberg, who offered many suggestions for improving the text, located errors, and also provided me with valuable advice on the preparation of figures by computer. Shaun Courtney's expert guidance in the mysteries of Unix, and his very willing assistance with a variety of LaTeX problems, are much appreciated. Most of the figures were prepared by Bruce Bassett and Jill Goode, while Diane Laugksch assisted me in the typing of drafts of sections of the book. I am most grateful to these individuals for their cheerful assistance.

I express my thanks to the staff at Springer-Verlag New York for their expert guidance and assistance with editorial aspects, as well as their advice on the the preparation of the manuscript using LaTeX.

Finally, I acknowledge with gratefulness the moral support and forbearance of my wife Shaada and son Jordi, who have had to spend many evenings and weekends without my company in order that I might bring this project to fruition.

B.D.R.
Cape Town
April 1997

# Contents

# Introduction

The usefulness of functional analysis may not be immediately evident to users of mathematics who have hitherto not encountered this branch of the subject. Indeed physicists, engineering scientists, and other applied mathematicians are often put off by what they perceive to be an unnecessarily high degree of abstraction inherent in functional analysis, and the conclusion is often reached that such a branch of mathematics could not possibly be of any use in an area of endeavor in which concrete solutions to concrete problems are sought.

However, there are many areas in which a knowledge of functional analysis is indispensable if one hopes to be able to probe deeply into the nature of a problem. In this book we try to convey some idea of the circumstances under which the student or researcher, equipped with not much more than the basics of functional analysis, can gain a great deal of insight into the properties of boundary value problems and their approximation. Not that what is being proposed is in any way a panacea: while learning about the power of functional analysis, it is equally important to be aware of its limitations; in other words, it is important to know which questions can conceivably be answered by adopting such an approach, and which cannot.

The introduction to functional analysis presented in this text is directed somewhat towards those aspects of the subject that are relevant to the qualitative treatment of boundary value problems and their approximation by finite elements. The precise manner in which one may call upon functional analysis as a useful tool in these applications is the subject of Parts II and III of this work. However, it is rather unsatisfactory to postpone until then an indication of how this branch of mathematics interacts with such applications, and in which ways it is useful. For this reason we present in this introductory chapter an overview of how boundary value problems are encountered, what kinds of mathematical questions arise in their treatment, and where functional analysis fits into the general scheme of things. The treatment is deliberately sketchy in its mathematical detail, since the aim here is to identify important beacons or landmarks, rather than to flesh out all their mathematical features; this latter task forms the bulk of this text.

Boundary value problems almost always arise as mathematical models of some real-life situation, whether of a physical, biological, economic, or other nature. We place the planned excursion in a concrete physical context in order to be able to show how the mathematics interacts with the physical dictates of the problem. The main vehicle chosen for the discussion in this chapter is the physical problem of *heat conduction* or, equivalently, of *diffusion* and, subsequently, its steady (that is, time-independent) variants. When arriving at the steady-state case we are able to make contact also with other problems that have the same mathematical formulation, viz. *electrostatics*, and the problem of *the deflection of an elastic membrane.*

We proceed now to examine the various stages that arise in the consideration of these physical problems, and their mathematical realizations.

## STAGE I: CONSTRUCTION OF A MATHEMATICAL MODEL

**Example 1: A model for heat conduction.** Consider a medium through which heat is flowing. The aim is first to construct a mathematical model of this physical problem which, on the one hand, is a sufficiently accurate representation of the situation, yet which is simple enough to yield to mathematical analysis.

Positions of points in the medium are denoted by the position vector $\boldsymbol{x}$ relative to some origin 0. A Cartesian coordinate system is chosen with origin at 0, so that the coordinates of the point $\boldsymbol{x}$ are $(x, y, z)$. Here we use the vector $\boldsymbol{x}$ and the triple $(x, y, z)$ interchangeably. If the time is denoted by $t$, then the aim of the exercise is to find the temperature distribution $u(\boldsymbol{x}, t)$ in the body, due to the presence of heat sources within the body and flux of heat through its surface (see Figure 1). Two equations suffice for a realistic model of this situation: an equation representing balance (or conservation) of energy, and a constitutive equation, which contains information about how heat flows in the medium itself.

Balance of energy states the following: assuming that there are no other types of energy present,

$$\begin{aligned}&\textit{The rate of change of thermal energy in a body} =\\ &\quad\textit{the heat generated by sources in the body} +\\ &\quad\textit{the flow of heat into the body from outside.}\end{aligned} \tag{1}$$

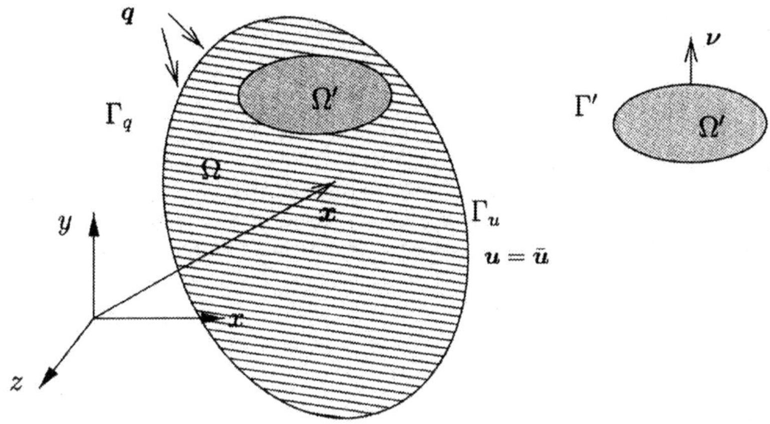


FIGURE 1. The problem of heat conduction

The next stage is to translate each of these terms into mathematical form. We do this by applying balance of energy to an arbitrary part $\Omega'$ of the body $\Omega$; the arbitrary region has a bounding surface $\Gamma'$ (see Figure 1). Now the thermal energy in a body is quantified by the *heat capacity* $c$, which is the amount of heat generated per unit mass, and per unit rise in temperature. If the mass density is denoted by $\rho$ and the temperature by $u$, then the total thermal energy in $\Omega'$ at a particular time is therefore given by

$$\int_{\Omega'} c(\boldsymbol{x})\rho(\boldsymbol{x})u(\boldsymbol{x},t)\; dV. \tag{2}$$

We emphasize here that the body is *nonhomogeneous*; that is, its properties vary with position in the body, so that $c$ and $\rho$ are functions of position. In the preceding expression, and henceforth, $dV$ denotes the volume element $dx\,dy\,dz$ and $\int_{\Omega'}$ is shorthand for the triple integral $\iiint_{\Omega'}$. In this or other problems in which the domain is two-dimensional, $dV$ is interpreted as the area element $dx\,dy$, and $\int_{\Omega'}$ as the area integral $\iint_{\Omega'}$.

Heat may be generated inside the body as a result of a heat source (for example, a chemical reaction). This is given in the form of a function $f(\boldsymbol{x},t)$, which represents the amount of heat generated per unit volume, per unit time. Thus the total heat generated by such a source in $\Omega'$ is given by

$$\int_{\Omega'} f(\boldsymbol{x},t)\; dV. \tag{3}$$

Finally, the flux or flow of heat is represented by a vector $\boldsymbol{q}(\boldsymbol{x},t)$, called the *heat flux*, which specifies both the magnitude and direction of the flow of heat. The flow of heat across the surface $\Gamma'$ *into* the part $\Omega'$ of the body is given by

$$-\oint_{\Gamma'} \boldsymbol{q} \cdot \boldsymbol{\nu} \, dA, \tag{4}$$

since it is only the normal component of heat flux that will actually enter the body. Here $\boldsymbol{\nu}$ is the outward unit normal to the boundary and $dA$ denotes the element of surface area. This surface integral may be converted into a volume integral by using the divergence theorem of Gauss, according to which

$$-\oint_{\Gamma'} \boldsymbol{q} \cdot \boldsymbol{\nu} \, dA = -\int_{\Omega'} \operatorname{div} \boldsymbol{q} \, dV.$$

By putting together these various components of the equation of balance of energy (1) we therefore obtain the equation

$$\frac{d}{dt} \int_{\Omega'} c(\boldsymbol{x}) \rho(\boldsymbol{x}) u(\boldsymbol{x}, t) \, dV = \int_{\Omega'} f(\boldsymbol{x}, t) \, dV - \int_{\Omega'} \operatorname{div} \boldsymbol{q} \, dV.$$

Now the time derivative may be taken inside the integral, since the limits of integration are fixed; it then becomes the partial derivative $\partial/\partial t$, and we now have, upon rearrangement,

$$\int_{\Omega'} \left[ c(\boldsymbol{x}) \rho(\boldsymbol{x}) \frac{\partial u}{\partial t} + \operatorname{div} \boldsymbol{q} - f(\boldsymbol{x}, t) \right] \, dV = 0.$$

Since the volume under consideration is arbitrary, and the functions appearing in the integrand are assumed to be sufficiently smooth, the integrand must vanish in order for this equation to hold true. This observation leads to the preliminary form

$$c\rho \frac{\partial u}{\partial t} + \operatorname{div} \boldsymbol{q} = f \tag{5}$$

of the heat equation.

Clearly a further equation is required, since there are two unknowns: the temperature $u$ and the heat flux $\boldsymbol{q}$. Physically also, it is clear that we need another equation that will characterize the particular heat-conducting properties of the material under consideration. We choose a simple form of such an equation, viz. *Fourier's law*, which states that the heat flux is linearly related to the temperature gradient; that is,

$$\boldsymbol{q} = -K \nabla u, \tag{6}$$

where the positive scalar function $K$ is known as the *thermal conductivity*. The minus sign is introduced to accommodate the fact that heat flows from hot to cold. Now substitution of Fourier's law in the energy equation and division throughout by $c\rho$ give, finally,

$$\frac{\partial u}{\partial t} - \frac{1}{c\rho}\operatorname{div}(K\nabla u) = Q, \tag{7}$$

where we have set $Q = f/(c\rho)$. In full, equation (7) reads

$$\frac{\partial u}{\partial t} - \frac{1}{c\rho}\left[\frac{\partial}{\partial x}\left(K\frac{\partial u}{\partial x}\right) + \frac{\partial}{\partial y}\left(K\frac{\partial u}{\partial y}\right) + \frac{\partial}{\partial y}\left(K\frac{\partial u}{\partial y}\right)\right] = Q.$$

This partial differential equation (PDE) is the standard *heat equation* in its full unsteady, nonhomogeneous form.

To the PDE must be added information about the conditions on the boundary, and also initial conditions. There are many kinds of boundary conditions (BCs): for example, suppose that the temperature is prescribed on a part $\Gamma_u$ of the boundary, whereas on the remainder $\Gamma_q$ the heat flux is given (Figure 1). That is,

$$u = \bar{u}(\boldsymbol{x}, t) \text{ on } \Gamma_u \quad \text{and} \quad \boldsymbol{q} \cdot \boldsymbol{\nu} = \bar{q} \text{ on } \Gamma_q,$$

where $\bar{u}$ and $\bar{q}$ are prescribed functions. The second of these conditions can be simplified, using Fourier's law once again, so that it reads

$$\frac{\partial u}{\partial \nu} = g \text{ on } \Gamma_q,$$

where $g = -\bar{q}/K$ and $\partial/\partial\nu = \boldsymbol{\nu} \cdot \nabla$ denotes the normal derivative. In the event that the part $\Gamma_q$ of the boundary is *insulated*, $g = 0$.

Finally, we add an *initial condition* (IC), which specifies the temperature at time $t = 0$; that is,

$$u(\boldsymbol{x}, 0) = u_0(\boldsymbol{x}),$$

with $u_0$ being a given function. We have now arrived at an *initial boundary value problem* (IBVP) for heat conduction, which may be succinctly summarized as in Box 1.

**Box 1: The Initial Boundary Value Problem for Heat Conduction**

PDE: $$\frac{\partial u}{\partial t} - \frac{1}{c\rho}\operatorname{div}(K\nabla u) = Q \quad \text{in } \Omega,\ t > 0$$

BCs: $$u = \bar{u}(\boldsymbol{x}, t) \text{ on } \Gamma_u \quad \text{and} \quad \frac{\partial u}{\partial \nu} = g(\boldsymbol{x}, t) \text{ on } \Gamma_q$$

IC: $$u(\boldsymbol{x}, 0) = u_0(\boldsymbol{x}) \text{ in } \Omega$$

Later on, when discussing elliptic problems and their approximation by finite elements, we are more concerned with problems that are independent of time. This is the so-called *steady case* which is appropriate if, for example, the data such as the source term and the boundary terms are independent of time. In this case the time derivative disappears from the PDE, the initial condition is redundant, and we are left with a *boundary value problem* (BVP) for $u(\boldsymbol{x})$, summarized in Box 2.

> **Box 2: The Boundary Value Problem for Steady Heat Conduction**
>
> PDE: $$-\frac{1}{c\rho}\operatorname{div}(K\nabla u) = Q$$
>
> BCs: $u = \bar{u}(\boldsymbol{x})$ on $\Gamma_u$ and $\dfrac{\partial u}{\partial \nu} = g(\boldsymbol{x})$ on $\Gamma_q$

Finally, this BVP takes an even simpler form if the problem is *homogeneous*, that is, if the density, specific heat, and thermal conductivity are constant. The problem now becomes that shown in Box 3; the PDE there is known as the *Poisson equation* and the operator $\nabla^2$ on the left-hand side is the *Laplacian*, defined by

$$\nabla^2 u = \frac{\partial^2 u}{\partial x^2} + \frac{\partial^2 u}{\partial y^2} + \frac{\partial^2 u}{\partial z^2}. \tag{8}$$

The constant $k = K/(c\rho)$ is known as the *thermal diffusivity.* If there is no source term, so that the right-hand side of the PDE is zero, then the resulting equation is known as *Laplace's equation.*

> **Box 3: Poisson's Equation**
>
> PDE: $$-k\nabla^2 u = Q$$
>
> BCs: $u = \bar{u}(\boldsymbol{x})$ on $\Gamma_u$ and $\dfrac{\partial u}{\partial \nu} = g(\boldsymbol{x})$ on $\Gamma_q$

**Example 2: Electrostatics.** It is important to realize that PDEs such as those in Boxes 1 through 3 do not represent only one physical situation. As mentioned earlier, the heat equaton is also known as the diffusion equation because it serves as a model for diffusion. Likewise, the Poisson equation models a wide range of physical phenomena.

To make the point we consider as a further example the case of *electrostatics.* Suppose that we are given a distribution of stationary electric charges in a region $\Omega$ in space; this distribution may be specified by a scalar function $\rho$ which gives the charge per unit volume, or charge density, at any point. The charge density in turn gives rise to a vector force field known as the *electric field*, and denoted by $\boldsymbol{E}$. The electric field at a point $\boldsymbol{x}$ gives the force per unit charge acting on a charge located at $\boldsymbol{x}$.

Now, just as we considered in Example 1 the relationship between the flux of heat through the boundary $\Gamma'$ of an arbitrary region $\Omega'$ and the change of heat inside that region, in the same way we may consider the relationship between the flux of the electric field through $\Gamma'$ (using the same notation as in Example 1), and the total charge inside the region enclosed by $\Gamma'$. The result is *Gauss's law*, which states that

$$\int_{\Gamma'} \boldsymbol{E} \cdot \boldsymbol{\nu} \, dA = 4\pi \int_{\Omega'} \rho \, dV; \tag{9}$$

that is, the flux of the electric field $\boldsymbol{E}$ through any closed surface equals $4\pi$ times the total charge enclosed by that surface. By exploiting the divergence theorem of Gauss, the surface integral on the left-hand side of (9) can be converted to a volume integral, and this way we arrive at the counterpart of (5); that is,

$$\operatorname{div} \boldsymbol{E} = 4\pi q. \tag{10}$$

We require some additional information in order to solve this problem since (10) is a single equation involving an unknown vector-valued quantity. The requisite information is provided by the fact that the line integral of the electric field between any two points in space is *path-independent*; that is, given points $\boldsymbol{x}_1$ and $\boldsymbol{x}_2$ and a curve $\boldsymbol{x}(s)$ joining these two points, with $\boldsymbol{x}_1(s) = s_1$ and $\boldsymbol{x}_2(s) = s_2$, the value of the integral

$$\int_{s_1}^{s_2} \boldsymbol{E} \cdot \boldsymbol{\tau} \, ds$$

is independent of the curve chosen to join $\boldsymbol{x}_1$ and $\boldsymbol{x}_2$. Here $\boldsymbol{\tau}$ is the unit tangent vector along the curve (Figure 2). An immediate mathematical consequence is that it is possible to express the electric field as the gradient of an *electric potential* function $\phi$; that is,

$$\boldsymbol{E} = -\nabla \phi. \tag{11}$$

The minus sign takes care of the fact that the electric field points in the direction of decreasing potential. Figure 2 shows schematically the curves

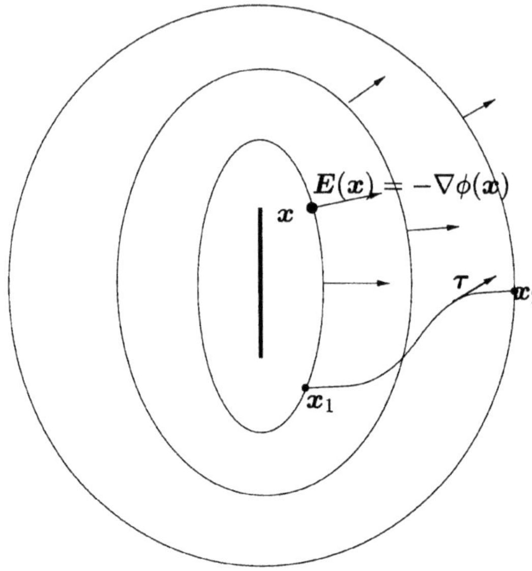


FIGURE 2. Curves of constant potential and electric field vectors in the plane normal to, and passing through the center of, a uniformly charged disk

of constant potential, and a few of the electric field vectors, in the vicinity of a uniformly charged disk.

Now we are ready to formulate the problem of determining the electric field: by substituting (11) in (10) we obtain again a *Poisson equation*

$$-\nabla^2\phi = 4\pi\rho. \tag{12}$$

Naturally this differential equation will have to be supplemented by suitable boundary conditions involving either the potential or the electric field; we then arrive at a problem exactly as that given in Box 3, with the obvious changes in notation.

Once $\phi$ has been found from (12), the electric field can then be determined from (11).

We see then that two problems which differ vastly in a physical sense have a common mathematical structure.

BOX 4: THE COMMON STRUCTURE OF THE HEAT CONDUCTION AND ELECTROSTATIC PROBLEMS

| | Heat Conduction | Electrostatics |
|---|---|---|
| Basic variable $u$ or $\phi$ | temperature | potential |
| Flux quantity $\boldsymbol{q}$ or $\boldsymbol{E}$ | heat flux | electric field |
| Balance law | thermal energy | Gauss's law |
| Constitutive law | Fourier's law | potential law |

This commonality is summarized in Box 4. Of course the relation (11) is not a constitutive law; indeed, the region in question is assumed to be a vacuum! But (11) plays the same role mathematically as does Fourier's law; so it is not inappropriate to place it alongside Fourier's law in the table.

**Example 3: A model for deformation of a membrane.** We conclude the set of examples in this chapter with one which is discussed again in later chapters: this is the problem of determining the shape at equilibrium of a thin elastic membrane. Such a membrane is initially planar, and occupies a two-dimensional domain $\Omega$, as shown in Figure 3. It is fixed along part of its boundary. The membrane is subjected to a transverse force $f$ per unit area, as a result of which it takes up a nonplanar shape. The problem is to find the deformed shape of the membrane, which is given by the function $u(\boldsymbol{x})$.

In this problem the main unknown is the transverse displacement, again represented by $u$ (see Figure 3). The terms and equations introduced for the heat problem remain valid, provided that they are interpreted correctly. First, all functions of time alone vanish, since we are dealing with a steady problem. Second, balance of energy is replaced by the principle of balance of forces (actually, this is the time-independent version of the principle of balance of momentum), which states that the net total force acting on any part of the membrane is zero.

A constitutive equation is required in order to characterize the behavior of the material comprising the membrane. This equation now expresses the fact that the vertical force depends not on the displacement $u$, but

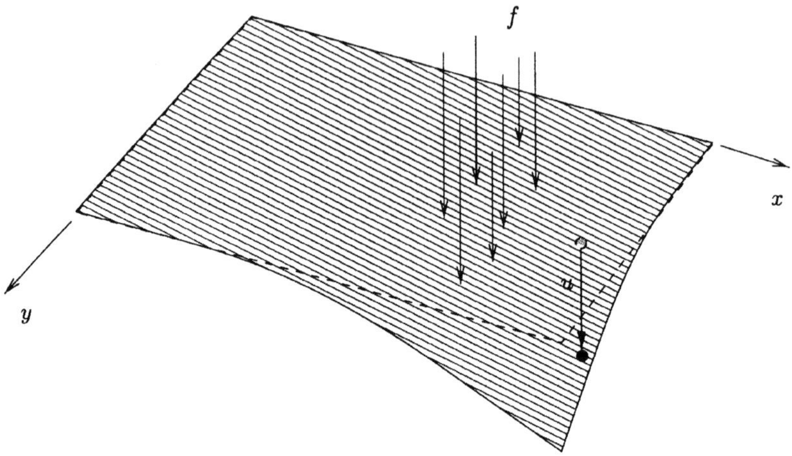


FIGURE 3. Deformation of a thin elastic membrane

rather on the *displacement gradient* $\nabla u$, which is what characterizes local deformation of the membrane.

By considering the balance of forces acting on an arbitrary section of the membrane, as shown in Figure 3, we eventually arrive at the PDE in Box 2 (if the membrane is nonhomogeneous) or the Poisson equation of Box 3 (if it is homogeneous).

BVPs, as opposed to IBVPs, are particularly relevant to later developments, since they are representative examples of elliptic equations (the notion of an elliptic problem is explored in some depth in Chapter 8). In order to focus the rest of this discussion on the main goals, we proceed further by taking one of the BVPs, rather than the original IBVP, as a representative example.

**An alternative formulation for BVPs: the variational problem.** The preceding developments lead to a particular kind of mathematical model, viz. one involving a PDE (plus boundary and possibly also initial conditions). But although this is a commonly adopted form of the model, it is not unique. There are other, more or less equivalent, ways of putting the model into mathematical form, and it may be that one of the alternatives would be more appropriate, depending on how we would want to pursue the mathematical investigation, and depending also on the types of approximations that we might wish to – or be forced to – consider.

An important alternative, and one which is featured heavily later on since it is at the heart of the finite element method of approximation, is that of the *variational problem*. Traditionally this meant that the problem was formulated as one in which it was required to mimimize a particular functional; however, the term "variational" has a wider significance that is explored in Chapter 9. We begin the discussion here with the original understanding of a variational problem as being the same thing as a minimization problem, and subsequently consider an alternative variational formulation.

This exploration is carried out in the context of the simple boundary value problem associated with the Poisson equation. We simplify matters even further by assuming that the temperature (or the potential in the case of electrostatics, or the displacement in the case of the membrane problem) is prescribed to be zero on the entire boundary.

The first stage is to introduce a *functional* $J$, that is, an operator that maps a function $v(\boldsymbol{x})$ to a real number, defined by

$$J(v) = \tfrac{1}{2}\int_\Omega |\nabla v|^2 \, dV - \int_\Omega fv \, dV. \tag{13}$$

In the context of the membrane problem, for example, this represents, to within a constant, the total potential energy of the membrane; the first term on the right-hand side represents the strain energy or stored energy due to deformation, and the second term represents the potential energy of the force.

Now it can be shown (and this is done in Chapter 9) that the problem of Box 3, with the modified boundary condition, is equivalent to the following.

*Find $u$ such that* $J(u) \leq J(v)$ *for all admissible $v$.*

Again, in the context of the membrane problem, this represents the statement of the principle of minimum potential energy. Exactly what is meant by an admissible function is a matter that takes up some time when BVPs are discussed in full detail, but it suffices for this preliminary overview that we consider functions which satisfy two properties:

(i) they are continuously differentiable; that is, the functions and their derivatives are continuous on $\overline{\Omega}$, where $\overline{\Omega}$ denotes the region $\Omega$ together with its boundary $\Gamma$. In this way the integrand in (13) makes sense;

(ii) they satisfy the boundary condition $u = 0$ on $\Gamma_u$.

Then the minimization problem may be summarized concisely as in Box 5.

**Box 5: Minimization Problem for the Poisson Equation**

Find a function $u$ in $X$ that satisfies

$$J(u) \leq J(v) \quad \text{for all functions } v \text{ in } X,$$

$$\text{where } J(v) = \tfrac{1}{2}\int_\Omega |\nabla v|^2 \, dV - \int_\Omega fv \, dV$$

and

$$X = \{\text{continuously differentiable functions on } \overline{\Omega} \text{ that vanish on } \Gamma_u\}$$

Now, when one wishes to find the minimum of a function of a single variable $h(x)$, say, then of course this minimum (assuming it exists) is characterized by the necessary condition $h'(x_0) = 0$, $x_0$ being the point at which the minimium is attained. The case such as that in Box 5, where it is required to find a *function* that minimizes a given *functional*, is not dissimilar, despite its greater generality. Indeed, suppose we assume that a minimum does exist, and that this minimum is achieved at the function $u$. If we replace $v$ by $u+\epsilon v$, where $v$ is arbitrary, although a member of $X$, then we may treat $J(u+\epsilon v)$ as a function of the single variable $\epsilon$, and write $J(u+\epsilon v) \equiv F(\epsilon)$, say. A minimum is then achieved at $\epsilon = 0$, so the condition for a minimum is therefore that

$$\frac{d}{d\epsilon}F(\epsilon)\bigg|_{\epsilon=0} = 0 \quad \text{or} \quad \frac{d}{d\epsilon}J(u+\epsilon v)\bigg|_{\epsilon=0} = 0.$$

This is very easy to work out, since

$$\begin{aligned} J(u+\epsilon v) &= \tfrac{1}{2}\epsilon^2 \int_\Omega |\nabla v|^2 \, dx + \epsilon \int_\Omega \nabla u \cdot \nabla v \, dx + \int_\Omega |\nabla u|^2 \, dx \\ &\quad -\epsilon \int_\Gamma fv \, dx - \int_\Gamma fu \, dx. \end{aligned}$$

Differentiation with respect to $\epsilon$, and evaluation at $\epsilon = 0$, result in the equation

$$\int_\Omega \nabla u \cdot \nabla v \, dx = \int_\Gamma fv \, dx. \tag{14}$$

Thus we have arrived at an alternative variational formulation. Because of its close association with the minimization problem, and because minimization problems are the subject of the calculus of variations, (14) is also known as a *variational boundary value problem* (VBVP). This problem is summarized in Box 6.

Box 6: The Variational Boundary Value Problem for the Poisson Equation

Find a function $u$ in $X$ that satisfies

$$\int_\Omega \nabla u \cdot \nabla v \, dV = \int_\Gamma fv \, dV$$

for all $v \in X$

We show in Chapter 9, the subject of which is VBVPs, that although minimization problems may be formulated as VBVPs, it is possible to derive the VBVP corresponding to a PDE and boundary conditions without going via the minimization problem. Indeed, in some cases there may in fact *be no minimization problem*, so the VBVP is in this sense a more fundamental formulation.

As an example of how the VBVP is derived directly from the classical formulation, we return to the Poisson equation in Box 3; multiplying this equation throughout by an arbitrary function $v$ in $X$, where $X$ is as defined previously, and then integrating over $\Omega$, we obtain

$$-\int_\Omega v\nabla^2 u \, dV = \int_\Omega fv \, dV.$$

Now the left-hand side may be transformed by using the divergence theorem, according to which

$$-\int_\Omega v\nabla^2 u \, dV = -\int_\Gamma v\frac{\partial u}{\partial n} \, dA + \int_\Omega \nabla u \cdot \nabla v \, dV.$$

The boundary integral $\int_\Gamma$ may be broken up into two parts: an integral over $\Gamma_u$, and an integral over the complementary part $\Gamma_q$. Since $v$ belongs to $X$ by assumption, and since all members of $X$ have the property that they vanish on $\Gamma_u$, the integral over $\Gamma_u$ vanishes. Likewise, we have the boundary condition $\partial u/\partial \nu = 0$ on $\Gamma_q$, so that the integral over $\Gamma_q$ also vanishes. Thus the entire boundary integral is zero, and we arrive in this way at the problem in Box 6.

## STAGE II: CONSTRUCTION OF A SOLUTION (IF POSSIBLE)

Having formulated the problem, one obvious second step would be to solve it in closed form. If an exact solution is available, then most people would agree that the task has been essentially completed, and all that remains is to use the solution to obtain information about the system that has been modeled. Certainly for the case of a homogeneous medium, that is, one whose properties are independent of position, the problem can often be solved in closed form if the data are reasonably "nice", and if the geometry of the boundary $\Gamma$ is very simple: for example, a rectangle (or parallelipiped in three dimensions), circular cylinder, or sphere. In this case the method of separation of variables can be used to obtain a series solution of the heat equation; this is a method that most students encounter in elementary courses on PDEs or Fourier series.

But such a happy situation is exceptional; in general *it is not possible* to solve even relatively simple BVPs, such as those given previously, in closed form, if the geometry of the domain is complex, for example. Thus we are left with the problem of trying to obtain information about the solution, even though there is little prospect of finding that solution. It is at this stage that tools such as functional analysis come into their own.

## STAGE III: WELL-POSEDNESS OF THE PROBLEM

Now, rather than pursuing the possibly fruitless goal of an exact solution, we first attempt to gain *qualitative information* about the problem. What kind of qualitative information is required? Well, it is generally agreed that above anything else it is necessary to know the answers to the following questions.

Does a solution *exist*?

If so, is this solution *unique*?

Does the solution *depend continuously* on the data?

The first two questions need no clarification; in the case of the third we are asking whether the solution changes by only a small amount if the data are changed by a small amount (of course, what is meant by "small amount" must be made clear). A problem for which small changes in data cause wild fluctuations in the solution is clearly unstable, and one would be tempted in such a case to reconsider whether the mathematical model does indeed represent reality adequately and, if so, exactly how to interpret such sensitivity.

If the answer to all three questions is in the affirmative, then the problem is said to be *well-posed*. Armed with such knowledge, which can only properly be obtained with the aid of the tools of functional analysis, the process of seeking approximate solutions can then proceed from a firm base.

It is possible in some circumstances to construct counterexamples which demonstrate that a problem does *not* have a unique solution. It is also possible in such cases to obtain *necessary* conditions for existence of a solution. That is, with a minimum of manipulation we can establish conditions satisfied either by the solution or by the data, assuming that a solution does exist. Take as an example the BVP in Box 3, but assume that the boundary condition takes the form

$$\frac{\partial u}{\partial \nu} = g \quad \text{on } \Gamma; \tag{15}$$

in other words, if the physical problem is heat conduction, the heat flux is given on the entire boundary.

Now in fact it is easy to see that this problem does not have a unique solution; indeed, if $u$ is a solution, then so is $u + c$, where $c$ is any constant, since

$$-\nabla(u + c) = -\nabla u = f \quad \text{on } \Omega$$

and

$$\frac{\partial}{\partial \nu}(u + c) = \frac{\partial u}{\partial \nu} = g \quad \text{on } \Gamma.$$

Physically, we may add a constant temperature to the body, and the resulting temperature distribution would still be consistent with the mathematical model.

It is possible to go even further, and to show that a solution will exist *only if* the data satisfy a particular condition; to see this we integrate Laplace's equation over $\Omega$ and make use of the divergence theorem and the boundary condition (15), to find that

$$\begin{aligned} -\int_\Omega f \, dV = \int_\Omega \nabla^2 u \, dV &= \int_\Omega \operatorname{div}(\nabla u) \, dV \\ &= \oint_\Gamma \frac{\partial u}{\partial \nu} \, dA \\ &= \oint_\Gamma g \, dA. \end{aligned}$$

That is,

$$\int_\Omega f \, dV + \oint_\Gamma g \, dA = 0. \tag{16}$$

Thus it is not possible to find a solution to the problem unless the data $f$ and $g$ satisfy the *compatibility condition* (16). For the problem of heat conduction this asserts that the net amount of heat generated in the body must be zero; such a condition makes perfect sense since otherwise we could not expect to have a steady problem.

## STAGE IV: CONSTRUCTION OF APPROXIMATE SOLUTIONS

Assuming that a closed-form solution is out of the question, but that the problem has nevertheless been shown to be well-posed, the following stage involves that of constructing the next best thing: a good approximation. Perhaps the two most well-known procedures for achieving this are the *finite difference* and *finite element methods*. The finite element method has been a great success, particularly since the advent of high-speed computers, and it is the main topic of Part III of this book. We therefore focus on this particular method of approximation in this section.

The first point to bear in mind about the finite element method is that it takes as a point of departure the *variational* formulation of the problem, for example, that given in Boxes 5 or 6, rather than the formulation (such as that in Box 3) as a PDE. The method is in fact a special case of the *Ritz–Galerkin method*, whose essence can be described succinctly. We could start with Box 5 or 6; choosing the minimization problem, we pose this not on the whole of the space $X$ – remember that this problem in general will not be solvable in closed form – but instead on a *finite-dimensional subspace* $X_n$ of $X$. In other words, functions $\phi_1, \phi_2, \ldots, \phi_n$, all of which belong to $X$, are selected as basis functions, and $X_n$ is defined by

$$\begin{aligned} X_n &= \{\text{all functions of the form } v_n = b_1\phi_1 + \ldots + b_n\phi_n, \\ &\qquad b_1, \ldots, b_n \text{ constant}\}. \end{aligned}$$

Such a representation of $v$ is substituted in the functional $J$, which now reads

$$\begin{aligned} J(v_n) &= \tfrac{1}{2}\int_\Omega \nabla v_n \cdot \nabla v_n \, dV - \int_\Omega f v_n \, dV \\ &= \tfrac{1}{2}\int_\Omega (b_1\nabla\phi_1 + \cdots + b_n\nabla\phi_n)\cdot(b_1\nabla\phi_1 + \cdots + b_n\nabla\phi_n)\, dV \\ &\quad - \int_\Omega f(b_1\phi_1 + b_2\phi_2 + \cdots b_n\phi_n)\, dV \\ &= \tfrac{1}{2}\sum_{i,j=1}^{n} K_{ij} b_i b_j - \sum_{j=1}^{n} F_j b_j, \end{aligned}$$

or, more concisely,

$$J(v) = \tfrac{1}{2}\boldsymbol{b}^T\boldsymbol{K}\boldsymbol{b} - \boldsymbol{b}^T\boldsymbol{F},$$

where $\boldsymbol{b}^T = [b_1, b_2, \ldots, b_n]$ and $\boldsymbol{K}$ and $\boldsymbol{F}$ are, respectively, the $n \times n$ symmetric matrix and $n \times 1$ vector with components

$$K_{ij} = \int_\Omega \nabla\phi_i \cdot \nabla\phi_j \, dV \quad \text{and} \quad F_j = \int_\Omega f\phi_j \, dV.$$

So the essence of the Ritz–Galerkin method is that the problem of minimizing a functional over an infinite-dimensional space $X$ of candidate functions has been reduced to the very simple – and solvable! – problem of minimizing a quadratic function of $n$ variables. If the minimizing vector is denoted by $\boldsymbol{a}$, then this vector is the solution to the set of simultaneous equations

$$\boldsymbol{K}\boldsymbol{a} = \boldsymbol{F}; \tag{17}$$

this set of equations has a unique solution since $\boldsymbol{K}$ is positive-definite and therefore has an inverse. Once solved, the approximate solution may be found from $u_n = a_1\phi_1 + \ldots + a_n\phi_n$.

The finite element method, as has been mentioned already, is a special case of the Ritz–Galerkin method in which the basis functions $\phi_i$ are constructed by a particular process which has the virtues that, first, it is systematic, and second, the quality of the approximation improves with increase of the dimension $n$ of the basis. The method is introduced in Part III.

## STAGE V: QUALITY OF APPROXIMATE SOLUTIONS

Although many practitioners would regard Stage IV as a suitable point at which to conclude proceedings, it is nevertheless of the utmost importance to obtain some information about the *quality of the approximation.* Without such information the approximate solution itself is of little use, since it could conceivably bear very little resemblance to the exact solution. We need to know whether in some sense the approximation is a good one, and also how it could be improved.

Once again a knowledge of functional analysis is required in order to pursue this line of enquiry satisfactorily. Suppose that the exact solution to one of the problems discussed previously is denoted by $u$, and the approximate solution, obtained using the Ritz–Galerkin finite element method, is denoted by $u_n$. What we first of all require is some qualitative information about the *error* $u - u_n$. Now since $u$ is unknown, it is of course not possible to evaluate the error exactly. So we have instead to be satisfied with an *estimate* of the error. To complete the picture, one would hope to obtain from such an estimate not only an idea about the size of the error, but also some information about whether the approximate procedure gives a family of solutions that *converge* to the exact solution and, if so, what the *rate of convergence* is.

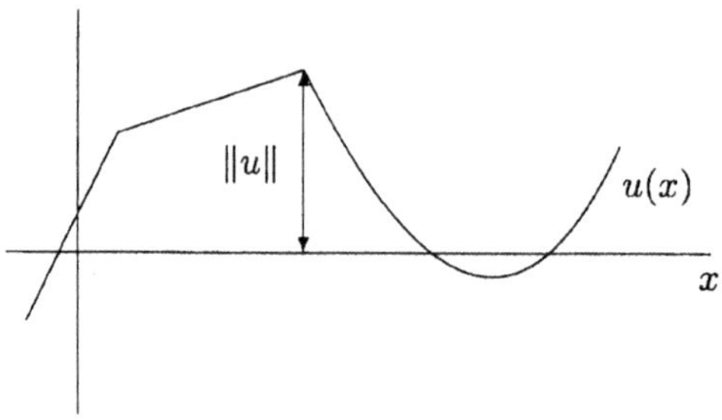


FIGURE 4. The norm of a continuous function

The finite element method is ideally suited to such a convergence analysis. In the framework of functional analysis one has to be precise first of all about the set, or space, of functions $X$ to which the solution belongs. We gave a rather simple example earlier, but it is shown that it is necessary to be a lot more careful about the choice of space. Such a space is generally a *normed space*, meaning that there is an operation $\|\cdot\|$, called a norm, defined on the space. The norm measures the magnitude of the function in a manner analogous to that in which the length $|\boldsymbol{a}|$ of a vector $\boldsymbol{a}$ is defined by $|\boldsymbol{a}| = \sqrt{\boldsymbol{a}\cdot\boldsymbol{a}}$. For example, if $X$ is the space of all bounded continuous functions, then one way of defining a norm on $X$ is according to

$$\|u\| = \max\{|u(\boldsymbol{x})| \text{ for all points } \boldsymbol{x} \text{ in } \Omega\}.$$

(see Figure 4). With the aid of a norm, the procedure of obtaining an error estimate and the rate of convergence can be phrased in more concrete form. For example, the Ritz–Galerkin procedure entails the replacement of $X$ by an $n$-dimensional approximation $X_n$ which is constructed in such a way that the *error estimate*

$$\|u - u_n\| \leq Cn^{-p} \tag{18}$$

holds, where $C$ is a positive constant not related to $n$, and $p$ is a positive constant, also independent of $n$. There is a wealth of information contained in this inequality, if it can be derived for the problem at hand. First, it confirms that we may improve the approximation by increasing the size of $X_n$; that is, $\|u - u_n\|$ decreases with an increase in $n$. Second, the sequence of approximations obtained by a progressive increase in the dimension $n$ of $X_n$ converges to $u$, in the sense that $\|u - u_n\|$ approaches zero as $n$ gets larger. Third, we get to know not only that the method converges, but also about the *rate* of convergence . This is given by the number $p$: the larger $p$ is, the faster will be the approach to the exact solution.

So we see that, even though an exact solution may be elusive, the combination of a careful qualitative analysis, using the tools of functional analysis, together with an approximate solution procedure, will yield extremely useful information about the problem. The finite element method in particular provides a systematic way of defining the finite-dimensional spaces $X_n$, for increasing values of $n$.

In summary, then, the overall impression gained is that although functional analysis will not in general provide a means or technique for actually finding closed-form solutions to boundary value problems, it can help us to develop theories that will throw light on the *nature* of solutions to problems. Given that closed-form solutions may be impossible to achieve, by whatever method, the need for an understanding of some of these qualitative properties of solutions is compelling.

## Outline of the rest of this book

The next seven chapters constitute Part I of this book and provide a systematic development of the basic functional analysis required for a reasonably in-depth study of BVPs and their approximation.

In Part II, which comprises Chapters 8 and 9, we present a detailed study of elliptic BVPs in both the classical (that is, PDE) as well as the variational forms. This part of the book can therefore be thought of as supplying the technical detail that was missing in the heuristic overview of BVPs presented earlier. The opportunity is also taken in these chapters to introduce further physical examples, taken mostly from mechanics, and to use these examples to illustrate the theory.

Finally, Part III is concerned with methods of obtaining approximate solutions. The focus is on the Galerkin and finite element methods, and both the computational and mathematical aspects are studied so that the steps leading to the equation (17) are presented in detail and the analysis necessary to arrive at error estimates of the form (18) is also developed.

# Part I

# Linear Functional Analysis

# 1
# Sets

Functional analysis conventionally takes as its starting point the idea of the existence of collections of mathematical objects, for example, numbers, vectors, or functions. Such collections, which are known as sets, are endowed with additional structure, and when this is done it becomes possible to elaborate on their properties and build up a coherent theory.

Sets are clearly basic to a proper study of mathematics, and for this reason we start the study of functional analysis with some introductory aspects of set theory. After a review of the algebra of sets in Section 1, we go on to take a closer look at sets of numbers and of $n$-tuples, since these are fundamental to much of the later developments in this work. In Chapter 2 we discuss some important sets whose members are *functions*. This chapter ends, in Section 1.4, by introducing further set-theoretic and logical notions which are required in the proofs of theorems.

## 1.1 The algebra of sets

A *set* is any well-defined collection of objects. These objects – in the present context mainly numbers, vectors, or functions – are called *members* or *elements* of the set. A set is usually denoted by a capital letter, for example, $A$, and if the object $x$ is a member of $A$ we write

$$x \in A$$

which is read "*$x$ is an element of $A$*" or "*$x$ belongs to $A$*". Likewise, the expression

$$x \notin A$$

reads "*$x$ is not an element of $A$*". Various ways of defining sets are given in the following examples.

**Examples**

1. The set $A$ of all positive integers less than or equal to 5 is given by

$$A = \{1, 2, 3, 4, 5\}. \tag{1.1}$$

Here we have defined $A$ simply by listing its elements. A more sophisticated, but concise, way of defining the set would be to write

$$A = \{n : \; n \text{ is an integer}, \; 1 \leq n \leq 5\}$$

or, if we denote by $\mathbb{Z}$ the set of all integers,

$$\mathbb{Z}^+ = \{n : \; n \in \mathbb{Z}, \; 1 \leq n \leq 5\}. \tag{1.2}$$

The expression in brackets is read "*the set of all $n$, $n$ being an integer, such that* $1 \leq n \leq 5$."

2. The preceding example was of a *finite set*, that is, a set comprising a finite number of elements. An *infinite set* is one that does not have a finite number of elements; for example, the set $\mathbb{Z}^+$ of all nonnegative integers, that is,

$$A = \{n : \; n \in \mathbb{Z}, \; n \geq 0\},$$

is an infinite set (as is $\mathbb{Z}$ itself).

3. The *empty* or *null* set is the set with no elements, and is denoted by $\varnothing$. For example,

$$\{n : \; n \in \mathbb{Z}, \; n < 1, \text{ and } n > 1\} = \varnothing.$$

**Subsets, equal sets.** If $A$ and $B$ are two sets, we say that $A$ is a *subset* of $B$ if each element of $A$ is an element of $B$. This is denoted by

$$A \subset B.$$

According to this definition every set is, of course, a subset of itself, and so in order to distinguish subsets that do not coincide with the set in question, we say that $A$ is a *proper subset* of $B$ if $A$ is indeed a subset of $B$ and if, furthermore, $B$ also contains elements that do not belong to $A$. If it is desirable to indicate that $A$ is a subset of $B$ which is possibly the set $A$ itself, we write

$$A \subseteq B.$$

If $A$ is not a subset of $B$, this is indicated by

$$A \not\subset B.$$

For example, if $A$ is given by (1.1) or (1.2), then $C = \{1,2,3\}$ is a proper subset of $A$, but $D \not\subset A$, where $D = \{1,2,6\}$.

Two sets $A$ and $B$ are *equal* if they contain exactly the same elements. When this is the case, we write

$$A = B.$$

According to the definition of a subset, it is clear that two sets $A$ and $B$ are equal if, and only if,

$$A \subset B \quad \text{and} \quad B \subset A.$$

For example, if

$$A = \{x : \; x^2 = 4\} \quad \text{and} \quad B = \{2, -2\},$$

then $A = B$.

**Union, intersection, difference.** We make the assumption from now on that all sets under discussion are subsets of a single fixed set called the *universal set*, which is denoted by $U$. The definition of $U$ varies from one context to another.

The *union* of two sets $A$ and $B$, written $A \cup B$, is the set consisting of all elements that are in $A$ or in $B$. That is,

$$A \cup B = \{x : \; x \in A \text{ or } x \in B\}.$$

This is shown graphically in Figure 1.1(a), in the form of a *Venn diagram*; the universal set is represented by the rectangle and its subsets are points or areas within the rectangle.

The *intersection* of two sets $A$ and $B$, written $A \cap B$, is the set of all elements that belong to both $A$ *and* $B$ (Figure 1.1(b)). In other words,

$$A \cap B = \{x : \; x \in A \text{ and } x \in B\}.$$

The *difference* of two sets $A$ and $B$, written $A - B$, is the set of all elements of $A$ that do not belong to $B$ (Figure 1.2(a)):

$$A - B = \{x : x \; \in A, \; x \notin B\}.$$

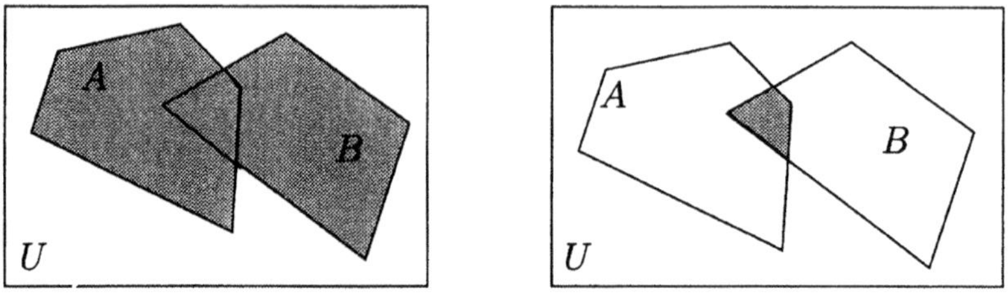


FIGURE 1.1. (a) The union and (b) the intersection of two sets

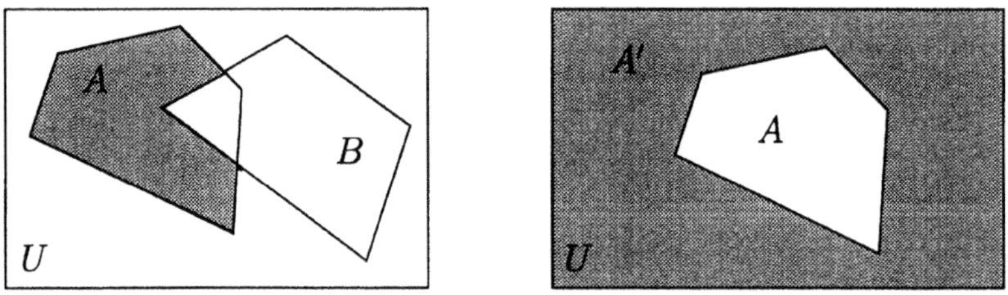


FIGURE 1.2. (a) The difference $A - B$ of two sets $A$ and $B$; (b) the complement $A'$ of a set $A$

The *complement* of a set $A$, denoted by $A'$, is the set of all elements not in $A$ (Figure 1.2(b)). That is,

$$A' = \{x \in U : x \notin A\}.$$

**Example**

4. Let $A = \{x : x \in \mathbb{Z},\ 1 \leq x \leq 10\}$ and $B = \{9, 10, 11, 12\}$. Then

$$\begin{aligned} A \cup B &= \{x : x \in \mathbb{Z},\ 1 \leq x \leq 12\}, \\ A \cap B &= \{9, 10\}, \\ A - B &= \{x : x \in \mathbb{Z},\ 1 \leq x \leq 8\}, \end{aligned}$$

and

$$A' = \{\dots - 3, -2, -1, 0, 11, 12, 13, \dots\}$$

(the universal set is taken here to be $\mathbb{Z}$).

**Countable sets.** It is necessary at times to know whether the elements of an infinite set can be "labelled" by positive integers. In other words, we wish to distinguish infinite sets of the form

$$A = \{a_1, a_2, a_3, \dots\} \tag{1.3}$$

from those that cannot be labeled as in (1.3). A set that can be put in one-to-one correspondence, in other words, labelled, with positive integers is called a *countable set.* Of course, any finite set $A$ is countable, for if $A$ has $m$ members we may label them $a_1, a_2, ..., a_m$.

**Examples**

5. The set of *functions*

$$P = \{f_k \ (k = 1, 2, \dots) : \ f_k(x) = x^k\}$$

is countable.

6. As shown in the next section, sets such as the set of points

$$A = \{x : \ 0 \le x \le 1\}$$

on the real line are not countable. That is, the set of real numbers between 0 and 1 cannot be labeled in the form $a_1, a_2, \dots$.

**Cartesian products.** Given two sets $A$ and $B$, their Cartesian product $A \times B$ is the set of *all ordered pairs* $(a, b)$, where $a \in A$ and $b \in B$. That is,

$$A \times B = \{(a, b) : \ a \in A, \ b \in B\}.$$

For example, if

$$A = \{1, 2, 3\} \ \text{ and } \ B = \{7, 8\}, \tag{1.4}$$

then

$$A \times B = \{(1,7), \ (1,8), \ (2,7), \ (2,8), \ (3,7), \ (3,8)\}.$$

If we think of the members of $A$ as lying on a horizontal axis and the members of $B$ as lying on the vertical axis of a pair of Cartesian coordinate axes, then $A \times B$ may be represented by a set of points in the plane, as shown in Figure 1.3. The idea of the Cartesian product may be extended to products of more than two sets. For example, the Cartesian product $A_1 \times A_2 \times A_3 \times \dots A_n$ is defined to be the set of *all ordered n-tuples*

$$(a_1, a_2, \dots, a_n) \ \text{ where } \ a_1 \in A_1, \ a_2 \in A_2, \dots, \ a_n \in A_n.$$

Note that, in general, $B \times A \ne A \times B$. For example, if $A$ and $B$ are as in (1.4), then

$$B \times A = \{(7,1), (7,2), (7,3), (8,1), (8,2), (8,3)\} \ne A \times B.$$

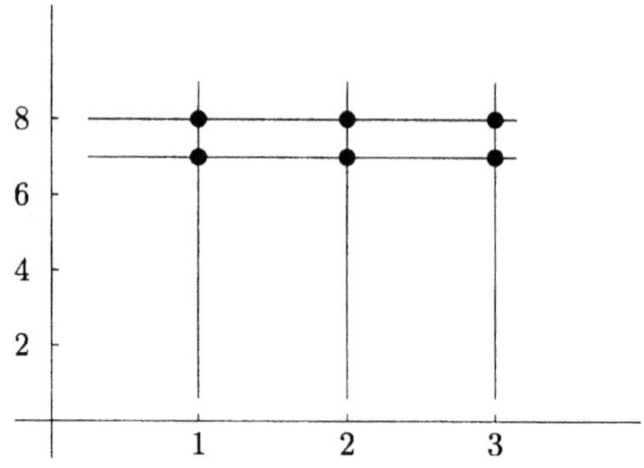


FIGURE 1.3. The Cartesian product $A \times B$ of the two sets in (1.4)

## 1.2 Sets of numbers

Although we encounter a wide variety of sets in this work, sets of numbers are of particular importance, and pervade the developments that follow. For this reason we set out in this section some of the salient definitions and properties of sets of numbers.

We have already come across the set $\mathbb{Z}$ of all integers, defined by

$$\mathbb{Z} = \{\dots, -3, -2, -1, 0, 1, 2, 3, \dots\}.$$

Occasionally we also make use of the set $\mathbb{N}$ of *natural numbers* or positive integers:

$$\mathbb{N} = \{1, 2, 3, \dots\}.$$

A *rational number* is a number that can be expressed as the ratio of two integers. We denote the set of rational numbers by $\mathbb{Q}$, so that

$$\mathbb{Q} = \{x : x = p/q,\ p, q \in \mathbb{Z},\ q \neq 0\}.$$

An important property of $\mathbb{Q}$, which we record here, is the following: *the set $\mathbb{Q}$ of rational numbers is a countably infinite set.* In other words, the rational numbers can be put in one-to-one correspondence with integers. You are asked to show this in Exercise 1.6 at the end of this Chapter.

Real numbers that do not belong to $\mathbb{Q}$ are called *irrational numbers.* For example, $\sqrt{2}$ cannot be written in the form $p/q$ for $p, q \in \mathbb{Z}$, and so is irrational (see later, Section 5). This brings us to $\mathbb{R}$, the set of *real numbers;* $\mathbb{R}$ consists of all rational as well as irrational numbers. It is convenient to think of $\mathbb{R}$ as being represented by an infinitely long line, called the real line. The origin is chosen to be at some point on this line, and once the location of the number 1 is fixed, the scale will be determined, and every real number then corresponds to a point on the real line. It should be clear that $\mathbb{R}$ is an uncountable set.

Assuming the rational numbers to be known, the existence of the real numbers is not, mathematically speaking at any rate, a fact that can be deduced in an obvious way. The construction of $\mathbb{R}$ is a process that was carried out in accordance with modern notions of rigor as recently as the late nineteenth century. We omit the detailed arguments that are required for a proper treatment of this subject, and appeal instead to well-known and intuitively obvious properties of the real numbers.

**Subsets of $\mathbb{R}$.** Very often we deal not with the whole real line but only a portion of it, called an *interval*. Thus, if $a$ and $b$ are two points on $\mathbb{R}$ such that $a \leq b$, then we define

the *open interval* $(a, b) = \{x : \ x \in \mathbb{R}, \ a < x < b\}$;

the *closed interval* $[a, b] = \{x : \ x \in \mathbb{R}, \ a \leq x \leq b\}$;

the *half-open intervals* $(a, b] = \{x : \ x \in \mathbb{R}, \ a < x \leq b\}$ and $[a, b) = \{x : \ x \in \mathbb{R}, \ a \leq x < b\}$.

Thus the terms "open" and "closed" indicate, respectively, that the endpoints of the interval are excluded from or included in the set. There are more technical definitions of open and closed sets however, which, although consistent with the preceding definitions, are mathematically more sound. We discuss these shortly.

**Complex numbers.** The set of *complex numbers* is denoted by $\mathbb{C}$, and is defined to be the set of numbers of the form $z = a + bi$, where $i = \sqrt{-1}$ and $a$ and $b$ are real numbers. The number $a$ is called the *real part* of $z$, and is written $\mathrm{Re}\,(z)$, whereas $b$ is called the *imaginary part* of $z$ and is written $\mathrm{Im}\,(z)$. The *modulus* of a complex number $z$ is defined as $(a^2 + b^2)^{1/2}$, and is written $|z|$. The *complex conjugate* of $z$ is the complex number $\overline{z} = a - bi$, so that $|z|^2 = z\overline{z}$. Complex numbers are conveniently represented graphically with respect to a pair of axes that correspond to the real and imaginary parts, respectively, as shown in Figure 1.4; this is referred to as the complex plane. Clearly one may set up a correspondence between $\mathbb{C}$ and the Cartesian product $\mathbb{R} \times \mathbb{R}$.

Continuing the geometrical interpretation, the *argument* $\theta$ of a complex number $z = a + bi$ is the angle defined by

$$\theta \equiv \arg z = \arctan(b/a), \qquad -\pi < \theta \leq +\pi.$$

The last condition ensures that $\theta$ is uniquely defined.

There is a close relationship, in the complex plane, between the exponential function exp and trigonometric functions. Indeed, we have

$$\exp(i\theta) \equiv e^{i\theta} = \cos\theta + i \sin\theta,$$

and so every complex number $z = a + bi$ can be written in *polar form* as

$$z = re^{i\theta},$$

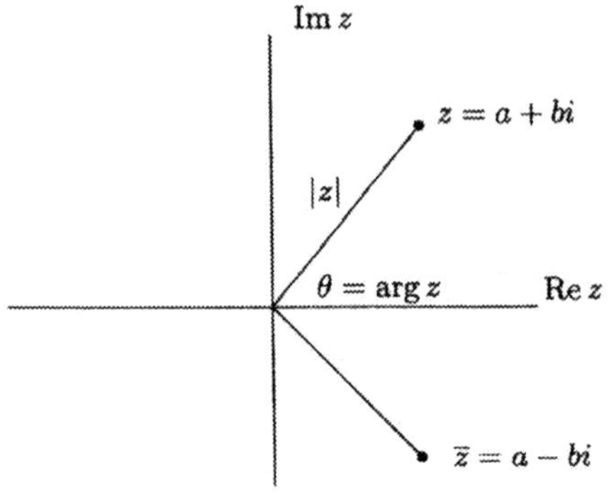


FIGURE 1.4. Graphical representation of complex numbers

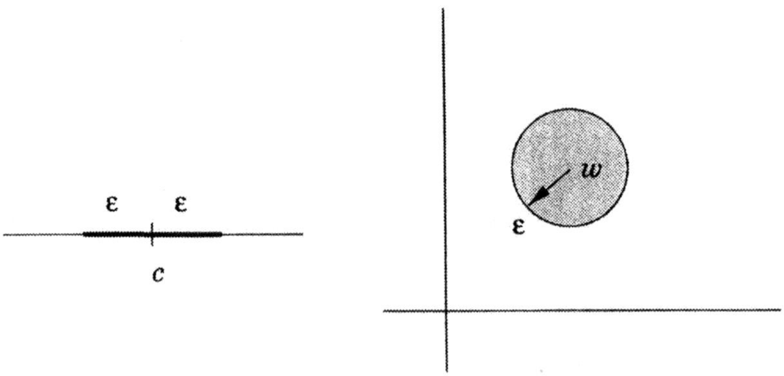


FIGURE 1.5. Neighborhood of a point $c$ in $\mathbb{R}$ and of a complex number $w$

where $r = |z|$.

**Open sets.** Given any point $c$ on the real line and a positive number $\epsilon$, the open interval $(c - \epsilon, c + \epsilon) = \{x : c - \epsilon < x < c + \epsilon\}$ is called a *neighborhood of* $c$. Likewise, if $w$ is any complex number and $\epsilon$ a positive real number, a neighborhood of $w$ is the set $\{z \in \mathbb{C} : |w - z| < \epsilon\}$. Neighborhoods are illustrated in Figure 1.5.

Now, let $\mathbb{K}$ represent either $\mathbb{R}$ or $\mathbb{C}$, and let $X$ be a subset of $\mathbb{K}$. Then $c$ is called an *interior point of* $X$ if we can find a neighborhood of $c$, all of whose points belong to $X$. A set $X \subset \mathbb{K}$ is called an *open set* if every point of $X$ is an interior point.

### Examples

7. The open interval $(a, b)$ is an open set: for any point $c$ in $(a, b)$ we can define a neighborhood lying entirely in $(a, b)$ by choosing $\epsilon$ to be less than $|c - a|$ and $|c - b|$. Thus every point in $(a, b)$ is an interior point.

   On the other hand, the closed interval $[a, b]$ is *not* an open set: the points $a$ and $b$ are such that, no matter how small we choose $\epsilon$, it is not possible to find neighborhoods of $a$ and $b$, all of whose points lie in $[a, b]$. Thus $a$ and $b$ are not interior points and so $[a, b]$ is not open. Similar considerations apply to the half-open intervals $[a, b)$ and $(a, b]$; the points $a$ and $b$, respectively, are not interior points.

8. The real line $\mathbb{R}$ is an open set since every point in $\mathbb{R}$ has a neighborhood that lies in $\mathbb{R}$.

9. A simple example of an open set in $\mathbb{C}$ is the disk of radius $r$ and center $z_0$, defined by $D(z_0; r) = \{z \in \mathbb{C} : |z - z_0| < r\}$.

**Points of accumulation and closed sets.** In order to give a rigorous definition of closed sets in $\mathbb{R}$ and $\mathbb{C}$ we define first a *point of accumulation.* Let $X$ be a set in $\mathbb{K}$, where $\mathbb{K}$ is either $\mathbb{R}$ or $\mathbb{C}$, and $c$ a point in $\mathbb{K}$ ($c$ does not necessarily belong to $X$). Then $c$ is called a point of accumulation of $X$ if *every* neighborhood of $c$ contains at least one point of $X$ distinct from $c$. Furthermore, a set $X \subset \mathbb{K}$ is a *closed set* if it contains all of its points of accumulation. It is possible to show also (see Exercise 1.9) that a set $X$ is closed if and only if its complement $X' = \mathbb{K} - X$ (here the universal set is $\mathbb{K}$) is *open.*

Finally, we define the *closure* $\overline{X}$ of a set $X \subset \mathbb{K}$ to be the union of $X$ and all its points of accumulation. Clearly $\overline{X}$ is a closed set; however, if $X$ is not closed, the operation of closure is a means of constructing the closed set that is nearest to $X$.

### Examples

10. The set $\{1, \frac{1}{2}, 1, \frac{1}{3}, 1, \frac{1}{4}, \dots\}$ has two points of accumulation, namely, 1 and 0. Since these do not belong to the set, it is not closed. On the other hand, the closed set $\{1, 2, 3, \dots\}$ has no points of accumulation.

11. Consider the interval $(a, b)$: according to the preceding definition, every point in $(a, b)$ is a point of accumulation. Furthermore, $a$ and $b$ are also points of accumulation of $(a, b)$ since every neighborhood of these two points contains members of $(a, b)$. But $a$ and $b$ do not belong to the set, and so it is not closed. The closure of $(a, b)$, on the other hand, is $[a, b]$.

12. The *closed interval* $[a, b]$ is a closed set.

13. The unit circle $S = \{z \in \mathbb{C} : |z| = 1\}$ is a closed set in $\mathbb{C}$; every member of $S$ is a point of accumulation of $S$.

**Sequences of numbers.** The concept of a sequence is central to analysis. In due course we deal with sequences in fairly arbitrary spaces; to set the stage for these later developments, and in order to be able to deal here with some further ideas which are pertinent to real or complex numbers, we introduce sequences in this elementary setting. First, however, we clarify the manner in which the symbol $\infty$ is employed.

The nature of $\infty$, which represents infinity, is often misunderstood. To begin with, $\infty$ is *not* a number: it is merely a means of representing unboundedness. For example, the set of positive even integers can be written as $\{2, 4, \dots\}$ or as $\{2n\}_{n=1}^{\infty}$, the second representation indicating that $n$ increases indefinitely. Another example concerns the real line: since $(a, b)$ denotes an open interval in $\mathbb{R}$, using this same notation we may write

$$\mathbb{R} = (-\infty, \infty)$$

to indicate that the "interval" corresponding to $\mathbb{R}$ is not bounded.

Let $\mathbb{K}$ be, as before, either $\mathbb{R}$ or $\mathbb{C}$, and let $X$ be any subset of $\mathbb{K}$. A sequence $\{u_1, u_2, \dots, u_n, \dots\}$ in $X$ is a countable set of elements of $X$ with the general element $u_n$ of $X$ being associated with a positive integer $n$. If the sequence has a *finite* number of elements, it is called a *finite sequence*; otherwise it is called an *infinite sequence.*

Most of the time we deal with infinite sequences, and we generally use the notation $\{u_n\}_{n=1}^{\infty}$ or simply $\{u_n\}$ to denote the infinite sequence $\{u_1, u_2, \dots, u_n, \dots\}$.

**Example**

14. Sequences may be described either by displaying them or by giving a formula for determining the general element. For example, let $X$ be the closed interval $[0, 1]$; the sequence

$$\{1, 0, \tfrac{1}{3}, 0, \tfrac{1}{5}, 0, \dots\}$$

is defined by actually displaying the first few elements; alternatively, it could be defined by stating that the $n$th term $x_n$ of the sequence is

$$x_n = \begin{cases} 0 & \text{for even } n, \\ 1/n & \text{for odd } n. \end{cases}$$

Ultimately what is of most interest about sequences is the way in which they behave as $n$ gets progressively larger; this brings us to the next topic, namely, that of convergence of sequences.

**Convergence of sequences.** We begin with sequences in $\mathbb{R}$. Consider then the sequence of real numbers $\{x_n\} = \{1/n\}_{n=1}^{\infty}$. As $n$ gets larger the term $x_n$ gets closer to zero and we say, loosely at this stage, that the sequence *converges* to 0. On the other hand, the sequence $\{2^n\} = \{2, 4, 8, \ldots\}$ increases indefinitely: no matter what number $N$ we specify, it will always be possible to choose a value of $n$ such that $2^n$ will be greater than $N$. This sequence is said to *diverge.*

The preceding examples behave in a fairly obvious way, and so they could be discussed without recourse to a rigorous definition of convergence. Later on it is necessary to discuss convergence of arbitrary sequences in *normed spaces*, and the definitions that are introduced in that context are fairly obvious generalizations of the definitions used in the rather specialized case of real and complex numbers.

Let $X$ be a subset of $\mathbb{K}$, and $\{u_n\}$ a sequence in $X$. Let $u$ be a number in $X$, and form the sequence $\{|u_1 - u|, |u_2 - u|, \ldots, |u_n - u|, \ldots\}$ (in the case $\mathbb{K} = \mathbb{C}$, $|\cdot|$ of course represents the modulus of a complex number). If the number $|u_n - u|$ approaches 0 as $n$ gets larger, we agree to call the sequence convergent. Another, more formal, way of stating this is as follows. Pick *any positive number* $\epsilon$; then $\{u_n\}$ is said to converge to $u$ if it is always possible to make $|u_n - u|$ smaller than $\epsilon$ simply by choosing $n$ large enough, larger than some number $N$, say. That is, a sequence $\{u_n\}$ in a subset $X$ of $\mathbb{K}$ is *convergent* if, given *any* $\epsilon > 0$, there is a member $u \in X$ for which a number $N$, possibly depending on $\epsilon$, can be found such that

$$|u_n - u| < \epsilon \quad \text{for all } n > N. \tag{1.5}$$

If this is the case, we write $u_n \to u$ (which is read "$u_n$ converges to $u$"), and $u$ is called the *limit* of the sequence. Yet another way of stating (1.5) informally is

$$\lim_{n\to\infty} |u_n - u| = 0 \quad \text{or} \quad \lim_{n\to\infty} u_n = u, \tag{1.6}$$

which is read "the limit as $n$ tends to $\infty$ of $u_n$ is $u$". Note, however, that by (1.6) we mean (1.5); it is also worth noting that the symbol $\lim_{n\to\infty}$ is synonymous with "$n$ becomes arbitrarily large".

### Examples

15. Consider the sequence $\{\alpha_n\} = \{(3n^2 - 1)/(n^2 - 5n)\}_{n=6}^{\infty}$. As $n$ gets very large we would expect this sequence to approach the limit 3 (since the terms $3n^2$ and $n^2$ dominate the numerator and denominator, respectively). We check this by asking whether, for any $\epsilon > 0$, a number $N$ can be found such that

$$|\alpha_n - 3| = \frac{15n - 1}{n^2 - 5n} < \epsilon$$

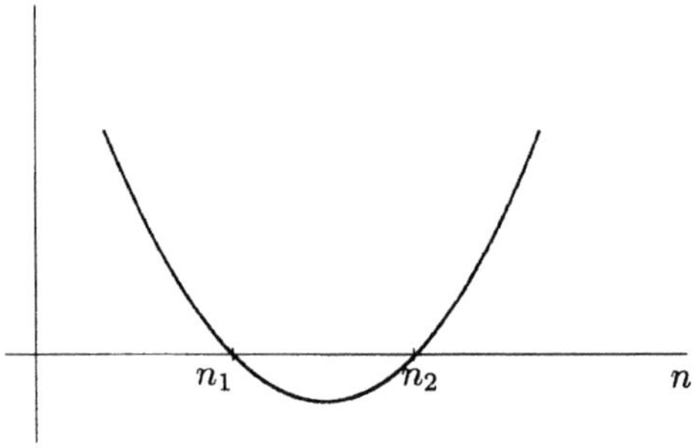


FIGURE 1.6. The function $f(n)$ in Example 15

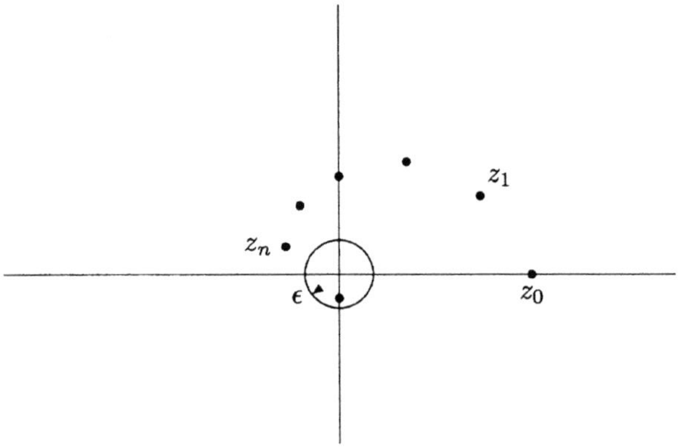


FIGURE 1.7. The sequence in Example 16

whenever $n > N$. This is equivalent to seeking $N$ such that

$$\epsilon(n^2 - 5n) > 15n - 1 \quad \text{or} \quad \epsilon n^2 - 5(\epsilon + 3)n + 1 > 0 \tag{1.7}$$

for $n > N$. Denote the left-hand side of (1.7) by $f(n)$ and treat $n$ as a real number; then the graph of $f(n)$ is as shown in Figure 1.6, and $f$ has roots $n_1$ and $n_2$, with $n_2 \geq 6$. If we choose $N = n_2$, then clearly $f(n) > 0$ for $n > N$, or $|\alpha_n - 3| < \epsilon$ for $n > N$, so that $\alpha_n \to 3$.

16. Consider the sequence $\{z_n \in \mathbb{C} : z_n = a^n e^{in\theta}\}_{n=0}^{\infty}$ in which $a$ and $\theta$ are fixed, and $0 < a < 1$. The sequence is shown in Figure 1.7. To confirm that $z_n \to 0$ as $n \to \infty$, we choose $\epsilon > 0$ and consider whether a number $N$ can be found such that $|z_n| < \epsilon$ when $n > N$. Since $|z_n| = a^n$, this amounts to checking whether there exists $N$ such that $a^n < \epsilon$ for all $n > N$. The answer is affirmative; indeed, by taking the logarithms of both sides we see that $n > \log \epsilon / \log a$, so it suffices to choose $N = \log \epsilon / \log a$.

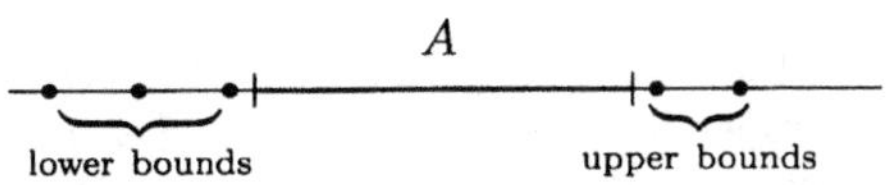


FIGURE 1.8. Upper and lower bounds of a set $A$

**Supremum and infimum.** Suppose that the set $A$ is a subset of $\mathbb{R}$: if there is a real number $p$ such that $p \geq x$ for all points $x$ in $A$, then we call $p$ an *upper bound* of $A$, and say that $A$ is bounded above by $p$. Similarly, if there is a real number $q$ such that $q \leq x$ for all $x$ in $A$, then $q$ is called a *lower bound* of $A$, and we say that $A$ is bounded below by $q$. Note that $A$ can have many upper and lower bounds. If $A$ has both an upper and a lower bound, then it is said to be *bounded*. Thus another way of characterizing a bounded set $A$ is as one for which there exists a positive number $M$ such that

$$|x| \leq M \quad \text{for all } x \in A.$$

Now suppose that there is a number $m$ which belongs to $A$ and that also is an upper bound of $A$. We call $m$ the *maximum* of the set $A$ and we write

$$\max A = m.$$

Similarly, if there is a number $n$ that belongs to $A$ and which, furthermore, is a lower bound of $A$, then this number is called the *minimum* of $A$ and we write

$$\min A = n.$$

## Examples

17. Let $A$ be the closed unit interval $[0,1] = \{x : \; x \in \mathbb{R}, \; 0 \leq x \leq 1\}$. Then any number $a \geq 1$ is an upper bound, any number $b \leq 0$ is a lower bound, and

$$\max A = 1, \qquad \min A = 0.$$

18. Let $A = (0,1) = \{x : \; X \in \mathbb{R}, \; 0 < x < 1\}$. In this case $A$ has no maximum or minimum, although it is bounded; the numbers 0 and 1 are upper and lower bounds, respectively, but do not belong to $A$.

The preceding examples illustrate once again the essential difference between closed and open intervals: closed intervals have minima and maxima whereas open intervals do not. Still, we would like to be able to express the fact that, from the point of view of boundedness, a set such as $(0,1)$ is not that different from $[0,1]$ in that it does have a *least upper bound*, which is the smallest of all its upper bounds, and a *greatest lower bound*, which is the largest of all its lower bounds, even though these bounds do not belong to the set.

In general the *supremum* or *least upper bound* of a set $A$ is a number $p'$ which is an upper bound of $A$, and which satisfies $p' \leq p$ for all upper bounds $p$. When $p'$ exists, we write

$$p' = \sup A.$$

Similarly, the *infimum* or *greatest lower bound* of $A$ is a number $q'$ which is a lower bound of $A$, and which satisfies $q' \geq q$ for all lower bounds $q$. We normally write this as

$$q' = \inf A.$$

We note that when $\max A$ exists then clearly

$$\max A = \sup A.$$

Likewise, if $\min A$ exists, then

$$\min A = \inf A.$$

**Examples**

19. Let $A = (0,1]$. Then $\max A = \sup A = 1$, and $\inf A = 0$ although $\min A$ does not exist.

20. Let $A$ be the positive real line $\mathbb{R}^+ = \{x : x \in \mathbb{R}, x \geq 0\}$. Then $\inf \mathbb{R}^+ = \min \mathbb{R}^+ = 0$ and $\sup \mathbb{R}^+$ does not exist, since $\mathbb{R}^+$ is not bounded above.

**The Bolzano–Weierstrass theorem.** The question naturally arises as to whether it is possible to characterize those subsets for which all sequences contain a point of accumulation in the subset. The answer, in the case of subsets of $\mathbb{R}$, is surprisingly straightforward: all that is required is for the set to be closed and bounded. This result is the substance of the following theorem.

THEOREM 1 (THE BOLZANO–WEIERSTRASS THEOREM). *Let $[a,b]$ be a closed and bounded interval on the real line, and $\{x_n\}$ a sequence in $[a,b]$. Then this set has a point of accumulation in $[a,b]$.*

PROOF. Let $c_1$ be the infimum or greatest lower bound of the sequence

$\{x_1, x_2, \ldots\}$. Next, let $c_2$ be the infimum of the sequence $\{x_2, x_3, \ldots\}$. Continuing in this way, we denote by $c_n$ the infimum of the sequence starting with $x_n$, that is, $\{x_n, x_{n+1}, \ldots\}$.

Clearly $\{c_1, c_2, \ldots\}$ is a monotone increasing sequence, and furthermore this sequence is bounded above (by $b$). It follows (Exercise 1.14) that this sequence $\{c_n\}$ converges to a limit $c$, say, which lies in $[a, b]$. We show next that $c$ is in fact a point of accumulation of the sequence $\{x_n\}$.

To do this, choose any $\epsilon > 0$, and choose also a positive integer $N$; then there exists $m \geq N$ such that

$$|c_m - c| < \epsilon.$$

Now $c_m$ is the infimum of the set of numbers $\{x_m, x_{m+1}, \ldots\}$, so that there exists $k \geq m$ such that

$$|x_k - c_m| < \epsilon.$$

Hence it follows, using the triangle inequality, that

$$|x_k - c| = |x_k - c_m + c_m - c| \leq |x_k - c_m| + |c_m - c| < 2\epsilon.$$

The assertion is thus proved. □

The Bolzano–Weierstrass theorem is in fact valid for arbitrary closed and bounded subsets in $\mathbb{R}$, and in $\mathbb{C}$. This may be shown by modifying the preceding proof appropriately, and is left as an exercise.

**Compactness.** A set $X \subset \mathbb{K}$ is said to be *compact* if every sequence of elements in $X$ has a point of accumulation in $X$. In other words, a set is called compact if it has a property which by Theorem 1 is possessed by all closed and bounded intervals. It is in fact a key property of $\mathbb{K}$ that subsets of $\mathbb{K}$ are compact if and only if they are closed and bounded:

$$\text{compact} \equiv \text{closed} + \text{bounded}.$$

## 1.3 $\mathbb{R}^n$ and its subsets

In the previous section we dealt in some detail with subsets of the real line or intervals. Here we extend some of the concepts to higher-dimensional regions. We start with a description of $\mathbb{R}^2$, which is defined by $\mathbb{R}^2 = \mathbb{R} \times \mathbb{R}$ so that members of $\mathbb{R}^2$ are *ordered pairs of real numbers:*

$$\mathbb{R}^2 = \{(x, y) : \; x, \, y \in \mathbb{R}\}.$$

Just as $\mathbb{R}$ may be represented geometrically by a line, with members of $\mathbb{R}$ being points on the line, in the same way $\mathbb{R}^2$ may be thought of as a *plane*

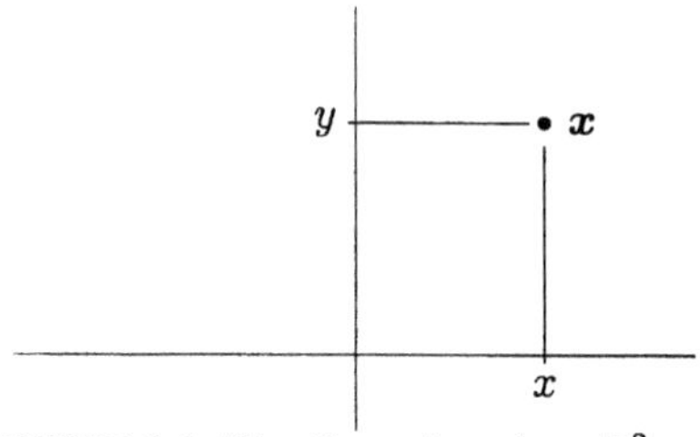


FIGURE 1.9. The Cartesian plane $\mathbb{R}^2$

extending indefinitely in all directions. If we use the notation $\boldsymbol{x} = (x, y)$ to denote a typical member of $\mathbb{R}^2$, then clearly $\boldsymbol{x}$ represents a point in the plane with coordinates $(x, y)$, as shown in Figure 1.9. This plane is known as the Cartesian plane.

The situation just described is easily generalized to higher dimensions: for example, the set $\mathbb{R}^3 \equiv \mathbb{R}\times\mathbb{R}\times\mathbb{R}$ is the set of all ordered *triples* $\boldsymbol{x} = (x, y, z)$ of real numbers; that is,

$$\mathbb{R}^3 = \{\boldsymbol{x} = (x, y, z) : \; x, y, z \in \mathbb{R}\}.$$

As with $\mathbb{R}^2$, we simply represent a typical member of $\mathbb{R}^3$ by $\boldsymbol{x}$. Geometrically we can regard $\mathbb{R}^3$ as being synonymous with three-dimensional space, any member $\boldsymbol{x} \in \mathbb{R}^3$ being a point in this space with coordinates $x$, $y$, and $z$.

Generally, we define $\mathbb{R}^n$ to be the set of all *ordered n-tuples* of real numbers:

$$\mathbb{R}^n = \{\boldsymbol{x} = (x_1, x_2, ..., x_n) : \; x_i \in \mathbb{R}, \; i = 1, ..., n\}.$$

When working in two or three dimensions it is often convenient to use the alternative notations

$$\boldsymbol{x} = (x, y) \text{ for } \boldsymbol{x} \in \mathbb{R}^2 \quad \text{and} \quad \boldsymbol{x} = (x, y, z) \text{ for } \boldsymbol{x} \in \mathbb{R}^3,$$

depending on circumstances.

**Open sets in $\mathbb{R}^n$.** The generalization to higher dimensions of the interval on the real line is the *domain* in $\mathbb{R}^n$; in order to describe exactly what we mean by a domain, however, it is necessary first of all to generalize to $\mathbb{R}^n$ the definition of an open set introduced earlier. Recall that a neighborhood of a point $c$ in $\mathbb{R}$ is an open interval of points $x$ satisfying $|x - c| < \epsilon$. Now we can read this inequality as "the distance from $x$ to $c$ is less than $\epsilon$", and so it follows that all we need in order to extend the idea of a neighborhood to $\mathbb{R}^n$ is a means of measuring distance. In $\mathbb{R}^2$ the distance between two points $\boldsymbol{x}$ and $\boldsymbol{y}$ is defined by

$$|\boldsymbol{x} - \boldsymbol{y}| = \sqrt{(x_1 - y_1)^2 + (x_2 - y_2)^2},$$

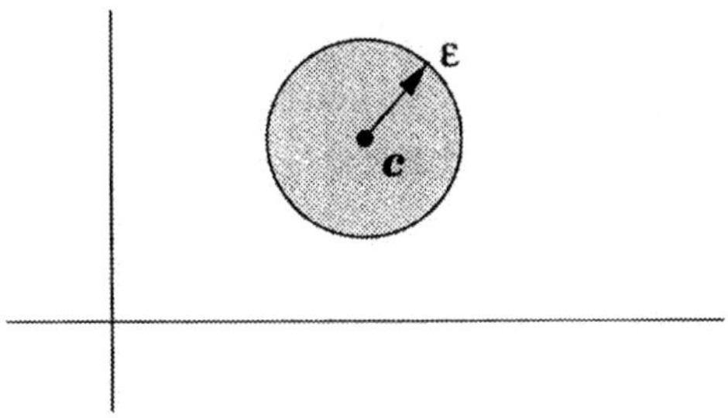


FIGURE 1.10. A neighborhood of the point $c$ in $\mathbb{R}^2$

where $(x_1, x_2)$ are the coordinates of $\boldsymbol{x}$ and $(y_1, y_2)$ the coordinates of $\boldsymbol{y}$; hence, we can define a neighborhood of a point $\boldsymbol{c}$ in $\mathbb{R}^2$ to be the set of points that are a distance less than $\epsilon$ away from $\boldsymbol{c}$, for some $\epsilon > 0$; that is, if we denote a neighborhood of $\boldsymbol{c}$ by $N(\boldsymbol{c}; \epsilon)$, then

$$N(\boldsymbol{c}; \epsilon) = \{\boldsymbol{x} : \ \boldsymbol{x} \in \mathbb{R}^2, \ |\boldsymbol{x} - \boldsymbol{c}| < \epsilon\},$$

and $\epsilon$ is called, for obvious reasons, the *radius* of the neighborhood (Figure 1.10). We immediately generalize to $\mathbb{R}^n$ and define a neighborhood of a point $\boldsymbol{c}$ in $\mathbb{R}^n$ to be the set

$$N(\boldsymbol{c}; \epsilon) = \{\boldsymbol{x} : \ \boldsymbol{x} \in \mathbb{R}^n, \ |\boldsymbol{x} - \boldsymbol{c}| < \epsilon\}$$

where $\epsilon$ is the radius of the neighborhood and the distance $|\boldsymbol{x} - \boldsymbol{c}|$ from $\boldsymbol{x}$ to $\boldsymbol{c}$ is defined by

$$|\boldsymbol{x} - \boldsymbol{c}| = \sqrt{(x_1 - c_1)^2 + (x_2 - c_2)^2 + \cdots + (x_n - c_n)^2}.$$

Now that we have at our disposal the concept of a neighborhood in $\mathbb{R}^n$, we can define open sets in $\mathbb{R}^n$ simply by modifying appropriately the definition given in Section 1.2 for subsets of $\mathbb{R}$; specifically, $\boldsymbol{c} \in \mathbb{R}^n$ is called an *interior point* of a set $\Omega$ of points in $\mathbb{R}^n$ if we can always find a neighborhood of $\boldsymbol{c}$, all of whose points belong to $\Omega$. The situation is depicted in Figure 1.11 for the case $n = 2$, where $\Omega$ is defined to be the set of all points lying *inside* but not on the curve $\Gamma$. We can define a neighborhood $N(\boldsymbol{c}; \epsilon)$ lying entirely inside $\Omega$ by choosing $\epsilon$ to be less than or equal to $d$, the shortest distance from $\boldsymbol{c}$ to the boundary $\Gamma$. Finally, a subset $\Omega$ of $\mathbb{R}^n$ is an *open set* if every point of $\Omega$ is an interior point.

**Points of accumulation and closed sets in $\mathbb{R}^n$.** As with open sets, closed sets in $\mathbb{R}^n$ are defined in much the same way as their counterparts in $\mathbb{R}$. Specifically, if $\Omega$ is a subset of $\mathbb{R}^n$ and $\boldsymbol{c}$ is a point in $\mathbb{R}^n$ (not necessarily in $\Omega$ though), then $\boldsymbol{c}$ is called a *point of accumulation* of $\Omega$ if every neighborhood of $\boldsymbol{c}$ contains at least one point of $\Omega$ distinct from $\boldsymbol{c}$.

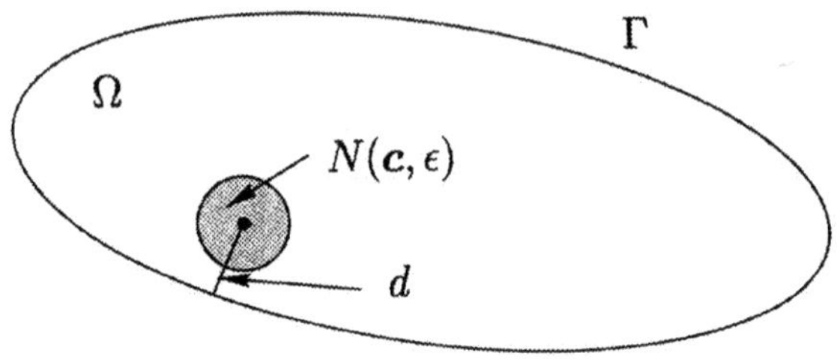


FIGURE 1.11. A neighborhood of $\boldsymbol{c}$, an interior point of $\Omega$

Then, if the set $\Omega$ contains all of its points of accumulation, it is called a *closed set*. The *closure* $\overline{\Omega}$ of a set $\Omega$ is defined to be the union of $\Omega$ and its points of accumulation. The *boundary* $\operatorname{bdy}\Omega$ of $\Omega$ is defined by $\operatorname{bdy}\Omega = \Omega - \{\text{points of accumulation}\}$.

**Example**

21. The unit square $\overline{\Omega} = \Omega \cup \Gamma = \{\boldsymbol{x} : \ \boldsymbol{x} \in \mathbb{R}^2, \ 0 \le x_1 \le 1, 0 \le x_2 \le 1\}$ is closed (see the previous example); however $\Omega$ is not closed since all the points lying on $\Gamma$ are limit points of $\Omega$ but do not belong to $\Omega$.

**Compactness.** A set $\Omega \subset \mathbb{R}^n$ is said to be *compact* if every sequence of elements in $\Omega$ has a point of accumulation in $\Omega$. As in the case of $\mathbb{R}$ and $\mathbb{C}$,

$\Omega$ is compact if and only if it is *closed* and *bounded.*

**Domains in $\mathbb{R}^n$.** We now describe the kinds of sets in $\mathbb{R}^n$ that are of greatest relevance. First, we define a *connected set* $\Omega$ in $\mathbb{R}^n$ to be a set which has the property that *every* pair of points in $\Omega$ can be connected by a curve that lies entirely in $\Omega$. Examples of connected and disconnected set are shown in Figure 1.12.

We define next a *domain* in $\mathbb{R}^n$ to be an *open connected* set in $\mathbb{R}^n$. Domains are central to the consideration of boundary value problems, as the examples in the Introduction indicate. Our interest is exclusively confined to domains in $\mathbb{R}$, $\mathbb{R}^2$, and occasionally in $\mathbb{R}^3$; in the case of $\mathbb{R}^2$ and $\mathbb{R}^3$ the boundary $\Gamma$ (that is, the curve (in $\mathbb{R}^2$) or surface (in $\mathbb{R}^3$) within which all points of the domain lie) is assumed to be sufficiently smooth, in the sense that it possesses no cusps or suchlike singularities. Examples of admissible and inadmissible domains are also shown in Figure 1.12.

Later, it is necessary to be more precise about what is meant by an admissible domain, and there we define what is called a Lipschitz domain; this is in a sense the standard "nice" domain with which one works in the context of boundary value problems.

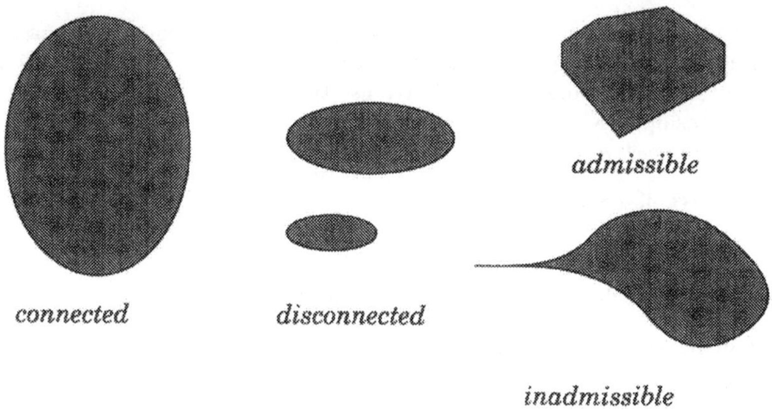


FIGURE 1.12. Connected and disconnected sets, and admissible and inadmissible domains

## 1.4 Relations, equivalence classes, and Zorn's lemma

Having acquired some familiarity with sets of a simple nature, we return now to abstract set theory, and develop some ideas that are useful later. We return to the notion of ordered pairs, and the concept of a relation.

**Relations.** Formally, if we have two sets $A$ and $B$ with Cartesian product $A \times B$, then a *relation* is a subset of $A \times B$. However, this formal definition obscures an intuitively simple interpretation of what constitutes a relation. Essentially we wish to formalize the notion that, given two sets, *some* members of the one set may be related to some members of the other in a special way. A few examples should help to clarify these ideas.

**Examples**

22. Let $A$ be the set of all men (in a given community, say), and $B$ the set of all women. Then for $a \in A$ and $b \in B$, "$a$ is the husband of $b$" defines a relation on $A \times B$.

23. Let $A = \{2, 3, 4\}$, $B = \{3, 4, 5, 6\}$, and consider the relation "$y$ is divisible by $x$", for $(x, y) \in A \times B$. Then the subset making up this relation is

$$\{(2,4), (2,6), (3,3), (3,6), (4,4)\}.$$

The kinds of relations that are particularly useful are those that have some well-defined structure built into them, and we now consider some of these special types of relations, with particular reference to the case in which the ordered pairs come from a single set; that is, given a set $A$, we consider relations on $A \times A$. We use the notation "$\sim$" to indicate the relationship between two elements of a set $A$.

Given a set $A$, a relation $\sim$ on $A$ is

**reflexive** if $a \sim a$,

**symmetric** if $a \sim b$ implies that $b \sim a$,

**antisymmetric** if $a \sim b$ and $b \sim a$ imply that $a = b$,

**transitive** if $a \sim b$ and $b \sim c$ imply that $a \sim c$,

for all $a, b, c \in A$.

**Equivalence relations.** A relation that is *reflexive, symmetric, and transitive* is called an equivalence relation.

**Partial and linear orderings.** A relation $\sim$ on a set $A$ is a partial ordering if it is *reflexive, antisymmetric, and transitive.* It is conventional to denote a partial ordering by the suggestive symbol "$\leq$" rather than the generic "$\sim$", since the standard operation "$\leq$" on the set $\mathbb{R}$ in fact defines a partial ordering on that set.

Finally, a *partially ordered set* is a pair $(A, \leq)$, where $A$ is a set and $\leq$ is a partial ordering on $A$.

If $(A, \leq)$ is a partially ordered set and $\leq$ also satisfies the condition

$$x \leq y \quad \text{or} \quad y \leq x \quad \text{for every } x, y \in A,$$

then the set $A$ is said to be *linearly ordered*, and $\leq$ is called a *linear ordering on $A$.*

### Examples

24. Consider the relation "$<$" on the real line. This is not reflexive since, for any real number $x$, $x \not< x$. It is also not symmetric, though it is transitive: $x < y$ and $y < z$ imply that $x < z$.

25. As mentioned earlier, the operation "$\leq$" defines a partial ordering on $\mathbb{R}$; it is reflexive ($x \leq x$), antisymmetric ($x \leq y$ and $y \leq x$ imply that $x = y$), and transitive ($x \leq y$ and $y \leq z$ imply that $x \leq z$). Note that it is not, however, an equivalence relation, since it is not symmetric ($x \leq y$ does *not* imply that $y \leq x$).

26. Let $\mathcal{F}$ be a family of sets; that is, $\mathcal{F}$ is a set whose members are themselves sets. Then set inclusion $\subset$ is a partial ordering on $\mathcal{F}$; note in particular that for any two sets $A$ and $B$ in $\mathcal{F}$, $A \subset B$ and $B \subset A$ imply that $A = B$.

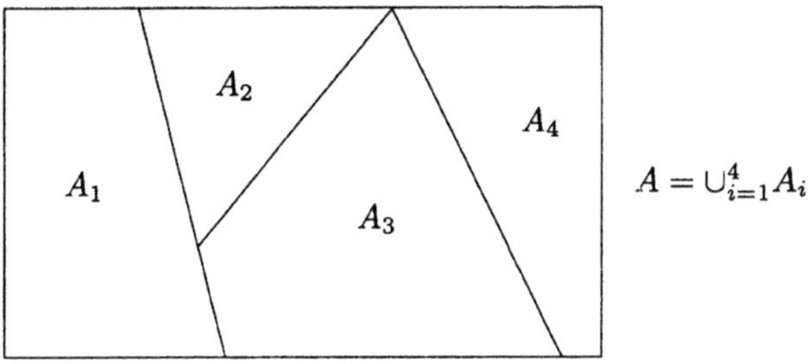


FIGURE 1.13. Illustration of the concept of a partition

27. Let $A$ be the set of triangles in the plane, and let "$\sim$" be the relation defined by "is similar to". Then this defines an equivalence relation on $A$.

**Partitions and equivalence classes.** Let $A$ be any set, and suppose that it is possible to define subsets $A_1$, $A_2$, ... of $A$ which have the properties that

(i) the sets $A_i$ are pairwise disjoint; that is, $A_i \cap A_j = \varnothing$ for all $i$, $j = 1, 2, \ldots$ such that $j \neq i$;

(ii) $A_1 \cup A_2 \cup \ldots = A$.

Then the family of sets $\{A_1, \ldots\}$ is called a *partition of* $A$. The motivation for this name is easily understood if one considers Figure 1.13, which illustrates the concept for the case of a set $A$ in $\mathbb{R}^2$.

**Examples**

28. Let $X = \{1, 2, 3, \ldots, 9\}$, $A = \{1, 4, 7\}$, $B = \{2, 3, 5, 6\}$, and $C = \{7, 8, 9\}$. Then $\{A, B, C\}$ is *not* a partition of $X$; $X = A \cup B \cup C$ but $A \cap C = \{7\} \neq \varnothing$.

29. Consider the plane $\mathbb{R}^2$ and the family of subsets $A_a$ defined by $A_a = \{\boldsymbol{x} \in \mathbb{R}^2 : x_2 = a\}$. Thus $A_a$ is the set of points lying on the horizontal line $x_2 = a$. Then $\{A_a : a \in \mathbb{R}\}$ defines a partition on $\mathbb{R}^2$.

It turns out that there is a close relation between partitions and equivalence relations, and we explore this next. Suppose that $\sim$ defines an equivalence relation on $A$, and for each $a \in A$, define the set $A_a$ by

$$A_a = \{x \in A : x \sim a\}.$$

Then $A_a$ is called an *equivalence class* determined by $a$. In Example 29, the equivalence relation $\boldsymbol{x} \sim \boldsymbol{y}$ may be defined on $\mathbb{R}^2$ by $x_2 = y_2$; then the horizontal line $A_a$ passing through $a$ is the equivalence class defined by $a$. We show that the family of equivalence classes in fact defines a partition of $A$.

First, we note the result that if $b \in A_a$, then $A_b = A_a$. To see this, observe that by definition, $b \sim a$. Now take any $x \in A_b$ such that $x \sim b$; then since the relation is transitive, $x \sim a$. Thus $A_b \subset A_a$. A repetition of the argument, starting with $x \in A_a$, yields the result that $A_a \subset A_b$, so that $A_b = A_a$, as desired.

We note also, by way of a preliminary result (see Exercise 1.23), that if two equivalence classes $A_a$ and $A_b$ contain at least one element in common, then they are in fact equal; that is,

$$\text{if } A_a \cap A_b \neq \varnothing, \text{ then } A_a = A_b.$$

We are now ready to prove the main result.

THEOREM 2. *Let $\sim$ be an equivalence relation on a set $A$. Then the equivalence classes defined by $\sim$ constitute a partition of $A$.*

PROOF. We must show that the set of equivalence classes $\{A_a : a \in A\}$ satisfies

(i) $\cup\{A_a : a \in A\} = A$;
(ii) $A_a \cap A_b = \varnothing$ for $b \not\sim a$.

To prove (i), let $B = \cup\{A_a : a \in A\}$. Then any $b \in B$ belongs to some $A_a$, for a suitable choice of $a$, and hence $b$ belongs also to $A$. Thus $B \subseteq A$. Next, take any $c \in A$. By reflexivity we have $c \sim c$, or $c$ belongs to $A_c$, and hence $c$ belongs also to $B$. It follows that (i) holds.

The proof of (ii) follows directly from the result of Exercise 1.23, which implies that if $A_a \neq A_b$, then $A_a$ and $A_b$ must be disjoint. □

**Upper and lower bounds, maximal and minimal elements.** The analogy between the operation $\leq$ on the real line, and the notion of a partially ordered set, may be exploited further by extending to the more general situation, in an appropriate way, properties on the real line that arise from the use of $\leq$. Thus, suppose that $(P, \leq)$ is a partially ordered set, and $A$ a subset of $P$. Then a member $u$ in $P$ is called an *upper bound* of $A$ if $x \leq u$ for all $x \in A$. An element $m$ in $P$ is called a *maximal element* of $P$ if $x \in P$ and $m \leq x$ imply that $m = x$. The dual concepts of a lower bound and a minimal element are defined similarly (cf. the definitions of infimum and minimum in Section 1.2).

The notions of upper bounds and maximal elements bring us to a fundamental axiom of mathematics, known as Zorn's Lemma. Very often one finds in functional analysis, and also in other branches of mathematics such as algebra and topology, that the elementary notions of set theory, as have been presented in this chapter, are not sufficient to allow proofs or definitions to be constructed satisfactorily, or at all. Just as it is necessary in elementary mechanics to invoke a number of self-evident truths, or axioms

– for example, Newton's laws of motion – in order to construct a theory of moving bodies, in the same way it becomes necessary to introduce into the general mathematical framework axioms of set theory that are deemed to be generally acceptable, in order to be able to proceed with the business of constructing theories. Zorn's Lemma is one such axiom.

ZORN'S LEMMA. *Let $A$ be a nonempty partially ordered set. If every linearly ordered subset of $A$ has an upper bound, then $A$ has a maximal element.*

We encounter an application of Zorn's Lemma in Chapter 6, in order to prove a theorem about orthonormal bases in Hilbert spaces.

It is possible, instead of invoking Zorn's lemma in a particular situation, to make use of one of many alternative axioms, should one of these alternatives prove to be more appropriate to the situation at hand. We do not get into a detailed discussion here about the various alternative axioms since these are rather peripheral to further developments; but we mention for completeness one such alternative, the *Axiom of Choice*, which is often encountered. First, we introduce the notion of a choice function.

**Choice functions.** Let $\mathcal{F} = \{A_1, A_2, \ldots\}$ be a family of sets, and suppose that we choose from each set $A_i$ a member $a_i$, say. The resulting set of choices $\{a_1, a_2, \ldots, a_i, \ldots\}$ is known as a choice function. We deal with functions or maps in detail in Chapter 5, and it suffices for now to recall that a function $f$ from a set $X$ to a set $Y$ is a rule that associates with each member $x \in X$ exactly one member $y \in Y$. In the present context the choice function acts on the family $\mathcal{F}$, and associates with each member $A_i$ exactly one member $a_i$ of that set. Note that as we run through the various permutations or choices, we actually set up the Cartesian product $A_1 \times A_2 \times \ldots$.

Now one may ask whether, for a given family of sets, there are any choice functions; in other words, we should like to know whether it is *always* possible to select one member from each of the sets in $\mathcal{F}$. The question is trivial for finite sets, but for infinite sets it is not, and it turns out that this question cannot be answered in general using only the usual axioms of set theory. This motivates the introduction of the Axiom of Choice.

**Axiom of Choice.** *Let $\mathcal{F} = \{A_1, A_2, \ldots\}$ be a family of nonempty sets. Then there exists at least one choice function for the family. That is, the Cartesian product of any nonempty family of nonempty sets is a nonempty set.*

This rather innocent-looking axiom has as a consequence some very important and deep results in mathematics, as the following shows.

THEOREM 3. *Zorn's Lemma and the Axiom of Choice are equivalent axioms.*

We omit the rather lengthy proof of this theorem.

## 1.5 Theorem proving

We end this chapter with a collection of notions and procedures that are central to proving results in mathematics. No doubt these will have been previously encountered in various guises, but it is well to reiterate these notions in one place, in order for there to be clarity about what constitutes a proof, and about some of the ideas that are used in constructing proofs.

**Necessity and sufficiency.** Suppose that we are given two mathematical statements, labeled A and B, and suppose we are told that if A holds true, then so does B. For example, A may be the statement "$a = 2$", and B the statement "$a^3 = 8$". We may put this another way by stating that statement A *implies* statement B, and by writing

$$A \Longrightarrow B.$$

Yet another way of expressing the relation between A and B is to assert that B holds *if* A holds, or that a *sufficient condition* for B to hold, is that A holds. These three different ways of stating the same fact are summarized in the following.

| |
|---|
| A $\Longrightarrow$ B |
| B holds *if* A holds |
| a *sufficient condition* for B to hold is that A holds |

Now consider the converse, that is, the case in which B implies A, or B $\Rightarrow$ A. Of course we could simply go back and transpose A and B in all the preceding statements; but it is useful to consider this relationship from a diffferent angle. Specifically, we may now state that B $\Rightarrow$ A is the same as stating that B holds *only if* A holds, which is to say that, if A does not hold, then neither does B. A third way of making this assertion is to state that *a necessary condition* for B to hold is that A holds. Going back to the simple example, we can state that a necessary condition for $a^3 = 8$ to hold is $a = 2$. We summarize again.

| |
|---|
| B $\Longrightarrow$ A |
| B holds *only if* A holds |
| a *necessary condition* for B to hold is that A holds |

This brings us to the third possibility, which is that statements A and B imply each other. In this case the two statements are equivalent, and we write $A \Leftrightarrow B$; furthermore, for B to hold it is now *necessary and sufficient* that A holds. We now have

| |
|---|
| B $\Longleftrightarrow$ A |
| B holds *iff* A holds |
| a *necessary and sufficient condition* for B to hold is that A holds |

The term "iff" is shorthand for "if and only if". In the context of proofs of theorems and the like, when faced with the task of showing that statement A is true if and only if B is true, the typical approach is a two-stage one:

- sufficiency (if): assume that B holds, and show that this implies A;
- necessity (only if): assume that A holds, and show that this implies B.

**Example**

30. Let A be the statement "$a^2 > 4$" and B the statement "$a > 2$". Assume first that B is true; then clearly A is true. Thus B is a sufficient condition for A to hold. Conversely, assume that A is true; this implies that $|a| > 2$; that is, $a < -2$ and $a > 2$. Thus A is *not* a sufficient condition for B to hold; alternatively, B is *not* a necessary condition for A to hold (since $a > -2$ would also be acceptable). The two statements are therefore not equivalent. On the other hand, if A is the statement "$a^2 > 4$ and $a > 0$", and B is the statement "$a > 2$", then A and B are equivalent.

**Reductio ad absurdum or proof by contradiction.** The method of reductio ad absurdum is an ancient strategy for constructing proofs. It exploits the fact that the statement "if A holds, then B holds" is equivalent to the statement "if B does not hold, then A does not hold". Faced with the task of proving that A implies B, the procedure starts off by assuming that B does not hold. The task is then to show that this implies that A is not valid, usually by obtaining a contradiction of the original assumption.

**Example**

31. A classical example of proof by contradiction is the proof that $\sqrt{2}$ is irrational. We begin by assuming that $\sqrt{2}$ is rational. Set $x = \sqrt{2}$; since this is rational by assumption, we may write $x = p/q$ for some integers $p$ and $q$ with $q$ nonzero. It may also be assumed that $p$ and $q$ have no common divisor (if they do, this may be divided out). Thus $x^2 = 2 = p^2/q^2$, or $p^2 = 2q^2$, which implies that $p^2$ is even. Therefore $p$ is even, and since it is divisible by 2, $p^2$ is divisible by 4. Since $q^2 = p^2/2$, $q^2$ is therefore even. But then 2 is a common divisor of $p$ and $q$, which constitutes a contradiction. Thus $\sqrt{2}$ is irrational.

## 1.6 Bibliographical remarks

There is a wide range of books that deal with the subject matter of this chapter. Very readable accounts of the real and complex numbers systems are to be found in the texts by Apostol [2], Binmore [6, 7], Lang [29], and Royden [44]. Oden [36] presents a fairly detailed account of the algebra of sets, with an applications-oriented readership in mind. The text by Lipschutz [31] in the Schaum's Outline Series provides an accessible account of set theory, replete with hundreds of examples and exercises, which would take the reader some way beyond the contents of this chapter. Finally, the first chapter of the monograph by Hewitt and Stromberg [19] provides a treatment that is more detailed and somewhat more advanced, although very well written, of set theory and of the real and complex numbers.

## 1.7 Exercises

**The algebra of sets**

1.1. Let $A = \{x \in \mathbb{Z} : x^2 - x - 6 = 0\}$ and $B = \{x \in \mathbb{Z} : x^2 < 10\}$. List the elements of $A$ and $B$. What are $A \cup B$, $A \cap B$, $A \cap \mathbb{Z}^+$, and $A - \mathbb{Z}^+$ ?

1.2. Let $A = \{1,2\}$, $B = \{7,8\}$, and $C = \{9,1\}$. Find $B \times (A \cup C)$ and $(A \cap C) \times B$.

1.3. Show that

$$A \cap (B \cup C) = (A \cap B) \cup (A \cap C),$$
$$A \cup (B \cap C) = (A \cup B) \cap (A \cup C).$$

Illustrate these identities.

1.4. Let $n(A)$ denote the number of elements of a finite set $A$. Prove that

$$n(A \cup B) = n(A) + n(B) - n(A \cap B).$$

How would you generalize this identity to $n(A \cup B \cup C)$?

1.5. The power set of a set $A$, denoted by $2^A$ or $\mathcal{P}(A)$, is the set of all subsets of $A$. What are $\mathcal{P}(A)$ and $\mathcal{P}(B)$ if $A = \{1, 2, 3\}$ and $B = \{\{1, 2\}, 3\}$?

**Sets of numbers**

1.6. Show that the set $\mathbb{Q}$ of rational numbers is countable. [Hint: Set up a table of the form

$$\begin{array}{cccc} 1/1 & 1/2 & 1/3 & \cdots \\ 2/1 & 2/2 & 2/3 & \cdots \\ 3/1 & \cdots & & \end{array}$$

].

1.7. Find all the points of accumulation of the following subsets of $\mathbb{R}$.

(i) $[a, b]$; (ii) $\mathbb{Q}$; (iii) $(0, 1) \cup \{2\}$.

1.8. Which of the following subsets of $\mathbb{R}$ are closed, open, or neither?

(i) $A = \{x :\ \sin(1/x) = 0\}$; (ii) $A = \{x :\ x \sin(1/x) = 0\}$; (iii) $A = \{x :\ \sin(1/x) > 0\}$.

1.9. Show that a set $I \subset \mathbb{R}$ is closed if and only if its complement is open.

1.10. Find all the points of accumulation of the set $A = \{z \in \mathbb{C} : z = x + iy,\ x^2 - y^2 < 1\}$, and determine whether this set is open or closed.

1.11. Write down the first few terms of the following sequences.

(i) $\{(-1)^n/n\}_{n=1}^{\infty}$;
(ii) $\{\frac{1}{2}(1 - (-1)^n)\}_{n=1}^{\infty}$;
(iii) $\{3n^2/(5n^2 - 6)\}_{n=1}^{\infty}$.

1.12. Determine which of the following sequences are convergent, and find their limits.

(i) $\dfrac{(4 - 2n - 3n^2)}{(2n^2 + n)}$; (ii) $\dfrac{(-1)^n n^4}{2 + n^4}$; (iii) $\dfrac{n}{1 + n}$.

1.13. The sequence $\{(3n+2)/(n-1)\}$ converges to 3 as $n \to \infty$. Find the smallest integer $N$ such that

$$\left|\frac{3n+2}{n-1} - 3\right| < \epsilon$$

whenever $n > N$, for the case $\epsilon = 0.001$.

1.14. A sequence $\{u_n\}$ is *bounded* if there are constants $M$ and $N$ such that $M \le u_n \le N$ for all $n$. Also, $u_n$ is *monotone increasing* if $u_{n+1} \ge u_n$ for all $n$, and *monotone decreasing* if $u_{n+1} \le u_n$ for all $n$. Show that every bounded monotone (increasing or decreasing) sequence converges, and that the limit is the supremum (or infimum).

1.15. Find $\max A$, $\min A$, $\sup A$, and $\inf A$ when

(i) $A = \{1/n : \ n = 1, 2, 3, \ldots\}$;

(ii) $A = \{x : 0 < x^2 < 1\}$;

(iii) $A = \{x : \ (x-a)(x-b)(x-c) < 0, \ a < b < c\}$;

(iv) $A = \{|z^2 + 1| : \ z \in \mathbb{C}, \ |z| \le 1\}$ .

1.16. Show that

$$\inf A = -\sup(-A).$$

1.17. Suppose that $A$ and $B$ are two sets of real numbers that are bounded above, with $\sup A = a$ and $\sup B = b$. Let $C$ be the set of real numbers formed by considering all products of the form $xy$, where $x \in A$ and $y \in B$. Give a counterexample to show that, in general, $\sup C \ne ab$.

1.18. Show that the supremum has the following properties.

(i) If $I \subset \mathbb{R}$ and $\alpha$ is any positive real number, then $\sup_{x \in I} \alpha x = \alpha \sup_{x \in I} x$;

(ii) if $I \subset \mathbb{R}$ and $\alpha$ is any real number, then $\sup_{x \in I}(\alpha + x) = \alpha + \sup_{x \in I} x$.

**Subsets of $\mathbb{R}^n$**

1.19. Determine the points of accumulation of the following sets and establish which of these sets are open, closed, or neither.

(i) $\Omega = \{\boldsymbol{x} : \ \boldsymbol{x} \in \mathbb{R}^2, \ 0 \le x \le 1, \ 0 < y \le 1\}$;

(ii) $\Omega = \{\boldsymbol{x} : \ \boldsymbol{x} \in \mathbb{R}^3, \ x^2 + y^2 + z^2 < a^2, \ z > 0\}$.

1.20. The *diameter* dia $(\Omega)$ of a set in $\mathbb{R}^n$ is defined by

$$\operatorname{dia}\Omega = \sup\{|\boldsymbol{x} - \boldsymbol{y}| : \boldsymbol{x}, \boldsymbol{y} \in \Omega\}.$$

Find dia $(\Omega)$ for the sets in Exercise 1.19.

**Relations, equivalence classes, and Zorn's lemma**

1.21. Which of the following statements defines (i) an equivalence relation; and (ii) a partial ordering on the set of natural numbers $\mathbb{N}$?

(a) $\beta$ is a multiple of $\alpha$; (b) $\alpha\beta$ is the square of a number;
(c) $\alpha + 2\beta = 6$; (d) $\alpha$ divides $\beta$.

1.22. Let $\sim$ be the relation on the set $A = \{2, 3, 4, 5, 6\}$ defined by the statement "$|a - b|$ is divisible by 3". Write $\sim$ as a set of ordered pairs, that is, as a subset of $A \times A$, and represent it graphically as a set of points in the plane.

1.23. If $\sim$ defines an equivalence relation on a set $A$, show that if $A_a \cap A_b \neq \varnothing$, then $A_a = A_b$.

1.24. Consider the relation on $\mathbb{Z}\times\mathbb{Z}$ in which $\boldsymbol{a} \sim \boldsymbol{b}$ is defined by $|a_1| + |a_2| = |b_1| + |b_2|$. Show that this is an equivalence relation, and illustrate the manner in which $\mathbb{Z} \times \mathbb{Z}$ is partitioned.

# 2

# Sets of functions and Lebesgue integration

In due course we endow sets with particular properties and on the basis of these assumed properties construct a theory for special kinds of sets such as Hilbert spaces. In the development of this theory it is not necessary to appeal to the precise character of a set: the basic axioms, and the theorems that follow from these axioms, apply equally to sets whose members are numbers or matrices or functions. Before embarking on the task of describing this general framework, however, we first introduce two important examples of sets, or *spaces* (as they are usually called when endowed with additional properties) of functions: these are the spaces of continuous functions, and the $L^p$ spaces of functions whose $p$th powers are integrable. With these at our disposal it is possible in subsequent chapters to illustrate aspects of the general theory, using as special examples sets such as $\mathbb{R}$ or $\mathbb{R}^n$ which were introduced in the last chapter, as well as spaces of functions.

In Section 2.1 the concept of continuity is introduced, and the space $C^m(\Omega)$ of $m$-times continuously differentiable functions is defined.

There are of course many well-behaved functions that are not continuous, and that also feature in the developments to follow; an example is the Heaviside step function. These functions need to be characterized in an alternative manner, and this is done by exploiting not the degree of continuity or smoothness of the function, but rather its *integrability*. This process leads naturally to the definition of the $L^p$ spaces. In order to discuss these adequately it is necessary first, however, to extend the definition of the integral encountered in elementary courses on calculus; this is the Riemann integral, and it is not adequate for our purposes. Its extension, known as the *Lebesgue* integral, in turn relies on an acquaintance with the

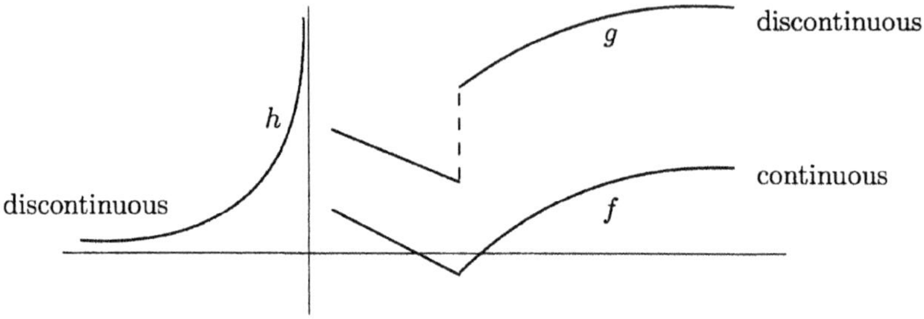


FIGURE 2.1. Examples of continuous and discontinuous functions

notion of Lebesgue measure, which is the subject of Section 2.2. Section 2.3 is then devoted to a discussion of the $L^p$ spaces.

## 2.1 Continuous functions

Continuous functions occur in great abundance in applied mathematics and engineering. This is not surprising, since many natural phenomena that are modeled mathematically involve quantities which may be represented (perhaps approximately) by continuous functions. Our aim in this section is to describe, rather informally at first, the concept of continuity, and subsequently to arrive at a mathematically suitable definition of a continuous function.

We begin by considering two arbitrary functions $f$ and $g$ whose graphs are shown in Figure 2.1. The function $f(x)$ is continuous, by which we mean that it is possible to draw its graph without lifting one's pen. On the other hand, the function $g(x)$ is discontinuous, in that its graph has a break. Roughly speaking, then, a continuous function of a single variable may be characterized as one whose graph is an uninterrupted curve. Another type of discontinuous function is one that is *unbounded* at some point. For example, the function $h$ shown in Figure 2.1 is not continuous at $x = 0$ since $h(x)$ "tends to infinity" as $x$ approaches zero. Again, it is not possible to represent $h(x)$ by an uninterrupted curve.

The preceding examples give a qualitative feel for what constitutes a continuous function. However, for subsequent work we need a definition of continuity that agrees with our intuition and which is also sufficiently robust to be used in any mathematical situation. The following is a suitable definition.

**Continuous functions of one variable.** Let $f$ be a function on an interval $I$ (open or closed) of the real line. Then $f$ is *continuous* at a point $x_0 \in I$ if, *given* any positive number $\epsilon$, no matter how small, it is possible

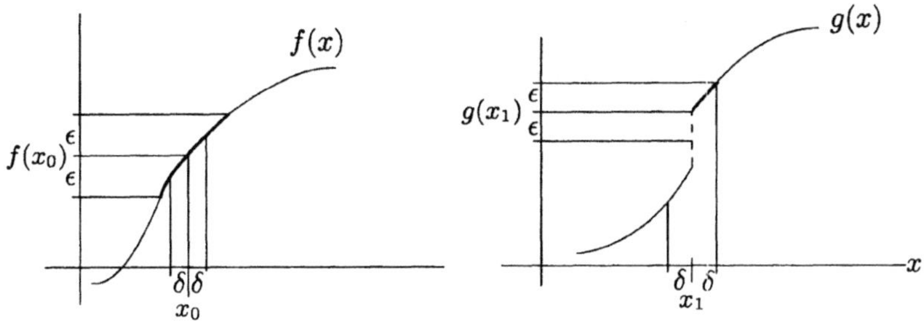


FIGURE 2.2. Graphical interpretation of the definition of continuity

to find a positive number $\delta$ (depending on $\epsilon$ and on the point $x_0$) such that

$$|f(x) - f(x_0)| < \epsilon \qquad \text{for all } x \text{ in } I \text{ with} \qquad |x - x_0| < \delta. \tag{2.1}$$

If $f$ is continuous *at all points in* $I$ then $f$ is said to be *continuous on* $I$. Generally the number $\delta$ will vary from point to point for a given value of $\epsilon$, but if it so happens that $\delta$ depends only on $\epsilon$ and not on $x$, then we say that $f$ is *uniformly continuous* on $I$1.

The preceding definition of continuity has the advantage of a very simple geometrical meaning. To see this, consider the graph of the function $f$ shown in Figure 2.2. Choose a positive number $\epsilon$ and draw lines parallel to the $x$ axis at heights $f(x_0) \pm \epsilon$. Then $f$ is continuous at $x_0$ if we can find a positive number $\delta$ such that the graph of $f(x)$ lies inside the horizontal band bounded by $f(x_0) \pm \epsilon$ (that is, $|f(x) - f(x_0)| < \epsilon$) for all values of $x$ in the vertical band $|x - x_0| < \delta$. In other words, the whole portion of the graph lying in the vertical band is contained in the horizontal band. For the function $f$ shown in Figure 2.2 this is clearly possible at any point $x_0$ lying in $I$, no matter how small a value of $\epsilon$ is chosen. Thus $f$ is continuous. On the other hand, the function $g$ in Figure 2.2 is continuous at all points in $I$ *except* at $x = x_0$; no matter how small we make $\delta$, the vertical band will always include a portion of the graph that lies outside the horizontal band.

## Examples

1. $f(x) = x^2$ is continuous on $[0, 1]$: to see this, consider

$$\begin{aligned} |f(x) - f(x_0)| &= |x^2 - x_0^2| = |(x - x_0)(x + x_0)| \\ &\le 2|x - x_0| \quad \text{for} \quad x, x_0 \in [0, 1]. \end{aligned}$$

If $|x - x_0| < \delta$, then $|f(x) - f(x_0)| < 2\delta$. Hence, if $\epsilon$ is given, it suffices to take $\delta = \frac{1}{2}\epsilon$ to guarantee that $|f(x) - f(x_0)| < \epsilon$ whenever $|x - x_0| < \delta$. Since $\delta$ does not depend on $x_0$, $f$ is also uniformly continuous on $[0, 1]$.

2. $f(x) = 1/x$ is continuous on the half-open interval $(0, 1]$. To show this, we begin by considering

$$|f(x) - f(x_0)| = \frac{|x_0 - x|}{|x|\,|x_0|}$$

for an arbitrary fixed $x_0$ in $(0, 1]$. Let $0 < \delta < x_0$; then every $x$ in the interval $|x - x_0| < \delta$ satisfies $x > x_0 - \delta > 0$ and we have

$$|f(x) - f(x_0)| < \frac{\delta}{x_0(x_0 - \delta)}.$$

Then if $\epsilon$ is given, we choose $\delta = \epsilon x_0^2/(1 + \epsilon x_0)$ (this is found by setting $\delta/x_0(x_0 - \delta) = \epsilon$). Thus $f$ is continuous for any $x_0$ in $(0, 1]$. Note that $\delta$ depends on $\epsilon$ and on $x_0$, so we have not been able to prove that $f$ is *uniformly* continuous. Indeed, it can be shown (see Exercise 2.3) that $f$ is *not* uniformly continuous.

For functions of more than one variable the preceding ideas are easily extended. For example, consider a function $f(\boldsymbol{x}) \equiv f(x, y)$ of two variables defined on an open subset $\Omega$ of $\mathbb{R}^2$, as shown in Figure 2.387q. To check for continuity at a point $\boldsymbol{x}_0 = (x_0, y_0)$ in $\Omega$ we choose a positive number $\epsilon$ and construct a pair of horizontal *planes* at heights $f(\boldsymbol{x}_0) \pm \epsilon$ above the $xy$ plane. Then $f(\boldsymbol{x})$ is continuous at $\boldsymbol{x}_0$ if it is always possible to construct a *cylinder* of radius $\delta$ (that is, the set of points $\boldsymbol{x}$ for which $|\boldsymbol{x} - \boldsymbol{x}_0| < \delta$), this radius depending on $\epsilon$, such that the part of the surface lying within the cylinder is contained in the horizontal band $|f(\boldsymbol{x}) - f(\boldsymbol{x}_0)| < \epsilon$. This is but a special case of the general definition of continuity defined for functions of any number of variables, which we now state.

**Continuity in $\mathbb{R}^n$.** A function $f(\boldsymbol{x})$ defined on a subset $\Omega$ of $\mathbb{R}^n$ is continuous at a point $\boldsymbol{x}_0$ in $\Omega$ if, for every positive number $\epsilon$, no matter how small, it is possible to find a positive number $\delta$ (depending on $\epsilon$ and $\boldsymbol{x}_0$) such that

$$|f(\boldsymbol{x}) - f(\boldsymbol{x}_0)| < \epsilon \quad \text{whenever} \quad |\boldsymbol{x} - \boldsymbol{x}_0| < \delta \text{ and } \boldsymbol{x} \in \Omega. \tag{2.2}$$

If (2.2) holds for every $\boldsymbol{x}_0$ in $\Omega$, then $f$ is said to be continuous on $\Omega$. Furthermore, if $\delta$ does not depend on $\boldsymbol{x}_0$, then $f$ is said to be uniformly continuous on $\Omega$.

**The space $C(\Omega)$.** For any domain $\Omega$ in $\mathbb{R}^n$ the collection of all continuous functions defined on $\Omega$ forms a set, or space, which is denoted by $C(\Omega)$.

For functions defined on a subset $\Omega = (a, b)$ of the real line, we simply write $C(a, b)$. The space of functions that are continuous on the closed set $\overline{\Omega} = \Omega \cup \Gamma$ ($\Omega$ and its boundary $\Gamma$) is denoted by $C(\overline{\Omega})$, and by $C[a, b]$ for functions on the closed interval. There is more than a mere technical

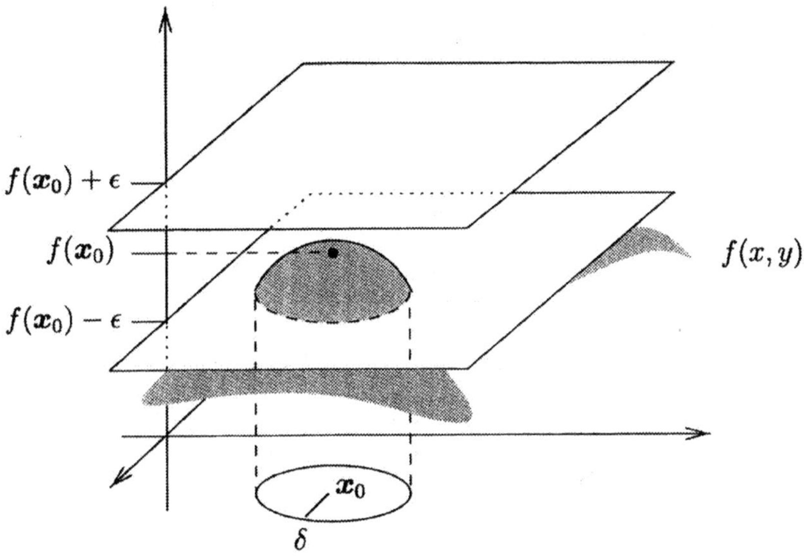


FIGURE 2.3. Continuity of a function of two variables

difference between continuous functions defined on open and closed sets; for example, $u(x) = 1/x$ is continuous on $(0,1)$ but not on $[0,1]$. We return to this point momentarily.

**The spaces $C^m(\Omega)$ and $C^\infty(\Omega)$.** Among all the continuous functions defined on a subset $\Omega$ of $\mathbb{R}^n$, some have the property that their first derivatives and possibly some derivatives of higher order are also continuous. It is important to identify such functions, and so we introduce the space $C^m(\Omega)$ of functions which, together with all of their derivatives up to and including those of order $m$, are continuous on $\Omega$. That is,

$$C^m(a,b) = \{u : u, du/dx, \ldots, d^m u/dx^m \text{ are all continuous functions}\}$$

for $\Omega = (a,b) \subset \mathbb{R}$,

$$C^m(\Omega) = \{u : u, \partial u/\partial x, \partial u/\partial y, \ldots, \partial^m u/\partial x^k \partial y^{m-k} (k = 0, \ldots, m) \text{ are all continuous functions }\}$$

for $\Omega \subset \mathbb{R}^2$, and so on.

We define $C^\infty(\Omega)$ to be the space of functions *all* of whose derivatives exist and are continuous on $\Omega$.

Clearly the inclusions

$$C^\infty(\Omega) \subset \cdots \subset C^m(\Omega) \subset C^{m-1}(\Omega) \subset \cdots \subset C^0(\Omega) = C(\Omega)$$

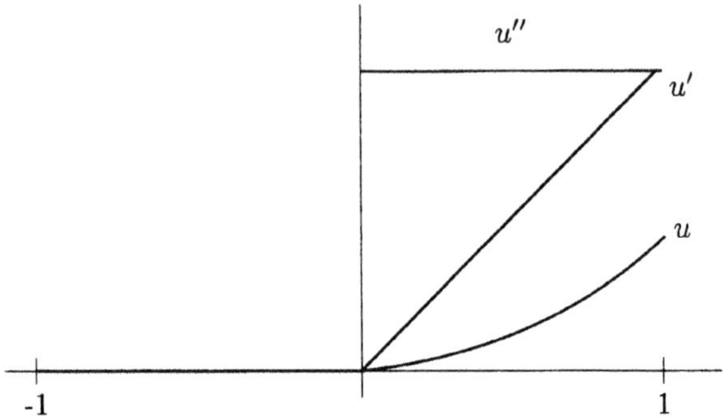


FIGURE 2.4. The function in Example 3, and its first and second derivatives

hold, so that the spaces $C^m$ constitute a gradation which permits continuous functions to be classified according to their degree of smoothness: for any function in $C^m(\Omega)$, the higher the value of $m$, the smoother the function.

## Examples

3. The function

$$u(x) = \begin{cases} 0, & -1 \leq x < 0, \\ x^2, & 0 \leq x \leq 1, \end{cases}$$

belongs to $C^1[-1,1]$ since $u$ and $du/dx$ are both continuous, but $d^2u/dx^2$ is discontinuous (Figure 2.4).

4. $u(x) = \sin x$ belongs to $C^\infty(-\infty, \infty)$ since $u$ and all of its derivatives are continuous on the whole real line.

5. The wedge-shaped function shown in Figure 2.5 is a member of $C(\Omega)$, where $\Omega = (0,1) \times (0,1)$ is the unit square in $\mathbb{R}^2$, but not of $C^1(\Omega)$, since $\partial u/\partial x$ is not a continuous function.

**Continuous functions on compact sets.** We saw earlier that the function $u(x) = 1/x$ is continuous on the open set $(0,1)$, but not on the closed set $[0,1]$, as a result of the singularity at $x = 0$. It turns out that continuous functions defined on compact sets (recall from Chapter 1 that these are closed and bounded sets in $\mathbb{R}^n$) may be characterized further, in the sense that they are necessarily *bounded* on such sets.

A function $f$ defined on a set $\Omega$ in $\mathbb{R}^n$ is said to be bounded if it is possible to find a number $M > 0$ such that

$$|f(\boldsymbol{x})| \leq M \quad \text{for all } \boldsymbol{x} \in \Omega;$$

in other words, the function does not "blow up" anywhere. Another way of characterizing boundedness is to consider the set $f(\Omega)$ of values of $f(\boldsymbol{x})$,

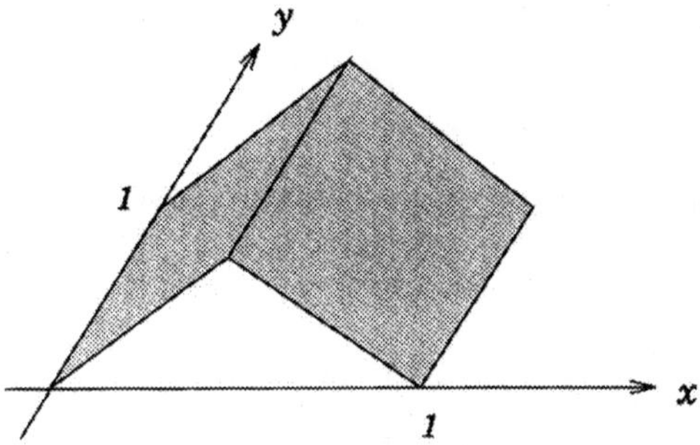


FIGURE 2.5. The wedge-shaped function of Example 5

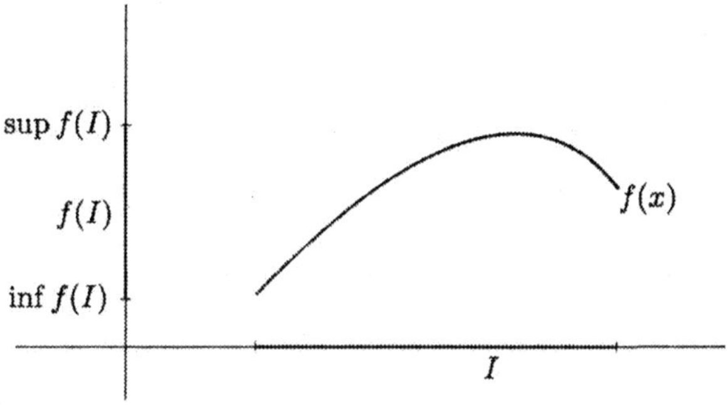


FIGURE 2.6. The image $f(I)$ of the function $f$ defined on an interval $I$ in $\mathbb{R}$

as $\boldsymbol{x}$ ranges over all points in $\Omega$; that is,

$$f(\Omega) = \{y \in \mathbb{R} : \ f(\boldsymbol{x}) = y \ \text{ for some } \boldsymbol{x} \in \Omega\}.$$

This situation is depicted in Figure 2.6, for a function of a single variable. $f(\Omega)$ is called the *range* or *image* of $f$, a concept which is explored in greater detail in Chapter 5. With the notion of the range at our disposal it is possible to characterize a bounded function $f$ as being one for which

$$\sup f(\Omega) \text{ and } \inf f(\Omega) \text{ exist (that is, are finite).}$$

Continuous functions behave in a special way on compact sets, as the next theorem shows.

THEOREM 1. *Let $\Omega$ be a bounded domain (that is, a bounded open, connected set) in $\mathbb{R}^n$, and $f$ a continuous function defined on the compact set $\overline{\Omega}$. Then*

*(a) $f$ is bounded on $\overline{\Omega}$ and, furthermore, $f$ achieves its supremum and infimum on $\overline{\Omega}$;*

*(b) $f$ is uniformly continuous on $\overline{\Omega}$.*

Note that part (a) of the theorem states that there is a point $\boldsymbol{z}$, say, in $\overline{\Omega}$, such that $\boldsymbol{z} = \sup f(\Omega) = \max f(\Omega)$; that is, $f(\boldsymbol{z}) \geq f(\boldsymbol{x})$ for all points $\boldsymbol{x} \in \overline{\Omega}$; a similar interpretation applies with respect to the infimum.

PROOF OF THEOREM 1. We prove (a).

We show first that the function has a maximum; thus we have to show that the function $f$ is indeed bounded above, that is, that $f(\boldsymbol{x}) \leq M$ for some $M$ in the interval.

If $f$ is not bounded from above, then corresponding to each positive integer $m$ it is possible to find a point $\boldsymbol{x}$ in the domain $\overline{\Omega}$ such that $f(\boldsymbol{x}) > m$. By the Bolzano–Weierstrass Theorem applied to sets in $\mathbb{R}^n$, the sequence $\{\boldsymbol{x}_m\}$ constructed in this way has a point of accumulation $\mathbf{c}$ that lies in $\overline{\Omega}$. Furthermore, since $f$ is continuous, given $\epsilon = 1$ there exists $\delta > 0$ such that

$$|f(\boldsymbol{x}) - f(\mathbf{c})| < 1 \quad \text{whenever} \quad |\boldsymbol{x} - \mathbf{c}| < \delta.$$

In particular, this applies to points in the sequence, so that

$$|f(\boldsymbol{x}_m)| - |f(\mathbf{c})| \leq |f(\boldsymbol{x}_m) - f(\mathbf{c})| \leq 1,$$

whence

$$m < |f(\boldsymbol{x}_m)| \leq 1 + |f(\mathbf{c})|.$$

For $m$ sufficiently large this is a contradiction. Thus $f$ is bounded from above.

Suppose then that the least upper bound of all the values of $f(\boldsymbol{x})$ is $\ell$. Then, given a positive integer $m$ we can find a point $\boldsymbol{y}_m$ in $\overline{\Omega}$ such that

$$|f(\boldsymbol{y}_m) - \ell| < 1/m.$$

Denote by $\boldsymbol{d}$ a point of accumulation of the sequence of points $\{\boldsymbol{y}_m\}$; then $f(\boldsymbol{d}) \leq \ell$. The proof of the theorem follows if we can show that in fact $f(\boldsymbol{d}) = \ell$. Given $\epsilon$, there exists $\delta$ such that

$$|f(\boldsymbol{y}_m) - f(\boldsymbol{d})| < \epsilon \quad \text{whenever} \quad |\boldsymbol{y}_m - \boldsymbol{d}| < \delta.$$

This is true for infinitely many values of $m$ since $\boldsymbol{d}$ is a point of accumulation. But

$$|f(\boldsymbol{d}) - \ell| \leq |f(\boldsymbol{d}) - f(\boldsymbol{y}_m)| + |f(\boldsymbol{y}_m) - f(\boldsymbol{d})| < \epsilon + 1/m.$$

This holds for every $\epsilon$ and $m$; hence $|f(\boldsymbol{d}) - \ell| = 0$, as was to be shown.

The proof for the minimum is carried out in much the same way. □

**Examples**

6. Consider the function $u(x) = \sin x$ defined on $[0, 2\pi]$, which is closed. The supremum of $u(x)$ is 1 which is achieved at $x = \pi/2$, whereas the infimum is $-1$ which is achieved at $x = 3\pi/2$. Theorem 1 tells us that $u$ is uniformly continuous.

7. Let $u(x) = 1/x$; we have seen earlier that this function is continuous on the open interval $(0, 1)$, but that it is *not* uniformly continuous there (see also Exercise 2.3). It is not continuous on $[0, 1]$; furthermore, $\inf u = 1$ (at $x = 1$), but $\sup u$ does not exist.

8. Note that Theorem 1 gives *sufficient* conditions for a function to be bounded and uniformly continuous. These are not *necessary* conditions, however; for example, if $u(x) = x^2$ on $(0, 1)$, then $\sup u = 1$, $\inf u = 0$, and the function $u$ is uniformly continuous, although it achieves its supremum and infimum (at $x = 0$ and $x = 1$, respectively) outside the set $(0, 1)$, which is open.

**Lipschitz continuous functions.** A function $f$ defined on a set $\Omega$ in $\mathbb{R}^n$ is said to be Lipschitz continuous (or simply Lipschitz) if there exists a constant $L > 0$ such that

$$|f(\boldsymbol{x}) - f(\boldsymbol{y})| \leq L|\boldsymbol{x} - \boldsymbol{y}| \quad \text{for all } \boldsymbol{x},\, \boldsymbol{y} \in \Omega. \tag{2.3}$$

It is straightforward to show (Exercise 2.10) that every Lipschitz function is uniformly continuous, although of course the converse is not true. This may be better appreciated by considering the interpretation of Lipschitz continuity for functions of a single variable (Figure 2.7): (2.3) states that the slope of the chord joining any two points on a Lipschitz function is bounded above by a constant $L$ which is independent of the two points. We see also that the definition of Lipschitz continuity does not require that the derivative exist at every point. However it is not difficult to show that, if $\Omega$ is a compact set, then every continuously differentiable function on $\Omega$ is Lipschitz.

## 2.2 Measure of sets in $\mathbb{R}^n$

Many functions that occur in practical applications are not continuous, and cannot therefore be accommodated in one of the spaces $C^m(\Omega)$. A simple example is the Heaviside step function, which has many applications in physics and engineering, and which is defined by

$$H(x) = \begin{cases} 0, & x \leq 0, \\ 1, & x > 0. \end{cases}$$

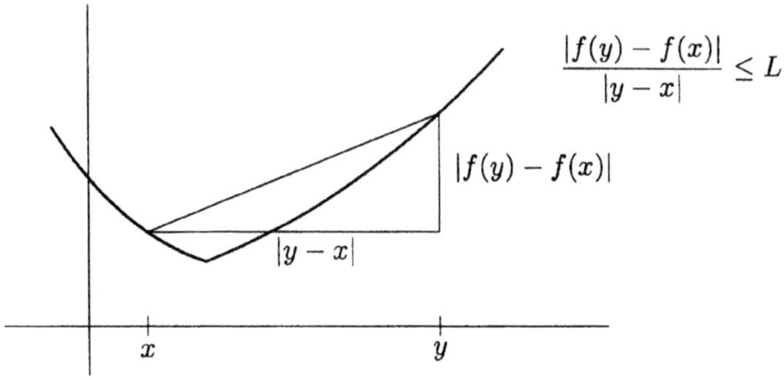


FIGURE 2.7. A Lipschitz continuous function of a single variable

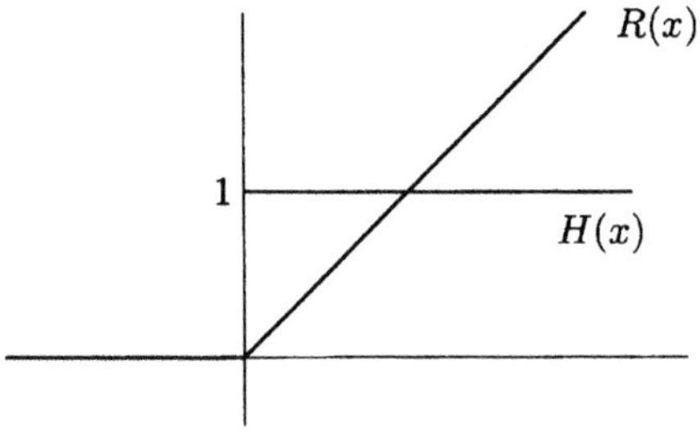


FIGURE 2.8. The Heaviside step function $H(x)$ and its integral, the ramp function $R(x)$

Though functions like $H(x)$ are not continuous, they do nevertheless possess the important property that they are *integrable*; that is, their integrals exist. For example, the integral of $H(x)$ is the *ramp* function $R(x)$ shown in Figure 2.8; clearly, $R(x) \in C(-\infty, \infty)$.

Our aim is to set up a space of functions that may be classified according to whether they, and their powers, are integrable. That is, for a given function $f$ we investigate the range of exponents $p$ for which the integral

$$\int_a^b |f(x)|^p \, dx$$

is meaningful (that is, finite), where $p \geq 1$ is a real number. This permits the introduction of the spaces $L^p(a, b)$ or, more generally, $L^p(\Omega)$. Now recall that in the case of the spaces $C^m$ it is possible to obtain a precise idea of the degree of smoothness of a function by determining the largest value of $m$ for which it belongs to $C^m$. The smoothness of two functions may then, for example, be compared by determining the largest numbers $m$ of the spaces $C^m$ of which they are members. In the same way, we will see that the $L^p$ spaces are also "nested", in the sense that $L^p \subset L^q$ for the

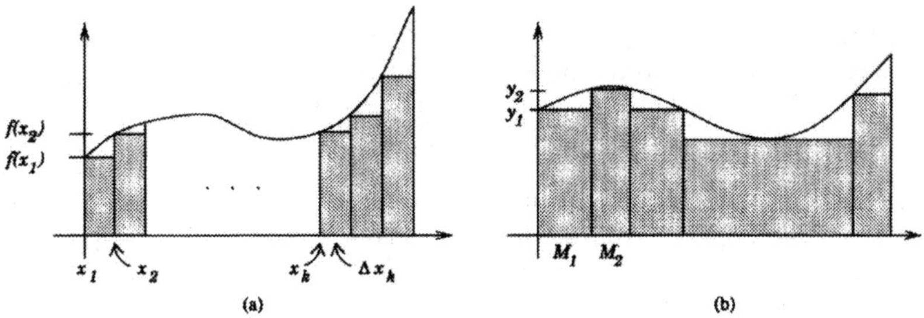


FIGURE 2.9. The basic idea behind (a) Riemann and (b) Lebesgue integration

case in which $p > q$; thus these spaces also provide a means of comparing functions, this time through their integrability.

In order to give such spaces a proper treatment it is necessary first of all to discuss the notion of Lebesgue measure. This in turn allows us to introduce the notion of Lebesgue integration, which is a generalization of the "standard" Riemann integration, and in so doing to go on to introduce the spaces $L^p(\Omega)$.

Measure theory is a well-established branch of mathematics, and Lebesgue measure is but one example of a measure. It is an important example, though, and it is also intuitively the easiest to grasp. There is no need to make reference subsequently to any other measure than that of Lebesgue, so rather than give a general treatment of the subject, this section is restricted to an overview of the theory of Lebesgue measure that is concise, but which nevertheless suffices for our purposes.

In order to appreciate the need to extend the notion of Riemann integration, we return first to the definition of the Riemann integral. Restricting the discussion for now to functions of a single variable, consider a function $f$ defined on the interval $[a, b]$. The Riemann integral is based on the idea of dividing $[a, b]$ into a finite number $N$ of subintervals, the $k$th subinterval having length $\Delta x_k$, and then considering sums of the form

$$f(x_1)\Delta x_1 + f(x_2)\Delta x_2 + \cdots + f(x_N)\Delta x_N,$$

as shown in Figure 2.9(a). This sum represents an approximation to the area under the graph of $f$. If the function is sufficently well-behaved – for example, piecewise continuous – then the approximation may be improved by increasing $N$, that is, by refining the subdivision of $[a, b]$, so that in the limit, as $N$ gets very large, we arrive at the Riemann integral, which is usually denoted by

$$\int_a^b f(x)\, dx.$$

The Riemann integral is the integral used in everyday applications, and it is generally adequate for most purposes, but it also suffers from certain deficiencies. For example, there are certain "nasty" functions that we are unable to deal with using the Riemann integral: an example is the function

$$u(x) = \begin{cases} 1, & x \text{ is rational}, \\ 0, & x \text{ is irrational}, \end{cases} \tag{2.4}$$

defined on the interval $[0, 1]$. With the more general *Lebesgue* integral we avoid these problems; the Lebesgue integral is able to handle functions like (2.4) and, furthermore, gives the same result as the Riemann integral if the function is Riemann-integrable. Also, limits of Lebesgue-integrable functions are always Lebesgue-integrable.

Although it might seem rather pedantic to abandon the Riemann integral for the preceding reasons – after all, how often are we required to integrate something like the function defined in (2.4)? – we demonstrate later that spaces of Lebesgue-integrable functions possess properties which allow them to be classified as Banach spaces or Hilbert spaces, with the fortunate consequence that it is then possible to draw on the vast reservoir of results for such spaces. From a practical point of view, Riemann and Lebesgue integrals coincide when the former exists, so all we will have done would be to broaden the class of functions that can be integrated.

Since the question of whether the integral of a function $f$ makes sense depends very much on the function, a suitable alternative approach to the Riemann integral might be to approximate $f$ by a very simple function, the integral of which can be computed without any difficulty. Then, in contrast to the Riemann integral, the approximation to the integral of $f$ can be progressively improved, not by further subdivisions of the domain, but by refining the approximation to $f$ (see Figure 2.9(b)). The approximating functions that serve this purpose are indeed known as simple functions, and are defined to be functions that take on a finite number of values. Provided that we have no problems with the subsets $M_k$ on which they take their constant values, the integral of $f$ can be approximated by a sum of the form

$$y_1\mu(M_1) + y_2\mu(M_2) + \cdots + y_N\mu(M_N),$$

in which $\mu(M_k)$ denotes the "size", or *measure* of $M_k$. By a process of refinement which leads to a progressive improvement in the approximation of $f$ (this is shown schematically in Figure 2.9(b)) in the limit as $N$ goes to infinity, we arrive at the integral of $f$.

Now for this strategy to work we have to have available a means of measuring the size of sets such as $M_k$, even for approximations of fairly nasty functions, in which case $M_k$ may take on a rather complex form. So the problem of constructing an adequate definition of the integral has been transferred to one of formulating a mathematically acceptable definition of the size of a set.

It can be shown that, rather surprisingly, not all subsets of $\mathbb{R}^n$ can be assigned a size that is independent of rotations and translations; this is known as the Banach–Tarski paradox, the essence of which is that it is possible to break up a ball of radius $r$ into a finite number of complex pieces, move them around, and reassemble them to get two balls of radius $r$! Those special sets that do not suffer from such shortcomings, and which will do for evaluations of integrals, are known as *measurable sets.* So we have to consider next the issue of identifying those subsets of a set $\Omega$ (which would be $\mathbb{R}^n$ or a subset of $\mathbb{R}^n$) that can be used in an appropriate definition of the integral.

Suppose that this family of subsets is denoted by $\mathcal{M}$; what properties do we want the members of this family to have? First, $\Omega$ itself should be a member of this family. Second, if $M$ belongs to $\mathcal{M}$, that is, $M$ is a measurable set, then we would like its *complement* $\Omega - M$ to be measurable as well. Next, this family should contain the open subsets of $\Omega$. And finally, the intuitive notion of size dictates that if $\{M_1, M_2, \ldots,\}$ is a countable family of measurable subsets that belong to $\mathcal{M}$ with the property that the sets $M_k$ are mutually disjoint, then the size of $M_1 \cup M_2 \cup \ldots$ may be evaluated by determining the size of each of the sets $M_k$, and adding the result; that is, we require that

$$\mu(M_1 \cup M_2 \cup \ldots) = \mu(M_1) + \mu(M_2) + \ldots .$$

This property of $\mu$ is known as *countable additivity.*

To summarize, we require that the following categories of sets all belong to $\mathcal{M}$:

1. $\Omega$ itself;
2. $\Omega - M$, for $M \in \mathcal{M}$;
3. all open sets in $\Omega$; and
4. $M_1 \cup M_2 \cup \ldots$, for any countable family $\{M_1, M_2, \ldots,\}$ of disjoint sets in $\mathcal{M}$.

The members $M_k$ of a family $\mathcal{M}$ of sets that satisfy the properties 1 through 4 are known as *measurable sets*, and $\Omega$ is called a *measurable space.* It is on such spaces that the Lebesgue measure is defined.

**Lebesgue measure.** In order to define the Lebesgue measure we start with the familiar: in $\mathbb{R}^n$, define an $n$-cell to be any set of the form

$$C = \{\boldsymbol{x} : \ a_i < x_i < b_i, \quad i = 1, 2, \ldots, n\}.$$

Thus a 1-cell is an open interval in $\mathbb{R}$, a 2-cell is the area bounded by a rectangle in $\mathbb{R}^2$, and so on. We further define the $n$-volume of $C$, denoted $\mathrm{vol}\,(C)$, by

$$\mathrm{vol}\, C = (b_1 - a_1)(b_2 - a_2) \cdots (b_n - a_n).$$

Of course, by the 1-volume of a 1-cell (interval) in $\mathbb{R}$ we understand the length of the interval, the 2-volume of a 2-cell (rectangle) in $\mathbb{R}^2$ is its area, and so on. Then the Lebesgue measure $\mu$ on $\mathbb{R}^n$ is a quantity (a function, to be precise) that measures the content of a set. It is defined to have the following properties.

1. $\mu(\varnothing) = 0$;
2. $\mu(C) = \operatorname{vol} C$ for any $n$-cell $C$;
3. if $\{M_1, M_2, \ldots\}$ is a collection of mutually disjoint sets in $\mathcal{M}$, then
$$\mu(M_1 \cup M_2 \cup \ldots) = \mu(M_1) + \mu(M_2) + \ldots ;$$
4. the measure of any measurable set can be approximated from above by *open* sets; that is, for any measurable $M$,
$$\mu(M) = \inf\{\mu(O) : M \subset), \ O \text{ is open}\};$$
5. the measure of any measurable set can be approximated from below by *compact* sets (recall that these are closed and bounded sets in $\mathbb{R}^n$); that is, for any measurable $M$,
$$\mu(M) = \sup\{\mu(C) : C \subset M, \ C \text{ is compact}\}.$$

Thus we see that Lebesgue measure is a very natural extension of the familiar notions of length, area, and volume, with the added quality that the careful definition also permits the measurement of the content of a very wide range of sets in $\mathbb{R}^n$.

This definition of measure also suffices for the construction of a theory of integration that works for a range of functions much wider than that accommodated by Riemann integration.

**Sets of measure zero.** Suppose that $P$ is a property that may or may not hold at a point $x$ in $\mathbb{R}^n$. For example, for a given function $f$, $P$ could be the property "$f(x) = 0$". If $\mu$ is the Lebesgue measure on $\mathbb{R}^n$, then we say that a property holds *almost everywhere* (abbreviated a.e.) if the only subset on which it does not hold is one with measure zero. For example, if $f$ and $g$ are two functions that are equal everywhere except at a number of isolated points in $\mathbb{R}$, then $f = g$ a.e. (Figure 2.10). The same would apply to two functions that are defined in $\mathbb{R}^2$, and differ only on a curve in $\mathbb{R}^2$, such as the boundary $\Gamma$ of a domain $\Omega$.

Sets of measure zero are readily identifiable, at least in the kinds of applications with which we are concerned. For example, any *countable* set has zero measure. So in particular, for the sets $\mathbb{Z}$ and $\mathbb{Q}$ of integers and rational numbers, respectively, we have
$$\mu(\mathbb{Z}) = \mu(\mathbb{Q}) = 0.$$

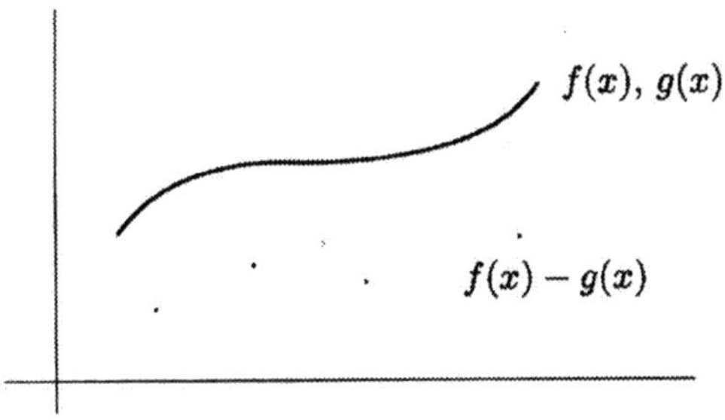


FIGURE 2.10. Functions $f$ and $g$ that are equal almost everywhere

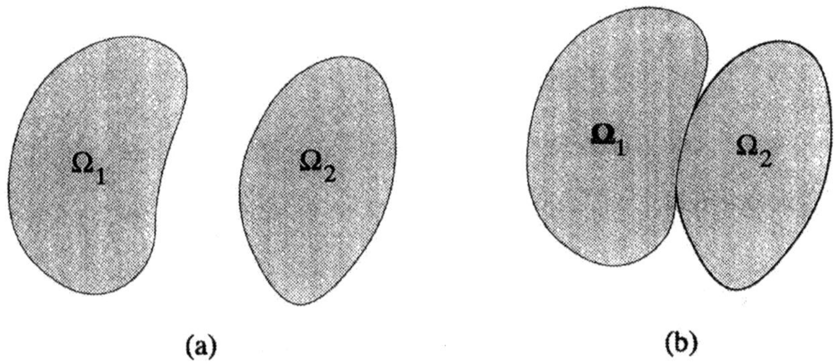


FIGURE 2.11. (a) Disjoint sets: $\overline{\Omega}_1 \cap \overline{\Omega}_2 = \varnothing$ and $\mu(\overline{\Omega}_1 \cap \overline{\Omega}_2) = 0$; (b) nonoverlapping sets: $\Omega_1 \cap \Omega_2 = \varnothing$, $\overline{\Omega}_1 \cap \overline{\Omega}_2 \neq \varnothing$, and $\mu(\overline{\Omega}_1 \cap \overline{\Omega}_2) = 0$.

Another example concerns disjoint or nonoverlapping sets: if $\Omega_1$ and $\Omega_2$ satsify $\overline{\Omega}_1 \cap \overline{\Omega}_2 = \varnothing$ or more generally, if $\Omega_1$ and $\Omega_2$ are nonoverlapping in the sense that $\overline{\Omega}_1 \cap \overline{\Omega}_2$ is a nonempty set of zero measure, then

$$\mu(\overline{\Omega}_1 \cap \overline{\Omega}_2) = 0.$$

This is illustrated in Figure 2.11.

## 2.3 Lebesgue integration and the space $L^p(\Omega)$

Before we begin discussion of the integral itself, we have to select a suitable family or space of functions for which integration can be usefully defined. Returning to the motivation given at the beginning of the previous section, we say that a function defined on a measurable set $\Omega$ in $\mathbb{R}^n$ is *measurable* if the inverse image $f^{-1}(M)$ of any measurable set $M$ (in $\mathbb{R}$, of course) is itself measurable. The idea is illustrated in Figure 2.12.

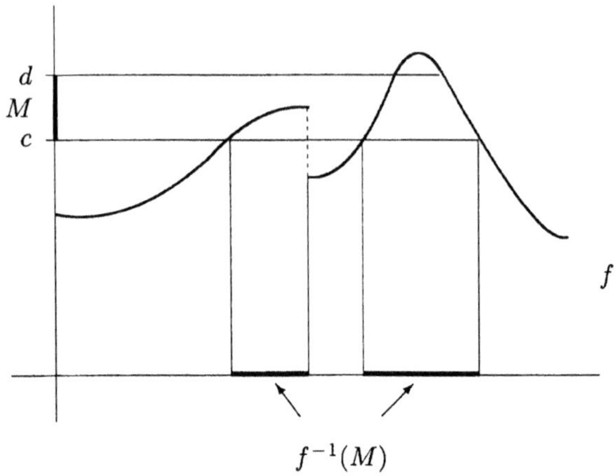


FIGURE 2.12. The definition of a measurable function

Before continuing further we consider a few examples of measurable functions.

**Examples**

9. Any *continuous* function is measurable: in particular if $M$ is the interval $(c, d)$ (Figure 2.12), then it is possible to show that $f^{-1}(c, d)$ is also open. To see this, take any point $y_0$ in $M$; then it is possible to choose $\epsilon$ such that the neighborhood $\{y : |y - y_0| < \epsilon\}$ lies entirely in $M$, $M$ being open. Now denote by $J$ the interval $f^{-1}(M)$; by definition there exist points $x$ and $x_0$ in $J$ such that $f(x) = y$ and $f(x_0) = y_0$, and so $|f(x) - f(x_0)| < \epsilon$. By the continuity of $f$ it follows that there exists $\delta > 0$ such that $|x - x_0| < \delta$. Thus $J$ is open.

10. Consider the Heaviside function $H$ defined by

$$H(x) = \begin{cases} 1 & \text{if } x \geq 0 \\ 0 & \text{if } x < 0, \end{cases}$$

and shown in Figure 2.13; if we choose $M$ as shown in the figure, then $H^{-1}(M) = \{x : x \geq 0\}$ which is measurable; on the other hand, if we choose the measurable set $L$, then $H^{-1}(L) = \{x : x < 0\}$ which again is measurable. Continuing in this way, we can verify that the sets $H^{-1}(M)$ for measurable $M$ are all measurable. Thus $H$ is a measurable function.

11. Let $\Omega$ be a measurable set, and $E$ a measurable subset of $\Omega$; then the *characteristic function* $\chi_E$ of $E$ is defined by

$$\chi_E(\boldsymbol{x}) = \begin{cases} 1 & \text{if } \boldsymbol{x} \in E \\ 0 & \text{if } \boldsymbol{x} \notin E. \end{cases} \tag{2.5}$$

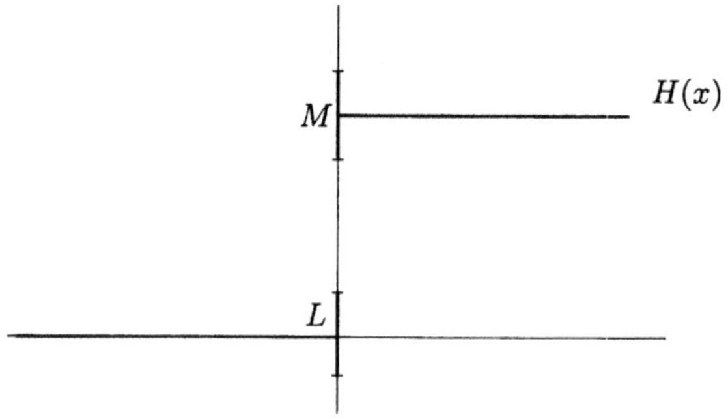


FIGURE 2.13. The Heaviside step function $H(x)$

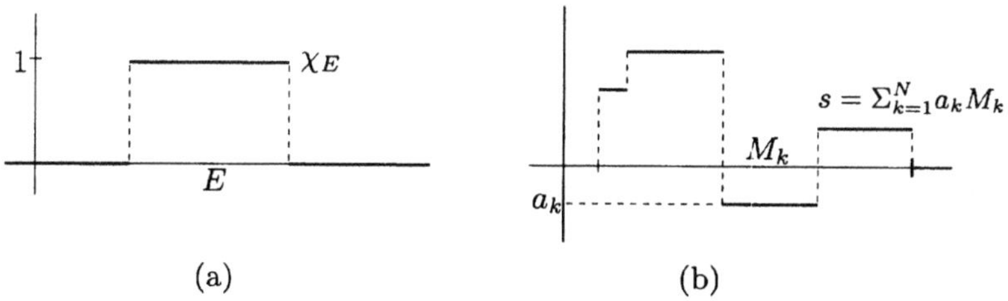


FIGURE 2.14. (a) The characteristic function $\chi_E$, and (b) a simple function $s$

This is illustrated in Figure 2.14. It can be shown that $\chi_E$ is a measurable function (see Exercise 2.12) provided that $E$ is measurable.

12. We return to the example given in (2.4), and observe that this can be written in the alternative form $u = \chi_Q$, where $Q$ is the set of rational numbers. Since $Q$ is a measurable set – with Lebesgue measure zero, since it is countable – it follows that the function $u$ is measurable, by Example 11.

With the characteristic function at our disposal we can now define *simple functions* $s$: these are functions on $\Omega$ that take on only a finite number of values. In other words, suppose that $M_1, M_2, \ldots, M_N$ is a partition of $\Omega$; then each simple function is a measurable function of the form

$$s = a_1\chi_{M_1} + a_2\chi_{M_2} + \ldots + a_N\chi_{M_N}, \tag{2.6}$$

where $a_k$ is the value of $s$ on $M_k$. These notions are illustrated in Figure 2.14. Since sums of measurable functions are measurable, we can conclude that every step function is measurable.

**The Lebesgue integral.** We are now in a position to define the Lebesgue integral, and we begin by doing so for simple functions, for which case the definition takes on an intuitively obvious character.

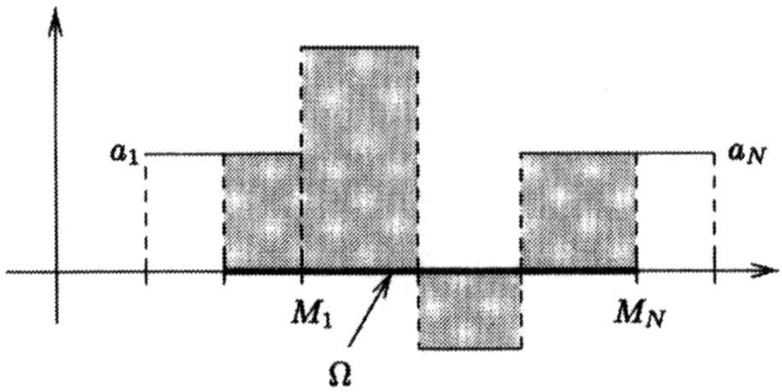


FIGURE 2.15. The Lebesgue integral of a simple function

*The (Lebesgue) integral of a simple function* $s$ on $\Omega$ is defined by

$$\int_\Omega s\,dx = a_1\mu(M_1 \cap \Omega) + a_2\mu(M_2 \cap \Omega) + \ldots + a_N\mu(M_N \cap \Omega),$$

where $M_k$ are measurable and pairwise disjoint (Figure 2.15). A special case is the integral over a measurable subset $E$ of $\Omega$ with finite measure; it suffices to put $s = \chi_E$, and we obtain

$$\int_E dx \equiv \int_\Omega \chi_E\,dx = \mu(E).$$

Returning to Example 12, we find now that it is a trivial matter to integrate the function $u$: indeed

$$\int_{\mathbb{R}} u\,dx = \int_{\mathbb{R}} \chi_Q\,dx = \mu(Q) = 0.$$

Here, then, is one example of a function that is not Riemann-integrable, but which is indeed Lebesgue-integrable.

We now extend the definition of the integral to more general classes of functions; this is achieved by approaching measurable functions as limits of sequences of simple functions.

We begin by introducing the notion of a *nondecreasing sequence* of simple functions: this is a sequence $S = \{s_1, s_2, \ldots, s_k, \ldots\}$ of simple functions which have the property that

$$s_1(\boldsymbol{x}) \leq s_2(\boldsymbol{x}) \leq \cdots \leq s_k(\boldsymbol{x}) \leq \cdots \quad \text{for all } \boldsymbol{x} \in \mathbb{R}^n.$$

An example is shown in Figure 2.16. A word about nomenclature: the sequence is termed "nondecreasing" rather than "increasing" since the latter would refer to functions having the property $s_1 < s_2 < \ldots$; in other words the possibility of equality would be excluded. To obtain the Lebesgue integral of a measurable function $f$ we first set up a sequence of nondecreasing

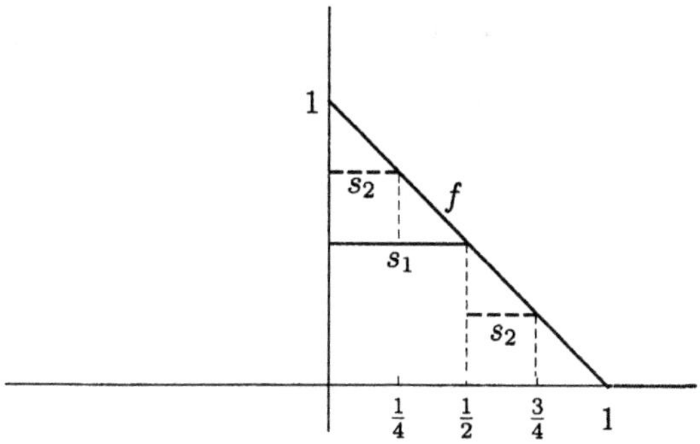


FIGURE 2.16. A nondecreasing sequence of simple functions that approximate a measurable function $f$

simple functions that approximate $f$; next, we evaluate the integrals of these simple functions – this is a well-defined procedure – and, finally, take the limit to obtain the integral of $f$. In order for this intuitively reasonable strategy to work we must first be sure, though, that it will always be possible to set up such sequences; this guarantee is given in the following result.

LEMMA 1. *If $f$ is a nonnegative measurable function on $\mathbb{R}^n$, then it is possible to find a nondecreasing sequence $S$ of simple functions on $\mathbb{R}^n$ such that*

$$\lim_{n\to\infty} s_n(\boldsymbol{x}) = f(\boldsymbol{x}) \text{ at all points } \boldsymbol{x} \text{ in } \mathbb{R}^n. \tag{2.7}$$

The meaning of (2.7) should be clear from the discussion of sequences in Chapter 1: for each $n$, $s_n(\boldsymbol{x})$ is a real or complex number, so (2.7) is a statement concerning the convergence of a sequence of real or complex numbers. Simply put, a rather arbitrary function may be approximated as closely as we wish by simple functions. Figure 2.16 illustrates this notion in one dimension.

We come at last to the definition of the integral, and begin by defining the integral for *nonnegative functions*, that is, functions that satisfy the condition $f(\boldsymbol{x}) \geq 0$ a.e. on their domain. Suppose that $f$ is a measurable function defined on a measurable set $\Omega$, and further that $f$ is nonnegative on $\Omega$: $f(\boldsymbol{x}) \geq 0$ a.e.; then *the Lebesgue integral of $f$ over $\Omega$* is defined by

$$\int_\Omega f\, dx = \lim_{k\to\infty} \int_\Omega s_k\, dx, \tag{2.8}$$

where $s_n$ are nondecreasing simple functions that approximate $f$ in the sense of Lemma 1. The definition (2.8) should be considered in the light of the discussion of sequences in Chapter 1. Assuming that we are dealing with real-valued functions (the same argument applies to complex-valued functions), $\int_\Omega f\,dx$ is a real number, and if we set $\alpha_k = \int_\Omega s_k\,dx$, then $\{\alpha_k\}$ is a sequence of real numbers. The definition (2.8) then states that $\alpha_k \to \int_\Omega f\,dx$ as $k \to \infty$. Thus, in contrast to the approach taken with the Riemann integral, the Lebesgue integral of $f$ may be obtained as the limit of integrals of simple functions that approximate $f$ more and more closely as the limit is approached.

As mentioned earlier, functions that are Riemann-integrable are also Lebesgue-integrable, and the two integrals coincide. Indeed, for well behaved functions – for example, piecewise continuous functions – it is clear that the Lebesgue integral, like the Riemann integral, amounts to the area under the graph of the function. However, as we indicated earlier, there are Lebesgue-integrable functions which are not Riemann-integrable.

### Example

13. Suppose that we wish to integrate the function $f$ shown in Figure 2.16. Now this function is piecewise continuous, and it is well known from elementary integration that its integral is the area under the triangle, and is equal to $\frac{1}{2}$. However, the purpose of this example is to show how the definition (2.8) may be deployed in practice, so we in fact construct a sequence of nondecreasing simple functions that converge to $f$.

    There are many different ways of constructing the requisite family of simple functions; we consider just one, in which the first member $s_1$ is as shown in Figure 2.16; the second member, $s_2$, is constructed in a similar manner, ensuring that it too satisfies $s_2 \le f$. The process may now be continued in a fairly obvious manner.

    Considering next the integrals of the simple functions, we see that

    $$\int_{\mathbb{R}} s_1(x)\,dx = \tfrac{1}{4} \quad \text{and} \quad \int_{\mathbb{R}} s_2(x)\,dx = \tfrac{3}{8}.$$

    In fact it is possible to show (see Exercise 14) that

    $$\int_{\mathbb{R}} s_k(x)\,dx = \frac{1}{2} - \frac{1}{2^{k+1}},$$

    so that as $k$ goes to $\infty$ we approach the value given by the Riemann integral.

To complete the theory of the Lebesgue integral we now extend the treatment to include functions that are not necessarily nonnegative. Suppose then that $f$ is any measurable function; then $f$ may be decomposed into its *positive part* $f^+$ and *negative part* $f^-$ (Figure 2.17), which are defined by

$$f^+(x) = \begin{cases} f(x) & \text{if } f(x) \ge 0 \\ 0 & \text{otherwise} \end{cases} \quad \text{and} \quad f^-(x) = \begin{cases} 0 & \text{if } f(x) \ge 0 \\ -f(x) & \text{otherwise.} \end{cases}$$

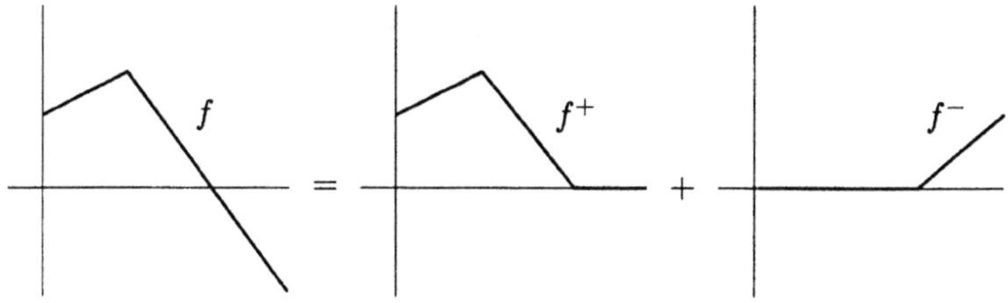


FIGURE 2.17. The positive and negative parts of a function

More concisely, we can write

$$f^+ = \tfrac{1}{2}(f + |f|), \quad f^- = \tfrac{1}{2}(|f| - f),$$

so that

$$f = f^+ - f^-.$$

We observe that both $f^+$ and $f^-$ are nonnegative functions, so that the preceding theory applies to these two components of $f$. It is possible to show that $f^+$ and $f^-$ are both measurable if $f$ is, and so we may define the Lebesgue integral of $f$ by

$$\int_\Omega f\ dx = \int_\Omega f^+\ dx - \int_\Omega f^-\ dx. \tag{2.9}$$

It is at this point that we can clarify the need to define the integral of a function in terms of its positive and negative parts; first we need to note that the integral, as defined in (2.8), need not be finite; that is, it is possible to have $\int_\Omega f\ dx = +\infty$ for a nonnegative function. Continuing this line of argument, it is quite conceivable that evaluation of the righ-hand side of (2.9) will give $\infty - \infty$, which is of course meaningless. It is therefore to be understood that the notation $\int_\Omega f\ dx$ makes sense only if one of the terms on the right-hand side of (2.9) is finite. We go one step further, and give a special name to those functions $f$ for which $f^+ + f^-$ has a finite integral.

**Integrable functions.** A measurable function $f$ is said to be *integrable* on a measurable set $\Omega$ in $\mathbb{R}^n$ if

$$\int_\Omega |f|\ dx < \infty.$$

A word about notation is in order. In multivariable calculus it is customary to write multiple integrals as

$$\int \cdots \int_\Omega u(\boldsymbol{x})\ dx_1 dx_2 \cdots dx_n$$

for functions of $n$ variables; so, for example, a double integral is written as

$$\iint_{\Omega} u(\boldsymbol{x})\ dx\, dy.$$

We adopt the convention throughout that multiple integrals are written in the concise form

$$\int_{\Omega} u(\boldsymbol{x})\ dx,$$

the context making clear the dimension of the domain over which the integral is taken. This convention has in fact been implicit in the developments leading to the Lebesgue integral; we made no distinction there between integrals taken over $\mathbb{R}$ and over $\mathbb{R}^n$, for any $n$.

It is also worth bearing in mind that since sets of measure zero are irrelevant in the evaluation of integrals, integrals may be defined over open sets or over their closures. So, for example, it makes no difference whether an integral is defined over an open interval $(a, b)$, or over $[a, b]$.

All the usual properties of Riemann integrals extend to Lebesgue integrals, and we summarize without proof some of these properties.

THEOREM 2. *Let $u(x)$ and $v(x)$ be Lebesgue-integrable functions on $\Omega \subset \mathbb{R}^n$. Then*

(a) $\displaystyle\int_{\Omega} [\alpha u(\boldsymbol{x}) + \beta v(\boldsymbol{x})]\ dx = \alpha \int_{\Omega} u(\boldsymbol{x})\ dx\ +\ \beta \int_{\Omega} v(\boldsymbol{x})\ dx$
*for constants $\alpha, \beta$;*

(b) if $u(\boldsymbol{x}) \leq v(\boldsymbol{x})$ *for almost all $\boldsymbol{x} \in \Omega$, then*

$$\int_{\Omega} u(\boldsymbol{x})\ dx \leq \int_{\Omega} v(\boldsymbol{x})\ dx;$$

(c) *if $u(\boldsymbol{x})$ is bounded above and below by numbers $m$ and $M$, then*

$$m\mu(\Omega) \leq \int_{\Omega} u(\boldsymbol{x})\ dx \leq M\mu(\Omega);$$

(d) *$|u|$ is also integrable, and*

$$\left| \int_{\Omega} u\ dx \right| \leq \int_{\Omega} |u|\ dx.$$

The following theorem is a powerful tool in functional analysis.

THEOREM 3 (THE LEBESGUE DOMINATED CONVERGENCE THEOREM). *Let $u_1, u_2, \ldots, u_k, \ldots$ be a sequence of measurable functions and suppose* *that* $u_k(x) \leq v(x)$ *a.e. for each* $k$*, where* $v$ *is an integrable function. Suppose that* $u(x) = \lim_{k\to\infty} u_k(x)$ *a.e. Then* $u$ *is integrable and*

$$\int_\Omega u\, dx = \lim_{k\to\infty} \int_\Omega u_k\, dx.$$

The usefulness of this theorem lies in the very mild conditions that are placed on $u_k$.

**The spaces $L^p(\Omega)$.** Let $p$ be a real number with $p \geq 1$. A function $u(x)$ defined on a subset $\Omega$ of $\mathbb{R}^n$ is said to belong to $L^p(\Omega)$ if $u$ is measurable and if the (Lebesgue) integral

$$\int_\Omega |u(x)|^p\, dx$$

exists (that is, is finite). The case $p = 2$ is special in many ways, as the developments in Chapter 3 and beyond make clear; functions in $L^2(\Omega)$ have the property that

$$\int_\Omega |u(x)|^2 dx < \infty,$$

and for this reason are referred to as *square-integrable* .

Of course, every bounded continuous function defined on a bounded set $\Omega$ belongs to $L^p(\Omega)$, but there are many other functions that have this property, as we show in the following.

**Examples**

14. The step function $H(x)$ defined by

$$H(x) = \begin{cases} 0, & x < 0 \\ 1, & 0 \leq x \end{cases}$$

belongs to $L^p(a, b)$ for any $p \geq 1$ and finite $a < 0$ and $b > 0$ since

$$\int_a^b |H(x)|^p\, dx = \int_0^b (1)^p dx = b < \infty.$$

15. The function $u(x) = x^{-1/3}$ belongs to $L^p(0, 1)$ for any $p < 3$, since

$$\int_0^1 |u(x)|^p\, dx = \int_0^1 x^{-p/3}\, dx = \frac{3}{3-p}\left[x^{(3-p)/3}\right]_0^1$$

which is finite for $p < 3$.

Some results that are frequently useful are embodied in the following theorem.

THEOREM 4. *Suppose that $\Omega$ is a bounded domain in $\mathbb{R}^n$. Then*

(a) $L^p(\Omega) \subset L^{p'}(\Omega)$ *for* $p \geq p'$;

(b) *if* $u \in L^p(\Omega)$, *then the integrals*

$$\int_\Omega |u(\boldsymbol{x})| \, dx \quad \text{and} \quad \int_\Omega u(\boldsymbol{x}) \, dx$$

*are finite;*

(c) *if* $u, v \in L^2(\Omega)$, *then the integral*

$$\int_\Omega u(\boldsymbol{x})v(\boldsymbol{x}) \, dx$$

*is finite.*

PROOF. The proof of (a) relies on the inequality

$$\left[\int_\Omega |u(\boldsymbol{x})|^{p'} dx\right]^{1/p'} \leq \left[\int_\Omega |u(\boldsymbol{x})|^p dx\right]^{1/p}$$

which holds if $p \geq p'$ (see Exercise 3.22 later for a derivation). If $u$ belongs to $L^p(\Omega)$, then the integral on the right is finite, and hence so is the integral on the left. Thus $u \in L^{p'}(\Omega)$ also.

Part (b) is a trivial consequence of (a): set $p' = 1$; then we have, for $u \in L^p(\Omega)$,

$$\left|\int_\Omega u(\boldsymbol{x}) \, dx\right| \leq \int_\Omega |u(\boldsymbol{x})| \, dx \leq \left[\int_\Omega |u(\boldsymbol{x})|^p \, dx\right]^{1/p} < \infty.$$

Part (c) is a result of the inequality

$$\left|\int_\Omega u(\boldsymbol{x})v(\boldsymbol{x}) \, dx\right| \leq \sqrt{\int_\Omega |u(\boldsymbol{x})|^2 dx} \sqrt{\int_\Omega |v(\boldsymbol{x})|^2 dx}$$

which arises again in Chapter 3 (Theorem 2) in the guise of the Cauchy-Schwarz inequality. □

There is a source of ambiguity in the definition of the space $L^p(\Omega)$ which we must remove in order to deal with it meaningfully. Suppose that $f(x)$

and $g(x)$ are two measurable functions that are equal a.e. (as in Figure 2.10); then

$$\int_a^b |f(x)|^p \, dx = \int_a^b |g(x)|^p \, dx.$$

It follows that $L^p(\Omega)$ can be partitioned into equivalence classes, each class comprising all those functions that are equal a.e. to a given one. In order to be able to define $L^p(\Omega)$ as a normed space (in the next chapter) it is necessary to regard the elements of this space not as functions, but rather as the equivalence classes of functions defined here. Notwithstanding this distinction, it is common practice to speak of the members of $L^p(\Omega)$ as functions; this is a harmless abuse of language provided that the precise nature of the space is properly understood.

**Complex-valued functions.** The theory presented here can be extended in a very obvious way to functions that are complex-valued. Given a function $f$ that is of the form $f(x) = u(x) + iv(x)$, we say that $f$ is measurable if $u$ and $v$ are. Furthermore, $f$ is integrable if $u$ and $v$ are, and

$$\int_\Omega f \, dx \equiv \int_\Omega u \, dx + i \int_\Omega v \, dx.$$

The definition of the $L^p$ spaces still stands, if the notation $|f|$ is interpreted as the modulus of a complex number: $|f|^2 = u^2 + v^2$.

**The space $L^\infty(\Omega)$.** If we let $p \to \infty$, then we may define the space $L^\infty(\Omega)$ to be the space of all measurable functions on $\Omega$ that are *bounded* almost everywhere on $\Omega$ (that is, except possibly on subsets of zero measure):

$$L^\infty(\Omega) = \{u : |u(x)| \le k \text{ a.e. on } \Omega \text{ for some } k \in \mathbb{R}\}.$$

Clearly for a bounded domain $\Omega$, $L^\infty(\Omega)$ is a subset of $L^p(\Omega)$ for all $p \ge 1$, since any $u \in L^\infty(\Omega)$ satisfies

$$\int_\Omega |u(x)|^p \, dx \le \int_\Omega k^p \, dx < \infty,$$

so that $u \in L^p(\Omega)$ also.

**Example**

16. The function

$$u(x) = \begin{cases} x^2, & 0 \le x < 1, \quad x \ne \frac{1}{2} \\ \to \infty, & x = \frac{1}{2}, \end{cases}$$

is bounded a.e. on $(0, 1)$ since $u(x) \to \infty$ only on a set of measure zero (the point $x = \frac{1}{2}$).

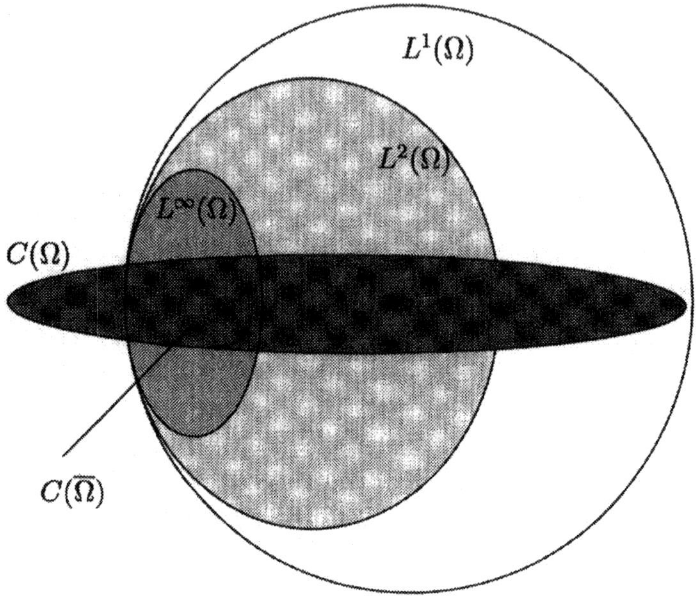


FIGURE 2.18. The relationship between the $L^p$ spaces and spaces of continuous functions

It is interesting to note that although we have $L^\infty(\Omega) \subset \cdots \subset L^p(\Omega) \subset \cdots \subset L^1(\Omega)$, the space $C(\Omega)$ of continuous functions is not a subset of any of the $L^p$ spaces. For example, the function $u(x) = x^{-1}$ belongs to $C(0,1)$ but not to $L^\infty(0,1)$ since it is not bounded. But the space of *bounded* continuous functions, (equivalently, the space $C(\overline{\Omega})$ of continuous functions defined on a compact set $\overline{\Omega}$) is a subset of $L^\infty(\Omega)$.

Figure 2.18 shows schematically how the spaces $C^m(\Omega)$ and $L^p(\Omega)$ are related.

## 2.4 Bibliographical remarks

A good treatment of the concept of continuity may be found in Apostol [2], Binmore [6], and Lang [29]. The treatment of measure and integration given here is somewhat superficial, although it suffices for subsequent needs. There exist many readable accounts of the Lebesgue theory, notable examples being Kolmogorov and Fomin [26], Reed and Simon [40], Royden [44], and Rudin [45], and Roman [42] gives an account that should be particularly accessible to nonspecialists in mathematics.

## 2.5 Exercises

### Continuous functions and the space $C^m(\Omega)$

2.1. Sketch and discuss the continuity of the functions

(a) $u(x) = x/(x^2 - 1), \quad -\infty < x < \infty.$

(b) $u(x) = \begin{cases} (x - |x|)/x, & x < 0 \\ 2, & x = 0 \end{cases}.$

2.2. Show that the following functions are continuous on the intervals given:

(a) polynomials of degree $k$ defined on the interval $[a, b]$;

(b) the function $u(x) = x^{1/2}$ on $[0, \infty)$.

Is either of these functions uniformly continuous?

2.3. (a) Show that $f(x) = 1/x$ is *not* uniformly continuous on $(0, 1)$. [Hint: recall that $f(x) - f(y) = (y - x)/xy$. Show, for example by choosing $x = 1/n$ and $y$ appropriately, that the distance $|x - y|$ can be shown arbitrarily small although $|f(x) - f(y)|$ is large.]

(b) Show, on the other hand, that $f(x) = 1/x$ is uniformly continuous on $[a, b]$, where $b > a > 0$.

2.4. Show that $f(\boldsymbol{x}) = x^2 + 2y$ is continuous at any point $\boldsymbol{x}$ in $\mathbb{R}^2$.

2.5. Let $E$ be a closed connected set in $\mathbb{R}^2$, and for any point $\boldsymbol{x}$ in $\mathbb{R}^2$ define the function $f$ by $f(\boldsymbol{x}) = d(\boldsymbol{x}, E)$, where $d(\boldsymbol{x}, E)$ is the *distance between $\boldsymbol{x}$ and $E$*, defined by

$$d(\boldsymbol{x}, E) = \inf\{|\boldsymbol{x} - \boldsymbol{y}| : \ \boldsymbol{y} \in E\}.$$

Draw a sketch that illustrates the function $f$, and show that $f$ is continuous.

2.6. If $f$ is continuous at a point $x_0$ in $[a, b]$ with $f(x_0) > 0$, show that there is a neighborhood $(x_0 - h, x_0 + h)$ about $x_0$ in which $f$ is positive.

2.7. Prove *Bolzano's Theorem*, which states that if $f(x)$ is a continuous function on $[a, b]$ with $f(a)f(b) < 0$ (that is, $f(a)$ and $f(b)$ have different signs), then there is at least one point $c$ in $[a, b]$ such that $f(c) = 0$. [Use the result in Exercise 2.6.]

2.8. To which spaces $C^m(\overline{\Omega})$ do the following functions belong ?

(a) $u(x) = \begin{cases} 0, & -1 \le x < 0 \\ x(1 + x), & 0 \le x \le 1 \end{cases}.$

(b) $u(\boldsymbol{x}) = (\sin x)(1-y), \quad (x,y) \in [0,\pi] \times [0,1]$.

(c) $u''(x) = \begin{cases} 0, & 0 \le x \le \frac{1}{2} \\ 1, & 0 < x < 1 \end{cases}$ .

2.9. Examine the continuity of $u(\boldsymbol{x}) = r$ on the unit disk, where $r^2 = x^2 + y^2$.

2.10. Show that every Lipschitz function is uniformly continuous.

## Measure of sets in $\mathbb{R}^n$

2.11. Let $I$ be an interval in $\mathbb{R}$, and consider the subset of all irrational numbers in $I$. Is this set measurable? If so, calculate its measure.

2.12. Show that the characteristic function $\chi_E$ of a set $E$ is a measurable function if and only if $E$ is itself measurable.

## Lebesgue integration and the spaces $L^p(\Omega)$

2.13. Prove Lemma 1.

2.14. Verify that the integral of the $n$th simple function approximating the function $f$ in Example 13 has the value $\frac{1}{2} - \frac{1}{2^{n+1}}$.

2.15. For the function $f$ defined by

$$f(x) = \begin{cases} -1 & & -1 \le x < 0 \\ +1 & \text{if} & 0 \le x \le 1 \\ 0 & & |x| > 1, \end{cases}$$

find $f^+$ and $f^-$ and determine the integral using (2.8). Repeat the exercise for the case in which

$$g(x) = \begin{cases} -1 & & -1 \le x < 0 \\ +1 & \text{if} & x \ge 0 \\ 0 & & x < -1 \end{cases} .$$

2.16. Show that the Lebesgue integral of $f$ exists if and only if that of $|f|$ exists, and that

$$\left| \int_\Omega f \, dx \right| \le \int_\Omega |f| \, dx.$$

2.17. What relationship must be satisfied by $a$ and $p$ in order that $u(x) = x^a$ belongs to $L^p(0,1)$? And for $u$ to belong to $L^p(1,\infty)$?

2.18. For what values of the real number $a$ does the complex-valued function $f$ defined by $f(x) = x^a(1 - xi)$ belong to $L^2(0,1)$?

2.19. Prove Theorem 4(c), which states that if $u$, $v \in L^2(\Omega)$, then $\int_\Omega u(\boldsymbol{x})v(\boldsymbol{x})\, dx$ is finite.

# 3

# Vector spaces, normed, and inner product spaces

From elementary courses in vector algebra and analysis we know that the idea of a vector as a directed line segment is not sufficient for us to build up a nontrivial theory, let alone be of use in concrete applications. Additional structure has to be added: we agree to add together vectors using the parallelogram law, and we define various forms of multiplication of vectors, for example, the scalar (dot) product and the vector (cross) product. Once these properties have been adopted, it becomes possible to construct a fairly sophisticated theory.

The same is true of sets in general. A set without structure is sterile, and not of much use from the point of view of the analyst. The question of what kinds of properties to assume is generally answered by looking at the properties of simple sets like $\mathbb{R}$ or the set of vectors, and by generalizing accordingly. This process of generalization is a recurrent theme in the next few chapters, and in this chapter we begin the process by defining first a vector space to be, broadly speaking, an arbitrary set whose members behave as vectors. Then we show how properties such as "length", "distance" and "scalar product" can be defined for vector spaces, leading to the notions of normed and inner product spaces.

## 3.1 Vector spaces and subspaces

We are familiar with the idea of a *set* being a collection of objects, all of which have a specified property. In most applications, though, it is useful

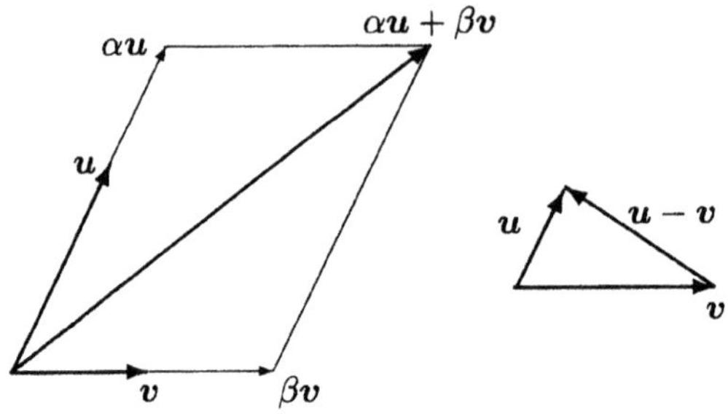


FIGURE 3.1. Vector addition and subtraction

to be able to add together multiples of members of a set and to have the assurance that the result of such an operation will yield something that is also a member of that set. This is the essence of a vector space.

Suppose we generalize from the behavior of vectors, starting by first reviewing some familiar properties of the set $V$ of all vectors in three-dimensional space. Given vectors $\boldsymbol{u}, \boldsymbol{v}, \boldsymbol{w}$ and real numbers $\alpha, \beta$ we know that:

1. $\alpha \boldsymbol{u} + \beta \boldsymbol{v}$ is also a vector (sums of multiples of vectors are also vectors, as shown in Figure 3.1);

2. $\boldsymbol{u} + \boldsymbol{v} = \boldsymbol{v} + \boldsymbol{u}$, and $\boldsymbol{u} + \boldsymbol{v} + \boldsymbol{w} = (\boldsymbol{u} + \boldsymbol{v}) + \boldsymbol{w}$ (when adding vectors together, the result does not depend on the order in which the addition is carried out);

3. there is a vector $\mathbf{0}$ called the *zero vector* that has the property $\boldsymbol{u}+\mathbf{0} = \boldsymbol{u}$ for all vectors $\boldsymbol{u}$;

4. there is a vector $-\boldsymbol{u}$, called the *negative* of $\boldsymbol{u}$, that has the property $\boldsymbol{u} + (-\boldsymbol{u}) = \mathbf{0}$ (we normally write this as $\boldsymbol{u} - \boldsymbol{u} = \mathbf{0}$). This in turn defines *subtraction*: by the difference $\boldsymbol{u} - \boldsymbol{v}$ we then mean the vector $\boldsymbol{u} + (-\boldsymbol{v})$ (Figure 3.1);

5. $(\alpha\beta)\boldsymbol{u} = \alpha(\beta\boldsymbol{u})$;

6. $(\alpha + \beta)\boldsymbol{u} = \alpha\boldsymbol{u} + \beta\boldsymbol{u}$, and $\alpha(\boldsymbol{u} + \boldsymbol{v}) = \alpha\boldsymbol{u} + \alpha\boldsymbol{v}$;

7. $1 \cdot \boldsymbol{u} = \boldsymbol{u}$ (this, with 6, tells us that $\boldsymbol{u} = (1 + 0)\boldsymbol{u} = 1 \cdot \boldsymbol{u} = 1 \cdot \boldsymbol{u} + 0 \cdot \boldsymbol{u} = \boldsymbol{u}$ so that $0 \cdot \boldsymbol{u} = 0$).

Now all of these properties of vectors are readily generalized to any set, and this is what we do next.

**Vector space.** Let $X$ be a set, and let $\mathbb{K}$ be either the set $\mathbb{R}$ of real numbers or the set $\mathbb{C}$ of complex numbers, either of these being referred to here as scalars, for convenience. Then $X$ is called a vector space (or linear space) if it has an operation + called *addition*, an operation of *multiplication by a scalar*, and satisfies the following *axioms*.

**VS1.** for all $u, v \in X$, and scalars $\alpha, \beta$, $\alpha u + \beta v$ is also a member of $X$;

**VS2.** $u + v = v + u$ and $u + (v + w) = (u + v) + w$ for all $u, v, w \in X$;

**VS3.** there is an element 0 of $X$ called the zero element that has the property

$$u + 0 = u \quad \text{for all } u \in X;$$

**VS4.** for every $u \in X$ there is an element $-u$ that satisfies $u + (-u) = 0$; then by the difference $u - v$ we understand $u + (-v)$;

**VS5.** $(\alpha\beta)u = \alpha(\beta u)$ for all scalars $\alpha, \beta$ and for all $u \in X$;

**VS6.** $(\alpha + \beta)u = \alpha u + \beta u$, and $\alpha(u + v) = \alpha u + \alpha v$ for all scalars $\alpha, \beta$ and for all $u, v \in X$;

**VS7.** $1 \cdot u = u$.

When $\mathbb{K}$ is chosen to be the real numbers, then $X$ is called a *real* vector space, whereas it is referred to as a *complex* vector space if $\mathbb{K}$ is chosen to be $\mathbb{C}$. These two sets do not exhaust the choices of scalars that may be made, but they more than suffice for our needs.

### Examples

1. We start with a trivial example: the set $V$ of vectors in $\mathbb{R}^3$ is a vector space. Indeed, $V$ served as a model for setting up the axioms of a vector space.

2. The set $\mathbb{R}^n$ of $n$-tuples is a real vector space, with addition defined by

$$\begin{aligned} \boldsymbol{x} + \boldsymbol{y} &= (x_1, x_2, \ldots, x_n) + (y_1, y_2, \ldots, y_n) \\ &= (x_1 + y_1, x_2 + y_2, \ldots, x_n + y_n) \quad \text{for } \boldsymbol{x}, \boldsymbol{y} \in \mathbb{R}^n \end{aligned}$$

   and scalar multiplication by

$$\alpha\boldsymbol{x} = \alpha(x_1, x_2, \ldots, x_n) = (\alpha x_1, \alpha x_1, \alpha x_2, \ldots, \alpha x_n) \quad \text{for } \alpha \in \mathbb{R}.$$

   The zero element is $\mathbf{0} = (0, \ldots, 0)$ and the element $-\boldsymbol{x}$ is given by $-\boldsymbol{x} = (-x_1, \ldots, -x_n)$.

3. The set $\mathbb{C}^n$ of $n$-tuples of complex numbers is a *complex* vector space, the operations of addition and scalar multiplication being defined as in the case of $\mathbb{R}^n$, with the scalars now being complex numbers.

4. The set $C^m(\Omega)$ of $m$-times continuously differentiable functions on $\Omega$ is a vector space. For, if $u$ and $v$ are two such functions, then so is the function $\alpha u + \beta v$ defined by $(\alpha u + \beta v)(\boldsymbol{x}) = \alpha u(\boldsymbol{x}) + \beta v(\boldsymbol{x})$. The zero element is simply the zero function and the function $-u$ is the function satisfying $(-u)(\boldsymbol{x}) = -1 \cdot u(\boldsymbol{x})$. It is a real or complex vector space, accordingly as the functions are real- or complex-valued.

5. The space $L^p(\Omega)$ is a vector space for $1 \le p < \infty$; this follows from the *Minkowski inequality for integrals*

$$\left[\int_\Omega |u \pm v|^p \, dx\right]^{1/p} \le \left[\int_\Omega |u|^p \, dx\right]^{1/p} + \left[\int_\Omega |v|^p \, dx\right]^{1/p}, \qquad (3.1)$$

the derivation of which is discussed in Exercise 3.6. If we replace $u$ by $\alpha u$ and $v$ by $\beta v$ in (3.1) then we see that

$$\begin{aligned} \left[\int_\Omega |\alpha u + \beta v|^p \, dx\right]^{1/p} &\le \left[\int_\Omega |\alpha u|^p \, dx\right]^{1/p} + \left[\int_\Omega |\beta v|^p \, dx\right]^{1/p} \\ &= \left[|\alpha|^p \int_\Omega |u|^p \, dx\right]^{1/p} + \left[|\beta|^p \int_\Omega |v|^p \, dx\right]^{1/p} \\ &= |\alpha| \left[\int_\Omega |u|^p \, dx\right]^{1/p} + |\beta| \left[\int_\Omega |v|^p \, dx\right]^{1/p} \end{aligned}$$

and this last expression is finite since $u$ and $v$ belong to $L^p(\Omega)$. Hence $\alpha u + \beta v \in L^p(\Omega)$. The remaining axioms are readily shown to be valid.

The space $L^\infty(\Omega)$ is likewise a vector space, as is easily verified.

As in the case of $C^m(\Omega)$, the spaces $L^p(\Omega)$ are real or complex vector spaces, accordingly as the functions are real- or complex- valued. If complex-valued, then $|\cdot|$ in the Minkowski inequality is interpreted as the modulus of a complex number.

6. The set $X$ of all nonnegative continuous functions, defined by

$$X = \{u : u(\boldsymbol{x}) \in C(\Omega), \ u(\boldsymbol{x}) \ge 0, \ \boldsymbol{x} \in \Omega\},$$

is *not* a vector space since, for example, $\alpha u$ is not a member of $X$ for negative values of $\alpha$.

Since all vector spaces are sets, it is natural to enquire whether subsets of vector spaces are also vector spaces. This is not always true, but in those cases in which it is true we give the subset a special name.

**Subspace.** A *subspace* $Y$ of a vector space $X$ is a subset of $X$ that is also a vector space.

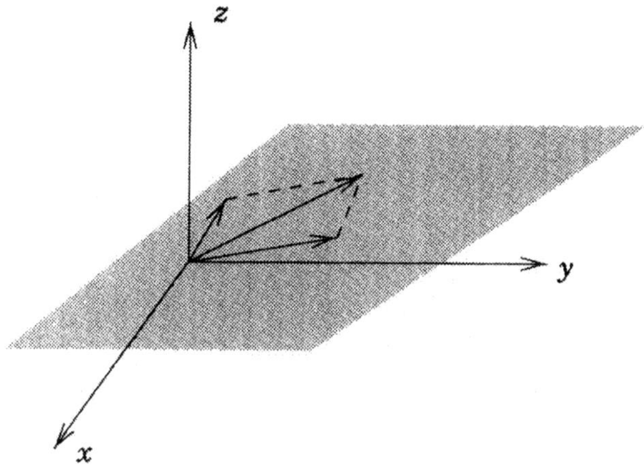


FIGURE 3.2. Planes passing through the origin are subspaces of $\mathbb{R}^3$

**Examples**

7. Consider the vector space $\mathbb{R}^3$; all points of the form $(x, y, 0)$ form a subspace of $\mathbb{R}^3$ – the $xy$ plane, in common parlance – since sums of multiples of points in the $xy$ plane also lie in this plane. Indeed, the set of points of any plane or line *passing through the origin* is a subspace of $\mathbb{R}^3$ (Figure 3.2).

8. The set $P_3[0,1]$ of polynomials of degree $\leq 3$ forms a subset of $C[0,1]$ and constitutes a subspace: for any polynomials $p(x), q(x) \in P_3[0,1]$,

$$\alpha p(x) + \beta q(x)$$

is also a polynomial of degree $\leq 3$, and therefore belongs to $P_3[0,1]$.

9. The set $C(\overline{\Omega})$ of bounded continuous functions forms a subspace of $L^p(\Omega)$ (see Section 2.3) for $1 \leq p \leq \infty$.

**Sum of subspaces.** Given two subspaces $V, W$ of a vector space $X$, we define the *sum* of $V$ and $W$, denoted by $V+W$, to be the set of all members of $X$ of the form $v + w$ with $v \in V$ and $w \in W$. In other words,

$$V + W = \{u \in X : u = v + w \quad \text{for } v \in V,\ w \in W\}.$$

The set $V + W$ is also a subspace of $X$ since if $u$ and $\overline{u}$ are members of $V + W$, so that $u = v + w$ and $\overline{u} = \overline{v} + \overline{w}$ with $v, \overline{v} \in V$ and $w, \overline{w} \in W$, then it follows that

$$\begin{aligned} \alpha u + \beta \overline{u} &= \alpha(v + w) + \beta(\overline{v} + \overline{w}) \\ &= \underbrace{(\alpha v + \beta \overline{v})}_{\in V} + \underbrace{(\alpha v + \beta \overline{w})}_{\in W}, \end{aligned}$$

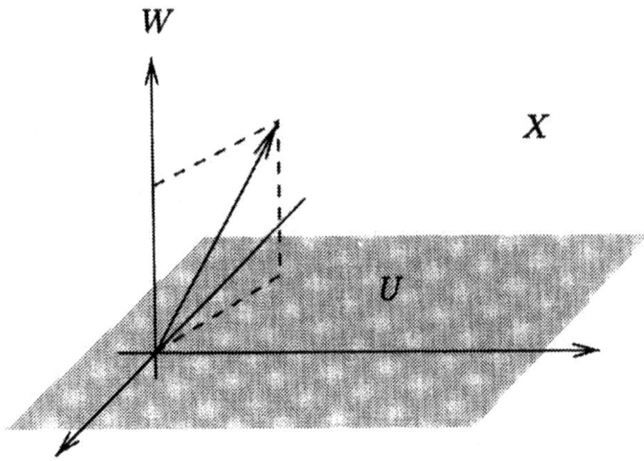


FIGURE 3.3. Direct sum of subspaces

so that $\alpha u + \beta \overline{u}$ is also in $V + W$.

**Example**

10. Let $X = \mathbb{R}^3$ and $V = \{\boldsymbol{x} \in \mathbb{R}^3 : \boldsymbol{x} = (\alpha, 0, 0),\ \alpha \in \mathbb{R}\}$, $W = \{\boldsymbol{x} \in \mathbb{R}^3 : \boldsymbol{x} = (\alpha, \beta, 0),\ \alpha,\ \beta \in \mathbb{R}\}$ (that is, $V$ is the $x$ axis and $W$ the $xy$ plane). Then the sum of $V$ and $W$ is the subspace of $X$ consisting of the points $\boldsymbol{x} = (\alpha, \beta, 0)$ for real numbers $\alpha, \beta$.

**Direct sum of subspaces.** If $V$ and $W$ are subspaces of a vector space $X$, then $X$ is said to be the *direct sum* of $V$ and $W$ if (i) it is the sum of $V$ and $W$; and (ii) $V$ and $W$ have only the zero element in common, that is, $V \cap W = \{0\}$. The direct sum is denoted by $V \oplus W$.

**Example**

11. Let $X = \mathbb{R}^3$, $U = \{\boldsymbol{x} \in \mathbb{R}^3 : \boldsymbol{x} = (\alpha, \beta, 0)\}$, $V = \{\boldsymbol{x} \in \mathbb{R}^3 : \boldsymbol{x} = (0, \beta, \gamma)\}$, and $W = \{\boldsymbol{x} \in \mathbb{R}^3 : \boldsymbol{x} = (0, 0, \gamma)\}$. Then clearly

$$X = U + V \text{ and } X = U + W.$$

But $U \cap V = \{\boldsymbol{x} : \boldsymbol{x} = (0, \beta, 0)\} \neq \{\mathbf{0}\}$, whereas $U \cap W = \{\mathbf{0}\}$. Thus $X = U \oplus W$ (Figure 3.3).

The question of when an arbitrary member of $u$ of a vector space $X$ has a *unique* representation $u = v + w$ for $v \in V,\ w \in W$ is easily resolved, as we show in the next result.

THEOREM 1. *Let $X$ be a vector space. Then $X = V \oplus W$ if and only if*

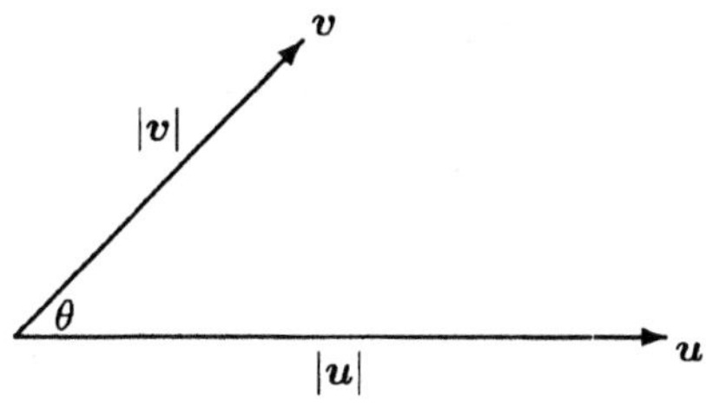


FIGURE 3.4. The inner product of two vectors

*for any $u \in X$ there are unique members $v \in V$ and $w \in W$ such that $u = v + w$.*

The proof of this theorem is treated in Exercise 3.4.

## 3.2 Inner product spaces

It is now possible to generalize to vector spaces many of the fundamental concepts of vector algebra and analysis, and we start with the concept of an *inner product* or *scalar* product. Recall that the scalar product $\boldsymbol{u} \cdot \boldsymbol{v}$ of two vectors $\boldsymbol{u}$ and $\boldsymbol{v}$ is a real number given by

$$\boldsymbol{u} \cdot \boldsymbol{v} = |\boldsymbol{u}|\,|\boldsymbol{v}| \cos\theta,$$

where $\theta$ is the angle between $\boldsymbol{u}$ and $\boldsymbol{v}$ (Figure 3.4). The inner product has the following properties. It is symmetric ($\boldsymbol{v} \cdot \boldsymbol{u} = \boldsymbol{v} \cdot \boldsymbol{u}$), linear ($(\alpha\boldsymbol{u} + \beta\boldsymbol{v}) \cdot \boldsymbol{w} = \alpha\boldsymbol{u} \cdot \boldsymbol{w} + \beta\boldsymbol{v} \cdot \boldsymbol{w}$), and positive-definite ($\boldsymbol{u} \cdot \boldsymbol{u} \geq 0$ and $\boldsymbol{u} \cdot \boldsymbol{u} = 0$ iff $\boldsymbol{u} = \boldsymbol{0}$). Furthermore, the scalar product in turn provides a means of measuring the *length* or *norm* of a vector: indeed, for any vector $\boldsymbol{u}$ $|\boldsymbol{u}| = (\boldsymbol{u} \cdot \boldsymbol{u})^{1/2}$. And finally, when equipped with the scalar product operation it is possible to measure the distance between two points $\boldsymbol{x}$ and $\boldsymbol{y}$ in $\mathbb{R}^3$: if this distance is denoted by $d(\boldsymbol{x}, \boldsymbol{y})$, then

$$\begin{aligned} d(\boldsymbol{x}, \boldsymbol{y}) &= \sqrt{(\boldsymbol{y} - \boldsymbol{x}) \cdot (\boldsymbol{y} - \boldsymbol{x})} \\ &= |\boldsymbol{y} - \boldsymbol{x}| \\ &= \sqrt{(y_1 - x_1)^2 + (y_2 - x_2)^2 + (y_3 - x_3)^2}\ . \end{aligned}$$

The function $d(\cdot\,,\cdot)$, being a device for measuring distances between points, is called the *metric*. The concepts of inner product, norm, and metric are defined in much the same way for arbitrary vector spaces. In this section we take the first step in this direction, and deal with the inner product.

**Inner product and inner product space.** Let $X$ be a complex vector space; then the inner product $(u, v)$ of $u, v \in X$ is an operation that satisfies the following *axioms*, for all $u, v, w \in X$ and $\alpha, \beta \in \mathbb{C}$.

**CIP1.** $(u, v) \in \mathbb{C}$ (the inner product is complex-valued).

**CIP2.** $(v, u) = \overline{(u, v)}$ (the operation is *Hermitian*).

**CIP3.** $(\alpha u + \beta v, w) = \alpha(u, w) + \beta(v, w)$ (it is linear in the first slot).

**CIP4.** $(u, u) \geq 0$ and $(u, u) = 0$ iff $u = 0$ (it is positive-definite).

A vector space $X$ endowed with an inner product $(\cdot\,, \cdot)$ is called an *inner product space*. Since it is the vector space together with the inner product that defines the inner product space, the conventional and more proper notation is $(X, (\cdot\,, \cdot))$; however, when it is clear which particular inner product is being used, or when the details of the inner product are not pertinent, the inner product space is denoted simply by $X$.

Although the inner product is in general a *complex* number, the inner product $(u, u)$ of any member with itself is always a *real* number, from Axiom CIP2, since this axiom tells us that $(u, u) = \overline{(u, u)}$. It follows then that the axiom of positive-definiteness CIP4 makes complete sense (it would not if $(u, u)$ were complex).

**Linearity.** Linearity is a notion that recurs on a regular basis, given that the focus of this work is on linear functional analysis and its applications. It is as well, therefore, to spend a moment considering its attributes. The Axiom CIP3 of linearity, which applies only to the first slot of the inner product, is actually made up of two parts: it states that the inner product is

$$\text{additive, that is, } (u + v, w) = (u, w) + (v, w), \tag{3.2}$$

and

$$\text{homogeneous, that is, } (\alpha u, v) = \alpha(u, v). \tag{3.3}$$

We thus have the relationship

$$\text{additivity} + \text{homogeneity} = \text{linearity}.$$

These two compnents of linearity can easily be deduced from CIP3, simply by setting $\alpha = \beta = 1$ (to obtain the property of additivity) and then by setting $\beta = 0$ (to obtain homogeneity). Conversely, the properties of additivity and homogeneity may be combined to give CIP3, by replacing $u$ and $v$ in (3.2) with $\alpha u$ and $\beta v$, and then by invoking homogeneity.

Whether linearity is defined with respect to the first or second slot in the inner product *does* make a difference for *complex* inner product spaces. To see this we note first of all that, with the use of the inner product axioms and the properties of complex conjugation,

$$(u, v + w) = \overline{(v + w, u)} = \overline{(v, u) + (w, u)} = \overline{(v, u)} + \overline{(w, u)} = (u, v) + (u, w)$$

so that *additivity* in the second slot follows. However, the inner product is not *homogeneous* in the second slot since, for any complex number $\alpha$, Axioms CIP2 and CIP3 give

$$(u, \alpha v) = \overline{(\alpha v, u)} = \overline{\alpha (v, u)} = \overline{\alpha}\,\overline{(v, u)} = \overline{\alpha}(u, v).$$

The property of linearity implies that the inner product of any member $u$ with the zero element is zero; that is,

$$(0, v) = 0 \quad \text{for all } v \in X.$$

This follows from the observation that, for any $u \in X$, $(u, v) = (u + 0, v) = (u, v) + (0, v)$. Comparison of the left- and right-hand sides gives the desired result. In the same way it follows that $(v, 0) = 0$ for any $v \in X$.

**Real inner product spaces.** It is possible to define an inner product on *real* vector spaces as well; indeed, this is a very important special case to which frequent reference is made. To do so, it suffices to change $\mathbb{C}$ to $\mathbb{R}$ in the preceding set of axioms, which then conveniently reduce to the set of axioms appropriate to real vector spaces. For convenience we summarize these here in one place:

for $u, v, w \in X$ and $\alpha, \beta \in \mathbb{R}$,

**RIP1.** $(u, v) \in \mathbb{R}$;

**RIP2.** $(v, u) = (u, v)$ (the operation is *symmetric*);

**RIP3.** $(\alpha u + \beta v, w) = \alpha(u, w) + \beta(v, w)$ (linearity);

**RIP4.** $(u, u) \geq 0$ and $(u, u) = 0$ iff $u = 0$ (positive-definiteness).

We note in particular that the property of Hermitian symmetry reduces to that of plain symmetry, for the real case, whereas the other axioms are unchanged. For real inner product spaces we have in addition the property of linearity in the second slot; this follows from (3.2) and (3.3), bearing in mind that $\alpha$ is a real number in this case.

In general we refer simply to inner product spaces, and the context makes clear whether the space is real or complex. General results are always proved for complex inner product spaces, since these then apply a fortiori to real spaces.

**Examples**

12. Let $X = \mathbb{R}^3$; then the conventional or Euclidean scalar product defined by

$$(\boldsymbol{x}, \boldsymbol{y}) = \boldsymbol{x} \cdot \boldsymbol{y} = x_1 y_1 + x_2 y_2 + x_3 y_3$$

for $\boldsymbol{x} = (x_1, x_2, x_3)$ and $\boldsymbol{y} = (y_1, y_2, y_3)$ satisfies the (real) inner product axioms.

13. Consider the space $L^2(a,b)$ of square-integrable functions defined on the interval $(a,b)$. An inner product for $L^2(a,b)$ may be defined by

$$(u, v) \equiv \int_a^b u(x)\overline{v(x)}\, dx \quad \text{for } u, v \in L^2(a,b). \tag{3.4}$$

We have, in particular,

$$\begin{aligned}
(v, u) = \int_a^b v(x)\overline{u(x)}\, dx &= \int_a^b \overline{u(x)}v(x)\, dx \\
&= \int_a^b \overline{\overline{v(x)}u(x)}\, dx = \overline{\int_a^b \overline{v(x)}u(x)\, dx} \\
&= \overline{\int_a^b u(x)\overline{v(x)}\, dx} = \overline{(u, v)}
\end{aligned}$$

using the properties of complex numbers, including the property that, for complex integrands, the complex conjugate of the integral equals the integral of the complex conjugate (can you see why?). Thus Axiom CIP2 is satisfied. Second,

$$\begin{aligned}
(\alpha u + \beta v, w) &= \int_a^b [\alpha u(x) + \beta v(x)]\overline{w(x)}\, dx \\
&= \alpha \int_a^b u(x)\overline{w(x)}\, dx + \beta \int_a^b v(x)\overline{w(x)}\, dx \\
&= \alpha(u, w) + \beta(v, w)
\end{aligned}$$

and so Axiom CIP3 is satisfied. Finally,

$$(u, u) = \int_a^b u(x)\overline{u(x)}\, dx = \int_a^b |u(x)|^2\, dx,$$

which is clearly positive since it is the integral of a positive function. The only function $u$ for which this integral vanishes is $u(x) = 0$ a.e. (recall the properties of the Lebesgue integral), and so Axiom CIP4 is satisfied.

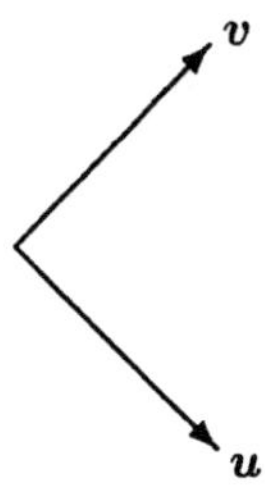

FIGURE 3.5. Orthogonal vectors

**Orthogonality.** We continue the abstraction of properties of vectors in three-dimensional space, and recall next the concept of orthogonality: two vectors $\boldsymbol{u}$ and $\boldsymbol{v}$ are orthogonal if $\boldsymbol{u} \cdot \boldsymbol{v} = 0$, that is, if they are at right angles to each other (Figure 3.5). Since we have at our disposal the concept of an inner product, it is a very simple matter to extend this notion of orthogonality to any inner product space, irrespective of whether the geometric interpretation applies. Thus two members $u, v$ of an inner product space $X$ are said to be *orthogonal* if

$$(u, v) = 0.$$

When this is the case we write, as in the case of vectors, $u \perp v$.

**Example**

14. Consider the functions $u(x) = \sin x$ and $v(x) = \cos x$, with $u, v \in L^2(-\pi, \pi)$. Making use of the inner product (3.4) (but bearing in mind that we are dealing with real-valued functions here) we find that

$$(u, v) = \int_{-\pi}^{\pi} \sin x \cos x \; dx = 0$$

and so $u$ and $v$ are orthogonal in $L^2(-\pi, \pi)$.

**The Cauchy–Schwarz inequality.** We return for a moment to vectors and observe another property of the dot product. Recall that for two vectors $\boldsymbol{u}$ and $\boldsymbol{v}$ with an angle $\theta$ between them,

$$\boldsymbol{u} \cdot \boldsymbol{v} = |\boldsymbol{u}|\,|\boldsymbol{v}| \cos\theta$$

or

$$\boldsymbol{u} \cdot \boldsymbol{v} = (\boldsymbol{u} \cdot \boldsymbol{u})^{1/2} (\boldsymbol{v} \cdot \boldsymbol{v})^{1/2} \cos\theta.$$

But $|\cos\theta| \leq 1$, and so it follows that

$$|\boldsymbol{u} \cdot \boldsymbol{v}| \leq (\boldsymbol{u} \cdot \boldsymbol{u})^{1/2} (\boldsymbol{v} \cdot \boldsymbol{v})^{1/2}.$$

This property in fact holds for any inner product space, as the next result shows.

THEOREM 2 (THE CAUCHY–SCHWARZ INEQUALITY). *If $u$ and $v$ are members of an inner product space $X$ with inner product $(\cdot, \cdot)$, then*

$$|(u,v)| \leq (u,u)^{1/2}(v,v)^{1/2}. \tag{3.5}$$

PROOF. We assume that neither $u$ nor $v$ is zero; for the case in which either of these is zero, (3.5) is satisfied trivially. The proof then follows from the observation that, for any complex number $\alpha$,

$$(u - \alpha v, u - \alpha v) \geq 0,$$

using Axiom CIP3. Upon expansion and use of the axioms of linearity and Hermitian symmetry this becomes

$$\begin{aligned} 0 &\leq (u,u) - (\alpha v, u) - (u, \alpha v) + (\alpha v, \alpha v) \\ &= (u,u) - (\alpha v, u) - \overline{(\alpha v, u)} + (\alpha v, \alpha v) \\ &= (u,u) - 2\text{Re}[\alpha(v,u)] + |\alpha|^2(v,v), \end{aligned}$$

where Re$z$ denotes the real part of a complex number $z$. Now $\alpha$ is arbitrary, so if we choose $\alpha$ to be equal to $\overline{(v,u)}/(v,v)$, then $|\alpha| = |(v,u)|/(v,v)$ (remember that $(v,v)$ is real) and

$$\begin{aligned} 0 &\leq (u,u) - 2\text{Re}[\overline{(v,u)}(v,u)/(v,v)] + |(v,u)|^2/(v,v) \\ &= (u,u) - |(v,u)|^2/(v,v). \end{aligned}$$

The desired result is then obtained by rearranging the terms, multiplying throughout by $(v,v)$, and taking the square root of both sides. □

## 3.3 Normed spaces

At the beginning of the previous section we introduced the concept of a norm by drawing an analogy with the notion of the length of a vector. As with the definition of an inner product, this notion may be abstracted in a natural way if we start from scratch with an arbitrary vector space $X$: a *norm* $\|\cdot\|$ on $X$ is an operation that satisfies the following axioms for any members $u, v$ of $X$, and scalars (real or complex, as appropriate) $\alpha$:

**N1.** $\|u\| \in \mathbb{R}$.

**N2.** $\|u\| \geq 0$ and $\|u\| = 0$ iff $u = 0$ (positive-definiteness).

**N3.** $\|\alpha u\| = |\alpha| \|u\|$ (positive homogeneity).

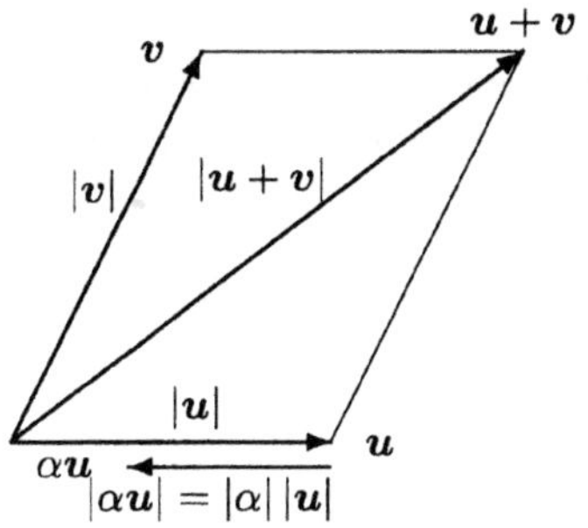


FIGURE 3.6. Axioms N3 and N4 as they apply to vectors

**N4.** $\|u+v\| \leq \|u\| + \|v\|$ (triangle inequality).

A few remarks about these axioms are in order. The first asserts that, by analogy with the length of a vector, the norm is expected to yield a real number; furthermore, Axiom N2 asserts that this quantity will always be positive, except for the case of the zero element, in which case it is zero.

The third axiom, that of positive homogeneity, is again a straightforward abstraction of a property of vectors, as shown in Figure 3.6. It is also clear from this figure why it is *positive* homogeneity, and not simply homogeneity as in the case of the inner product, that is required. Axiom N3 applies equally to real and complex spaces, with $|\alpha|$ being interpreted appropriately.

Finally, the triangle inequality abstracts the situation that is summarized in the parallelogram law for addition of vectors (Figure 3.6).

## Examples

15. Let $X = \mathbb{R}^3$; then the usual or *Euclidean* norm defined on $\mathbb{R}^3$ is

$$\|\boldsymbol{x}\| = (x_1^2 + x_2^2 + x_3^2)^{1/2} \quad \text{for} \quad \boldsymbol{x} = (x_1, x_2, x_3).$$

The extension to $\mathbb{R}^n$ is obvious.

16. Norms, like inner products, are not unique quantities. A case in point is $\mathbb{R}^n$, on which it is possible to define a whole family of norms: for each real number $p$ in the range $1 \leq p < \infty$ the quantity $\|\cdot\|_p$ defined by

$$\|\boldsymbol{x}\|_p = [|x_1|^p + |x_2|^p + \cdots + |x_n|^p]^{1/p}$$

is a norm on $\mathbb{R}^n$, the Euclidean norm corresponding to the case $p = 2$. Axioms N2 and N3 are seen by inspection to hold, and the triangle inequality is a consequence of the *Minkowski inequality for sums*; for $1 \leq p < \infty$,

$$\left[\sum_{i=1}^{n} |x_i \pm y_i|^p\right]^{1/p} \leq \left[\sum_{i=1}^{n} |x_i|^p\right]^{1/p} + \left[\sum_{i=1}^{n} |y_i|^p\right]^{1/p}.$$

The proof of this inequality is outlined in Exercise 3.19. The case $p = \infty$ may also be included in this family if we define $\|\cdot\|_\infty$ on $\mathbb{R}^n$ by

$$\|\boldsymbol{x}\|_\infty = \max_{1\le i\le n} |x_i|;$$

this is also a norm on $\mathbb{R}^n$.

17. Let $X = L^p(\Omega)$ with $1 \le p < \infty$. The standard norm on $L^p$ is defined by

$$\|u\|_{L^p} = \left[\int_\Omega |u(x)|^p \, dx\right]^{1/p} \tag{3.6}$$

which is of course a well-defined quantity for any $u \in L^p(\Omega)$. Axioms N1 through N3 are easily shown to hold, and the triangle inequality follows from the Minkowski inequality for integrals (3.1) which can be written, using the notation (3.6), as

$$\|u \pm v\|_{L^p} \le \|u\|_{L^p} + \|v\|_{L^p}.$$

For convenience, and when there is no danger of ambiguity, we denote the $L^p$-norm simply by $\|\cdot\|_p$ rather than the more cumbersome $\|\cdot\|_{L^p}$.

18. Consider the space $L^\infty(\Omega)$ of bounded measurable functions, that is, functions $u$ that satisfy $|u(x)| \le k$ a.e. on $\Omega$. Because we are dealing with equivalence classes of functions, the notion of the supremum is meaningless, and has to be replaced by the *essential supremum*, defined to be the greatest lower bound of the constants $k$ that bound $|u|$ almost everywhere:

$$\operatorname*{ess\,sup}_{x\in\Omega} |u(x)| = \inf\{k : |u(x)| \le k \text{ a.e.}\}.$$

Then $L^\infty(\Omega)$ is a normed space, with norm $\|\cdot\|_{L^\infty}$ defined by

$$\|u\|_{L^\infty} = \operatorname*{ess\,sup}_{x\in\Omega} |u(x)|.$$

The first three norm axioms obviously hold; to verify the triangle inequality we note that, for any two functions $u$ and $v$ in $L^\infty(a,b)$, $u(x)$ and $v(x)$ are simply real or complex numbers, so that

$$|u(x) + v(x)| \le |u(x)| + |v(x)|;$$

thus, recalling the properties of the supremum in Chapter 1, and bearing in mind that these properties carry over to the essential supremum, we have (Figure 3.7)

$$\begin{aligned}
\|u+v\|_{L^\infty} &= \text{ess sup}|u(x)+v(x)| \\
&\le \text{ess sup}(|u(x)| + |v(x)|) \\
&= \text{ess sup}|u(x)| + \text{ess sup}|v(x)| \\
&= \|u\|_{L^\infty} + \|v\|_{L^\infty}.
\end{aligned}$$

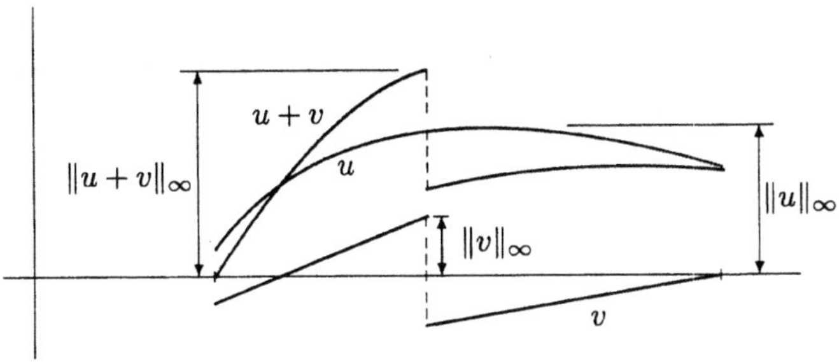


FIGURE 3.7. The triangle inequality for functions in $L^\infty(a,b)$

Here, too, it is convenient to denote the $L^\infty$-norm simply by $\|\cdot\|^\infty$ when this is unlikely to be ambiguous.

19. Since the space $C(\overline{\Omega})$ of bounded continuous functions is a subspace of $L_\infty(\Omega)$, it follows that we may endow $C(\overline{\Omega})$ with the sup-norm $\|\cdot\|_\infty$, thereby making it a normed space. In the case of continuous functions it suffices of course to make use of the supremum, or indeed the maximum, rather than the essential supremum. Under these circumstances it suffices also simply to denote this norm by $\|\cdot\|_\infty$; thus,

$$\|u\|_\infty = \max_{\boldsymbol{x}\in\overline{\Omega}} |u(\boldsymbol{x})| \quad \text{for} \quad u \in C(\overline{\Omega}). \tag{3.7}$$

**Normed space.** A vector space $X$ with a norm $\|\cdot\|$ defined on it is called a *normed space.* Since it is the vector space together with the norm that defines the normed space, the conventional and more proper notation is $(X, \|\cdot\|)$; however, when it is clear which particular norm is being used, or when the details of the norm are not pertinent, the normed space is denoted simply by $X$.

**The norm generated by an inner product.** The norm is a primitive concept that does not require for its definition the existence on an inner product. Indeed, a norm is any operation $\|\cdot\|$ that satisfies N1 through N3. But if we have an inner product space $(X, (\cdot\,,\cdot))$, then a norm $\|\cdot\|$ on $X$ may be defined according to

$$\|u\| = (u,u)^{1/2}, \tag{3.8}$$

and we say that $\|\cdot\|$ in (3.8) is *the norm generated* by the inner product. The analogy with vectors in $\mathbb{R}^3$ is clear: given the scalar (inner) product defined for vectors, the norm or length of a vector $\boldsymbol{u}$ is given by

$$\|\boldsymbol{u}\| = (\boldsymbol{u}\cdot\boldsymbol{u})^{1/2}.$$

The question now arises: is $\|\cdot\|$ defined in (3.8) really a norm? That is, does it satisfy all of the norm axioms? The answer, of course, is yes: first, the quantity $(u, u)$ is real, so that N1 is satisfied. Second, positive-definiteness of $\|u\|$ follows directly from the positive-definiteness of the inner product. Positive homogeneity is verified by considering that, for any complex $\alpha$,

$$\begin{aligned}\|\alpha u\|^2 &= (\alpha u, \alpha u) \\ &= \alpha\bar{\alpha}(u, u) = |\alpha|^2\|u\|^2\end{aligned}$$

using properties CIP2 and CIP3 of the inner product. Finally, in order to show that the triangle inequality is satisfied we consider

$$\begin{aligned}\|u+v\|^2 &= (u+v, u+v) \\ &= (u,u) + 2\mathrm{Re}(u,v) + (v,v) \\ &= \|u\|^2 + 2\mathrm{Re}(u,v) + \|v\|^2 \\ &\leq \|u\|^2 + 2|(u,v)| + \|v\|^2 \\ &\leq \|u\|^2 + 2\|u\|\,\|v\| + \|v\|^2 \\ &\quad \text{(using the Cauchy–Schwarz inequality)} \\ &= (\|u\| + \|v\|)^2.\end{aligned}$$

The desired result is now obtained by taking the square root of both sides.

With the understanding that the norm on an inner product space is that generated by the inner product, the Cauchy–Schwarz inequality may be written in the alternative form

$$|(u,v)| \leq \|u\|\,\|v\|.$$

**The parallelogram law.** The preceding discussion shows that it is true that *every inner product space is automatically a normed space*, so it is natural to enquire whether the converse is true: in other words, can a norm be used to generate an inner product? The answer, unfortunately, is negative. To see this, we introduce an identity that is valid on any normed space, and which may be used to provide a partial answer to the question. This identity is known as the parallelogram law as a result of its interpretation in $\mathbb{R}^2$ (Figure 3.8). Let $X$ be an *inner product space* with norm $\|u\| = (u,u)^{1/2}$; then

$$\|u+v\|^2 + \|u-v\|^2 = 2(\|u\|^2 + \|v\|^2) \tag{3.9}$$

for all $u, v \in X$. Comparison of (3.9) with Figure 3.8 should make clear the reason for referring to this identity as the parallelogram law. Indeed, from the cosine rule in $\mathbb{R}^2$, $\|\boldsymbol{u} - \boldsymbol{v}\|^2 = \|\boldsymbol{u}\|^2 + \|\boldsymbol{v}\|^2 - 2\|\boldsymbol{u}\|\,\|\boldsymbol{v}\|\cos\theta$ and $\|\boldsymbol{u} + \boldsymbol{v}\|^2 = \|\boldsymbol{u}\|^2 + \|\boldsymbol{v}\|^2 - 2\|\boldsymbol{u}\|\,\|\boldsymbol{v}\|\cos(180^o - \theta)$; adding, we obtain (3.9). The proof for the more general case is easily carried out (see Exercise 3.10). Since the parallelogram law holds for any norm generated by an

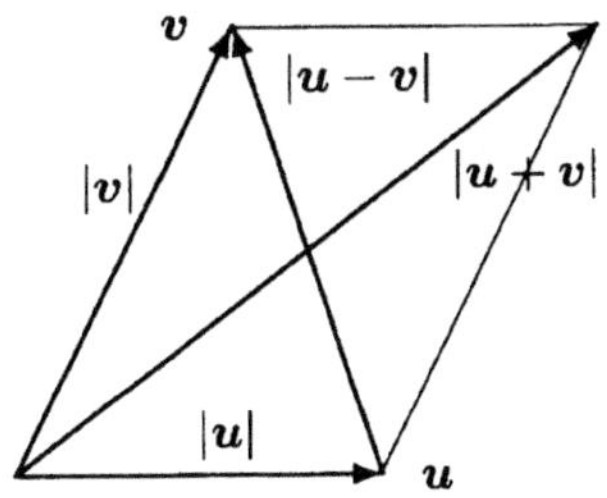


FIGURE 3.8. The parallelogram law in two-dimensional space

inner product it follows that, for any normed space $X$, *if the norm does not satisfy the parallelogram law, then there is no inner product that generates this norm.* When this is so, the space $X$ is not an inner product space.

**Example**

20. Consider $C[0,1]$ with the sup-norm $\|\cdot\|_\infty$. Then choosing

$$u(x) = 1 \quad \text{and} \quad v(x) = x$$

we have

$$\begin{aligned} \|u+v\|_\infty &= \sup|u(x)+v(x)| = \sup|1+x| = 2, \\ \|u-v\|_\infty &= \sup|1-x| = 1. \end{aligned}$$

Thus

$$\|u+v\|_\infty^2 + \|u-v\|_\infty^2 = 5.$$

On the other hand,

$$\|u\|_\infty = \sup|1| = 1 \quad \text{and} \quad \|v\|_\infty = \sup|x| = 1$$

and so

$$\|u\|_\infty^2 + \|v\|_\infty^2 = 2.$$

The parallelogram law does not hold, and so $C[0,1]$ with the sup-norm is *not* an inner product space. In the same way we can show that $C(\overline{\Omega})$ with the sup-norm is also not an inner product space.

**Equivalent norms.** It has already been pointed out that a norm is not a unique object, in the sense that a variety of norms may be defined on any given vector space. Suppose then that two alternative norms $\|\cdot\|_A$ and $\|\cdot\|_B$ are defined on a vector space $X$. These norms are said to be *equivalent* to each other if there are positive constants $m$ and $M$ such that

$$m\|u\|_A \le \|u\|_B \le M\|u\|_A \tag{3.10}$$

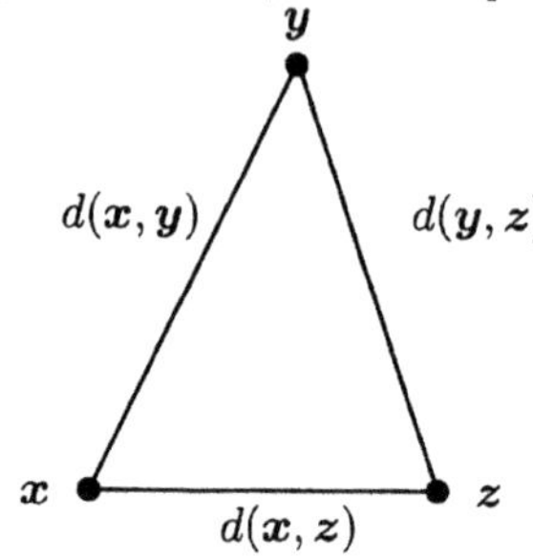


FIGURE 3.9. The triangle inequality in $\mathbb{R}^2$

for all $u \in X$.

**Example**

21. Consider the case $X = \mathbb{R}^2$, with the norms $\|\cdot\|_2$ and $\|\cdot\|_\infty$ defined in Examples 15 and 16. Since $|x_1| \leq \|\boldsymbol{x}\|_2$ and $|x_2| \leq \|\boldsymbol{x}\|_2$, it follows that $\max_i |x_i| \equiv \|\boldsymbol{x}\|_\infty \leq \|\boldsymbol{x}\|_2$. Furthermore, $|x_1| \leq \|\boldsymbol{x}\|_\infty$ and $|x_2| \leq \|\boldsymbol{x}\|_\infty$; squaring and adding, we find that $\|\boldsymbol{x}\|_2^2 \leq 2\|\boldsymbol{x}\|_\infty^2$. Thus

$$\|\boldsymbol{x}\|_\infty \leq \|\boldsymbol{x}\|_2 \leq \sqrt{2}\|\boldsymbol{x}\|_\infty$$

and $\|\cdot\|_2$ and $\|\cdot\|_\infty$ are equivalent norms.

## 3.4 Metric spaces

The final geometrical property that we wish to abstract is the notion of a metric. The motivation comes, as before, from the situation in $\mathbb{R}^3$, in which the *distance* $d(\boldsymbol{x}, \boldsymbol{y})$ between two points $\boldsymbol{x}$ and $\boldsymbol{y}$ is given by

$$d(\boldsymbol{x}, \boldsymbol{y}) = \sqrt{(x_1 - y_1)^2 + (x_2 - y_2)^2 + (x_3 - y_3)^2}\ .$$

A further property enjoyed by points in $\mathbb{R}^3$ is the triangle inequality (Figure 3.9): the length of one side of a triangle is less than or equal to the sum of the lengths of the other two sides. That is,

$$d(\boldsymbol{x}, \boldsymbol{z}) \leq d(\boldsymbol{x}, \boldsymbol{y}) + d(\boldsymbol{y}, \boldsymbol{z}).$$

As with the concepts of inner product and norm, we define a metric $d(\cdot\,,\cdot)$ on an abstract set to have properties analogous to those relevant to distances between points in $\mathbb{R}^n$. One important respect in which the metric differs from both the norm and inner product is that it does not require the structure of a vector space for its definition.

**Metric and metric space.** Let $X$ be a *set*. If $u$ and $v$ are two members of $X$, a *metric* on $X$ is a real number $d(u, v)$ with the following properties, for any $u, v, w \in X$.

**M1.** $d(u, v) \geq 0$ and $d(u, v) = 0$ if and only if $v = u$.

**M2.** $d(v, u) = d(u, v)$.

**M3.** $d(u, w) \leq d(u, v) + d(v, w)$.

A set $X$ with a metric $d(\cdot\,,\cdot)$ defined on it is called a *metric space.* In order to emphasize the particular metric defined on a set, a metric space may alternatively be denoted by $(X, d)$.

**The metric generated by a norm.** The metric requires less structure on the underlying set for its definition than do the norm and inner product. Rather than defining a metric from scratch, we always work with normed (including inner product) spaces, and *define* the corresponding metric by

$$d(u, v) = \|u - v\|. \tag{3.11}$$

When $d(\cdot\,,\cdot)$ is thus defined, we say that $d(\cdot,\cdot)$ is the metric *generated by* the norm $\|\cdot\|$. That (3.11) does indeed satisfy the axioms for a metric is easily shown, and is left as an exercise.

There are nevertheless many examples of metrics that are not generated by norms, as some of the following examples show.

**Examples**

22. The *discrete metric* on any set $X$ is defined by

$$d(u, v) = \begin{cases} 0 & \text{if } v = u \\ 1 & \text{otherwise.} \end{cases} \tag{3.12}$$

Provided that the set $X$ is nonempty, the definition of this metric makes sense, and it may be checked that it satisfies the axioms for a metric.

23. Let $X = \mathbb{R}^2$, and define $d(\cdot\,,\cdot)$ by

$$d(\boldsymbol{x}, \boldsymbol{y}) = \begin{cases} 0 & \text{if } \boldsymbol{y} = \boldsymbol{x} \\ |\boldsymbol{x}| + |\boldsymbol{y}| & \text{otherwise.} \end{cases}$$

## 3.5 Bibliographical remarks

The concept of a vector space is a purely algebraic one, requiring as it does only a set of rules for combining elements and multiplying them by scalars. Further details on vector spaces may be found in texts on linear algebra, and good sources are Hoffman and Kunze [20] and Strang [50], as well as the text by Lang [29].

Good accounts of metric, normed, and inner product spaces may be found in Kolmogorov and Fomin [25], Kreyszig [27], Naylor and Sell [33], Rektorys [41], Roman [43], Smirnov [49], and Zeidler [54].

## 3.6 Exercises

### Vector spaces and subspaces

3.1. Which of the following are vector spaces?

(a) the set of $m \times n$ matrices;

(b) the set of $m \times m$ matrices with determinant equal to 1;

(c) the set of points $X = \{\boldsymbol{x} : \boldsymbol{x} = (x_1, x_2) \in \mathbb{R}^2,\ x_2 \geq 0\}$ (that is, the upper half plane);

(d) the set of solutions to the differential equation

$$a(x)\frac{d^2u}{dx^2} + b(x)\frac{du}{dx} + c(x)u = 0, \quad 0 < x < 1;$$

(e) the set of solutions to the differential equation

$$a(x)\frac{d^2u}{dx^2} + b(x)\frac{du}{dx} + c(x)u + d(x) = 0, \quad 0 < x < 1.$$

3.2. Consider the vector space $\mathbb{R}^2$ of ordered pairs. Which of the following subsets of $\mathbb{R}^2$ are subspaces?

(a) $V = \{\boldsymbol{x} = (x, y) : x = 0\}$;

(b) $V = \{\boldsymbol{x} = (x, y) : x + y = 1\}$.

3.3. Which of the following subsets of $C[a, b]$ are subspaces?

(a) $V = \{u \in C[a, b] : u(a) = u(b) = 0\}$;

(b) $V = \{u \in C[a, b] : u(a) = u(b) = 1\}$;

(c) $V = \left\{u \in C[a, b] : \int_a^b u(x)\, dx = 0\right\}$.

(d) $V = \{u \in C[0, 1] : u(x) = u(y)$ for all $x, y$ such that $x + y = 1\}$.

3.4. Prove Theorem 1, which states that if $V$ and $W$ are subspaces of a vector space $X$, then $X = V \oplus W$ if and only if every $u \in X$ has the unique representation

$$u = v + w$$

for some $v \in V,\ w \in W$.

3.5. Let $X = C[0,1]$, $V = \{v \in C[0,1] : v(x) = v(-x)\}$ (the set of even functions), and $W = \{w \in C[0,1] : w(x) = -w(-x)\}$ (the set of odd functions). Verify that $X = V \oplus W$.

3.6. The purpose of this exercise is to prove the Minkowski inequality for integrals

$$\left[\int_\Omega |u \pm v|^p \, dx\right]^{1/p} \leq \left[\int_\Omega |u|^p \, dx\right]^{1/p} + \left[\int_\Omega |v|^p \, dx\right]^{1/p}.$$

(a) Show that $\alpha\beta \leq (\alpha^p/p) + (\beta^q/q)$, where $1/p + 1/q = 1$ and $\alpha, \beta \in \mathbb{R}$. [Consider the following sketch. Show that area $A = \alpha^p/p$, area $B = \beta^q/q$.] Set $\alpha = u(\boldsymbol{x})/\left[\int_\Omega |u(\boldsymbol{x})|^p \, dx\right]^{1/p}$ and $\beta = v(\boldsymbol{x})/\left[\int_\Omega |v(\boldsymbol{x})|^q \, dx\right]^{1/q}$, integrate and manipulate to get the *Hölder inequality*

$$\int_\Omega |uv| \, dx \leq \left[\int_\Omega |u|^p \, dx\right]^{1/p} \left[\int_\Omega |v|^q \, dx\right]^{1/q}. \tag{3.13}$$

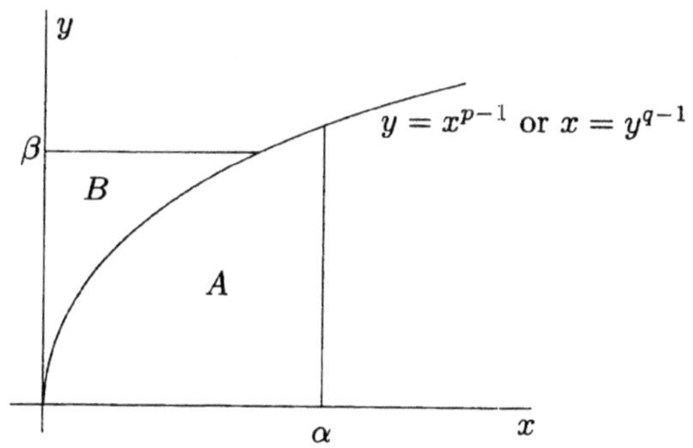


(b) Use the identity

$$(|\alpha| + |\beta|)^p = (|\alpha| + |\beta|)^{p-1}|\alpha| + (|\alpha| + |\beta|)^{p-1}|\beta|$$

to obtain the Minkowski inequality.

**Inner product spaces**

3.7. If $(u, w) = (v, w)$ for all $w \in X$, show that $u = v$.

3.8. Consider the space $C^m[0,1]$ with inner product $(\cdot\,,\cdot)_m$ defined by

$$(u, v)_m = \int_0^1 (uv + u'v' + \ldots + u^{(m)}v^{(m)}) \, dx.$$

Given $u(x) = x^3$ and $v(x) = 1 - (3x^2/2)$, show that $u$ and $v$ are orthogonal with respect to the inner product $(\cdot\,,\cdot)_0$. Are they orthogonal with respect to $(\cdot)_1$? Verify the Cauchy–Schwarz inequality using the inner product $(\cdot)_2$.

## Normed spaces

3.9. For a normed space $X$ show that

$$|\,\|u\| - \|v\|\,| \le \|u - v\| \quad \text{for all}\, u, v \in X.$$

3.10. Prove the parallelogram law

$$\|u + v\|^2 + \|u - v\|^2 = 2(\|u\|^2 + \|v\|^2) \quad \text{for}\, u, v \in X,$$

where $X$ is an inner product space.

3.11. Let $u$ and $v$ be nonzero elements in a real inner product space $X$. Show that

$$\|u + v\| = \|u\| + \|v\|$$

if and only if $v = \alpha u$ for some real number $\alpha > 0$.

3.12. If $X$ is a real inner product space, show that $\|x - y\| + \|y - z\| = \|x - z\|$ if and only if $y = \alpha x + (1 - \alpha)z$, where $0 \le \alpha \le 1$, and $\|\cdot\|$ is the norm generated by the inner product on $X$. Interpret this result for the case $X = \mathbb{R}^2$.

3.13. Show that the quantity

$$\|u\| = \left[\int_a^b \left(\frac{du}{dx}\right)^2 dx\right]^{1/2}, \quad u \in X$$

satisfies the norm axioms for the case in which $X$ is the space

$$X = \{u : \; u \in C^1[a, b],\; u(a) = u(b) = 0\}.$$

3.14. Is $\|\boldsymbol{x}\| = |x_1 x_2|^{1/2}$ a norm on $\mathbb{R}^2$?

3.15. Show that

$$(u, v) = \tfrac{1}{4}\{(\|u + v\|^2 - \|u - v\|^2) - i(\|u + iv\|^2 - \|u - iv\|^2)\},$$

where $u, v$ are members of a complex inner product space.

3.16. A subset $V$ of a linear space $X$ is said to be *convex* if, for every $u, v \in V$, $\alpha u + (1-\alpha)v$ is also in $V$, where $0 \le \alpha \le 1$. Show that the closed ball $B = \{u \in X : \|u\| \le 1\}$ is convex. What does $B$ look like when $X = C(0,1)$ with the sup-norm ?

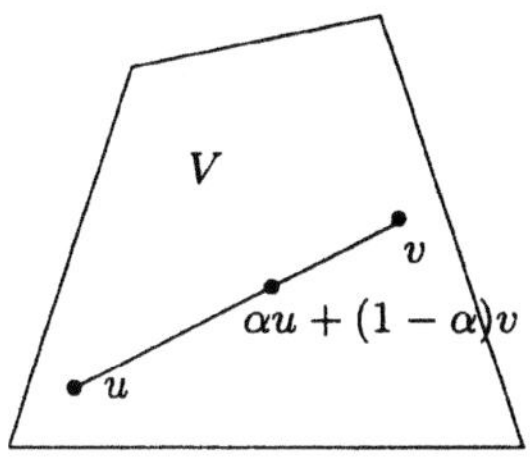


a convex set $V$

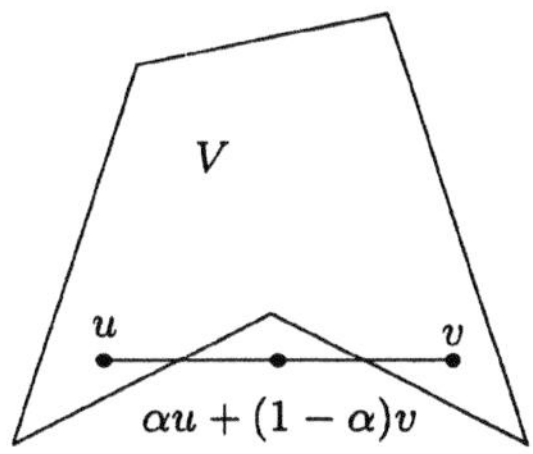


a nonconvex set $V$

3.17. Let $X$ be a real inner product space. Show that $u \perp v$ in $X$ if and only if $\|u + \alpha v\| = \|u - \alpha v\|$ for all real numbers $\alpha$, where $\|\cdot\|$ is the norm generated by the inner product on $X$. Illustrate this result in $\mathbb{R}^2$.

3.18. The distance from a point $x$ in a normed space $X$ to a closed and bounded subset $B$ of $X$ is defined by $d(x, B) = \inf\{\|x - y\| : y \in B\}$. Calculate $d(\boldsymbol{x}, B)$ if $X = \mathbb{R}^2$, $\boldsymbol{x} = (1,1)$, $B$ is the closed disk of radius $\frac{1}{2}$ and center $(\frac{1}{2}, 0)$, and $X$ has (i) the Euclidean norm; and (ii) the norm $\|\cdot\|_1$ (see Example 16).

3.19. The purpose of this exercise is to show that

$$\|\boldsymbol{x}\|_p = [|x_1|^p + \cdots + |x_n|^p]^{1/p}$$

defines a norm on $\mathbb{R}^n$, for $1 \le p < \infty$. In Exercise 3.6(a) set

$$\alpha = \frac{|x_i|}{\|\boldsymbol{x}\|_p}, \quad \beta = \frac{|y_i|}{\|\boldsymbol{y}\|_q},$$

sum over 1 to $n$, and manipulate to get the *Hölder inequality for sums*

$$\sum_{i=1}^{n} |x_i y_i| \le \|\boldsymbol{x}\|_p \, \|\boldsymbol{y}\|_q.$$

Use the identity in Exercise 3.6(b) to obtain the *Minkowski inequality for sums*

$$\left[\sum_{i=1}^{n} |x_i \pm y_i|^p\right]^{1/p} \le \|\boldsymbol{x}\|_p + \|\boldsymbol{y}\|_p,$$

which confirms that $\|\cdot\|_p$ is a norm for $\mathbb{R}^n$.

3.20. For any normed space $V$, the unit ball with center 0 and radius $r$ is defined by $B(0,r) = \{u \in V : \|u\| \leq r\}$. Sketch $B(0,r)$ for the case in which $V = \mathbb{R}^2$ with the norms $\|\cdot\|_p$ for $p = 1$, 2, and $\infty$.

3.21. Show that $\|\cdot\|_1$ and $\|\cdot\|_2$ are equivalent norms on $\mathbb{R}^2$.

3.22. The aim of this exercise is to show that, for a bounded domain $\Omega$,

$$\|u\|_{L^r} \leq [\mu(\Omega)]^{1/r-1/p}\|u\|_{L^p} \tag{3.14}$$

for $p > r \geq 1$, so that if $u \in L^p(\Omega)$, then $u \in L^r(\Omega)$ also. First, let $p$, $q$, $r$ be real numbers such that

$$\frac{1}{p} + \frac{1}{q} = \frac{1}{r} \quad \text{or} \quad \frac{1}{(p/r)} + \frac{1}{(q/r)} = 1. \tag{3.15}$$

Replace $u$ by $u^r$ and $v$ by $v^r$ in *Hölder's inequality* (3.13) and use (3.15) to obtain the generalization

$$\left[\int_\Omega |uv|^r\right]^{1/r} \leq \|u\|_{L^p}\|v\|_{L^q} \tag{3.16}$$

of *Hölder's inequality.* Then use (3.16) to obtain (3.14).

3.23. Show by means of a counterexample that the $L^1$-norm does not generate an inner product.

## Metric spaces

3.24. Let $D = \{z \in \mathbb{C} : |z| \leq 1\}$ be the closed unit disk in the complex plane, and define

$$d(z,w) = \begin{cases} |z-w| & \text{if } \arg(z) = \arg w \text{ or if one of } z \text{ and } w \text{ is zero,} \\ |z| + |w| & \text{otherwise.} \end{cases}$$

Verify that $d(\cdot\,,\cdot)$ defines a metric on $D$. This space is called the "French railroad space"; sketch a picture of the action of $d(\cdot\,,\cdot)$ to see why this is so.

3.25. Verify that (3.12) does indeed satisfy the axioms for a metric.

# 4
# Properties of normed spaces

Normed and inner product spaces possess a wealth of properties, and these in turn allow sophisticated theories to be developed and applied in a variety of contexts. Some of these properties are introduced in this chapter.

Arguably the most basic concept, and one which pervades most discussions involving normed spaces, is that of convergence of sequences. Sequences were introduced in Chapter 1, in the context of real and complex numbers. We show in Section 4.1 that the definition of convergence of a sequence in a normed space is a natural extension of that given in Chapter 1.

In Section 4.2 we focus attention on sequences in spaces of functions; these are a special case which occurs so often in the future as to warrant devoting some time to the elucidation of their characteristics.

The notion of completeness pervades functional analysis, and complete normed and inner product spaces are sufficiently important to be given special names: a complete normed space is called a Banach space and a complete inner product space is known as a Hilbert space. We describe completeness in Section 4.3, and then show in Section 4.4 how completeness of a space is related to the closedness of that space. We also discuss in this section the issue of how to complete a space that lacks this property.

Finally in Section 4.5, we discuss further properties of inner product spaces. In particular, we extend to arbitrary Hilbert spaces a property that is fairly obvious in three-dimensional space. $\mathbb{R}^3$ may be decomposed into two orthogonal subspaces (a simple example, once a set of Cartesian axes has been introduced, would be the $xy$-plane and the $z$-axis), and every vector may be written uniquely as the sum of orthogonal components in these

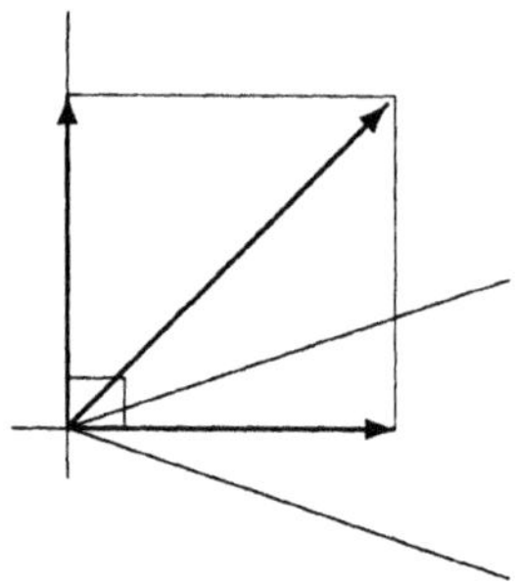

FIGURE 4.1. An example of orthogonal decomposition of a vector in $\mathbb{R}^3$

two subspaces, as shown in Figure 4.1. The generalization of this notion to arbitrary Hilbert spaces is known as the projection theorem, which also features later on.

## 4.1 Sequences

Sequences of numbers were defined in Chapter 1; here we look at sequences in normed spaces generally. A sequence in a normed space $X$ is an ordered set in $X$ whose members can be labeled with positive integers. We write $\{u_1, u_2, \dots\}$ or $\{u_k\}_{k=1}^{\infty}$.

**Example**

1. By way of moving away from sequences of numbers, consider the sequence of *functions* described by (Figure 4.2)

$$\{u_n\}_{n=1}^{\infty} \subset C[a,b], \quad u_n(x) = n(x-a).$$

Ultimately what is of most interest about sequences is the way in which they behave as $n$ gets progressively larger; this brings us to the next topic, namely, that of convergence.

**Convergence of sequences.** The notion of convergence of a sequence of elements in a normed space carries over in a natural way from the definition for sequences of numbers. Let $Y$ be a subset of a normed space $X$, then, and suppose that $\{u_n\}$ is a sequence in $Y$. Let $u$ belong to $Y$, and form the sequence of *real numbers* $\{\|u_1 - u\|, \|u_2 - u\|, \dots, \|u_n - u\|, \dots\}$. If the sequence of numbers $\|u_n - u\|$ converges to zero as $n$ gets larger, we agree to call the sequence convergent. Another, more formal way of stating this is as follows: pick *any positive number* $\epsilon$. Then $\{u_n\}$ is said to converge to

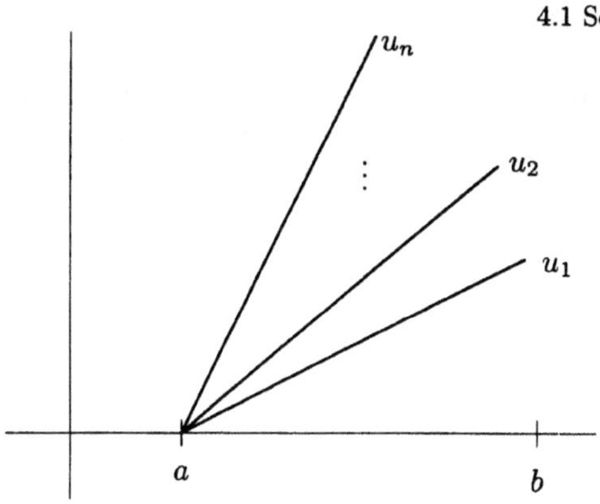


FIGURE 4.2. The sequence of functions with general member $u_n(x) = n(x-a)$

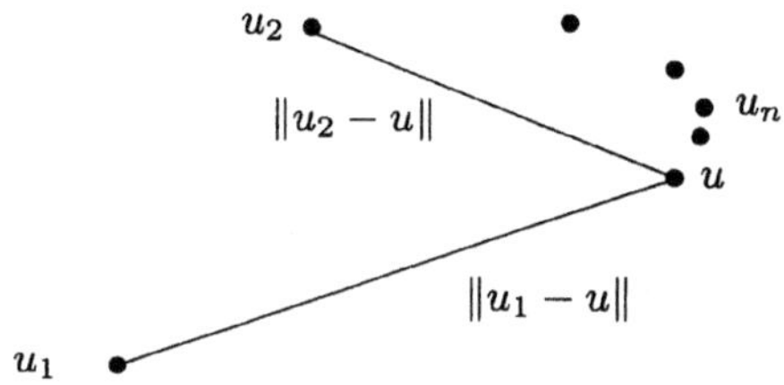


FIGURE 4.3. Convergence of a sequence to a point $u$

some element $u \in Y$ if, for any $\epsilon > 0$, it is always possible to make $\|u_n - u\|$ smaller than $\epsilon$ simply by choosing $n$ large enough, larger than some number $N$, say (Figure 4.3). The groundwork for a precise definition of convergence has now been laid.

**Convergence of a sequence in a normed space.** A sequence $\{u_n\}$ in a subset $Y$ of a normed space $X$ is *convergent* if there is a member $u \in Y$ for which, given *any* $\epsilon > 0$, a number $N$ can be found such that

$$\|u_n - u\| < \epsilon \quad \text{for all } n > N. \tag{4.1}$$

If this is the case, we write $u_n \to u$ (which is read "$u_n$ converges to $u$"), and $u$ is called the *limit* of the sequence. Yet another way of stating (4.1) informally is

$$\lim_{n\to\infty} \|u_n - u\| = 0 \quad \text{or} \quad \lim_{n\to\infty} u_n = u, \tag{4.2}$$

which is read "the limit as $n$ tends to $\infty$ of $u_n$, is $u$". Note, however, that by (4.2) we mean (4.1).

**Equivalent norms and convergence.** The notion of two equivalent norms $\|\cdot\|_A$ and $\|\cdot\|_B$ on a normed space $X$ was defined in (3.10). A useful attribute of equivalent norms is that properties of convergence carry over from one to the other. More precisely, if $\{u_n\}$ is a sequence in $X$ and $u_n \to u$ with respect to $\|\cdot\|_A$, in the sense that $\lim_{n\to\infty} \|u - u_n\|_A = 0$, then $u_n \to u$ with respect to $\|\cdot\|_B$ as well. To see this, we note from (3.10) that

$$\|u - u_n\|_B \leq M\|u - u_n\|_A \to 0 \text{ as } n \to \infty.$$

## 4.2 Convergence of sequences of functions

When discussing convergence of sequences in normed spaces whose members are functions, it is particularly important to specify which norm is being used, as convergence with respect to one norm does not necessarily imply convergence with respect to another. We are acquainted so far with two types of norms when dealing with spaces of functions: the sup-norm in Chapter 3, Examples 18 and 19, and the $L^p$-norm (3.6). As we show in this section, the type of convergence associated with the sup-norm (namely, uniform convergence) implies convergence in the $L^p$-norm, but not *vice versa*. We begin with a discussion of *pointwise* and *uniform* convergence.

Suppose that we know that a sequence $\{u_n(\boldsymbol{x})\}$ of continuous functions converges to a limit *at each point* $\boldsymbol{x} \in \Omega \subset \mathbb{R}^d$. This implies the following: if we *fix* $\boldsymbol{x}$, then the sequence of real numbers $u_n(\boldsymbol{x})$ $(n = 1, 2, \ldots)$ converges to a real number $u(\boldsymbol{x})$, say, and this in turn defines a function $u$. In other words, for every $\epsilon > 0$ there exists a number $N > 0$ such that

$$|u_n(\boldsymbol{x}) - u(\boldsymbol{x})| < \epsilon \quad \text{whenever } n > N. \tag{4.3}$$

Of course $N$ will depend on $\boldsymbol{x}$ and on the number $\epsilon$. If we now move to another value of $\boldsymbol{x}$ the statement (4.3) may not be true for the same $N$. However, if we can find a number $N$ independent of $\boldsymbol{x}$ such that (4.3) holds *for all* $\boldsymbol{x} \in \Omega$, then we say that $u_n$ *converges uniformly* to $u$ on $\Omega$. We now define these concepts formally.

**Pointwise and uniform convergence**. A sequence $\{u_n\}$ of functions defined on a subset $\Omega$ of $\mathbb{R}^d$ converges *pointwise* to $u(\boldsymbol{x})$ if for every $\epsilon > 0$ there exists a number $N$ depending on $\boldsymbol{x}$ and $\epsilon$ such that (4.3) holds. If $N$ does not depend on the value of $\boldsymbol{x}$, then $u_n$ is said to converge *uniformly* to $u$ on $\Omega$; this is written as $\lim_{n\to\infty} u_n = u$ (uniformly).

Note that we are using $\mathbb{R}^d$ rather than $\mathbb{R}^n$ here, for obvious notational reasons!

Uniform convergence has a very simple geometrical interpretation which is illustrated in Figure 4.4 for the case $\Omega = [a, b]$: according to the definition, for any given $\epsilon$ all the functions $u_n(x), u_{n+1}(x), \ldots$ lie in the "tube" of height $2\epsilon$ located symmetrically about the limit function $u(x)$, for $n$ greater than a number $N$ which of course depends on $\epsilon$, but *not* on $x$. Now that uniform convergence has been defined, one might ask how it is related to

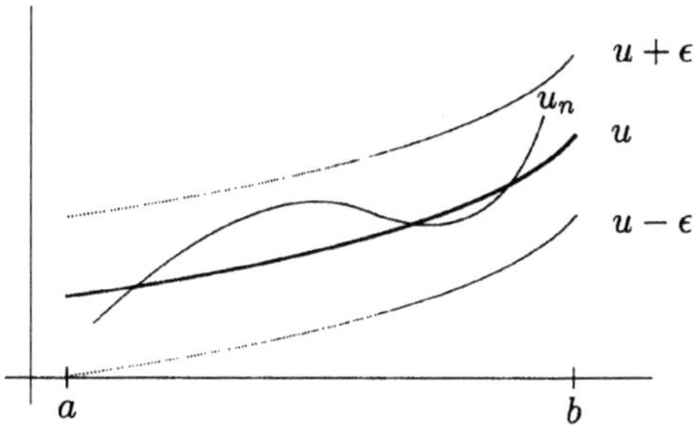


FIGURE 4.4. An illustration of the concept of uniform convergence

the formal definition (4.1) of convergence in terms of a norm. To answer this question, consider a sequence $\{u_n\}$ of functions that belong to the normed space $C[a,b]$ with the norm

$$\|u\|_\infty = \sup |u(x)|, \quad x \in [a,b].$$

Suppose that this sequence is convergent in the sup-norm; that is, given any $\epsilon > 0$ it is possible to find a number $N$ such that

$$\|u_n - u\|_\infty = \sup |u_n(x) - u(x)| < \epsilon \tag{4.4}$$

for all $x \in [a,b]$, whenever $n > N$. But since $|u_n(x) - u(x)| \leq \sup |u_n(x) - u(x)|$, it follows that (4.3) also holds.

In other words, *convergence in the sup-norm implies uniform convergence.* Conversely, suppose that $\{u_n\}$ is a uniformly convergent sequence, so that (4.3) holds. Then $\epsilon$ is an upper bound for $|u_n(x) - u(x)|$, for any $x$ in $[a,b]$. But this implies that the *least upper bound* or *supremum* of $|u_n(x) - u(x)|$ must also be less than $\epsilon$, so that

$$\|u_n - u\|_\infty \equiv \sup |u_n(x) - u(x)| < \epsilon \text{ for all } x \in [a,b],\ n > N,$$

or alternatively

$$\lim_{n\to\infty} [\sup |u_n(x) - u(x)|] = 0.$$

That is, *uniform convergence implies convergence in the sup-norm.* This useful result can be proved in much the same way for functions defined on domains $\Omega$ in $\mathbb{R}^d$ and so we simply record the general result.

THEOREM 1. *A sequence of functions $\{u_n\}$, where $u_n \in C(\overline{\Omega})$ and $\Omega$ is a domain in $\mathbb{R}^d$, converges uniformly to $u$ if and only if*

$$\lim_{n\to\infty} [\sup_{\boldsymbol{x}\in\Omega} |u_n(\boldsymbol{x}) - u(\boldsymbol{x})|] = 0. \tag{4.5}$$

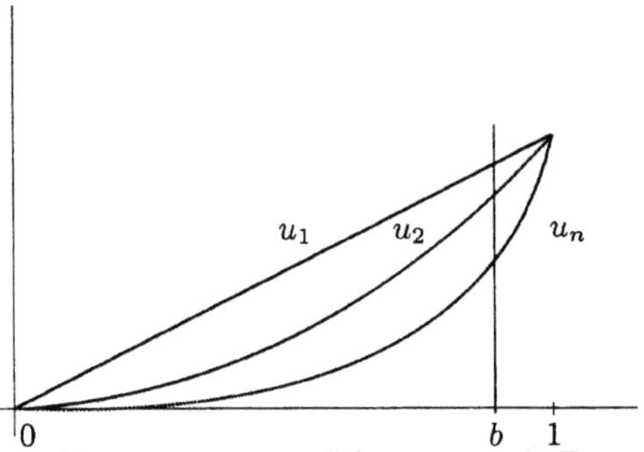


FIGURE 4.5. Nonuniform convergence of the sequence in Example 2

**Examples**

2. Let $u_n = x^n$, defined on $[0, 1]$. This sequence convergences *pointwise* to 0 for $0 \le x < 1$, and to 1 at $x = 1$. If we set $u(x) = 0,\ 0 \le x < 1$, and $u(x) = 1$ for $x = 1$, then

$$\sup |u_n(x) - u(x)| = 1 \quad \text{for all } n,$$

this supremum being attained at a value of $x$ "infinitesimally" close to $x = 1$ (Figure 4.5). Hence the sequence does not converge uniformly on $[0, 1]$. However, it does converge uniformly to zero on $[0, b]$, where $0 < b < 1$, since in this case $\sup |u_n(x) - u(x)| = b^n$ which goes to 0 as $n \to \infty$.

3. Consider the sequence $\{u_n(x) = n^2 x(1-x)^n\}$ defined on $[0, 1]$; the larger $n$ is, the larger and the closer to the $y$-axis the maximum value of $u_n(x)$ will be. For each fixed $x \in [0, 1]$ the sequence converges to zero; but as $n$ increases the supremum of $|u_n(x) - u(x)| = |u_n(x)|$, attained at $x = 1/(n+1)$, also increases (Figure 4.6). Condition (4.5) cannot be satisfied, and so we do not have uniform convergence. But convergence is uniform on any interval $[a, 1]$ where $0 < a < 1$; indeed, for sufficiently large $n$ the undesirable behavior of the maximum value of $u_n$ will fall outside the interval $[a, 1]$.

There is a close connection between the notions of continuity and convergence, in the context of functions. Continuous functions have of course been defined in Chapter 2, and this definition, which is encapsulated in (2.2), may be referred to as the $\epsilon - \delta$ definition of continuity, for obvious reasons. Depending on the context, it is often convenient to have available a definition of continuity that is based on sequential considerations. Such a definition does exist, and goes as follows. Suppose that we have a domain $\Omega$ in $\mathbb{R}^d$, and that $\{x_n\}_{n=1}^{\infty}$ is a *convergent* sequence of points in $\Omega$, with

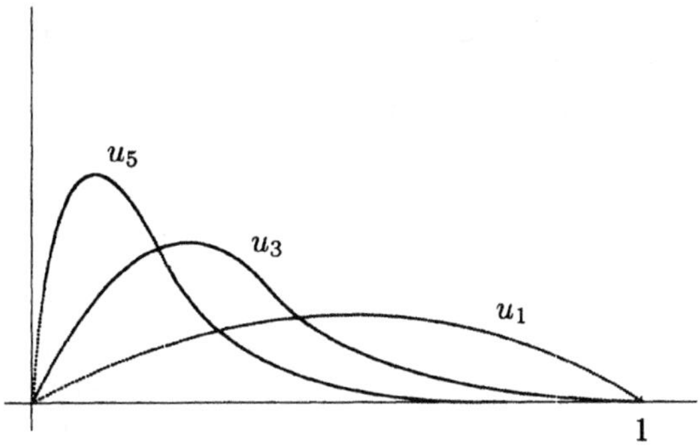


FIGURE 4.6. The sequence of functions in Example 3

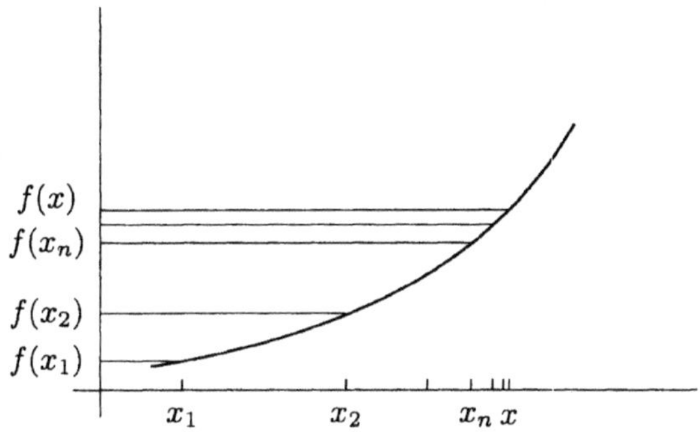


FIGURE 4.7. Illustration of the sequential definition of continuity

limit $\boldsymbol{x}$ in $\Omega$. Then a function $f$ defined on $\Omega$ is said to be continuous if

$$\lim_{n\to\infty} f(\boldsymbol{x}_n) = f(\boldsymbol{x}). \tag{4.6}$$

What this definition states is that if one takes a sequence of points that converges, then these points are mapped to a sequence of *numbers* (real or complex) $f(\boldsymbol{x}_1), f(\boldsymbol{x}_2), \ldots$ which, first, converges, and second, the limit of which coincides with $f(\boldsymbol{x})$. These ideas are illustrated in Figure 4.7. Now there is little point in having alternative definitions of the same concept unless these are equivalent, so it is essential that we establish the connection between the $\epsilon - \delta$ definition of continuity and the sequential definition. These are in fact equivalent, as the following theorem confirms.

THEOREM 2. *Let $\Omega$ be a domain in $\mathbb{R}^d$, and let $f$ be a function defined on $\Omega$. Let $\boldsymbol{x}$ be a point in $\Omega$. The function $f$ is continuous at $\boldsymbol{x}$ if and only if, for every sequence $\{\boldsymbol{x}_n\}$ of elements in $\Omega$ that converges to $\boldsymbol{x}$, (4.6) holds.*

PROOF. First assume that the $\epsilon - \delta$ definition (2.2) is valid. Then given any $\epsilon > 0$, there exists $\delta$ such that $|f(\boldsymbol{y}) - f(\boldsymbol{x})| < \epsilon$ when $\boldsymbol{y} \in \Omega$ and $\boldsymbol{y} - \boldsymbol{x}| < \delta$.

If we now take any sequence $\{\boldsymbol{x}_n\}$ of elements in $\Omega$ that converges to $\boldsymbol{x}$, then from the definition of convergence we know that $|\boldsymbol{x}_n - \boldsymbol{x}| < \delta$ for all $n \geq N$, for suitably large $N$. Thus, taking $\boldsymbol{y}$ to be the point $\boldsymbol{x}_n$, we see that $|f(\boldsymbol{x}_n) - f(\boldsymbol{x})| < \epsilon$ for $n \geq N$, which is just another way of stating (4.6). Thus the $\epsilon - \delta$ definition implies the sequential definition of continuity.

Conversely, assume that (4.6) is valid. It suffices to prove then that, given $\epsilon > 0$, there exists $N$ such that whenever $|\boldsymbol{y}-\boldsymbol{x}| < 1/N$ then $|f(\boldsymbol{y})-f(\boldsymbol{x})| < \epsilon$. We use the method of proof by contradiction. Suppose that this assertion is false. Then, for some $\epsilon$ and for every positive integer $n$ there exists $\boldsymbol{x}_n \in \Omega$ such that $|\boldsymbol{x}_n - \boldsymbol{x}| < 1/n$ but $|f(\boldsymbol{x}_n) - f(\boldsymbol{x})| > \epsilon$. This in turn implies that $\{\boldsymbol{x}_n\}$ is a convergent sequence, so that this statement contradicts (4.6). The theorem is thus proved. □

**$L^p$-convergence.** We continue the discussion of convergence of sequences of functions, and move on to the larger normed space $L^p(\Omega)$ with the usual $L^p$-norm defined in (3.6), and with $1 \leq p < \infty$. The definition (4.1) states that a sequence $\{u_n\} \subset L^p(\Omega)$ converges in the $L^p$-norm to an element $u \in L^p(\Omega)$ if for any given $\epsilon > 0$ it is possible to find a number $N$ such that

$$\|u_n - u\|_{L^p} < \epsilon \quad \text{whenever } n > N, \tag{4.7}$$

or

$$\left[\int_\Omega |u_n(\boldsymbol{x}) - u(\boldsymbol{x})|^p \, dx\right]^{1/p} < \epsilon \quad \text{whenever } n > N, \tag{4.8}$$

or

$$\lim_{n\to\infty} \int_\Omega |u_n(\boldsymbol{x}) - u(\boldsymbol{x})|^p dx = 0. \tag{4.9}$$

This type of convergence is referred to as *$L^p$-convergence*, and in the case $p = 1$ it is referred to as *convergence in the mean.* It is important to note that although uniform convergence implies $L^p$-convergence (if $\Omega$ is bounded; see Exercise 4.7), the converse is not true. The relationship between uniform, $L^p-$, and pointwise convergence can be summarized as follows.

UNIFORM CONVERGENCE

$\Downarrow$ $\qquad\qquad$ $\Downarrow$

POINTWISE CONVERGENCE $\not\Longleftrightarrow$ $L^p$ CONVERGENCE

**Example**

4. Let $\{u_n\} = \{(1+nx)^{-1}\}_{n=1}^{\infty}$. This sequence converges to 0 on $[0,1]$ in the $L^2$ norm since

$$\int_0^1 [u_n(x) - 0]^2 \, dx = \int_0^1 (1+nx)^{-2} \, dx = (1+n)^{-1}$$

which goes to zero as $n \to \infty$. It can be shown that $u_n \to 0$ in the $L^p$-norm for any $p > 1$.

## 4.3 Completeness

As we have seen, convergent sequences all have the property that the distance between successive members of a sequence, measured by means of some appropriate norm, becomes progressively smaller, and the sequence approaches a definite limit which is, moreover, a member of the normed space concerned. Unfortunately, the situation is not always so clear-cut: some normed spaces have the deficiency that, although it is possible to set up sequences in these spaces with the property that the distance between successive members becomes progressively smaller, the sequence does not in fact have a limit in this space. For example, suppose we take a look at the half-open interval $(0,1]$ with the norm $\|\cdot\| = |\cdot|$, and consider the sequence $\{u_n\} = \{1/n\}_{n=1}^{\infty}$. This sequence behaves in all respects as a convergent sequence, and converges to 0, but 0 is not in the space $(0,1]$!

This behavior is undesirable for a number of reasons, and we always make a strong distinction between spaces in which sequences that behave as convergent sequences do in fact converge to a limit and, on the other hand, those spaces in which the limits of such sequences are possibly "missing".

In order to proceed with the discussion, we first need to have a means of identifying sequences with the property that the distance between successive members decreases. These are called *Cauchy sequences*, and their definition makes no reference to the notion of convergence, or of a limit, since it is possible for such sequences not to converge.

**Cauchy sequence.** A sequence $\{u_n\}$ in a subset $Y$ of a normed space $X$ is called a *Cauchy sequence* if

$$\lim_{m,n\to\infty} \|u_m - u_n\| = 0 \tag{4.10}$$

or, more formally, if for any given $\epsilon > 0$ there exists a number $N$ such that

$$\|u_m - u_n\| < \epsilon \quad \text{whenever } m, n > N. \tag{4.11}$$

Every convergent sequence is a Cauchy sequence (see Exercise 4.13), but the point has been made that not every Cauchy sequence is convergent, for the simple reason that, although the members may be converging to a limit, the limit may not be part of the space. When this is so, then we say that the space is *incomplete*. The situation may be remedied, however, by adding to the space those elements that are the limits of Cauchy sequences but which were not originally in the space. This process is called *completion* of the space, which is then said to be *complete*. We discuss completions in more detail in the next section, but we first define formally a complete space, and then give some simple but important examples of complete spaces.

**Complete space.** A subset $Y$ of a normed space $X$ is *complete* if *every* Cauchy sequence in $Y$ converges to an element of $Y$.

**Example**

5. The set $\mathbb{R}$ of real numbers with the norm $\|\cdot\| = |\cdot|$ is complete, as is any *closed* interval of $\mathbb{R}$. The completeness of $\mathbb{R}$ is taken as a fundamental property of the real number system, whereas the completeness of closed intervals follows from the equivalence between closedness and completeness, in a sense made precise in Section 4.

6. The set $\mathbb{R}^n$ with any of the norms $\|\cdot\|_p$ defined by $\|\boldsymbol{x}\|_p = [\sum_{i=1}^n |x_i|^p]^{1/p}$ for $1 \le p < \infty$ is complete, as is $\mathbb{R}^n$ with the norm $\|\cdot\|_\infty$ defined by $\|\boldsymbol{x}\|_\infty = \max_{1\le i\le n} |x_i|$. This follows from the completeness of $\mathbb{R}$ (see Exercise 4.12).

7. The space $C[0,1]$ with the integral norm $\|u\|^2 = \int_0^1 u^2\, dx$ is *not* complete. To see this, consider the sequence $\{u_n\}$ defined by

$$u_n = \begin{cases} 0, & 0 \le x < \frac{1}{2}, \\ (x - \frac{1}{2})^{1/n}, & \frac{1}{2} \le x \le 1. \end{cases}$$

It is readily verified that $\{u_n\}$ is a Cauchy sequence; however, its "limit" $u(x)$ is the *discontinuous* function (Figure 4.8)

$$u(x) = \begin{cases} 0, & 0 < x < \frac{1}{2}, \\ 1, & \frac{1}{2} \le x \le 1. \end{cases}$$

Hence $C[0,1]$ (and in general $C[a,b]$) is not complete in the $L^2$-norm, and indeed it is not complete in the $L^p$-norm for any $p$ such that $1 \le p < \infty$. In a similar way we may show that $C(\overline{\Omega})$ with the $L^p$ norm $\|u\| = [\int_\Omega |u|^p dx]^{1/p}$ is not complete, for $1 \le p < \infty$.

8. The space $C[a,b]$ with the sup-norm $\|u\|_\infty = \sup\{|u(x)|,\ x \in [a,b]\}$, *is complete*. We may show this by demonstrating first of all that every Cauchy sequence converges uniformly to some function $u(x)$, and that this limiting function is necessarily continuous (see Exercise 4.11). Similarly, $C(\overline{\Omega})$ with the norm $\|u\|_\infty = \sup\{|u(\boldsymbol{x})|,\ \boldsymbol{x} \in \overline{\Omega}\}$,

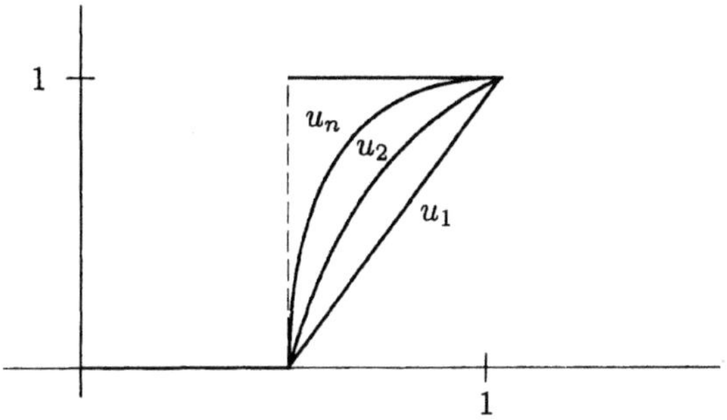


FIGURE 4.8. The sequence in Example 9

is complete. This example and the previous one demonstrate that completeness of a space depends crucially on the choice of norm.

9. The space $L^2(\Omega)$ with the usual $L^2$-norm is complete. We do not give the proof, as this would take us too far afield. It is important to note that the notion of *Lebesgue* integration is essential for the completeness of $L^2$; the space of functions that are Riemann-square integrable is *not* complete, since it is possible to construct Cauchy sequences of Riemann-integrable functions whose limits are not Riemann-integrable. Generally, for any $p \geq 1$ the space $L^p(\Omega)$ with the $L^p$-norm is complete (this holds for $L^\infty(\Omega)$ as well).

Completeness is an extremely important property, because complete spaces possess many useful characteristics that are absent from incomplete spaces. Fortunately, most of the spaces of functions with which we work are complete. Normed and inner product spaces that are complete have special names, which are introduced here.

**Banach and Hilbert spaces.** A complete normed space is called a *Banach* space; a complete inner product space is called a *Hilbert* space.

Since every inner product defines a norm, every Hilbert space is a Banach space.

## Examples

10. $\mathbb{R}^n$ with the norm $\|\cdot\|_p$ defined in Example 6, and with $1 \leq p \leq \infty$, is a Banach space, and $\mathbb{R}^n$ with the norm $\|\cdot\|_2$ is a Hilbert space.

11. The space $C[a,b]$ with the norm $\|\cdot\|_\infty$ is a Banach space, as are the spaces $L^p(a,b)$ with the $L^p$-norm. The space $L^2(a,b)$ with the $L^2$-norm is a Hilbert space.

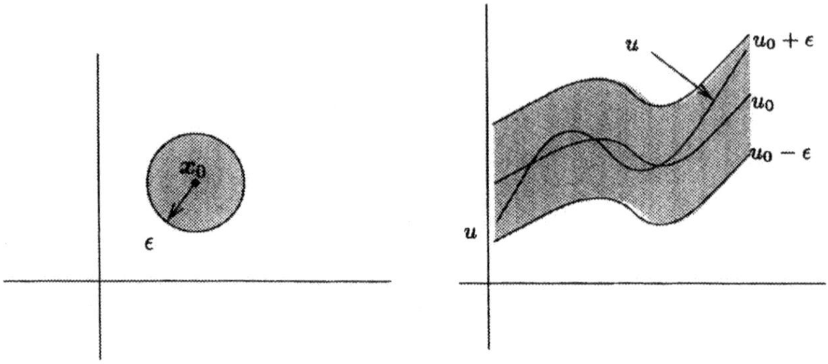


FIGURE 4.9. Open neighborhoods in $\mathbb{R}^2$ and in $C[0,1]$

## 4.4 Open and closed sets, completion

The notion of completeness, which we have just met, is an example of a *topological* property of a normed space. Topological properties of a space are those that are based on the concept of distances between points. Naturally we may use a norm to measure distances between points, although the broader concept of a topological space does not require a norm for its definition; a normed space is just a special case of a topological space.

In this section we discuss a few more topological concepts that are necessary for a proper understanding of later material. Common to all of these concepts – and indeed to topological considerations in general – is the idea of an *open set*, the definition of which in turn depends on the idea of a neighborhood. We have come across both neighborhoods and open sets in the context of the particular spaces introduced earlier; here we generalize to normed spaces.

**Neighborhood**. Let $X$ be a normed space and let $u_0$ be an arbitrary point in $X$. The set

$$N(u_0, \epsilon) = \{u \in U : \ \|u - u_0\| < \epsilon\}$$

is called an *open neighborhood* of $u_0$ with radius $\epsilon$, where $\epsilon > 0$ (Figure 4.9). Similarly, the set

$$\overline{N}(u_0, \epsilon) = \{u \in U : \|u - u_0\| \leq \epsilon\}$$

is called a *closed neighborhood* of $u_0$ with radius $\epsilon$.

### Examples

12. We have already met neighborhoods of points in $\mathbb{R}^n$ (see Chapter 1, Section 3); there, the norm used is of course the Euclidean norm.

13. Consider the space $C[0,1]$ with the sup-norm $\|u\|_\infty = \sup|u(x)|$. An open neighborhood of the function $u_0(x)$ of radius $\epsilon$ is the set of all continuous functions on $[0,1]$ for which (Figure 4.9)

$$\|u - u_0\|_\infty = \sup_{x\in[0,1]} |u(x) - u_0(x)| < \epsilon.$$

14. In the normed space $L^2(0,1)$ with the usual $L^2$-norm, an open neighborhood of $u_0$ is the set of all functions $u \in L^2(0,1)$ for which

$$\|u - u_0\|_{L^2} = \left[\int_0^1 [u(x_0) - u(x)]^2\,dx\right]^{1/2} < \epsilon.$$

This open neighborhood is unfortunately not easy to represent graphically.

As can be seen from the definition and the preceding examples, the idea of an open neighborhood is generalized in an almost trivial way from the concept in $\mathbb{R}^n$. This is a common feature of functional analysis in normed spaces, and we encounter it many more times. Indeed, we proceed now to the idea of open and closed subsets in normed spaces, and essentially generalize what was introduced in Chapter 1.

**Open and closed sets.** A subset $Y$ of a normed space $X$ is an *open set* if, for every point $v$ in $Y$, there is an open neighborhood $N(v,\epsilon)$ of $v$ that lies entirely in $Y$. A point $w$ in $Y$ is a *point of accumulation* of $Y$ if every open neighborhood of $w$, no matter how small, also contains at least one point $v$ in $Y$. Finally, $Y$ is a *closed set* if it contains all of its points of accumulation. We also define the *closure* $\overline{Y}$ of a set $Y$ to be the union of $Y$ and all of its points of accumulation.

## Examples

15. To start with, it may be worth reviewing some of the examples in Sections 1.2 and 1.3.

16. The set $B(u_0,r) := \{u : u \in X, \|u-u_0\| < r\}$, where $X$ is any normed space, is called the *open ball* with center $u_0$ and radius $r$, and is an open set. Indeed, for any point $v$ in $B(u_0,r)$ the open neighborhood $N(v,\epsilon)$ lies entirely in $B(u_0,r)$ provided that $\epsilon$ is less than $d$, the shortest distance from $v$ to the boundary of $B(u_0,r)$ (Figure 4.10). More formally, set

$$S(u_0,r) = \{u \in X,\ \|u - u_0\| = r\};$$

then

$$d = \inf\{\|v - u\|,\quad u \in S(u_0,r)\}$$

and we require $\epsilon < d$.

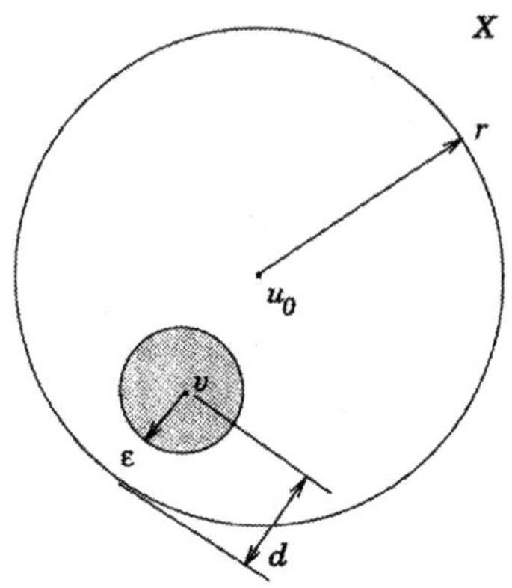


FIGURE 4.10. The open ball $B(u_0, r)$ with center $u_0$ and radius $r$

17. The set $\overline{B}(u_0, r) = \{u \in X, \ \|u - u_0\| \leq r\} = B(u_0, r) \cup S(u_0, r)$ is called the *closed ball* of radius $r$, and is a closed set since, for any $v \in \overline{B}(u_0, r)$, the neighborhood $N(v, \epsilon)$ contains points of $\overline{B}(u_0, r)$ other than $v$. Hence every member of $\overline{B}(u_0, r)$ is a point of accumulation. Furthermore, for any point $u_1$ not in $\overline{B}(u_0, r)$ we can always construct a neighborhood $N(u_1, \epsilon)$ that contains no point of $\overline{B}(u_0, r)$. Indeed, let $l = \inf \|u_1 - v\|, v \in \overline{B}(u_0, r)$; then we simply choose $\epsilon < l$. Thus $\overline{B}(u_0, r)$ contains all its points of accumulation and is therefore closed. Although these concepts all apply to normed spaces in general- - and in particular apply to open and closed balls that often look nothing like balls – the basic ideas may nevertheless be more readily assimilated by considering their interpretation in $\mathbb{R}^2$, endowed with the Euclidean norm.

18. Consider the space $C[a, b]$ with the sup-norm, and let $V = \{u \in C[a, b], |u(x)| \leq 1\}$. Then $V$ is closed, since $V$ is in fact the closed ball of radius 1, centered at $u_0(x) = 0$, as can be seen in Figure 4.11.

There is yet another way of characterizing closed sets, namely, by looking at the limits of convergent sequences. This characterization is described in the following theorem.

THEOREM 3. *A subset $Y$ of a normed space $X$ is closed if and only if every convergent sequence of points in $Y$ has its limit in $Y$.*

PROOF. First assume that $Y$ is closed, and consider the convergent se-

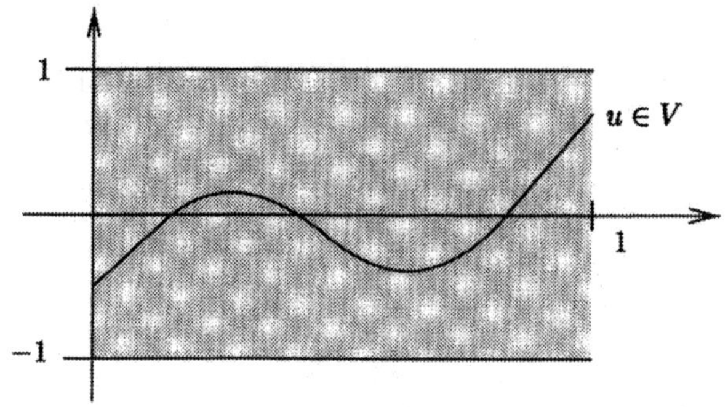


FIGURE 4.11. The closed ball of unit radius in $C[a,b]$

quence $\{u_n\} \subset Y$; when viewed as a sequence in $X$, $\{u_n\}$ has a limit $u$, say, in $X$. We want to show that $u \in Y$.

Now by assumption, given $\epsilon > 0$, there exists a number $N$ such that

$$\|u_n - u\| < \epsilon \quad \text{whenever} \quad n > N.$$

Stated otherwise, for every $\epsilon > 0$ the neighborhood $N(u, \epsilon)$ contains at least one member of $Y$ (that is, a member of the sequence $\{u_n\}$) distinct from $u$. Thus $u$ is a point of accumulation of $Y$ and, since $Y$ is closed, it follows that $u \in Y$.

Conversely, assume that every convergent sequence has its limit in $Y$. Let $u_0$ be a point in $\overline{Y}$; then there is at least one member of the sequence, $u_n$, say, in the neighborhood $N(u_0, 1/n)$. This holds for all values of $n$, so that $\lim_{n\to\infty} u_n = u_0$ (Figure 4.12). But since every convergent sequence has its limit in $Y$ by assumption, it follows that $u_0 \in Y$ and so $Y = \overline{Y}$; that is, $Y$ is closed. □

It appears from the foregoing result that closed sets, like complete spaces, have no "holes" in them, in the sense that convergent sequences have their limits in the sets. On the other hand, it would seem that open sets have the same deficiencies as incomplete spaces; for example, the interval $(0,1) \subset \mathbb{R}$ is an open set and is also incomplete, whereas $[0,1]$ is both closed and complete. Still, it is not clear under what circumstances a closed subset of a normed space is complete; we clarify this in the next result.

THEOREM 4. *Let $X$ be a complete normed space and $Y$ a subset of $X$. Then $Y$ is complete if and only if $Y$ is closed in $X$.*

PROOF. This is not too difficult, and is left as an exercise (see Exercise 4.21). □

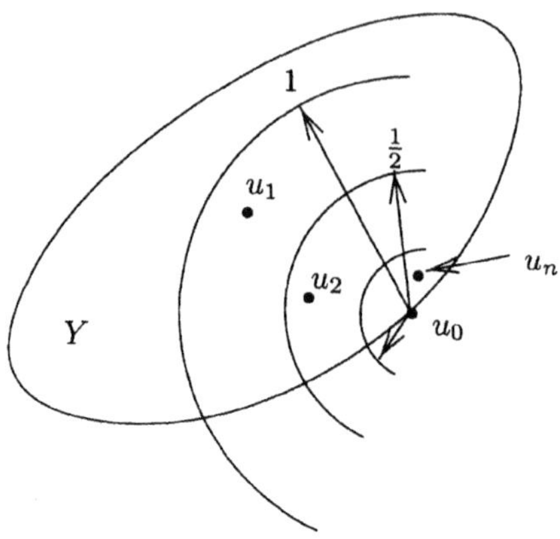


FIGURE 4.12. An illustration of the argument used in the proof of Theorem 3

## Example

19. Let $X = C[0,1]$ and $Y = P[0,1]$, the set of all polynomials on the unit interval. First, we recall that $C[0,1]$ is complete in the sup-norm $\|\cdot\|_\infty$. Now $P[0,1]$ is a subspace of $C[0,1]$, but $P$ is *not closed* in $C$. To see this, consider the fact that $u(x) = e^x$ is a point of accumulation of $P$: for any $\epsilon > 0$ we can always find at least one polynomial $p(x) \in P$ lying in a neighborhood of $u$; indeed, given any $\epsilon > 0$, it is possible to find a polynomial $p(x) = 1 + x + x^2/2! + \cdots + x^n/n!$ such that

$$\|u - p\|_\infty = \sup |e^x - p(x)| < \epsilon$$

for sufficiently large $n$. But the point of accumulation $e^x$ does not belong to $P[0,1]$ and so, since $P[0,1]$ does not contain all its points of accumulation, it is not closed.

**Compact sets.** We met compact sets in $\mathbb{R}^n$ earlier, in Chapter 1; there, a set $S$ in $\mathbb{R}^n$ was defined to be compact if every sequence in $S$ has a point of accumulation in $S$. That definition of compactness carries over without modification to arbitrary normed spaces: if $X$ is a normed space and $S$ a subset of $X$, then $S$ is said to be *compact* if every sequence in $S$ has a point of accumulation in $S$.

Although it is reassuring to observe that the definition of compactness is valid for any normed space, a word of caution is in order, in that not all of the properties of compact sets in $\mathbb{R}^n$ carry over to arbitrary normed spaces. One of these is worth pointing out here: for subsets of $\mathbb{R}^n$ compactness is equivalent to closedness + boundedness, but *the result does not hold in arbitrary normed spaces.* Certainly every compact set is closed and bounded (and hence complete), *but the converse is not true.* This observation emphasizes the need to exercise caution when generalizing from the particular; such generalizations, or illustrations of abstract concepts in $\mathbb{R}$ or $\mathbb{R}^n$, are often very helpful, but there are times when one's intuition can be misleading.

Apart from the results given in Chapters 1 and 2, compactness does not play much of a role in subsequent developments, and we do not pursue the topic further here.

Earlier we described in a vague fashion how an incomplete set $Y$ may be made complete by adding to it those limits of Cauchy sequences that were not originally in $Y$. The resulting set $\tilde{Y}$, say, is then called the completion of $Y$. We conclude this section by recording some properties of the completion $\tilde{Y}$ of an incomplete set $Y$, but in order to do this it is first of all necessary to define a few more topological concepts.

**Dense sets.** If $X$ and $Y$ are two subsets of a normed space, then $Y$ is said to be *dense in* $X$ if the closure of $Y$ is $X$, that is, if $\overline{Y} = X$. This definition implies of course that *every* member of $X$ is either a member of $Y$ or a point of accumulation of $Y$, so that a neighborhood $N(u_0, \epsilon)$ of any point $u_0$ in $X$ contains at least one member of $Y$. This in turn leads to an alternative definition of a dense set: $Y$ is dense in $X$ if and only if there are points in $X$ arbitrarily close to points in $Y$, or, given any point $u_0 \in X$ and any number $\epsilon > 0$, it is possible to find $v \in Y$ such that

$$\|u_0 - v\| < \epsilon.$$

**Example**

20. The set $\mathbb{Q}$ of rational numbers is dense in $\mathbb{R}$; the closure $\overline{\mathbb{Q}}$ $(= \mathbb{R})$ of $\mathbb{Q}$ consists of all rational and irrational numbers.

21. *The Weierstrass theorem* states that, for any $u \in C[a,b]$ and for every $\epsilon > 0$, it is possible to find a polynomial $p \in P[a,b], p(x) = a_0 + a_1 x + \cdots$, such that

$$\|u - p\|_\infty < \epsilon.$$

That is, every bounded continuous function can be approximated arbitrarily closely by a polynomial; we refer to this as *uniform approximation.* Stated otherwise, the space $P[a,b]$ is dense in $C[a,b]$.

We go on now to the issue of characterizing some of the subsets of $L^p$ that are dense in this space. The discussion is confined to a subset of the space of bounded continuous functions comprising those continuous functions that, roughly speaking, are zero in a neighborhood of the boundary.

**Functions with compact support.** Suppose that a function $u$ defined

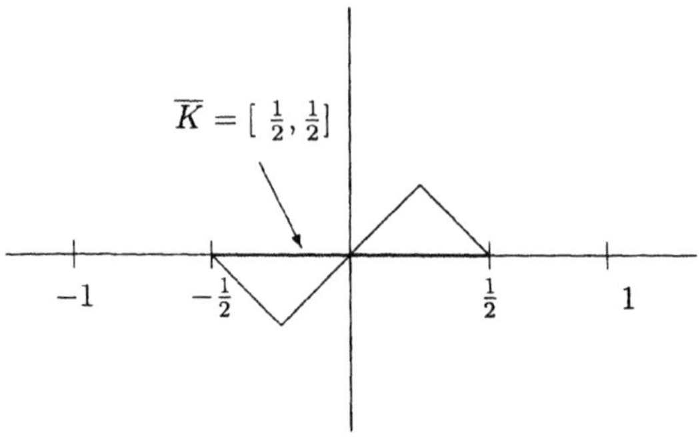


FIGURE 4.13. The support of the function in Example 24

on a domain $\Omega$ is nonzero only for points belonging to a proper subset $K$ of $\Omega$. Let $\overline{K}$ be the *closure* of $K$. Then $\overline{K}$ is called the *support* of $u$. We say that $u$ has *compact support* on $\Omega$ if its support $\overline{K}$ is a compact – that is, a closed and bounded – subset of $\Omega$. The set of continuous functions with compact support is denoted by $C_0(\Omega)$.

**Example**

22. Let $\Omega = (-1, 1)$ and define $u$ to be the function

$$u(x) = \begin{cases} x, & 0 \le |x| \le \frac{1}{4}, \\ \frac{1}{2} - x, & \frac{1}{4} < x < \frac{1}{2}, \\ -\frac{1}{2} - x, & -\frac{1}{2} < x < -\frac{1}{4}, \\ 0, & \frac{1}{2} \le |x| < 1. \end{cases}$$

Then $K$ will be the open set given by $K = (-\frac{1}{2},\ 0) \cup (0,\ \frac{1}{2})$ and $\overline{K} = [-\frac{1}{2},\ \frac{1}{2}]$ (note that $x = 0$ and $x = \pm\frac{1}{2}$ are points of accumulation of $K$). Since $\overline{K}$ is a closed bounded subset of $\Omega$, it follows that $u(x)$ has compact support on $\Omega$ (see Figure 4.13).

23. Consider next the function $u(x) = \sin \pi x$ on $\Omega = (-1, 1)$. Here $K = (-1, 0) \cup (0, 1)$ and $\overline{K} = [-1, 1]$ which is *not* a subset of $\Omega$. Hence $u(x)$ does not have compact support on $\Omega$.

**Dense sets in $L^p$.** We are now able to show that the space $C_0(-\infty, \infty)$ of continuous functions having compact support is dense in $L^p(-\infty, \infty)$ for $1 \le p < \infty$. This implies that any set $X$ which contains $C_0(-\infty, \infty)$ is itself dense in $L^p(-\infty, \infty)$.

Take any $u \in L^p(-\infty, \infty)$ and define the sequence of functions $u_n$ by

$$u_n(x) = \begin{cases} n, & u(x) > n \\ u(x) & -n \le u(x) \le n \\ -n, & u(x) < -n. \end{cases}$$

For each value of $n$ we have $|u_n|^p \le |u|^p$, so that $u_n \in L^p(-\infty, \infty)$. Furthermore, $u_n \to u$ (a.e.) as $n \to \infty$, and

$$|u - u_n|^p \le (|u| + |u_n|)^p \le 2^p |u|^p.$$

Now we apply the Dominated Convergence Theorem (Theorem 3 of Chapter 2) to deduce that $\|u - u_n\|_{L^p} \to 0$ as $n \to \infty$. We thus see that *bounded functions are dense in* $L^p(-\infty, \infty)$ *for* $1 \le p < \infty$.

Now take a bounded function $v$ in $L^p(-\infty, \infty)$ and set $v_n(x) = v(x)$ for $-n \le x \le n$, and $v_n(x) = 0$ otherwise. By repeating the argument in the previous paragraph we can conclude that *bounded functions with compact support are dense in* $L^p(-\infty, \infty)$ *for* $1 \le p < \infty$.

With these preliminary results, the following theorem can now be proved (see Exercise 4.23).

THEOREM 5. *The space* $C_0(-\infty, \infty)$ *is dense in* $L^p(-\infty, \infty)$ *for* $1 \le p < \infty$.

In Chapter 7 we come across a rather special class of functions with compact support; this is the space $C_0^\infty(\Omega)$ which consists of functions that, together with *all* their derivatives, are continuous and have compact support. It is in fact possible, by an appropriate modification and extension of the proof of Theorem 5, to prove the following.

THEOREM 6. *The space* $C_0^\infty(\Omega)$ *is dense in* $L^p(\Omega)$ *for* $1 \le p < \infty$, *where* $\Omega$ *is any open set in* $\mathbb{R}^n$.

The proof may be found in some of the texts referred to at the end of this chapter.

**Separable space.** A normed space $X$ is said to be separable if it contains a *dense* set $Y$ that is *countable*. Recall from Chapter 1 that a countable set is one whose elements can be put in one-to-one correspondence with integers. It follows then that a separable space is one with the following property: there exists a countable subset $\{\{v_1, v_2, \dots\}$ such that for each $\epsilon > 0$ and for each $u$ in $X$, there is a member $v_n$, say, with $\|u - v_n\| < \epsilon$.

**Example**

24. We have seen that the set of rationals $\mathbb{Q}$ is dense in $\mathbb{R}$; since $\mathbb{Q}$ is countable, it follows that $\mathbb{R}$ is separable. In the same way, $\mathbb{R}^n$ is separable: a countable dense subset is the set $\mathbb{Q}^n$ of $n$-tuples of rational numbers.

25. From the Weierstrass theorem (Example 21) we see that the *countable* set $Q[a,b]$ of polynomials with rational coefficients is dense in $C[a,b]$. Furthermore, from Theorem 6 we know that $C[a,b]$ is dense in $L^p(a,b)$. It follows that $Q[a,b]$ is dense in $L^p(a,b)$ for $1 \le p < \infty$, and that $L^p(a,b)$ is therefore separable.

**Completion of a set.** We return now to the idea of completion of a set. Recall from Theorem 4 that an arbitrary subset $Y$ of a Banach space $X$ is itself complete if and only if $Y$ is closed. It follows that the *closure* $\overline{Y}$ of $Y$ is *complete*; furthermore, according to the definition of a dense set, $Y$ is *dense* in $\overline{Y}$. What we have described is of course a way of completing an incomplete set $Y$. In future, by the completion $\widetilde{Y}$ of a set $Y$, we always mean the *closure* of $Y$, so that $\widetilde{Y} = \overline{Y}$. This definition obviously relies on the fact that $Y$ must be a subset of a complete space $X$.

Recall from Example 7 that the space $C(\overline{\Omega})$ with the norm

$$\|u\|_{L^p} = \left[\int_\Omega |u(\boldsymbol{x})|^p \, dx\right]^{1/p}$$

is not complete. The completion of this space is actually $L^p(\Omega)$, which is obtained by adding to $C(\overline{\Omega})$ those limits of Cauchy sequences (for example, the function $u(x)$ in Example 7) that are not in $C(\overline{\Omega})$. Thus $C(\overline{\Omega})$ is dense in $L^p(\Omega)$. In Chapter 7 we show how this result is obtained as a special case of a much broader result, the essence of which is that the space $C^m(\overline{\Omega})$ is dense in the Sobolev space $H^m(\Omega)$ of functions that, together with their derivatives of order $\le m$, are square-integrable. The result quoted here is for the special case $m = 0$.

## 4.5 Orthogonal complements in Hilbert spaces

In this section attention is confined to inner product spaces. We exploit the concept of orthogonality in Hilbert spaces and present results that are generalizations of well-known geometrical situations in $\mathbb{R}^3$.

Let $X$ be an *inner product space* and $Y$ any subspace of $X$; the *orthogonal complement* $Y^\perp$ of $Y$ is defined to be the set

$$Y^\perp = \{w \in X : \ (w,v) = 0 \quad \text{for all} \quad v \in Y\}; \tag{4.12}$$

that is, $Y^\perp$ consists of all those members of $X$ that are orthogonal to every member of $Y$. If $w$ belongs to $Y^\perp$, we say that $w$ is orthogonal to $Y$ and write $w \perp Y$. Since $(v,v) = 0$ implies that $v = 0$, it is clear that the only member of both $Y$ and $Y^\perp$ is the zero element: $Y \cap Y^\perp = \{0\}$.

**Example**

26. A canonical example of orthogonal complements is provided in $\mathbb{R}^3$. Let $Y$ be the real line – the $x_3$-axis, say. Then $Y^\perp$ is the $x_1x_2$-plane since points in the $x_1x_2$-plane are orthogonal to every member of $Y$, as shown in Figure 4.1.

Our main aim in this section is to show that if $H$ is any *Hilbert* space and $M$ is a *closed subspace* of $H$, then $H = M \oplus M^\perp$; that is, every $u \in H$ has the unique representation

$$u = v + w \quad \text{with} \quad v \in M \quad \text{and} \quad w \in M^\perp$$

(recall the discussion of direct sums in Section 3.1). Before doing so, however, it is necessary to prove another intuitively obvious result, which is embodied in the following theorem.

THEOREM 7. (a) *Let $H$ be a Hilbert space and $M$ a closed subspace of $H$. Then for every $u \in H$ it is possible to find a unique member $v_0$ in $M$ such that*

$$d \equiv \|u - v_0\| = \inf\{\|u - v\|,\ v \in M\}.$$

*Moreover, $u - v_0 \perp v$ for all $v \in M$.*
(b) *If $S$ is also a closed subspace of $H$ with $S \subset M$ and $S \neq M$, then there is a member $\tilde{v}$ in $M$ such that $\tilde{v} \neq 0$ and $\tilde{v} \perp S$.*

REMARK. Part (a) of the theorem says that, provided $M$ is a closed subspace, we can always find a unique point $v_0$ in $M$ that is closer to $u$ than any other point in $M$. Furthermore, this point may be found by "dropping a perpendicular" from $u$ on to $M$. Part (b) is illustrated in Figure 4.14 for the case in which $H = \mathbb{R}^3$, $M$ is the $xy$-plane, and $S$ the $x$-axis.

PROOF. (a) By definition of the infimum, there exists a sequence $\{v_n\}$ in $M$ such that

$$d_n = \|u - v_n\| \to d. \tag{4.13}$$

If we can show that $v_n$ is a Cauchy sequence, then it follows from the completeness of $M$ (see Theorem 4) that the limit $v_0$, say, of the sequence will lie in $M$. To show that $v_n$ is Cauchy, consider

$$\begin{aligned} \|v_n - v_m\|^2 &= \|(v_n - u) - (v_m - u)\|^2 \\ &= -\|(v_n - u) + (v_m - u)\|^2 + 2(\|v_n - u\|^2 + \|v_m - u\|^2) \end{aligned}$$

using the parallelogram law, Exercise 3.10. Now

$$\|(v_n - u) + (v_m - u)\| = \|v_n + v_m - 2u\| = 2\|\tfrac{1}{2}(v_n + v_m) - u\| \geq 2d$$

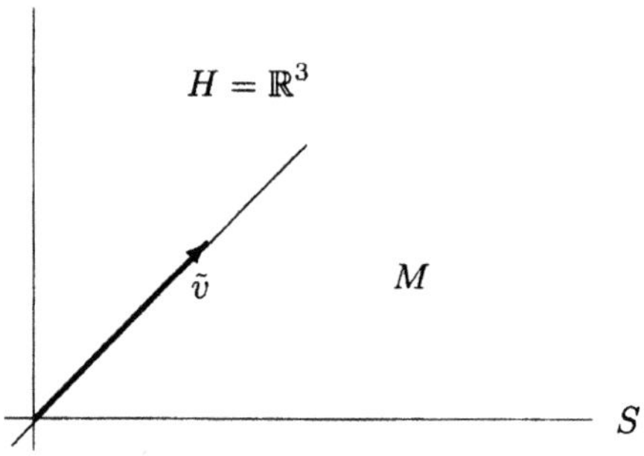


FIGURE 4.14. Illustration of part (b) of Theorem 7

and hence

$$\|v_n - v_m\|^2 \le -(2d)^2 + 2(d_n^2 + d_m^2) \to 0 \quad \text{as } m, n \to \infty,$$

from (4.13). Hence $v_n$ is a Cauchy sequence, and $v_n \to v_0$ in $M$. But since $v_0$ is in $M$ we must have

$$\|u - v_0\| \ge d;$$

furthermore,

$$\begin{aligned} \|u - v_0\| &= \|u - v_n + v_n - v_0\| \le \|u - v_n\| + \|v_n - v_0\| \\ &= d_n + \|v_n - v_0\| \to d. \end{aligned}$$

Hence $\|u - v_0\| = d$. The proof that $v_0$ is unique is left as an exercise (see Exercise 4.24).

To show that $(u - v_0, v) = 0$ for all $v \in M$, consider any point $v_0 + \alpha v$ in $M$; clearly,

$$d^2 \le \|u - (v_0 + \alpha v)\|^2 = \|u - v_0\|^2 - 2\alpha(v, u - v_0) + \alpha^2\|v\|^2. \quad (4.14)$$

Now suppose that $u - v_0$ is not orthogonal to $v$, and that $(u - v_0, v) = \beta \neq 0$. Also, take $\alpha$ to be equal to $\beta/\|v\|^2$ (recall that $\alpha$ is arbitrary). Then we obtain, from (4.14),

$$d^2 \le \|u - v_0\|^2 - \beta^2/\|v\|^2; \quad (4.15)$$

but the right-hand side of (4.15) is less than $d^2$ since $\|u - v_0\|^2 = d^2$ and $\beta^2/\|v\|^2 > 0$. This leads to a contradiction, and so we must have $\beta = 0$.

(b) Choose $\overline{v} \in M$ such that $\overline{v} \notin S$, and let $\overline{w}$ be the point in $S$ closest to $\overline{v}$ (we are applying part (a) of the theorem to $S$). Then $\tilde{v} = \overline{v} - \overline{w}$ is such a point. □

We are now ready to prove the following theorem.

THEOREM 8 (THE PROJECTION THEOREM.) *Let $M$ be a closed subspace of a Hilbert space $H$. Then every $u \in H$ can be uniquely written in the form*

$$u = v + w, \quad v \in M, \quad w \in M^{\perp};$$

that is, $H = M \oplus M^{\perp}$.

PROOF. First of all, $M$ is complete by Theorem 4. Second, according to Theorem 7, for each $u \in H$ there is a unique $\tilde{v}$ in $M$ such that $u - \tilde{v}$ is orthogonal to $M$. That is,

$$u - \tilde{v} = w \quad \text{or} \quad u = \tilde{v} + w, \quad w \in M^{\perp}. \tag{4.16}$$

To prove that there is only one such $w$, suppose that (4.16) holds for two elements $w_1$ and $w_2$ in $M^{\perp}$. Then

$$u = \tilde{v} + w_1, \quad u = \tilde{v} + w_2$$

and so $w_1 - w_2 = 0$, or $w_1 = w_2$. This proves the theorem. □

We conclude this section with a result that will prove useful later on.

THEOREM 9. *Let $M$ be a subspace of a Hilbert space $H$. Then $M^{\perp} = \{0\}$ if and only if $M$ is dense in $H$.*

In order to prove Theorem 9 we introduce the following property of orthogonal complements.

LEMMA 1. $M^{\perp\perp} = \overline{M}$, *where $\overline{M}$ denotes the closure of $M$ and $M^{\perp\perp} = (M^{\perp})^{\perp}$.*

PROOF OF LEMMA 1. First, we note that $M \subset M^{\perp\perp}$, since if $v$ is any point in $M$, then $v \perp M^{\perp}$ so that $v$ also belongs to $M^{\perp\perp}$. Furthermore, since $M^{\perp\perp}$ is closed (see Exercise 4.25), clearly $\overline{M} \subset M^{\perp\perp}$. All that remains is to show that $\overline{M} = M^{\perp\perp}$. Suppose that $\overline{M} \neq M^{\perp\perp}$; from Theorem 7(b), there is a nonzero point $w \in M^{\perp\perp}$ such that $w \perp \overline{M}$. Since $M \subset \overline{M}$ this means that $w \in M^{\perp}$. But $w \in M^{\perp} \cap M^{\perp\perp}$ implies that $w = 0$, a contradiction. Hence $M^{\perp\perp} = M$. □

PROOF OF THEOREM 9. First assume that $M$ is dense in $H$; then for any $u \in H$ and given any $\epsilon > 0$, we can always find a point $v$, say, in $M$ such that $\|v - u\| < \epsilon$. In particular, if $u \in M^{\perp}$ then

$$\epsilon^2 > \|v - u\|^2 = (v - u, v - u) = \|v\|^2 + \|u\|^2 \geq \|u\|^2.$$

Since $\epsilon$ is arbitrary we must have $u = 0$. Conversely, assume that $M^{\perp} = \{0\}$; then $M^{\perp\perp} = \{0\}^{\perp} = H$, so that from Lemma 1, $\overline{M} = H$. That is, $M$ is dense in $H$. □

**Example**

27. Since $C(\overline{\Omega})$ is dense in $L^2(\Omega)$ with the $L^2$-norm, it follows from Theorem 9 that if $u \in L^2(\Omega)$ and $(u, v) = 0$ for all $v \in C(\overline{\Omega})$, then $u = 0$. That is,

$$\int_\Omega u(\boldsymbol{x})v(\boldsymbol{x})\,dx = 0 \quad \text{for all } v \in C(\overline{\Omega}) \Rightarrow u(\boldsymbol{x}) = 0.$$

## 4.6 Bibliographical remarks

The results covered in this chapter are usually given a detailed treatment in books on functional analysis. Good references include Binmore [6], Kreyszig [27], Lang [29], Naylor and Sell [33], Oden [36], Roman [43], and Smirnov [49]; these texts contain a wealth of information on normed and inner product spaces. Naylor and Sell [33], in particular, treat in detail the issue of dense subspaces of $L^p$. Lang [29] has a simple and elegant proof of the Weierstrass Approximation Theorem.

## 4.7 Exercises

**Sequences**

4.1. Calculate $d(x, B)$ (see Exercise 3.18) if $X = \mathbb{R}$, $x = \frac{1}{2}$ and $B = \{x_n = 3 + (-1)^n n/(n^2+1) : n = 1, 2, 3, \ldots\} \cup \{3\}$.

4.2. Let $X$ be an inner product space and suppose that $\{u_n\}$ and $\{v_n\}$ are convergent sequences in $X$ with limits $u$ and $v$, respectively, convergence being defined via the norm generated by the inner product on $X$. Show that $(u_n, v_n) \to (u, v)$, and deduce that $(u_n, v) \to (u, v)$ and that $\|u_n\| \to \|u\|$.

4.3. If $u_n \to u$ in a normed space $X$ and $\|u_n - w\| \leq \alpha$ for some $w \in X$ and $\alpha \in \mathbb{R}$, show that $\|u - w\| \leq \alpha$.

**Convergence of sequences of functions**

4.4. Determine intervals on which the following sequences of functions converge pointwise: (a) $u_n(x) = x^n$; (b) $u_n(x) = 1/(1 + n^2x^2)$.

4.5. Show that the following sequences converge pointwise to 0 on the intervals given, but that they do not converge in the mean.

(a) $u_n(x) = \begin{cases} 0, & 0 \le x \le 1/n, \\ n, & 1/n < x < 2/n, \\ 0, & 2/n \le x \le 1; \end{cases}$

(b) $u_n(x) = n^{3/2} x e^{-n^2x^2}$ on $[-1, 1]$.

4.6. Does the sequence $u_n(x) = nx/(1 + n^2x^2), n = 1, 2, \ldots$ converge uniformly in $[0, 1]$? in $(a, 1]$ (for $0 < a < 1$)?

4.7. Show that uniform convergence of a sequence of functions implies $L^p$ convergence. Give an counterexample to show that the converse does not hold.

4.8. Let $a > 0$ be a fixed real number, and define $\|u\| = \sup\{|u(x)| : |x| \le a\}$ and $|||u||| = \min(1, \|u\|)$ on the space $C(-\infty, \infty)$. Why is $\|\cdot\|$ not a norm? Is $|||\cdot|||$ a norm?

**Completeness**

4.9. Show that the sequence $u_n(x) = x^{1/n}$ is a Cauchy sequence in $L^2(0, 1)$.

4.10. Consider the sequence $u_n(x) = x^n$ in the space $L^1(0, 1)$. Is this a Cauchy sequence?

4.11. The purpose of this exercise is to show that $C[a, b]$ is complete with respect to the sup-norm. Let $u_n(x)$ be a Cauchy sequence; show that $u_n(x_0)$ is a Cauchy sequence of real numbers for every fixed $x_0$ in $[a, b]$ and deduce that $u_n(x_0)$ converges to a number $u(x_0)$, say. Next, show that $u_n(x)$ converges uniformly to the function $u(x)$. Finally, since $u_n \to u$ uniformly, we have

$$|u_n(x) - u(x)| < \epsilon \text{ for all } n > N;$$

use the triangle inequality to show that

$$|u(x) - u(y)| \le 2\epsilon + |u_n(x) - u_n(y)|$$

and deduce from this result and the continuity of $u_n$ that $u$ is continuous.

4.12. Show that $\mathbb{R}^n$ with the norm $\|\cdot\|_p$ $(1 \le p \le \infty)$ is complete.

4.13. Show that every convergent sequence is a Cauchy sequence.

4.14. Consider $C[0, 1]$ with the $L^2$-norm. Show that the sequence $\{u_n\}$ is a Cauchy sequence, where $u_n(x)$ is as shown in the following figure. Next, show that if $u_n$ converges to $u(x)$, then we should have

$$u(x) = \begin{cases} 0, & 0 \le x < \frac{1}{2}, \\ 1, & \frac{1}{2} < x \le 1, \end{cases}$$

so that $C[0,1]$ with the $L^2$-norm is not complete.

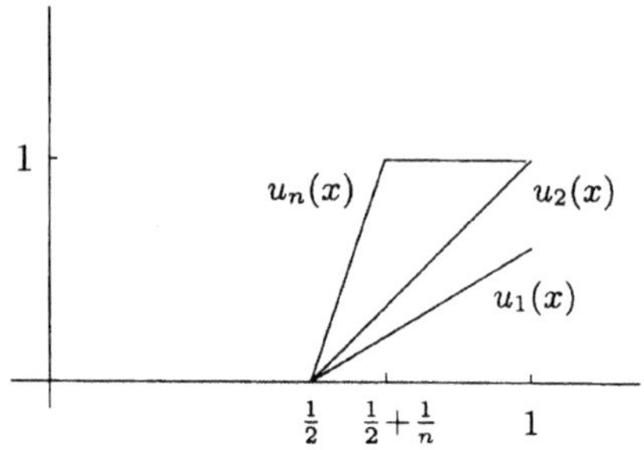


4.15. Show that the set $Y = \{v \in L^2(0,1) : \int_0^1 |v(x)|\, dx = 1\}$ is complete.

**Open and closed sets, completion**

4.16. Show that the function

$$u(x) = \begin{cases} -1, & -1 \leq x < 0, \\ +1, & 0 \leq x \leq 1, \end{cases}$$

is a point of accumulation of $C[-1,1]$ with respect to the $L^2$-norm.

4.17. Consider the space $C[a,b]$ with the sup-norm, and let $M$ be the subset consisting of functions $v$ satisfying $v(a) = 0$ and $|v(x)| < 1$. Is the function $u(x) = 1$ a point of accumulation of $M$?

4.18. Find the smallest value of $r$ such that the function $v(x) = \cos 2\pi x$ lies in the closed ball with center $u_0$ and radius $r$ in the space $C[0,1]$ with the sup-norm, where $u_0(x) = \sin 2\pi x$.

4.19. Show that a set $Y$ in a normed space $X$ is closed if and only if its complement $Y' = X - Y$ is open.

4.20. Prove Theorem 4, which states that if $X$ is a Banach space and $Y$ a subset of $X$, then $Y$ is complete if and only if $Y$ is closed in $X$.

4.21. Let $W$, $X$, and $Y$ be normed spaces, and suppose that $W$ is dense in $X$ and $X$ is dense in $Y$. Show that $W$ is dense in $Y$.

4.22. Prove Theorem 5.

**Orthogonal complements and the projection theorem**

4.23. Show that the element $v_0$ in Theorem 7 is unique.

4.24. If $Y$ is a subset of an inner product space $X$, show that $Y^\perp$ is a closed subspace of $X$. [Hint: let $\{u_n\}$ be a convergent sequence in $Y^\perp$ with limit $u_0$.]

4.25. Where in Lemma 1 is the completeness of $H$ used?

4.26. If $X$ and $Y$ are subsets of an inner product space $W$ and $X \subset Y$, show that $Y^\perp \subset X^\perp$.

# 5

# Linear operators

In the preceding chapters we have acquainted ourselves with some of the basic structures of normed and inner product spaces. We come now to another fundamental concept in functional analysis, namely, that of a mapping or operator from one space to another. At the most primitive level one requires only two sets in order to define an operator from one of them to the other, and these sets need not have any algebraic or topological structure for the definition to make sense. Obviously, though, the really interesting and useful properties of operators come to the fore when the two sets are given additional structure: if the two sets are vector spaces, we can introduce the concept of a linear operator, and if the sets are normed spaces as well, then it is possible to construct a rich theory of linear operators on such spaces. After a general introduction to operators in Section 5.1, we discuss the theory of linear operators on normed spaces in Section 5.2.

Projections are a class of operators that feature strongly in later chapters when we discuss approximations of boundary value problems. Apart from this, much of the geometrical structure of Hilbert spaces is laid bare with the aid of projection operators acting on these spaces. For these reasons we devote Section 5.3 to a discussion of projection operators on Hilbert spaces.

Operators that map members of a specified space into the real or complex numbers are special, and are given a special name: these are called functionals, and are discussed in Section 5.4. Finally, we discuss in Section 5.5 operators that map pairs of elements into the real or complex numbers in a linear fashion; these are known as bilinear forms. Linear functionals

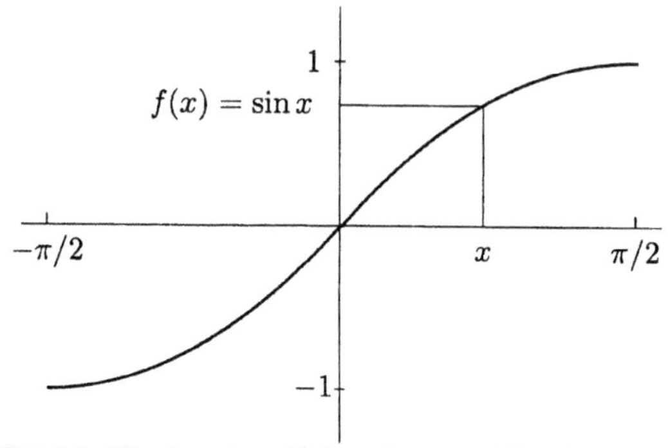


FIGURE 5.1. The function $f(x) = \sin x$ considered as a mapping

and bilinear forms play a central role in the study of linear boundary value problems, as we show in Chapter 9 and subsequently.

## 5.1 Operators

The subject of this chapter is not entirely unfamiliar; we have all come across both linear and nonlinear operators in earlier courses on linear algebra, differential equations, and so on. Here we continue the process of generalizing from the familiar. Consider the function $f(x)$, defined on the interval $I = [-\pi/2, \pi/2]$, as shown in Figure 5.1. This familiar situation is really just an example of the action of an operator: specifically, we have defined $f$ to be something that acts on any member $x$ in $I$, and produces a real number $\sin x$. Furthermore, the *image* $\sin x$ lies in the set $J = [-1, 1]$. More formally we write all of this as follows:

$$f : I \to \mathbb{R}, \quad f(x) = \sin x.$$

Here the first expression reads "$f$ maps elements of $I$ to elements of $\mathbb{R}$" and the second expression tells how $f$ does this: $f$ acts on $x$ to produce $\sin x$. The set $I$ is called the *domain* of the operator $f$, written $D(f)$. The set $\mathbb{R}$ in which $f(x)$ takes its values is called the *image* space, whereas the subset $J \subset \mathbb{R}$ consisting of all real numbers that are images of $I$ under the mapping $f$ is called the *range* of $f$, written $R(f)$. We now generalize.

Let $X$ and $Y$ be two sets, and suppose that a rule is given whereby an element $u$ of $X$ is mapped or transformed to an element $v$ of $Y$. This rule is called an *operator or transformation* or *mapping* and we write, for an operator $T$,

$$T : X \to Y, \quad Tu = v \quad (\text{or } T(u) = v), \quad u \in X, v \in Y.$$

The first expression reads "$T$ maps elements of $X$ to elements of $Y$" while the second reads "$T$ acts on $u$ to produce $v$". We refer to $Y$ as the *image*

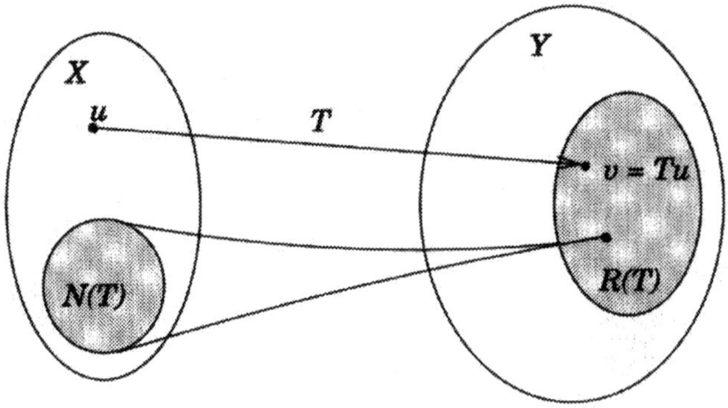


FIGURE 5.2. Illustration of concepts associated with a mapping $T : X \to Y$

*space*, $X$ is called the *domain* of $T$, written $D(T)$, and we write $R(T)$ for the *range* of $T$, which consists of all those elements of $Y$ that are images of members of $X$. In other words,

$$R(T) = \{v : \ v \in Y, \ Tu = v \text{ for some } u \in X\}.$$

Finally, the element $v$ is called the *image* of $u$ under the mapping $T$. These concepts are illustrated in Figure 5.2.

If the range of $T$ happens to be all of $Y$, then $T$ is called a *surjective* operator, and we say that $T$ maps $X$ *onto* $Y$. Otherwise $T$ maps $X$ *into* $Y$.

Assume that the image space of $T$ contains the zero element; then the *null space* $N(T)$ of $T$ is the set of all elements of $D(T)$ whose image is zero:

$$N(T) = \{u \in X : \ Tu = 0\}.$$

The *inverse image* of a member $v \in Y$ is denoted by $T^{-1}(v)$, and is the set of all $u \in X$ such that $Tu = v$:

$$T^{-1}(v) = \{u \in X : \ Tu = v\}.$$

Likewise, the inverse image of a subset $W$ of $Y$ is denoted by $T^{-1}(W)$, and is the set of all $u \in X$ such that $Tu \in W$ (Figure 5.3):

$$T^{-1}(W) = \{u \in X : \ Tu \in W\}.$$

**Examples**

1. All functions of a real variable are operators from a subset of $\mathbb{R}$ to $\mathbb{R}$, for example, the operator or function $f(x) = \sin x$ discussed at the beginning of this section. In the same way, the function

$$f : \mathbb{R}^2 \to \mathbb{R}, \quad f(\boldsymbol{x}) = f(x, y) = x^2 + y^2$$

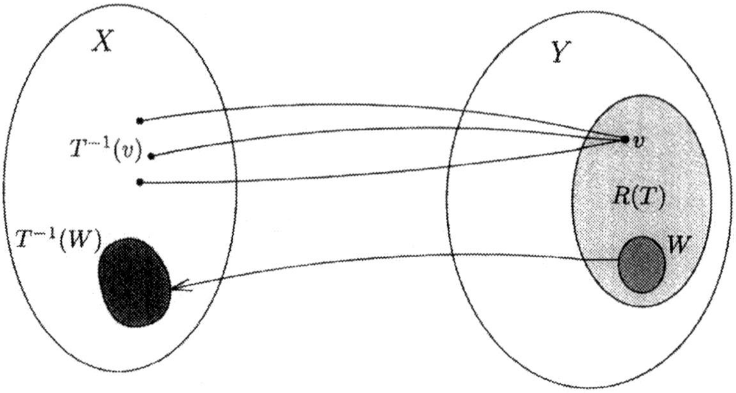


FIGURE 5.3. The inverse images of an element and of a set

is an operator that maps the point $\boldsymbol{x}$ in $\mathbb{R}^2$ to the real number $x^2+y^2$. This number is never negative, and indeed corresponding to any real number $r$ there are numbers $x$ and $y$ such that $f(\boldsymbol{x}) = r$ (think about a circle with radius $r$). Thus $R(f)$ = the set of all nonnegative real numbers, and hence $f$ is not surjective. The inverse image of the set $(-1, 0)$ is the empty set, whereas the inverse image of the point $+1$ is the set of all points on the unit circle.

2. An $n \times m$ matrix is an operator from $\mathbb{R}^m$ to $\mathbb{R}^n$. For example, the operator

$$\boldsymbol{T} : \mathbb{R}^3 \to \mathbb{R}^2, \quad \boldsymbol{T}\boldsymbol{x} = \begin{pmatrix} T_{11} & T_{12} & T_{13} \\ T_{21} & T_{22} & T_{23} \end{pmatrix} \begin{pmatrix} x_1 \\ x_2 \\ x_3 \end{pmatrix},$$

is a $2 \times 3$ matrix that transforms a member of $\mathbb{R}^3$ to a member of $\mathbb{R}^2$. Whether $\boldsymbol{T}$ is surjective depends on the entries $T_{ij}$ in $\boldsymbol{T}$; for example, if $T_{11} = T_{21} = 1$ and all other $T_{ij} = 0$, then $\boldsymbol{T}\boldsymbol{x} = (x_1, x_1)$ and the range of $\boldsymbol{T}$ is the subset of $\mathbb{R}^2$ described by the straight line running through the origin at 45°. The null space of $\boldsymbol{T}$ consists of all points $\boldsymbol{x}$ for which $\boldsymbol{T}\boldsymbol{x} = (x_1, x_1) = (0, 0)$: thus $N(T) = \{\boldsymbol{x} \in \mathbb{R}^3 : x_1 = 0\}$, which is the $x_2x_3$-plane.

3. Various examples of *differential* operators were presented in the Introduction, and reappear later; recall that these are operators that consist of combinations of ordinary or partial derivative operators. For example, if $u \in C^2(\Omega)$ with $\Omega$ a domain in $\mathbb{R}^2$, then the *Laplacian* operator $\Delta$ is defined by

$$\nabla^2 : C^2(\Omega) \to C(\Omega), \quad \nabla^2 u = \frac{\partial^2 u}{\partial x^2} + \frac{\partial^2 u}{\partial y^2} \equiv f$$

for a problem involving two variables $x$ and $y$ only. Thus $f$, the image of $u$, is a continuous function. To be specific, if $\Omega \subset \mathbb{R}^2$ and $u(\boldsymbol{x}) = x^2y^3$, then the image of $u$ is the function $f$ defined by

$$f(x,y) = 2y^3 + 6x^2y.$$

The question of whether $\Delta$ is a surjective operator is a question that is taken up in Chapter 8; this is equivalent to asking whether there exists a solution to the equation $\Delta u = f$.

Two operators $S : X \to Y$ and $T : X \to Y$ are said to be *equal* if for every $u \in X$ we have

$$Su = Tu.$$

When this is the case, we write $S = T$.

The *sum* of two operators $S : X \to Y$ and $T : X \to Y$ is defined to be the operator satisfying

$$(S+T)u = Su + Tu, \quad u \in X,$$

where $Y$ is a vector space. That is, $T + S$ has the same effect on any member of $u$ as would be obtained by applying $T$ and $S$ separately, and then adding together the result. In order for the definitions of the sum of operators, and of equality of operators, to make sense, the domains of the two operators $S$ and $T$ must be equal, as must the image spaces.

The *composition* or *product* $TS$ of two operators $S : X \to Y$ and $T : Y \to W$ is defined to be the operator satisfying

$$TS : X \to W, \quad (TS)u = T(Su) \quad \text{for all} \quad u \in X.$$

That is, the element $(TS)u \in W$ is found by first obtaining the element $Su \in Y$, and then by the action of $T$ on $Su$. Note that the composition $TS$ is meaningless if the element $Su$ does not belong to the domain of $T$. Furthermore, in general $TS \neq ST$; in fact, $ST$ may be quite meaningless.

**Example**

4. Let $X = \mathbb{R}^3$, $Y = \mathbb{R}^2$, $W = \mathbb{R}$, and let $\boldsymbol{T} : X \to Y$ and $\boldsymbol{S} : Y \to W$ be the matrices

$$\boldsymbol{T} = \begin{pmatrix} 1 & 2 & 1 \\ 2 & 3 & 2 \end{pmatrix}, \qquad \boldsymbol{S} = [1 \;\; 2].$$

Then for any $\boldsymbol{x} = (x, y, z)$ in $\mathbb{R}^3$,

$$\begin{aligned} (\boldsymbol{ST})\boldsymbol{x} &= \boldsymbol{S}\begin{pmatrix} 1 & 2 & 1 \\ 2 & 3 & 2 \end{pmatrix}\begin{pmatrix} x \\ y \\ z \end{pmatrix} = [1 \;\; 2]\begin{pmatrix} x + 2y + z \\ 2x + 3y + 2z \end{pmatrix} \\ &= 5x + 8y + 5z. \end{aligned}$$

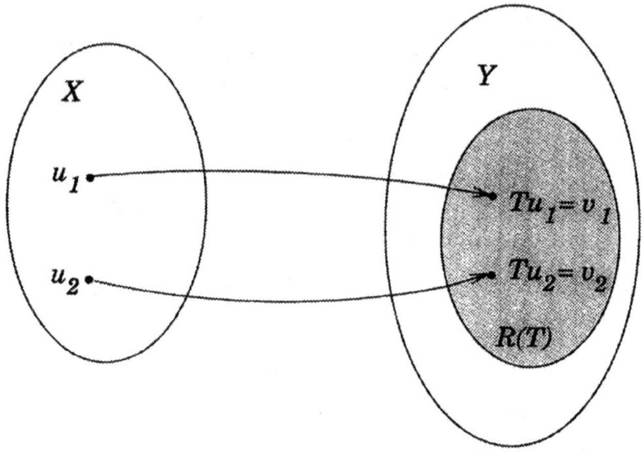


FIGURE 5.4. An injective operator and its inverse

It follows that the operator $\boldsymbol{TS}$ is meaningless since for any $\boldsymbol{x}$ in $\mathbb{R}^2$ we have $\boldsymbol{Sx} \in \mathbb{R}$ and $\boldsymbol{T}(\boldsymbol{Sx})$ makes no sense.

The *identity* operator is an operator from a set $X$ into itself, which maps each element of $X$ to the same element. That is,

$$I : X \to X, \quad Iu = u \quad \text{for all} \quad u \in X.$$

The *zero* operator 0 is an operator $0 : X \to Y$ which maps every element of $X$ to the zero element in $Y$ (we assume of course in this definition that $Y$ has a zero element):

$$0 : X \to Y, \quad 0u = 0 \quad \text{for all } u \in X.$$

**Example**

5. Let $X = Y = \mathbb{R}^3$; then the identity operator $\boldsymbol{I} : \mathbb{R}^3 \to \mathbb{R}^3$ is simply the $3 \times 3$ identity matrix. The zero operator from $\mathbb{R}^3$ to $\mathbb{R}^2$ is the $2 \times 3$ matrix containing all zeros.

**Injective (one-to-one) and invertible operators.** An operator $T : X \to Y$ is *one-to-one* or *injective* if no two distinct elements of $X$ are mapped to the same element in $Y$. That is, $T$ is *one-to-one* if

$$u_1 \neq u_2 \text{ implies that } Tu_1 \neq Tu_2$$

or, equivalently, if $Tu_1 = Tu_2$ implies that $u_1 = u_2$, for all $u_1, u_2 \in X$ (Figure 5.4). From this definition it is evident that each $v$ in the range of $T$ is *the image of exactly one element* $u$ in $X$. We may accordingly define an

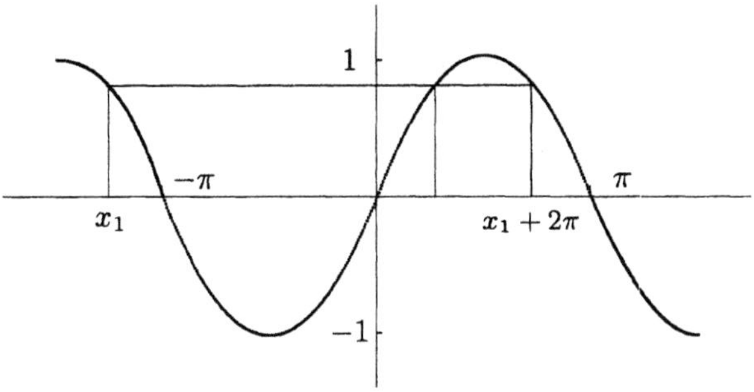


FIGURE 5.5. The function $f(x) = \sin x$

operator $T^{-1}$, called the *inverse* of $T$, which maps $v$ back to $u$. The inverse is then defined by

$$T^{-1} : R(T) \to X, \quad T^{-1}(Tu) = u. \tag{5.1}$$

In view of the definition of the composition of two operators, (5.1) indicates that

$$T^{-1}T = I.$$

In the same way, by starting with $T^{-1}$ we find that

$$TT^{-1} = I,$$

that is, $(T^{-1})^{-1} = T$. If the range of $T$ is all of $Y$ (that is, $T$ is surjective) and $T$ is also one-to-one, then $T$ is said to be *bijective*; $T^{-1}$ is a one-to-one operator from $Y$ onto $X$, and we say that $T$ is *invertible.*

**Examples**

6. The function $f : [-\pi/2,\ \pi/2] \to [-1, 1], f(x) = \sin x$, is one-to-one since to each value of $f(x) = \sin x$ there corresponds only one point $x$. However, if the domain of $f$ is the whole real line, then we see from Figure 5.5 that $f$ is *not* one-to-one since, for any $x_1$,

$$f(x_1 + 2n\pi) = f(x_1), \quad n = 1, 2, \ldots.$$

Returning to the case in which $D(f) = [-\pi, /2, \pi/2]$, the inverse function $f^{-1} : [-1, 1] \to [-\pi/2, \pi/2]$ is defined by $f^{-1}(y) = \arcsin y$.

7. Any nonsingular $n \times n$ matrix $\boldsymbol{T} : \mathbb{R}^n \to \mathbb{R}^n$ is one-to-one, with inverse $\boldsymbol{T}^{-1}$ being the usual matrix inverse.

8. The operator $T = d/dx : C^1[a, b] \to C[a, b]$ is *not* one-to-one since there are infinitely many functions, all differing from each other by

a constant, which have the same image or derivative. However, if we choose the domain of $T$ to be $X = \{u \in C^1[a,b] : u(a) = 0\}$, then $T$ is invertible with inverse $T^{-1}$ defined by

$$T^{-1}(v)(x) = \int_a^x v(y)dy.$$

**Restriction and extension.** Suppose that we are given an operator $T$ from $X$ to $Y$, so that $D(T) = X$. Let $U$ be a subset of $X$. Then the *restriction* of $T$ to $U$ is the operator $T|_U$ defined by

$$T|_U : U \to Y, \quad T|_U u = Tu \quad \text{for all} \quad u \in U.$$

Thus $T|_U$ is an operator with domain $U$ which has the same action on members of $U$ as does $T$.

Suppose next that $X$ is a subset of a bigger set $V$. Then an *extension* of $T$ to $V$ is an operator $\overline{T}$ with the property

$$\overline{T} : V \to Y, \quad \overline{T}|_X = T.$$

That is, $\overline{T}u = Tu$ for $u \in X$, so that $T$ is the restriction of $\overline{T}$ to $X$.

## 5.2 Linear operators, continuous, and bounded operators

**Linear operators.** A linear operator $T$ is an operator whose domain $X$ is a *vector space,* and which is

(a) additive: $T(u+v) = T(u) + T(v)$ for all $u, v \in X$; and
(b) homogeneous: $T(\alpha u) = \alpha T(u)$ for all $u \in X, \ \alpha \in \mathbb{K}$.

Here $\mathbb{K}$ is the field (either $\mathbb{R}$ or $\mathbb{C}$) over which the vector space is defined. We may summarize (a) and (b) in one statement by defining a linear operator to be one that satisfies

$$T(\alpha u + \beta v) = \alpha T(u) + \beta T(v) \quad \text{for all } u, v \in X, \ \alpha, \beta \in \mathbb{K}.$$

**Examples**

9. The differential operator $d^n/dx^n : C^n[a,b] \to C[a,b]$ is linear since

$$\frac{d^n}{dx^n}(\alpha u + \beta v) = \alpha \frac{d^n u}{dx^n} + \beta \frac{d^n v}{dx^n}.$$

Similar considerations apply to partial differential operators of all orders, which are also linear operators.

10. The operator $f : \mathbb{R} \to \mathbb{R}, f(x) = \sin x$, is not a linear operator since, if either $x$ or $y$ is nonzero, $f(x+y) = \sin(x+y) \neq f(x) + f(y) = \sin x + \sin y$.

We note that if $X$ and $Y$ are vector spaces and $T : X \to Y$ is linear, then $N(T)$ and $R(T)$ are *subspaces* of $X$ and $Y$, respectively.

Suppose that $T : X \to Y$ is a linear one-to-one operator, and that $Tu_0 = 0$ for some $u_0 \in X$. Since $T$ is linear we have, for any element $u$, $T(u) = T(u+0) = T(u) + T(0)$ which implies that $T(0) = 0$. But since $T$ is one-to-one, the inverse image of $0 \in Y$ must be a unique element in $X$, from which it follows that $u_0 = 0$.

Conversely, suppose that $T : X \to Y$ is a linear operator with the property that $Tu_0 = 0$ only for $u_0 = 0$. Then for any two distinct elements $u, v \in X$, $T(u) - T(v) = T(u - v) \neq 0$ since $u - v \neq 0$ by hypothesis, and so two distinct elements do not have the same image. Hence $T$ is one-to-one. We summarize all of this in the following important theorem.

THEOREM 1. *A linear operator $T$ is one-to-one if and only if $N(T) = \{0\}$.*

## Example

11. Let $\boldsymbol{T} : \mathbb{R}^n \to \mathbb{R}^n$ be the operator defined by an $n \times n$ matrix. It is easily shown that $\boldsymbol{T}$ is a linear operator; the question of whether $\boldsymbol{T}$ is one-to-one is equivalent to asking whether the equation

$$\boldsymbol{T}\boldsymbol{x} = \boldsymbol{y}$$

has a unique solution $\boldsymbol{x}$ for a given $\boldsymbol{y}$. According to Theorem 1 this question may be answered by considering the equation

$$\boldsymbol{T}\boldsymbol{x}_0 = \boldsymbol{0};$$

if the only element $\boldsymbol{x}_0$ satisfying this equation is $\boldsymbol{x}_0 = \boldsymbol{0}$, then $\boldsymbol{T}$ is one-to-one. Equivalently, we may check whether the matrix is nonsingular. For example, if $\boldsymbol{T} : \mathbb{R}^2 \times \mathbb{R}^2$ is given by

$$\boldsymbol{T} = \begin{pmatrix} 1 & 2 \\ 2 & 4 \end{pmatrix},$$

then

$$\boldsymbol{T}\boldsymbol{x}_0 = \boldsymbol{0} \Rightarrow \begin{pmatrix} 1 & 2 \\ 2 & 4 \end{pmatrix} \begin{pmatrix} \alpha \\ \beta \end{pmatrix} = \begin{pmatrix} 0 \\ 0 \end{pmatrix}$$

which has the solution $\boldsymbol{x}_0 = \gamma(-2, 1)$ for $\gamma \in \mathbb{R}$:

$$N(\boldsymbol{T}) = \{\boldsymbol{x}_0 \in \mathbb{R}^2 : \ \boldsymbol{x}_0 = \gamma(-2,1) \quad \text{for all } \gamma \in \mathbb{R}\}.$$

It follows that the equation $\boldsymbol{T}\boldsymbol{x} = \boldsymbol{y}$ will not have a unique solution; in fact, if $\boldsymbol{x}_1$ is any solution, then $\boldsymbol{x}_1 + \gamma(-2, 1)$ is also a solution. We observe also that $\boldsymbol{T}$ is singular, in that its determinant is zero.

**Isomorphisms.** Two vector spaces $X$ and $Y$ are said to be isomorphic, or more precisely, algebraically isomorphic, to each other if there exists a *linear bijective* map $T$ from $X$ onto $Y$; this map is then called an isomorphism. It follows that the inverse operator $T^{-1}$ is also an isomorphism from $Y$ onto $X$.

Isomorphisms are useful, in that they provide information about whether it is possible to put elements of one space $X$ in one-to-one correspondence with elements of another space $Y$. This in turn establishes that the two spaces concerned are "alike", in a rough sense. It should be borne in mind, though, that there are many attributes of a space, such as its topological properties, which cannot be inferred simply from its isomorphic relationship to another.

## Example

12. To emphasize the point that two isomorphic spaces can be quite different in nature, consider the case in which $X = \mathbb{R}^n$ and $Y = P_{n-1}[a, b]$, the space of polynomials of degree less than or equal to $n-1$. An arbitrary member of $Y$ is of the form $p(x) = a_0 + a_1 x + \cdots + a_{n-1}x^{n-1}$, and is therefore defined uniquely by the $n$ numbers $a_0, \ldots, a_{n-1}$. Let $T : X \to Y$ be the operator that associates with the point $\boldsymbol{a} = (a_0, \ldots, a_{n-1})$ the polynomial $p(x)$ introduced earlier; then clearly $T$ is linear and bijective, and is hence an isomorphism. Thus $X = \mathbb{R}^n$ and $Y = P_{n-1}[a, b]$ are isomorphic to each other.

All of the previous considerations depend only on the algebraic structure of $X$ and $Y$; in other words, we have required no more of $X$ and $Y$ than that they be linear spaces. When an operator maps elements from one *normed* space to another, though, many further interesting properties emerge. We take a look first at *continuous operators.*

Before giving a general definition of a continuous operator, it may be helpful to recall the discussion of continuous functions in Section 4 of Chapter 1. We defined a function $f : \mathbb{R} \to \mathbb{R}$ to be continuous at $x_0$ if, given any $\epsilon > 0$, it is always possible to find a $\delta > 0$ such that $|f(x_0) - f(x)| < \delta$ whenever $|x_0 - x| < \delta$. Now suppose we rephrase this in the language of open sets: $f$ is continuous at $x_0$ if, given any $\epsilon > 0$, it is always possible to find $\delta > 0$ such that the image of any point in the neighborhood $N(x_0, \delta)$ lies in the neighborhood $N(f(x_0), \epsilon)$. This is precisely how we define a continuous operator in any normed space.

**Continuous operator.** Let $T : X \to Y$ be an operator from a normed space $X$ to a normed space $Y$: then $T$ is continuous at $u_0 \in X$ if for every

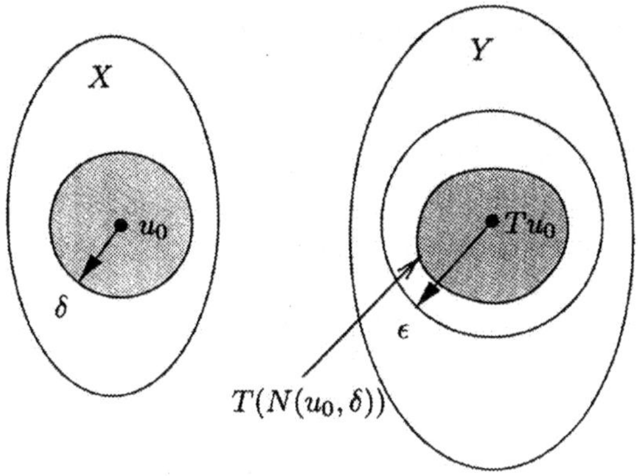


FIGURE 5.6. A neighborhood and its image

$\epsilon > 0$ there is a positive number $\delta$, possibly depending on $u_0$ and $\epsilon$, such that

$$\|Tu_0 - Tu\| < \epsilon \quad \text{whenever} \quad \|u_0 - u\| < \delta. \tag{5.2}$$

If (5.2) holds for every $u_0 \in X$, then we simply say that $T$ is continuous on $X$. Furthermore, if $\delta$ does not depend on $u_0$, then $T$ is said to be *uniformly continuous* on $X$.

The situation is shown schematically in Figure 5.6. Choose some point $u_0$ and a number $\epsilon > 0$; then $T$ is continuous if a number $\delta$ can be found such that the image of the points lying inside the neighborhood $N(u_0, \delta)$ is contained in the open ball of radius $\epsilon$ and center $Tu_0$. At this point we draw attention to the norms used in (5.2); since $u_0$ and $u$ are in $X$, the norm used when evaluating $\|u_0 - u\|$ is the norm defined on $X$; on the other hand, the norm used in the evaluation of $\|Tu_0 - Tu\|$ must be that defined on $Y$. When wishing to emphasize the distinction we write, for example, $\|u_0 - u\|_X$ and $\|Tu_0 - Tu\|_Y$. Generally, though, it is expected that there will be no confusion about which norm should be used.

### Examples

13. Let $X = Y = \mathbb{R}$ and let $f : \mathbb{R} \to \mathbb{R}$. Then the definition of continuity given previously coincides with that given in Section 2.1 if we use the norm $\|\cdot\| = |\cdot|$ on $\mathbb{R}$.

14. Let $X = Y = C[0,1]$ with the sup-norm, and define $T : C[0,1] \to C[0,1]$ by

$$Tu(x) = \int_0^x u(y)\, dy, \quad x \in [0,1]$$

(for example, if $u(x) = \cos x$, then $Tu(x)$ is the continuous function $\sin x$). Now

$$\|Tu_0 - Tu\| = \sup_{x\in[0,1]} \left| \int_0^x u_0(y)\, dy - \int_0^x u(y)\, dy \right|$$

and so we have to estimate the term inside $|\ldots|$. We find that

$$\begin{aligned} \left| \int_0^x (u_0(y) - u(y))\, dy \right| &\leq \int_0^x |u_0(y) - u(y)|\, dy \\ &\leq (x-0) \sup_{y\in[0,1]} |u_0(y) - u(y)| \\ &\qquad \text{(using Theorem 2, Chapter 2)} \end{aligned}$$

and so

$$\begin{aligned} \|Tu_0 - Tu\|_\infty &\leq \sup_{x\in[0,1]} x \sup_{y\in[0,1]} |u_0(y) - u(y)| \\ &\leq 1 \cdot \sup |u_0(y) - u(y)| = \|u_0 - u\|_\infty . \end{aligned}$$

Hence if $\|u_0 - u\|_\infty < \delta$, then $\|Tu_0 - Tu\|_\infty < \epsilon$ and so, given any $\epsilon > 0$ we simply choose $\delta = \epsilon$ to show that $T$ is continuous. Since $\delta$ does not depend on $u_0$, $T$ is in fact uniformly continuous.

15. Let $X = \mathbb{R}^n$ and $Y = \mathbb{R}^m$ and consider the linear operator $\boldsymbol{T}$ from $X$ to $Y$ represented by an $m \times n$ matrix. We endow both $\mathbb{R}^n$ and $\mathbb{R}^m$ with the Euclidean norm $\|\boldsymbol{x}\|_2$ introduced in Example 15, Chapter 3.

    Now consider the image $\boldsymbol{a}$ of $\boldsymbol{x}$ under the mapping $\boldsymbol{T}$. That is,

$$\boldsymbol{Tx} = \begin{pmatrix} T_{11}x_1 + \cdots + T_{1n}x_n \\ \vdots \\ T_{m1}x_1 + \cdots + T_{mn}x_n \end{pmatrix} = \boldsymbol{a} = \begin{pmatrix} a_1 \\ \vdots \\ a_m \end{pmatrix}$$

or

$$\sum_{j=1}^n T_{ij}x_j = a_i, \qquad 1 \leq i \leq m.$$

If $\boldsymbol{y}$ is another point in $\mathbb{R}^n$ with image $\boldsymbol{b} \in \mathbb{R}^m$, then

$$\begin{aligned} \|\boldsymbol{b} - \boldsymbol{a}\|_2^2 &= \sum_{i=1}^m \left( \sum_{j=1}^n T_{ij}(y_j - x_j) \right)^2 \\ &\leq \sum_{i=1}^m \left( \sum_{j=1}^n T_{ij}^2 \right) \sum_{j=1}^n (y_j - x_j)^2 \end{aligned}$$

(using the Cauchy–Schwarz inequality in $\mathbb{R}^n$)

$$\leq \quad k^2\|\boldsymbol{y}-\boldsymbol{x}\|^2, \quad \text{where } k^2 = \sum_{i=1}^{m}\sum_{j=1}^{n} T_{ij}^2.$$

Hence

$$\|\boldsymbol{T}\boldsymbol{y}-\boldsymbol{T}\boldsymbol{x}\| \leq k\|\boldsymbol{y}-\boldsymbol{x}\|,$$

so for given $\epsilon > 0$ we simply choose $\delta = \epsilon/k$; then $\|\boldsymbol{T}\boldsymbol{y}-\boldsymbol{T}\boldsymbol{x}\| < \epsilon$ whenever $\|\boldsymbol{y}-\boldsymbol{x}\| < \delta$ and $\boldsymbol{T}$ is thus continuous on $X$.

Continuous operators can also be characterized in terms of open sets, as the following theorem shows.

THEOREM 2. *An operator $T$ (not necessarily linear) from a normed space $X$ into a normed space $Y$ is continuous if and only if the inverse image $S_0$ of any open subset $S$ of $Y$ is an open subset of $X$.*

PROOF. Figure 5.7 illustrates the assertion of the theorem. Suppose that $T$ is continuous, and for any $u_0 \in S_0$ let $v_0 = Tu_0$. Since $S$ is open, there is a neighborhood $N(v_0, \epsilon)$ of $v_0$ contained entirely in $S$. By the continuity of $T$, $u_0$ has a neighborhood $N_0(u_0, \delta)$ that is mapped into $N(v_0, \epsilon)$. Thus, $N_0 \subset S_0$ since $N_0$ is part of the inverse image of $S$; so $S_0$ is open. Conversely, assume that the inverse image of every open set in $Y$ is an open set in $X$. Then in particular for every $u_0 \in X$ and any neighborhood $N(Tu_0, \epsilon)$ of $Tu_0$, the inverse image $N_0$, say, of $N$ is open. Hence $N_0$ also contains a neighborhood of center $u_0$ and, by definition of $N_0$, the image of this neighborhood lies in $N$. Since $u_0$ was arbitrary, $T$ is continuous. □

**Isometries.** Spaces that are isomorphic to each other were introduced earlier. With the notion of a norm available it is possible to take one step further the idea of two spaces being alike in some sense. Let $X$ and $Y$ be normed spaces, and suppose that there exists an operator $T : X \to Y$ which is linear and bijective – in other words, an isomorphism – and which, furthermore, has the property that

$$||Tu||_Y = ||u||_X \quad \text{for any } \ u \in X.$$

Then $T$ is called an isometry, and $X$ and $Y$ are said to be *isometrically isomorphic.* This in turn implies, from the linearity of $T$, that

$$||Tu - Tv|| = ||T(u-v)|| = ||u-v||$$

for any members $u$ and $v$ of $X$; that is, $T$ preserves the distances between elements. The situation is depicted schematically in Figure 5.8.

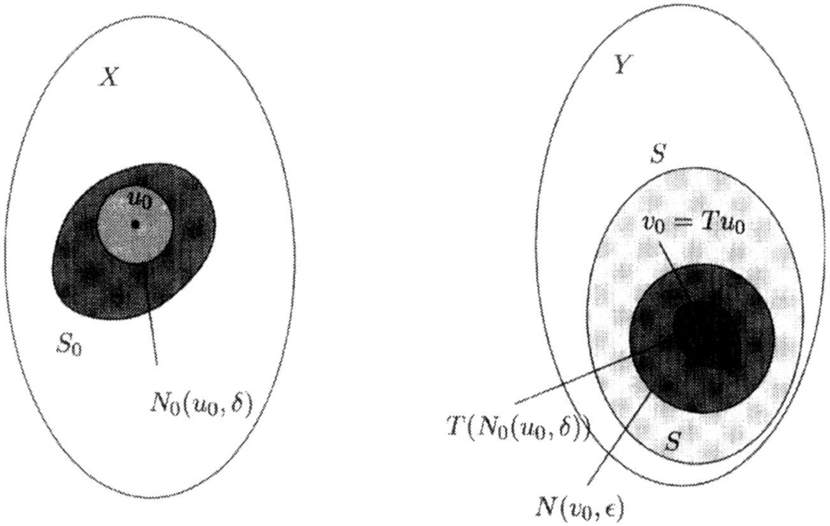


FIGURE 5.7. An equivalent definition of continuity

**Completions revisited.** Recall that we discussed in Section 4.4 the notion of completion of an incomplete subset $S$ of a normed space $X$, and defined the completion of such a set to be the closure $\overline{S}$ of $S$ or, equivalently, the union of $S$ with the limits of all Cauchy sequences in $S$ (these may of course have their limits either in $S$, or in the larger space $X$). It turns out that this treatment of completion is a somewhat simplified version of the full story, which can be given now that we have acquired some background in operator theory.

In general, by the completion of a subset $S$ of a normed space $X$ with norm $\|\cdot\|_X$ is meant any complete subset $S^*$ of a normed space $X^*$ with norm $\|\cdot\|_{X^*}$, with the property that $S^*$ has a dense subset $Y$ that is isometrically isomorphic to $S$.

We restricted the definition in Section 4.4 to the special case in which $S = Y$, and in which the isometric mapping is just the identity. The general procedure given here allows one to find a completion even when the incomplete space is not given as a subset of a complete normed space, but for our purposes the description given in Section 4.4 suffices.

**Bounded operators.** The concept of a bounded operator is closely connected with that of a continuous operator. Let $T : X \to Y$ be a *linear* operator; we say that $T$ is *bounded* if it is possible to find a number $K > 0$ such that

$$\|Tu\| \leq K\|u\| \quad \text{for all} \quad u \text{ in } X.$$

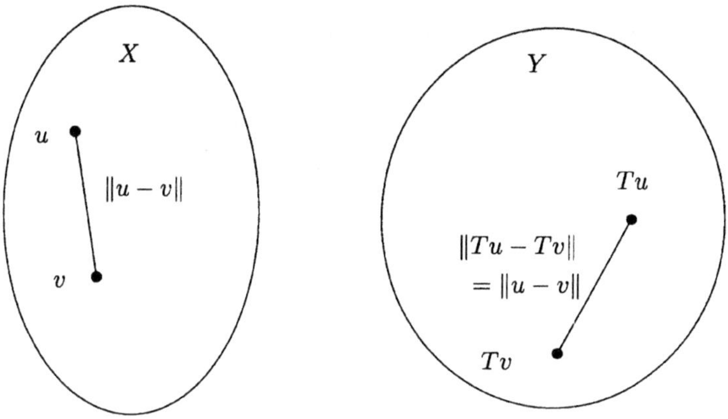


FIGURE 5.8. Two isometrically isomorphic spaces

For $u \neq 0$ we see that $K \geq \|Tu\|/\|u\|$. So the set $\{K : K \geq \|Tu\|/\|u\|, u \neq 0\}$ is bounded below, and the least upper bound, taken over all members $u$ of $X$, is called the *norm of* $T$, and is written $\|T\|$. That is,

$$\|T\| = \sup\{\|Tu\|/\|u\|, \quad u \neq 0\} \tag{5.3}$$

and so we can write

$$\|Tu\| \leq \|T\| \, \|u\|$$

for a bounded linear operator.

We now show that what we have called the "norm" of $T$ does indeed satisfy the norm axioms. First, there is the question of the space on which this quantity is a norm; this is the *set* $\mathcal{L}(X,Y)$ *of all bounded linear operators from* $X$ *to* $Y$, which forms a vector space; to see this we note that if $T$ and $S$ belong to $\mathcal{L}(X,Y)$, then $\alpha T + \beta S$, defined by

$$(\alpha T + \beta S)u = \alpha Tu + \beta Su$$

for all $u \in X$ and $\alpha, \beta \in \mathbb{R}$ or $\mathbb{C}$, also belongs to $\mathcal{L}(X,Y)$. Also, $I$ is the identity element and the zero operator the zero element.

The quantity defined in (5.3) does indeed obey all of the norm axioms: $\|T\| \geq 0$ and $\|T\|$ is zero if and only if $T = 0$ (the zero operator). The triangle inequality follows from

$$\begin{aligned} \|T+S\| &= \sup_{u\neq 0} \frac{\|(T+S)u\|}{\|u\|} = \sup_{u\neq 0} \frac{\|Tu+Su\|}{\|u\|} \\ &\leq \sup_{u\neq 0} \left( \frac{\|Tu\|}{\|u\|} + \frac{\|Su\|}{\|u\|} \right) = \|T\| + \|S\|. \end{aligned}$$

Thus $\mathcal{L}(X,Y)$ is a normed space with its norm defined by (5.3).

**Examples**

16. Let $\boldsymbol{T} : \mathbb{R}^2 \to \mathbb{R}^3$; then the space $\mathcal{L}(\mathbb{R}^2, \mathbb{R}^3)$ of linear operators from $\mathbb{R}^2$ to $\mathbb{R}^3$ is equivalent to the space of all real $3 \times 2$ matrices. If $\mathbb{R}^2$ and $\mathbb{R}^3$ are equipped with the 1-norm

$$\|\boldsymbol{x}\|_1 = \sum_{i=1}^{n} |x_i|$$

in which $n = 2$ or 3, respectively, then

$$\begin{aligned} \|\boldsymbol{T}\| &= \sup\left\{ \frac{\|\boldsymbol{Tx}\|_1}{\|\boldsymbol{x}\|_1}, \quad \boldsymbol{x} \neq 0 \right\} \\ &= \sup\{\|\boldsymbol{Tx}\|_1, \quad \|\boldsymbol{x}\|_1 = 1\} \quad \text{(see Exercise 5.11).} \end{aligned}$$

Now

$$\begin{aligned} \|\boldsymbol{Tx}\|_1 &= \sum_{i=1}^{3} |T_{i1}x_1 + T_{i2}x_2| \leq \sum_{i=1}^{3} (|T_{i1}|\,|x_1| + |T_{i2}|\,|x_2|) \\ &= |x_1| \sum_{i=1}^{3} |T_{i1}| + |x_2| \sum_{i=1}^{3} |T_{i2}| \\ &\leq \left( \max_{j=1,2} \sum_{i=1}^{3} |T_{ij}| \right) (|x_1| + |x_2|) \\ &\leq \max_{j=1,2} \sum_{i=1}^{3} |T_{ij}| \qquad (5.4) \end{aligned}$$

if $|x_1| + |x_2| = \|\boldsymbol{x}\|_1 = 1$. Hence

$$\|\boldsymbol{T}\|_1 = \max \|\boldsymbol{Tx}\|_1 \leq \max_j \sum_{i=1}^{3} |T_{ij}|.$$

Suppose that this maximum is attained for $j = 1$, and choose $\boldsymbol{x} = (1, 0)$; for this choice of $\boldsymbol{x}$

$$\|\boldsymbol{Tx}\|_1 = \sum_{i=1}^{3} |T_{i1}|,$$

so that we have equality in (5.4). Generally, $\|\boldsymbol{Tx}\|_1 = \sum_{i=1}^{3} |T_{ip}|$, where $p$ denotes the column in which $\sum_{i=1}^{3} |T_{ip}|$ is a maximum. Hence

$$\|\boldsymbol{T}\|_1 = \max_j \sum_{i=1}^{3} |T_{ij}|;$$

this quantity is known, for obvious reasons, as the *maximum column sum of* $\boldsymbol{T}$.

This special result is easily generalized to the case in which $\boldsymbol{T}$ is an $m \times n$ matrix; in other words, when $\boldsymbol{T} : \mathbb{R}^n \to \mathbb{R}^m$. When both $\mathbb{R}^n$ and $\mathbb{R}^m$ have the norms $\|\cdot\|_1$ then, following the steps in this example, we find that

$$\|\boldsymbol{T}\|_1 = \max_{1 \le j \le n} \sum_{i=1}^{m} |T_{ij}|.$$

Other matrix norms are also possible, depending on the choices of norms for the domain and image spaces. In Exercise 5.12, for example, you are asked to repeat this example for the case in which the two spaces are endowed with the max-norm.

17. Let $T = d/dx : C^1[0,1] \to C[0,1]$ with the sup-norm defined on $C^1$ and $C$. $T$ is *not* a bounded operator; to show this, we need only consider $u(x) = \sin nx$. Then $\|u\| = 1$ and $\|du/dx\| = \|n \cos nx\| = n$. It follows that $\|Tu\|$ can take on arbitrarily large values (for any chosen constant $K$, we simply choose $n$ big enough to invalidate the statement $\|Tu\| = n < K$). This result may be extended in an obvious way to show that all ordinary and partial differential operators are unbounded in the sup-norm.

The connection between bounded and continuous linear operators is one that is exploited very often. Suppose that $T : X \to Y$ is a bounded linear operator; then there exists $K > 0$ such that

$$\|Tu - Tv\| = \|T(u - v)\| \le K\|u - v\|, \quad u, v \in X.$$

Given any $\epsilon > 0$, we set $\delta = \epsilon/K$ to obtain

$$\|Tu - Tv\| < \epsilon \quad \text{whenever} \quad \|u - v\| \le \delta.$$

Thus $T$ is continuous if it is bounded. Now suppose instead that $T$ is continuous; then, with $u_0 = 0$ and $\epsilon = 1$ in (5.2) we can always find a $\delta > 0$ such that

$$\|Tu\| < 1 \quad \text{whenever} \quad \|u\| < \delta.$$

In particular, assume that $u \neq 0$ (the case $u = 0$ is trivial), and set $z = \delta u / 2\|u\|$; then $\|z\| = \delta/2$, hence $\|Tz\| < 1$, and so

$$1 > \|Tz\| = \tfrac{1}{2}\|T(\delta u/\|u\|)\|.$$

That is,

$$\|Tu\| < \delta^{-1}\|u\|.$$

For the case $u_0 = 0$ we have $Tu_0 = 0$ and so $\|Tu_0\| \leq \delta^{-1}\|u_0\|$. Thus $\|Tu\| \leq \delta^{-1}\|u\|$, and so $T$ is bounded. We thus have the following theorem.

THEOREM 3. *A linear operator $T$ from a normed space $X$ to a normed space $Y$ is continuous if and only if it is bounded.*

Theorem 3 is very useful when it needs to be shown that an operator is continuous; it is frequently more convenient to show boundedness, which in turn implies continuity.

## Example

18. Let $T : C[a,b] \to C[a,b]$ be defined by $Tu(x) = \int_a^b (x+\xi)u(\xi)\, d\xi$. Using the sup-norm we have, assuming $b > a > 0$,

$$\begin{aligned} |Tu(x)| &= \left| x\int_a^b u(\xi)\, d\xi + \int_a^b \xi u(\xi)\, d\xi \right| \\ &\leq \left| x\int_a^b u(\xi)\, d\xi \right| + \left| \int_a^b \xi u(\xi)\, d\xi \right| \\ &\leq |b| \sup |u(\xi)|\, |b-a| + |b| \sup |u(\xi)|\, |b-a| \end{aligned}$$

(see Theorem 2.2). Hence $\|Tu\|_\infty \leq 2b(b-a)\|u\|_\infty$ and so $T$ is bounded and consequently continuous.

It is a simple but nonetheless extremely important fact that if a linear operator $T$ is continuous, then for any convergent sequence $\{u_n\}$ in its domain, $T(\lim_{n\to\infty} u_n)$ and $\lim_{n\to\infty} T(u_n)$ yield the same result. Note that this says two things: if $u_n \to u$, then the sequence $\{T(u_n)\}$ in the range of $T$ converges, and the limit is the same as $T(u)$.

THEOREM 4. *Let $A : X \to Y$ be a bounded linear operator and let $\{u_n\}$ be a convergent sequence in $X$ with limit $u$. Then $Au_n \to Au$ in $Y$.*

PROOF. We have

$$\|Au_n - Au\|_Y = \|A(u_n - u)\|_Y \leq \|A\|\, \|u_n - u\|_X.$$

Hence

$$\lim_{n\to\infty} \|Au_n - Au\|_Y \leq \|A\| \lim_{n\to\infty} \|u_n - u\|_X = 0;$$

that is, $Au_n \to Au$. □

**Open mappings.** In Theorem 2 it was seen that a continuous operator $T$ may be characterized by the property that the inverse image of an open set in $R(T)$, the range of $T$, is itself an open set. This does not of course imply that $T$ maps open sets to open sets; for example, the operator $T$ defined by

$$T : (0, 2\pi) \to \mathbb{R}, \quad T(x) = \sin x \tag{5.5}$$

is continuous, but it maps the *open set* $(0, 2\pi)$ onto the *closed set* $[-1, 1]$.

Those special operators that do map open sets to open sets are called open mappings; more formally, an operator $T : X \to Y$, where $X$ and $Y$ are normed spaces, is called an *open mapping* if the image of every open set in $X$ is an open set in $Y$.

The question remains: under what conditions is an operator an open mapping? The answer is provided by one of the important theorems in functional analysis, the open mapping theorem. We now state this theorem, omitting the rather technical and lengthy proof.

THEOREM 5 (THE OPEN MAPPING THEOREM). *Let $X$ and $Y$ be Banach spaces, and let $T : X \to Y$ be a bounded linear operator from $X$ onto $Y$. Then $T$ is an open mapping.*

Note the essential ingredients: $X$ and $Y$ have to be Banach spaces, and $T$ must be bounded and surjective. In the example (5.5) the first requirement was met, but not the second, in that $T$ was not surjective.

The open mapping theorem has as a consequence a result that proves very useful when we study the existence of solutions to boundary value problems in Chapter 8. This is the so-called *Banach theorem.*

THEOREM 6 (THE BANACH THEOREM). *A bounded linear one-to-one operator $T$ from a Banach space $X$ onto a Banach space $Y$ has a continuous inverse $T^{-1}$.*

PROOF. Since $T$ is one-to-one with range all of $Y$, it remains to show that $T^{-1} : Y \to X$ is continuous. Let $S$ be any open set in $Y$; then its inverse image under the mapping $T^{-1}$ is $(T^{-1})^{-1}(S) = T(S)$, since $T$ is bijective. But from the Open Mapping Theorem we know that $T(S)$ is open. Hence, by Theorem 2, $T^{-1}$ is continuous. □

## 5.3 Projections

Consider the following situation in $\mathbb{R}^3$, shown in Figure 5.9. Given any vector $\boldsymbol{x}$ we define an operator $P$ which has the property that

$$P\boldsymbol{x} = (x_1, x_2, 0).$$

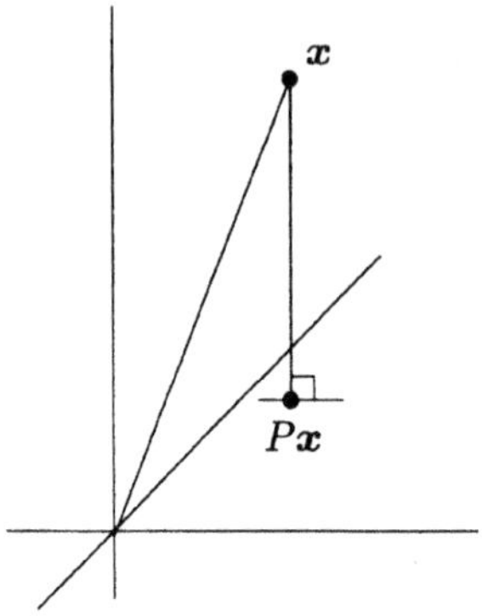


FIGURE 5.9. Projection of a point $\boldsymbol{x}$ onto the $x_1x_2$-plane

That is, $P$ projects any vector onto the $x_1x_2$-plane. It follows that if $\boldsymbol{y}$ is a vector of the form $(y_1, y_2, 0)$, then $P\boldsymbol{y} = \boldsymbol{y}$, so that $R(P) = \{\boldsymbol{y} : \boldsymbol{y} = (\alpha, \beta, 0)\}$ and $N(P) = \{\boldsymbol{y} : \boldsymbol{y} = (0, 0, \gamma)\}$, where $\alpha, \beta$, and $\gamma$ are real numbers. Furthermore, the only vector common to $R(P)$ and $N(P)$ is the zero vector. More generally, $P$ has the property that $P^2\boldsymbol{x} \equiv P(P\boldsymbol{x}) = P\boldsymbol{x}$ for all points $\boldsymbol{x}$ in $\mathbb{R}^3$. This is a simple and standard example of a projection operator on $\mathbb{R}^3$; we now generalize to arbitrary vector spaces.

**Projection operators.** A linear operator $P : X \to X$, where $X$ is a vector space is called a projection operator, or simply a projection, if

$$P^2 = P; \quad \text{that is,} \quad P(Pu) = Pu \quad \text{for all } u \in X.$$

**Example**

19. Let $X = C[0,1]$ and define the operator $P : C[0,1] \to C[0,1]$ by

$$Pu = u(0)(1-x) + u(1)x.$$

That is, $P$ maps a continuous function to its *linear interpolate*, as shown in Figure 5.10. To see that $P$ is a projection operator, note that

$$\begin{aligned} P(Pu) &= Pv, \quad \text{where } v = Pu = u(0)(1-x) + u(1)x \\ &= v(0)(1-x) + v(1)x = u(0)(1-x) + u(1)x = Pu. \end{aligned}$$

The characterization of the range and null space of a projection operator carries over in an obvious way from projections on $\mathbb{R}^3$ to those on vector spaces in general, and the main ideas are embodied in the following result.

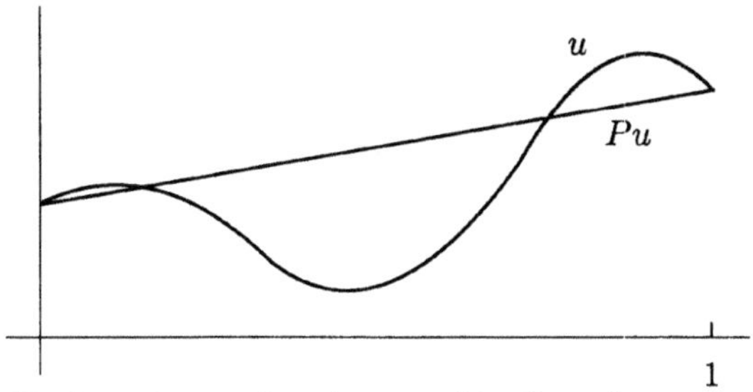


FIGURE 5.10. A continuous function $u$ and its linear interpolate $Pu$

THEOREM 7. *Let $P : X \to X$ be a linear projection operator on a vector space $X$. Then $R(P) \cap N(P) = \{0\}$ and every member of $X$ has the unique representation $u = v + w$ for some $v \in R(P)$ and $w \in N(P)$. That is, $X = R(P) \oplus N(P)$.*

PROOF. We recall from Section 3.1 the definition of the direct sum $U \oplus V$ of two subspaces $U, V$ of a vector space $X$: $X = U + V$ and $U \cap V = \{0\}$. Suppose then that $u \in R(P) \cap N(P)$. Then since $u \in R(P)$ there is a $v \in X$ such that $Pv = u$. Hence $Pv = P^2 v = Pu = 0$, since $u \in N(P)$ as well. Thus $u = 0$.

To show that $R(P) + N(P) = X$, let $u$ be any member of $X$, and let $Pu = v$. If we set $w = u - v$, then $Pw = Pu - Pv = P(Pu - Pv) = P(v - Pv) = Pv - Pv = 0$. Hence $u = v + w$ with $v \in R(P)$ and $w \in N(P)$. Thus $X = R(P) \oplus N(P)$. The uniqueness of the representation follows from Theorem 3.1. □

### Example

20. Let $X = C[-1, 1]$ and define $P$ to be the projection that maps any $X \in C[-1, 1]$ to its *even part* (Figure 5.11):

$$Pu = v, \text{ where } v(x) = \tfrac{1}{2}[u(x) + u(-x)].$$

The range of $P$ is then

$$R(P) = \{v \in C[-1, 1] : v(x) = v(-x)\},$$

that is, the space of all *even* functions, whereas the null space of $P$ is the space of all *odd* functions:

$$N(P) = \{v \in C[-1, 1] : v(x) = -v(-x)\}.$$

Clearly $X = R(P) \oplus N(P)$, since every continuous function can be represented as the sum of an odd and an even function, and furthermore the only function that is in $R(P) \cap N(P)$ is the zero function.

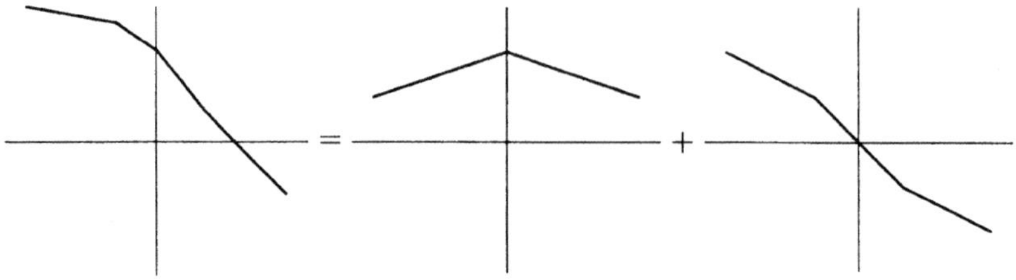

the function $v$ even part $Pv$ odd part $v - Pv$

FIGURE 5.11. Decomposition of a continuous function into its odd and even parts

**Orthogonal projections.** There is a further property of projection operators on $\mathbb{R}^n$ that is easily generalized if the vector space is also an *inner product space*: this is the concept of an orthogonal projection. We define an *orthogonal projection operator* to be a projection on an inner product space $X$ with the property that

$$R(P) \perp N(P);$$

that is, $(u, v) = 0$ for $u \in R(P)$ and $v \in N(P)$. The situation in $\mathbb{R}^3$ is obvious, as Figure 5.9 shows; if $P$ is the projection operator that maps vectors onto the $xy$-plane, then $R(P)$ is the $xy$-plane, $N(P)$ is the $z$ axis, and $R(P) \perp N(P)$.

Since we now have at our disposal a normed space (the norm being generated by the inner product), it is natural to enquire into the continuity of projection operators. We have the following result.

THEOREM 8. *An orthogonal projection $P : X \to X$ on an inner product space $X$ is continuous.*

PROOF. Any $u \in X$ can be expressed in the form $u = v + w$, where $v \in R(P)$ and $w \in N(P)$. Furthermore, $(v, w) = 0$, so it follows (why?) that $\|u\|^2 = \|v\|^2 + \|w\|^2$, hence $\|Pu\|^2 = \|v\|^2 \leq \|u\|^2$. Thus $P$ is bounded, hence continuous. □

Up to now we have discussed the situation that obtains when we are given a projection operator. What of the converse situation? Suppose we are given a subspace $Y$ of an inner product space $X$. Is it possible to define an orthogonal projection $P$ with the property that $R(P) = Y$? The answer lies in a logical extension to Theorem 8 of Chapter 4, the Projection Theorem, as we now show.

THEOREM 9. *Let $Y$ be a closed linear subspace of a Hilbert space $H$. Then* *there is exactly one orthogonal projection $P: H \to H$ with $R(P) = Y$. Moreover, $N(P) = Y^\perp$.*

PROOF. We know that $H = Y + Y^\perp$ and that every $u \in H$ has the unique representation

$$u = v + w, \quad \text{where } v \in Y, \ w \in Y^\perp.$$

Now define $P: H \to H$ by

$$P(v + w) = v;$$

that is, $P$ is a projection with range $R(P) = Y$. It is not difficult to show that $P$ is an orthogonal projection; so all that remains is to show that $P$ is unique. Let $Q$ be another orthogonal projection with $R(Q) = Y$. Now, by Theorem 8 of Chapter 4, $N(Q) = Y^\perp$, and

$$\begin{aligned} Pv = v = Qv \quad & \text{for } \ v \in Y, \\ Pw = 0 = Qw \quad & \text{for } \ w \in Y^\perp. \end{aligned}$$

Hence $P(v + w) = Q(v + w)$ for all $u = v + w \in H$; in other words, $P = Q$. □

So provided that $H$ is a *Hilbert* space, for any given closed subspace $Y$ it is possible to set up a unique orthogonal projection $P$ onto $Y$. By Theorem 8 of Chapter 4, we then have

$$H = R(P) \oplus N(P) = Y \oplus Y^\perp.$$

Note the close relationship between Theorem 9 and the Projection Theorem.

This section is concluded with a similar extension of Theorem 7, also from Chapter 4.

THEOREM 10. *Let $Y$ be a closed subspace of a Hilbert space $H$, and let $P$ be the orthogonal projection from $H$ onto $Y$. Then for any $u_0 \in H$*

$$\|u_0 - Pu_0\| \le \|u_0 - v\| \quad \textit{for all} \ \ v \in Y;$$

*that is,*

$$\|u_0 - Pu_0\| = \min_{v \in Y} \|u_0 - v\|.$$

PROOF. Since $H = R(P) \oplus N(P)$ from Theorem 7, clearly $u_0 - Pu_0 \in N(P) = Y^\perp$. Hence

$$(u_0 - Pu_0, Pu_0 - v) = 0 \quad \text{for all} \ \ v \in Y$$

since $Pu_0 - v \in Y$ also, and so

$$\begin{aligned} \|u_0 - v\|^2 &= \|u_0 - Pu_0 + Pu_0 - v\|^2 \\ &= \|u_0 - Pu_0\|^2 + \|Pu_0 - v\|^2 \\ &\geq \|u_0 - Pu_0\|^2, \end{aligned}$$

which proves the theorem. □

**Example**

21. Consider the space $L^2(-1,1)$. Let $V$ be the subspace of $L^2(-1,1)$ consisting of all *even* functions; that is,

$$V = \{v \in L^2(-1,1) : \ v(-x) = v(x) \text{ a.e. on } (-1,1)\}.$$

Then $V$ is a subspace of $L^2(-1,1)$, and in fact $V$ is closed (show this). According to Theorem 9 there is an orthogonal projection from $P$ onto $V$; in other words, $R(P) = V$. This projection is defined by

$$P : L^2 \to L^2, \quad Pv = \tfrac{1}{2}[v(x) + v(-x)].$$

$P$ is clearly a projection; it is linear and $P^2 = P$. The null space of $P$ is the set of odd functions, as characterized in Example 20.

We easily verify that $P$ is an orthogonal projection: if $u \in R(P)$ and $v \in N(P)$, then

$$(u, v) = \int_{-1}^{1} u(x)v(x)\, dx = 0$$

since the product $uv$ is an odd function. Hence $R(P) \perp N(P)$. So from Theorem 7,

$$L^2(-1,1) = R(P) \oplus N(P);$$

that is, every function in $L^2(-1,1)$ can be represented as the unique sum of an even and an odd function that also belong to $L^2(-1,1)$.

## 5.4 Linear functionals

**Linear functionals and dual spaces.** Let $\mathbb{K}$ denote either $\mathbb{R}$ or $\mathbb{C}$; then a linear functional $\ell$ on a vector space $X$ is defined to be any linear operator that maps elements of $X$ to $\mathbb{K}$; that is, $\ell : X \to \mathbb{K}$. Now we have seen that the set $\mathcal{L}(X,Y)$ of all bounded linear operators from a normed space $X$ to a normed space $Y$ is itself a normed space, with norm defined by (5.3). When $Y = \mathbb{K}$, $\mathcal{L}(X,\mathbb{K})$ is the space of *bounded linear functionals* on $X$, and is given a special name: this is called the *dual space* of $X$, and is denoted by $X'$. That is,

$$X' = \mathcal{L}(X, \mathbb{K}),$$

and for any $\ell \in X'$,

$$\|\ell(u)\| = |\ell(u)| \leq K\|u\| \quad \text{for all} \quad u \in X.$$

The second expression states that $\ell$ is bounded and hence continuous.

It is customary, when dealing with bounded linear functionals, to denote the action of such a functional $\ell$ on an element $u$ by

$$\langle \ell, u \rangle \quad \text{instead of the usual} \quad \ell(u);$$

we adopt this custom.

Using the definition (5.3) of an operator norm we see that the norm $\|\ell\|_{X'}$ of a member of $X'$ is given by

$$\|\ell\|_{X'} = \sup \frac{|\langle \ell, v \rangle|}{\|v\|}, \quad v \neq 0.$$

Most of the time we deal with the case $\mathbb{K} = \mathbb{R}$, and it is this case that is the focus in examples.

**Examples**

22. Let $f : L^2(a,b) \to \mathbb{R}$ be defined by

$$\langle f, u \rangle = \int_a^b u(x)\ dx.$$

Then $f$ is a linear functional: $\langle f, \alpha u + \beta v \rangle = \alpha\langle f, u\rangle + \beta\langle f, v\rangle$. Furthermore, using the Cauchy–Schwarz inequality on $L^2$,

$$|\langle f, u \rangle| = \left| \int_a^b 1 \cdot u(x)\ dx \right| \leq \|1\|_{L^2}\|u\|_{L^2} = |b - a|\ \|u\|_{L^2},$$

and so $f$ is bounded, and is thus a member of the dual space $[L^2(a,b)]'$.

23. One reason for the importance of functionals is exemplified by the Dirac delta "function" $\delta$, which occurs in many branches of physics and engineering. This quantity is commonly defined to be zero everywhere, with a "spike" at the origin; that is, $\delta(x) = 0$ for $x \neq 0$, $\delta(x) \to \infty$ at $x = 0$. Furthermore, $\delta$ is assumed to have the property

$$\int_{-\infty}^{\infty} \delta(x)u(x)\ dx = u(0)$$

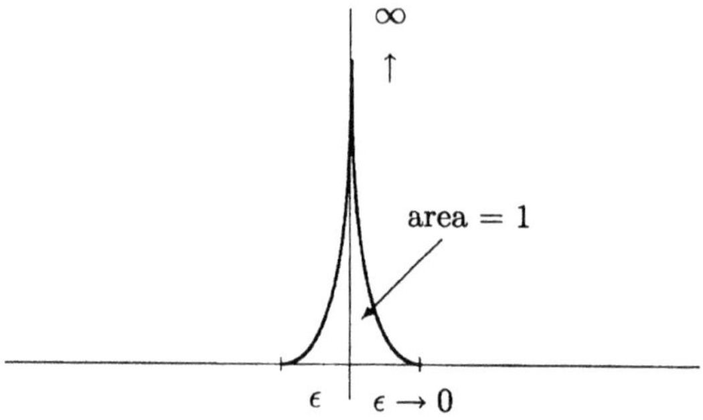


FIGURE 5.12. The Dirac delta

for any continuous function $u$ (Figure 5.12). However, it is not possible to construct a function in the ordinary sense having the properties just described. Rather, $\delta$ is more correctly defined as a *bounded linear functional* on the space of continuous functions $C[a, b]$ (here $b > 0 > a$), with

$$\delta : C(a, b) \to \mathbb{R}, \quad \langle \delta, u \rangle = u(0).$$

Thus the Dirac delta has a sampling property; it acts on a continuous function in such a way as to produce the value of that function at $x = 0$ (or at any other chosen point, with a small modification). The boundedness of $\delta$ follows from

$$|\delta(u)| = |u(0)| \leq \sup_{a \leq x \leq b} |u(x)| = \|u\|_\infty.$$

When approached in this way, there is no difficulty whatsoever in dealing with the Dirac delta; it is simply an operator that acts on a continuous function to produce its value at the origin.

**The Riesz Representation Theorem.** Suppose that $X$ is an inner product space, and $u$ a member of $X$. The inner product $(u, v)$ of $u$ with an arbitrary member $v$ of $X$ can be interpreted as the action of a functional; indeed, for given $u$ we have simply to define the functional $\ell$ by

$$\langle \ell, v \rangle = (u, v) \tag{5.6}$$

for a given $u$, and for any $v$. Linearity of $\ell$ is obvious; in addition, $\ell$ is bounded, with $\|\ell\| = \|u\|_X$. To see this, first observe that

$$|\langle \ell, v \rangle| = |(u, v)| \leq K\|v\|$$

using the Cauchy–Schwarz inequality, with $K = \|u\|$, so that $\|\ell\| \leq \|u\|$ (cf. equation (5.3)). Secondly, $|\langle \ell, u \rangle| = \|(u, u)\| = \|u\|^2$ so that $\|\ell\| \geq |(u, u)|/\|u\| = \|u\|$. Thus $\|\ell\| = \|u\|$.

A result of crucial importance in functional analysis states that for a bounded linear functional on a *Hilbert* space $H$, the converse also holds true: given $\ell$, it is always possible to find an element $u$ in $H$ such that (5.6) holds. This is the essence of the following theorem, which is regarded as one of the "big" theorems in linear functional analysis.

THEOREM 11 (RIESZ REPRESENTATION THEOREM). *Let $H$ be a Hilbert space and $\ell$ a bounded linear functional on $H$. Then there exists a unique element $u$ in $H$ such that*

$$\langle \ell, v\rangle = (v, u) \quad \textit{for all} \quad v \in H. \tag{5.7}$$

*Furthermore,*

$$\|\ell\| = \|u\|. \tag{5.8}$$

PROOF. Assume that $\ell \neq 0$ since, if $\ell = 0$, then (5.7) and (5.8) hold with $u = 0$. Also, observe that if a representation (5.7) exists, then the element $u$ must be nonzero. Second, for any $v$ in $H$ for which $\langle \ell, v\rangle = 0$ we must have $(u, v) = 0$. This implies that $u$ must be orthogonal to any member of the null space of $\ell$; that is, $u \in N(\ell)^{\perp}$. We are thus led to show the existence of $u$ by considering $N(\ell)$ and $N(\ell)^{\perp}$.

Now $N(\ell)$ is a *closed* subspace of $H$ (see Exercise 5.16). Furthermore, since $\ell \neq 0$ by assumption, $N(\ell) \neq H$ (if $N(\ell)$ were equal to $H$ this would imply that $\langle \ell, v\rangle = 0$ for all $v \in H$). Thus $N(\ell)$ is a proper subset of $H$ and so, by the Projection Theorem (Theorem 8, Chapter 4), $N(\ell)^{\perp} \neq \{0\}$. Hence there must be at least one nonzero element, $u_0$, say, in $N(\ell)^{\perp}$. Set

$$z = \langle \ell, v\rangle u_0 - \langle \ell, u_0\rangle v \quad \text{for any } v \in H;$$

then

$$\langle \ell, z\rangle = \langle \ell, v\rangle\langle \ell, u_0\rangle - \langle \ell, u_0\rangle\langle \ell, v\rangle = 0$$

and so $z \in N(\ell)$. Also, since $u_0 \in N(\ell)^{\perp}$ we have

$$0 = (u_0, z) = (u_0, \langle \ell, v\rangle u_0 - \langle \ell, u_0\rangle v) = \langle \ell, v\rangle(u_0, u_0) - \langle \ell, u_0\rangle(u_0, v)$$

which implies that

$$\langle \ell, v\rangle = \frac{\langle \ell, u_0\rangle(u_0, v)}{\|u_0\|^2}. \tag{5.9}$$

Finally, we set

$$u = \frac{\langle \ell, u_0\rangle}{\|u_0\|^2} u_0, \tag{5.10}$$

and from (5.9) we see that the element $u$ defined by (5.10) satisfies (5.7). The existence of $u$ has been proved.

The proof that $u$ is unique, and the derivation of (5.8), are more straightforward than the existence proof, and are left as exercises (see Exercise 5.30). □

In practice it is in many instances a difficult task to construct the element $u$ related to a linear functional $\ell$ by (5.7); but there are situations in which this can be done, and we give some examples.

### Examples

24. Let $f$ be a linear functional on $\mathbb{R}^n$; then $\langle f, \boldsymbol{x}\rangle$ is a real number and according to Theorem 11 we can always find a point $\boldsymbol{y} \in \mathbb{R}^n$ such that

$$\langle f, \boldsymbol{x}\rangle = \boldsymbol{x} \cdot \boldsymbol{y}.$$

For example, if $f$ is defined by

$$\langle f, \boldsymbol{x}\rangle = x_1 + \cdots + x_n \quad \text{for } \boldsymbol{x} = (x_1, \dots, x_n) \in \mathbb{R}^n,$$

then we may find $\boldsymbol{y} = (y_1, \dots, y_n)$ from

$$y_i = \boldsymbol{y} \cdot \boldsymbol{e}_i = \langle f, \boldsymbol{e}_i\rangle, \quad i = 1, \dots, n,$$

where $\mathbf{e}_i$ is the standard $i$th basis element of $\mathbb{R}^n$. Hence

$$y_i = 0 + \cdots + 1 + \cdots + 0 = 1 \text{ and so } \boldsymbol{y} = (1, 1, \dots, 1).$$

25. Let $\ell$ be a linear functional on $L_2(0,1)$ defined by

$$\ell : L^2 \to \mathbb{R}, \quad \langle \ell, v\rangle = \int_0^{1/2} v(x)\, dx.$$

Then according to Theorem 11 there exists a unique $u \in L^2(0,1)$ with the property that

$$\langle \ell, v\rangle = (u, v) \quad \text{or} \quad \int_0^1 u(x)v(x)\, dx = \int_0^{1/2} v(x)\, dx.$$

Clearly $u(x)$ is the function

$$u(x) = \begin{cases} 1, & 0 < x \le \frac{1}{2} \\ 0, & \frac{1}{2} < x < 1. \end{cases} \tag{5.11}$$

We have seen from the Riesz Representation Theorem that there is a *unique one-to-one correspondence* between bounded linear functionals on a Hilbert space $H$ and members of $H$. In other words, it is possible to set up a *bijective linear map* (in other words, one that is one-to-one with range all of $H$) $J : H \to H'$, called the *Riesz map* of $H$. In addition, from (5.8), $\|u\| = \|f\| = \|Ju\|$. Thus any Hilbert space $H$ and its dual $H'$ are *isometrically isomorphic* to each other, and may be identified with each other in this sense. This correspondence is written

$$H \sim H',$$

meaning that $H$ is identified with its dual.

**Example**

26. In view of the preceding remarks the correspondence

$$[L^2(\Omega)]' \sim L^2(\Omega) \tag{5.12}$$

exists. For example, if $\ell : L^2(0,1) \to \mathbb{R}$ with

$$\langle \ell, v \rangle = \int_0^{1/2} v(x)\, dx,$$

then according to Example 24, $u$ given by (5.11) is the unique member of $L^2(0,1)$ corresponding to $\ell$; that is, $\ell$ is identified with $u$, and $\|\ell\| = \|u\|_{L^2}$.

**The dual space of $L^p(\Omega)$.** The equivalence (5.12) has its counterpart in the case of the spaces $L^p(\Omega)$ which, with the exception of $p = 2$, are not Hilbert spaces. For any $p$ such that $1 \le p < \infty$, set $q = p/(p-1)$ (that is, $1/p + 1/q = 1$, with the usual rule $1/0 = \infty$). Note that the case $p = \infty$ is excluded.

Now let $g$ be any function in $L^q$ and define a functional $l_g$ on $L^p$ according to

$$\langle \ell_g, f \rangle = \int_\Omega f\bar{g}\, dx \quad \text{for all} \quad f \in L^p.$$

Then, using Hölder's inequality (3.13),

$$\langle \ell_g, f \rangle \le \|f\|_p \|g\|_q$$

(here the $L^p$-norm is denoted by $\|\cdot\|_p$) so that $\ell_g$ is a bounded linear functional on $L^p$, with

$$\|\ell_g\| \le \|g\|.$$

It is possible to go one step further, and to show that in fact

$$\|\ell_g\| = \|g\|;$$

this step is treated in Exercise 5.31.

Now it is clear from this argument that it is possible to associate with each $g$ in $L^q$ a bounded linear functional $l_g$ on $L^p$. What of the converse? Is it the case that *every* bounded linear functional on $L^p$ has the form $\ell_g$ for some $g \in L^q$? The answer is affirmative, and is embodied in a result known as Riesz's Theorem.

THEOREM 12 (RIESZ'S THEOREM). *Let $\ell$ be a bounded linear functional on $L^p(\Omega)$, for $1 \le p < \infty$. Then there is a function $g$ in $L^q(\Omega)$ such that*

$$\langle \ell, f\rangle = \int_\Omega f\bar{g}\, dx \quad \text{for all} \quad f \in L^p. \tag{5.13}$$

Furthermore,

$$\|\ell\|_{[L^p]'} = \|g\|_{L^q}.$$

Though Theorem 12 can be proved using techniques that are elementary, the proof is rather lengthy and is omitted. What is important, though, is to note that $[L^p]'$ and $L^q$ are *isometrically isomorphic*; that is,

$$[L^p(\Omega)]' \sim L^q(\Omega), \quad \frac{1}{p} + \frac{1}{q} = 1, \quad 1 \le p < \infty.$$

Although in particular $[L^1(\Omega)]' \sim L^\infty(\Omega)$, this relationship does not commute; in fact it can be shown that $[L^\infty(\Omega)]'$ is larger than $L^1(\Omega)$.

**Weak and weak* convergence.** The existence of bounded linear functionals on a normed space $X$ makes it possible to introduce an alternative form of convergence of sequences in $X$. Let $\{u_n\}$ be a sequence in $X$; then this sequence is said to *converge weakly* to a member $u$ of $X$ if

$$\langle \ell, u_n\rangle \to \langle \ell, u\rangle \quad \text{as } n \to \infty, \quad \text{for all } \ell \in X'.$$

When this is the case, $u$ is referred to as the weak limit of $\{u_n\}$, and we write $u_n \rightharpoonup u$.

When wishing to distinguish between weak convergence and the conventional notion of convergence defined in (4.2), the latter is referred to as *strong* or *norm* convergence. There is a difference between strong and weak convergence; on the one hand *strong convergence implies weak convergence*: indeed, if $u_n \to u$ (strongly), then for any bounded linear functional $\ell$

$$|\langle \ell, u_n - u\rangle| \le \|\ell\|\, \|u_n - u\| \to 0 \text{ as } n \to \infty.$$

However, the converse is not true. Take, for example, the function $u(x) = 0$ and the sequence defined by $u_n(x) = \cos nx$, on the interval $[0, 2\pi]$; then $u_n \in L^2(0, 2\pi)$ (for example), and it can be shown that $u_n \rightharpoonup 0$. On the other hand, $\|u_n - 0\|_{L^2} = \pi$ for all values of $n$, so $u_n$ does not converge strongly to 0.

It follows from the Riesz Representation Theorem that in a Hilbert space $H$, $u_n \rightharpoonup u$ if and only if $(u_n, v) \to (u, v)$ for all $v \in H$.

In the context of the spaces $L^p$, the notion of weak convergence takes on a more concrete form in the light of Riesz's Theorem. Indeed, the correspondence (5.13) implies that a sequence $\{u_n\}$ in $L^p$ $(1 \leq p < \infty)$ is weakly convergent with limit $u$ if and only if

$$\lim_{n\to\infty} \int_\Omega u_n \bar{g}\, dx = \int_\Omega u \bar{g}\, dx \quad \text{for all } g \in L^q(\Omega).$$

We mention briefly also the concept of weak* convergence, which applies to sequences of bounded linear functionals. Suppose that $X$ is a normed space with dual $X'$; then a sequence $\{\ell_n\}$ in $X'$ is said to converge weakly* to an element $\ell$ in $X'$ if

$$\langle \ell_n, u \rangle \to \langle \ell, u \rangle \quad \text{as } n \to \infty,$$

for all $u \in X$. In this case we write $\ell_n \overset{*}{\rightharpoonup} \ell$.

## 5.5 Bilinear forms

Another special type of operator that occurs very frequently in the study of boundary value problems is one that maps a *pair* of elements to the real or complex numbers, and which is linear in each of its slots. This called a *bilinear form*. Since we deal exclusively with real-valued bilinear forms, the definition and examples are restricted to this case. From the discussion of linear functionals it should be clear, though, that the extension to complex-valued forms is immediate.

If $X$ and $Y$ are vector spaces, a bilinear form $a : X \times Y \to \mathbb{R}$ is defined to be an operator with the properties

$$\begin{aligned} a(\alpha u + \beta w, v) &= \alpha a(u, v) + \beta a(w, v), \quad u, w \in X,\ v \in Y, \\ a(u, \alpha v + \beta w) &= \alpha a(u, v) + \beta a(u, w), \quad u \in X,\ v, w \in Y, \end{aligned} \tag{5.14}$$

where $\alpha$ and $\beta$ are real numbers.

**Examples**

27. Let $X = Y = \mathbb{R}^3$, and let $\boldsymbol{A}$ be any $3 \times 3$ matrix; then the operator defined by $a(\boldsymbol{x}, \boldsymbol{y}) = \boldsymbol{x} \cdot \boldsymbol{A}\boldsymbol{y}$ is a bilinear form. In particular, for any

inner product space $X$ the inner product $(\cdot,\cdot) : X \times X \to \mathbb{R}$ is a bilinear form. An example of a *nonlinear* form is

$$a : \mathbb{R}^3 \times \mathbb{R}^3 \to \mathbb{R}, \quad a(\boldsymbol{x}, \boldsymbol{y}) = |\boldsymbol{x}| + |\boldsymbol{y}|;$$

here, in general,

$$a(\alpha \boldsymbol{x} + \beta \boldsymbol{z}, \boldsymbol{y}) = |\alpha \boldsymbol{x} + \beta \boldsymbol{z}| + |\boldsymbol{y}| \neq \alpha a(\boldsymbol{x}, \boldsymbol{y}) + \beta a(\boldsymbol{z}, \boldsymbol{y})$$

(this last expression being equal to $\alpha(|\boldsymbol{x}| + |\boldsymbol{y}|) + \beta(|\boldsymbol{z}| + |\boldsymbol{y}|)$).

28. Let $X = Y = C^1[a,b]$. Then the operator defined by

$$a : C^1[a,b] \times C^1[a,b] \to \mathbb{R}, \quad a(u,v) = \int_a^b (uv + u'v')\, dx$$

is a bilinear form.

**Continuous bilinear forms.** Suppose that we are given a bilinear form $a : X \times Y \to \mathbb{R}$, where now $X$ and $Y$ are *normed* linear spaces. Consider the expression $a(u,v)$; if there is a positive number $K$ such that

$$|a(u,v)| \leq K\|u\|\,\|v\| \quad \text{for all} \quad u \in X, v \in Y, \tag{5.15}$$

then $a$ is called a *continuous* bilinear form (this definition should be compared with that of bounded operators, or bounded linear functionals). Later on it is shown that differential equations have associated bilinear forms, and in order for problems to be well-posed in a certain sense it is essential that these bilinear forms be continuous.

### Example

29. We discuss here an example that is typical of a class of problems that appears later. Denote by $H^1(c,d)$ the vector space of functions that together with their first derivatives, are square-integrable on the interval $(c,d)$. The reason for the notation $H^1$ becomes clear in Chapter 7, where it is also shown that $H^1(c,d)$ is in fact a Hilbert space if it is endowed with the inner product $(u,v)_{H^1} = \int_c^d [uv + u'v']\, dx$; the associated norm is then given by

$$\|u\|_{H_1}^2 = (u,u)_{H^1} = \int_c^d [u^2 + (u')^2]\, dx = \|u\|_{L^2}^2 + \|u'\|_{L^2}^2. \tag{5.16}$$

Now define the bilinear form $a(\cdot,\cdot)$ by

$$a : H^1(c,d) \times H^1(c,d) \to \mathbb{R}, \quad a(u,v) = \int_c^d [u'v' + \kappa uv]\, dx,$$

where $\kappa(x)$ is a bounded continuous function satisfying $\kappa_1 \geq \kappa(x) \geq \kappa_2 > 0$ for $x \in (c, d)$ and constants $\kappa_1$ and $\kappa_2$. To show that $a$ is continuous, consider

$$
\begin{aligned}
|a(u,v)| &= \left| \int_c^d [u'v' + \kappa uv] \, dx \right| \\
&\leq \left| \int_c^d u'v' \, dx \right| + \int_c^d |\kappa uv| \, dx \\
&\leq \left| \int_c^d u'v' \, dx \right| + \kappa_1 \int_c^d |u| \, |v| \, dx \\
&= |(u', v')_{L^2}| + \kappa_1 (|u|, |v|)_{L^2} \\
&\leq \|u'\|_{L^2} \|v'\|_{L^2} + \kappa_1 \|u\|_{L^2} \|v\|_{L^2} \\
&\quad \text{(using the Cauchy–Schwarz inequality).}
\end{aligned}
$$

Now from (5.16) it is clear that $\|u\|_{L^2} \leq \|u\|_{H^1}$, and likewise $\|u'\|_{L^2} \leq \|u\|_{H^1}$; thus

$$
\begin{aligned}
|a(u,v)| &\leq \|u\|_{H^1} \|v\|_{H^1} + \kappa_1 \|u\|_{H^1} \|v\|_{H^1} \\
&= (1 + \kappa_1) \|u\|_{H^1} \|v\|_{H^1}
\end{aligned}
$$

so that $a(\cdot, \cdot)$ is continuous, with constant $K = 1 + \kappa_1$.

In practice it is always desirable to find the *smallest* constant $K$ for which the inequality (5.15) holds. In this example it is possible to find a better constant $K = \max(1, \kappa_1)$. This is discussed further in Example 33.

**H-elliptic bilinear forms.** Given a bilinear form $a : H \times H \to \mathbb{R}$, where $H$ is an inner product space, we say that $a$ is *$H$-elliptic* if there exists a constant $\alpha > 0$ such that

$$
a(v, v) \geq \alpha \|v\|_H^2 \quad \text{for all} \quad v \in H.
$$

Thus an $H$-elliptic form is one that is always nonnegative, and takes the value 0 only for the case in which $v = 0$. In other words, it is *positive-definite.*

**Example**

30. Consider Example 29 again; this bilinear form is $H^1$-elliptic since

$$\begin{aligned} |a(v,v)| &= \left| \int_c^d [(v')^2 + \kappa v^2]\, dx \right| \\ &\geq \left| \int_c^d [(v')^2 + \kappa_2 v^2] dx \right| \\ &\geq \alpha \left| \int_c^d [(v')^2 + v^2] dx \right| = \alpha \|v\|_{H^1}^2, \end{aligned}$$

where $\alpha = \min(1, \kappa_2)$.

The Riesz Representation Theorem for linear functionals has a counterpart for bilinear forms that proves useful later. Suppose that we are given a real inner product space $H$ and a continuous, $H$-elliptic bilinear form $a$ on $H$; then for any given $u \in H$ it is possible to define a bounded linear functional $\ell$ on $H$ according to the rule

$$a : H \times H \to \mathbb{R}, \quad a(u,v) = \langle \ell, v \rangle \quad \text{for all } v \in H. \tag{5.17}$$

From the continuity of $a$ we find that

$$\|\ell\| \leq K \|u\|, \tag{5.18}$$

in which $K$ is the constant appearing in (5.15). Furthermore, from the $H$-ellipticity of $a(\cdot,\cdot)$ we have

$$\alpha \|u\|^2 \leq a(u,u) \leq \langle \ell, u \rangle$$

or

$$\|u\| \leq (1/\alpha) \|\ell\| \tag{5.19}$$

In this sense, then, $u$ and $a$ generate the bounded linear functional $\ell$. We now prove the converse assertion: namely, given a bilinear form $a$ and a linear functional with suitable properties, there exists a unique $u \in H$ satisfying (5.17). This is the *Lax-Milgram theorem.*

THEOREM 13 (THE LAX–MILGRAM THEOREM). *Let $H$ be a Hilbert space and let $a : H \times H \to \mathbb{R}$ be a continuous, $H$-elliptic bilinear form defined on $H$. Then, given any continuous linear functional $\ell$ on $H$, there exists a unique element $u$ in $H$ such that (5.17) and (5.19) hold for all $v \in H$.*

The proof of this theorem is rather lengthy, and is made more digestible by breaking it up into a series of five lemmas.

LEMMA 1. *Given any $u$ in $H$ there is a unique element $w$ in $H$ such that*

$$a(u,v) = (w,v) \quad \text{for all} \quad v \in H. \tag{5.20}$$

PROOF. Given any $u \in H, a(u,\cdot)$ is a bounded linear functional on $H$ since

$$a(u,\cdot) : H \to \mathbb{R}, \quad |a(u,v)| \le K'\|v\|,$$

where $K' = K\|u\|$. Hence, according to the Riesz Representation Theorem there is a unique element $w$ in $H$ such that $a(u,v) = (w,v)$. □

LEMMA 2. *Let $A$ be the operator that associates $u$ with $w$:*

$$A : H \to H, \quad Au = w. \tag{5.21}$$

*Then $A$ is a bounded linear operator.*

PROOF. Let $w_1, w_2 \in H$; then according to Lemma 1 there are elements $u_1$ and $u_2$ in $H$ that satisfy

$$a(u_1,v) = (w_1,v), \quad a(u_2,v) = (w_2,v) \quad \text{for all} \quad v \text{ in } H.$$

Since $a$ is bilinear,

$$\begin{aligned} a(\alpha u_1 + \beta u_2, v) &= \alpha a(u_1,v) + \beta a(u_2, v) \\ &= (\alpha w_1 + \beta w_2, v). \end{aligned} \tag{5.22}$$

Furthermore, from the definition of $A$ we have

$$Au_1 = w_1, \quad Au_2 = w_2 \text{ so that } \alpha Au_1 + \beta Au_2 = \alpha w_1 + \beta w_2. \tag{5.23}$$

But from (5.20) through (5.22) we see that $A$ maps $\alpha u_1 + \beta u_2$ to $\alpha w_1 + \beta w_2$:

$$A(\alpha u_1 + \beta u_2) = \alpha w_1 + \beta w_2.$$

The linearity of $A$ thus follows. $A$ is bounded since, choosing $u \neq 0$, setting $v = Au$ in (5.20), and using the continuity of $a$ and (5.21),

$$K\|u\|\|Au\| \ge a(u, Au) = (w, Au) = \|Au\|^2 \Rightarrow \|Au\| \le K\|u\|.$$

If $u = 0$, we simply choose $w = 0$. □

LEMMA 3. *$A$ is one-to-one with bounded inverse $A^{-1}$.*

PROOF. Let $R(A)$ denote the range of $A$ (of course, $R(A) \subset H$). We show that $Az = 0$ only for $z = 0$ and use Theorem 1 to show that $A$ is one-to-one. Let $z$ be such that $Az = 0$. Then, since $A$ by definition maps $z$ to a member $Az$ of $H$ such that $a(z, v) = (Az, v)$ we have

$$a(z, v) = (0, v) = 0 \quad \text{for all} \quad v \in H.$$

In particular, for $v = z$,

$$0 = a(z, z) \geq \alpha \|z\|^2$$

so that $\|z\| = 0$ or $z = 0$. Hence $A$ is one-to-one, and its inverse $A^{-1} : R(A) \to H$ exists. Furthermore, $A^{-1}$ is linear since $A$ is linear (see Exercise 5.9), and $A^{-1}$ is bounded since

$$\alpha \|u\|^2 \leq a(u, u) = (w, u) \leq \|w\| \, \|u\| \quad \text{(using the Cauchy–Schwarz inequality)}$$

whence $\|A^{-1}w\| = \|u\| \leq \alpha^{-1}\|w\|$. □

LEMMA 4. *$R(A)$ is a complete space.*

PROOF. Let $\{w_k\}$ be a Cauchy sequence in $R(A)$. Since $R(A)$ is a subset of $H$, $\{w_k\}$ is a Cauchy sequence in $H$ too, and so it converges in $H$; that is,

$$\lim_{k \to \infty} \|w_k - w\| = 0 \text{ in } H.$$

It is necessary to show that $w$ is in $R(A)$. To do this, let $u_k$ be defined by $Au_k = w_k$. Then

$$\begin{aligned} \|u_k - u_l\| &= \|A^{-1}w_k - A^{-1}w_l\| = \|A^{-1}(w_k - w_l)\| \\ &\leq \|A^{-1}\| \, \|w_k - w_l\| \end{aligned}$$

so that

$$\lim_{k,l \to \infty} \|u_k - u_l\| \leq \|A^{-1}\| \lim_{k,l \to \infty} \|w_k - w_l\| = 0$$

($\{w_k\}$ is a Cauchy sequence in $H$). Hence $\{u_k\}$ is also a Cauchy sequence in $H$, with limit $u$ in $H$. Furthermore, since $Au_k = w_k$ we have

$$\lim_{k \to \infty} Au_k = \lim_{k \to \infty} w_k = w \quad \text{or} \quad w = A(\lim u_k) = Au$$

(using Theorem 4). Hence $w$ is in the range of $A$, and since $w$ is the limit of an arbitrary Cauchy sequence, $R(A)$ is complete. □

LEMMA 5. *$R(A) = H$; that is, $A$ is bijective.*

PROOF. Suppose that $R(A)$ is a proper subspace of $H$, so that there is a nonzero element $u_0$ that lies in $R(A)^\perp$ (recall the Projection Theorem, Theorem 8 of Chapter 4). Then

$$(u_0, z) = 0 \quad \text{for all} \quad z \in R(A).$$

Using Lemma 2 we set $w_0 = Au_0$ so that $w_0 \in R(A)$. Then from Lemma 1 we have $a(u_0, v) = (w_0, v)$ for all $v$ in $H$. In particular, if we set $v = u_0$, then

$$\alpha\|u_0\|^2 \le a(u_0, u_0) = (w_0, u_0) = 0 \text{ since } w_0 \in R(A),\ u_0 \in R(A)^\perp.$$

Hence $u_0 = 0$, which is a contradiction, so that $R(A)^\perp = \{0\}$ and $R(A) = H$. □

Finally, we gather together all the pieces of information to give the following.

PROOF OF THE LAX–MILGRAM THEOREM. Lemma 1 shows that for any given $u \in H$ there is a unique $w \in H$ defined by (5.20). This lemma does *not* prove the converse; indeed, we define the operator $A$ by (5.20), and in order to prove that the converse is true it is necessary to show that $A$ is *bijective*. This is done in Lemmas 3, 4, and 5. Hence we conclude that given any $w \in H$ there exists a unique $u \in H$ such that

$$a(u, v) = (w, v) \text{ for all } v \in H. \tag{5.24}$$

By the Riesz Representation Theorem, every bounded linear functional $\ell$ can be expressed in the form

$$\langle \ell, v\rangle = (w, v) \quad \text{for all} \quad v \in H, \tag{5.25}$$

with $\|\ell\| = \|w\|$. Thus (5.24) and (5.25) imply (5.17), and (5.19) follows from the $H$-ellipticity of $a$ and the continuity of $\ell$. This proves the theorem. □

## 5.6 Bibliographical remarks

Operators from one set to another are usually known as functions when the sets have no structure. Much of the material covered in Section 5.1 falls into this category, and can usually be found in texts that discuss set theory. Good accounts of the material in Section 5.1 may be found in Naylor and Sell [33], Oden [36], Kreyszig [27], and Roman [42].

Section 5.2 makes use of both algebraic and topological properties of sets. The definition of a linear operator requires only the algebraic notion of a linear space, whereas the definition of a continuous operator obviously needs a normed space. The above-mentioned texts by Naylor and Sell, Oden, and Kreyszig are good references for further reading as is Volume 2 of the pair of texts by Roman [43], and Zeidler [54]. The same applies to the material of Sections 5.3 and 5.4; these texts are all good references. A detailed account, including proofs, of the equivalence of $[L^p(\Omega)]'$ and $L^q(\Omega)$ may be found in the book by Hewitt and Stromberg [19].

A good source for discussions of bilinear forms, including the Lax–Milgram Theorem, is Rektorys [41]. This theorem has been generalized to the case of a bilinear form $B : H \times Y \to \mathbb{R}$, where $H$ and $Y$ are distinct Hilbert spaces, by Babuška (see Babuška and Aziz [3] for this result); an account of this generalization is also given by Oden [36].

## 5.7 Exercises

### Operators

5.1. Describe the range and null space of the following operators.

(a) $M : (-1, 1) \to \mathbb{R}^2, \quad M(x) = (x, \sqrt{1 - x^2})$.

(b) $K : L^2(0, 1) \to \mathbb{R}, \quad Ku = \displaystyle\int_0^1 [u(x)]^2 \, dx$.

(c) $f : (0, \pi/2) \to \mathbb{R}, \quad f(x) = \tan x$.

5.2. Find the null space of $S = \begin{pmatrix} 1 & 2 \\ 2 & 3 \end{pmatrix}$ and of $T = \begin{pmatrix} 1 & 2 & 0 \\ 2 & 3 & 4 \\ 0 & -1 & 4 \end{pmatrix}$.

5.3. Which of the following operators is one-to-one? surjective?

(a) $K : C[0, 1] \to C[0, 1], \quad Ku = \displaystyle\int_0^x u(y) \, dy$.

(b) $T : \mathbb{R}^2 \to \mathbb{R}^2, \quad T(\boldsymbol{x}) = (y, x)$.

5.4. The operator $f : M \to \mathbb{C}$ is defined by $f(z) = z^2$, where $M$ is the subset of $\mathbb{C}$ defined by $M = \{z = x + iy \in \mathbb{C} : \ xy \geq 1, \ x > 0, \ y > 0\}$. Sketch the domain of $f$ and show that the image of the curve $xy = 1$ under the mapping $f$ is the line $\operatorname{Im} f(z) = 2$. Hence illustrate the range of $f$.

5.5. Describe the compositions $ST$ and $TS$ for the operators

(a) $T : \mathbb{R}^2 \to \mathbb{R}^2, \quad S : \mathbb{R}^2 \to \mathbb{R}^2$; $T(\boldsymbol{x}) = (x, -y)$, and $S(\boldsymbol{x}) = (2y, x)$.

(b) $T : \mathbb{R} \to \mathbb{R}$, $T(x) = \sin x$ and $S : \mathbb{R} \to \mathbb{R}$, $S(x) = x^2 - 1$

5.6. Suppose that $S : U \to V$ and $T : V \to W$ are invertible operators. Show that $TS$ is invertible and that $(TS)^{-1} = S^{-1}T^{-1}$.

**Linear operators, bounded, and continuous operators**

5.7. Which of the following are linear operators?

(a) $T : L^2(-1,1) \to L^2(-1,1),\quad Tu = \int_{-1}^{1} K(x,y)u(y)\,dy$;

(b) $T : C^1[a,b] \to C[a,b],\quad Tu = x^2\partial u/\partial x + 2u$;

(c) $M : \mathbb{R}^2 \to \mathbb{R},\quad M(\boldsymbol{x}) = xy$.

5.8. An operator $\boldsymbol{T} : \mathbb{R}^n \to \mathbb{R}^n$ is called an affine transformation if $\boldsymbol{Tx} = \boldsymbol{Ax} + \boldsymbol{b}$, where $\boldsymbol{A}$ is an $n \times n$ matrix and $\boldsymbol{b}$ an $n \times 1$ vector. Find the affine transformation in $\mathbb{R}^2$ that takes the triangle with vertices at $(0,0)$, $(0,1)$, and $(1,0)$ to the triangle with vertices at $(4,5)$, $(-1,2)$, $(3,0)$.

5.9. If $T : U \to V$ is an invertible linear operator, where $U$ and $V$ are vector spaces, show that $T^{-1}$ is linear.

5.10. If $d(\cdot, B)$ in Exercise 3.18 is regarded as an operator on $\mathbb{R}^n$, is this a linear operator? What is its null space?

5.11. Show that the norm of a bounded linear operator can equivalently be defined by

$$\|T\| = \sup\{\|Tu\| : \|u\| = 1\}$$

or by

$$\|T\| = \sup\{\|Tu\| : \|u\| \leq 1\}.$$

5.12. Let $X$ be the space $\mathbb{R}^n$ with the norm $\|\boldsymbol{x}\|_\infty = \max_{1\leq j\leq n} |x_j|$. If $\boldsymbol{A} : X \to X$ is a linear operator represented by an $n \times n$ matrix, show that

$$\|\boldsymbol{A}\|_\infty = \max_i \sum_{j=1}^{n} |A_{ij}|.$$

Determine $\|\boldsymbol{A}\|_\infty$ if $\boldsymbol{A} = \begin{bmatrix} -2 & 4 \\ 1 & 3 \end{bmatrix}$.

5.13. The $2 \times 2$ matrix $\boldsymbol{A}$ has elements $A_{11} = A_{22} = a$, $A_{12} = A_{21} = b$, with $a > 0$ and $b > 0$. Show that $\|\boldsymbol{A}\|_2 = a + b$.

5.14. Show that the identity operator $I : X \to X$ is continuous, where $X$ is *any* normed space. If $V$ is the normed space $C^1[a,b]$ with the sup-norm, and $W$ is the space $C^1[a,b]$ with the norm $\|u\|_W = \|u\|_\infty + \|u'\|_\infty$, produce an example to show that $I : V \to W$ is not continuous.

5.15. If $T : U \to V$ and $S : V \to W$ are bounded linear operators, show that $ST : U \to W$ is bounded with $\|ST\| \leq \|S\| \, \|T\|$.

5.16. Show that the null space $N(T)$ of a linear operator $T : U \to V$ is closed if $T$ is a bounded linear operator.

5.17. An operator $T : U \to V$ is *bounded below* if there exists a constant $K$ such that

$$\|Tu\|_V \geq K\|u\|_U, \quad u \in U.$$

If $T$ is a bounded below linear operator, show that $T$ is one-to-one, and that $T^{-1} : R(T) \to U$ is a bounded operator.

5.18. Show that the operator $D : C_0^1[0,1] \to C[0,1]$, $Du = du/dx$ is bounded below, where $C_0^1$ is the space of functions in $C^1$ that are zero at $x = 0$ and $x = 1$. Use the sup-norm. [Hint: consider $u(x) = \int_0^x u'(y)\, dy$.]

**Projections**

5.19. If $P : U \to U$ is a projection operator, show that $(I - P)$ is also a projection. How are $R(I - P)$ and $N(I - P)$ related to $R(P)$ and $N(P)$?

5.20. Show that $\|P\| = 1$ if $P$ is an orthogonal projection.

5.21. Give an example of a nonlinear operator $P$ that satisfies $P^2 = P$.

5.22. Show that $N(P) = R(P)^\perp$ if $P$ is an orthogonal projection on an inner product space.

5.23. Let $T$ be the transformation defined by

$$T : L^2(\mathbb{R}) \to L^2(\mathbb{R}), \quad Tu = \begin{cases} u(x) & \text{if } |x| < 1, \\ 0 & \text{otherwise.} \end{cases}$$

Show that $T$ is an orthogonal projection. What are the range and null space of $T$?

5.24. Show that the operator $P : L^2(-1,1) \to L^2(-1,1)$ defined by

$$Pu(x) = \tfrac{1}{2} \int_{-1}^{1} e^{i(x-y)} u(y)\, dy$$

is a projection. Is $P$ an orthogonal projection?

### Linear functionals and the Riesz Representation Theorem

5.25. Let $\boldsymbol{A}$ be a positive-definite symmetric $n \times n$ matrix; that is, $\boldsymbol{x}^T \boldsymbol{A} \boldsymbol{x} > 0$ for all nonzero vectors $\boldsymbol{x}$. Then the space $\mathbb{R}^n$ is a Hilbert space when endowed with the inner product

$$(\boldsymbol{x}, \boldsymbol{y}) = \sum_{i,j=1}^{n} A_{ij} x_i y_j.$$

Given a functional $\ell : \mathbb{R}^n \to \mathbb{R}$, find the element $\boldsymbol{x}$ such that $\langle \ell, \boldsymbol{y} \rangle = (\boldsymbol{x}, \boldsymbol{y})$ when $\ell$ is defined by:

(a) $\langle \ell, \boldsymbol{y} \rangle = y_1 + y_2 + \cdots + y_n$;
(b) $\langle \ell, \boldsymbol{y} \rangle = y_1$.

5.26. For each $f \in L^2(0,1)$ let $u(x)$ be the solution of $u'' + u' - 2u = f$ with $u(0) = u(1) = 0$. Define the functional $\ell$ by

$$\ell : L^2(0,1) \to \mathbb{R}, \quad \langle \ell, f \rangle = \int_0^1 u(x)\, dx.$$

Show that $\ell$ is a bounded linear functional, and find the function $u$, the value of $\langle \ell, f \rangle$, and the element $g$ such that $\langle \ell, f \rangle = (g, f)$, when $f(x) = 2x$.

5.27. Repeat Exercise 5.26 for the differential equation $u'' - 2u' + u = f$.

5.28. If $X$ is a normed space (not necessarily complete), prove that $X'$ is a Banach space.

5.29. Where in the proof of the Riesz Representation Theorem is the completeness of the Hilbert space $H$ first used, and where is it used subsequently?

5.30. Complete the proof of Theorem 11 by showing that $u$ is unique and that $\|\ell\| = \|u\|$.

5.31. For any $p$ and $q$ such that $1 \leq p < \infty$ and $1/p + 1/q = 1$, let $g$ be a function in $L^q$ and define a functional $l_g$ on $L^p$ according to

$$\langle \ell_g, f \rangle = \int_\Omega f \bar{g}\, dx \quad \text{for all} \quad f \in L^p.$$

Show that $\|\ell_g\| = \|g\|$. [Hint: in the prelude to Theorem 12 it is shown that $\|\ell_g\| \leq \|g\|$; choose $f = |g|^{q-1} \operatorname{sgn} g$, show that $|f|^p = |g|^q$, and hence that $\|\ell_g\| \geq \|g\|$.]

5.32. If $Y$ is a dense subset of a normed space $X$ and $\ell$ is a member of $X'$, show that

$$\langle \ell, v \rangle = 0 \quad \text{for all } v \text{ in } Y \text{ implies that } \ell = 0.$$

## Bilinear forms and the Lax–Milgram Theorem

5.33. Show that the constant $K$ in Example 29 can be improved upon, in that $K = \max(1, \kappa_1)$.

5.34. If $a : X \times X \to \mathbb{R}$ is a continuous bilinear form on an inner product space $X$, show that

$$\lim_{n\to\infty} a(u_n, v_n) = a(u, v)$$

if $u_n \to u$ and $v_n \to v$.

5.35. Let $\ell : H_0^1(0,1) \to \mathbb{R}$ and $a : H_0^1(0,1) \times H_0^1(0,1) \to \mathbb{R}$ be defined by

$$\langle \ell, v\rangle = \int_0^1 (-1-4x)v\,dx, \quad a(u,v) = \int_0^1 (x+1)u'v'\,dx,$$

where

$$H_0^1(0,1) = \{v \in L^2(0,1) : \ v' \in L^2(0,1),\ v(0) = v(1) = 0\};$$

this is a Hilbert space (see Chapter 7) with the inner product

$$(u,v)_{H_0^1} = \int_0^1 (uv + u'v')\,dx = (u,v)_{L^2} + (u',v')_{L^2}.$$

Show that $\ell$ is continuous, that $a$ is continuous and $H_0^1$- elliptic, and verify that the unique element $u$ satisfying

$$a(u,v) = \langle \ell, v\rangle \quad \text{for all } v \in H_0^1(0,1)$$

is $u(x) = x^2 - x$. [Hint: it may be necessary to use integration by parts. You may assume that a constant $C > 0$ exists such that $\|v\|_{L^2} \le C\|v'\|_{L^2}$.]

5.36. Let $a : H \times H \to \mathbb{R}$ be a continuous, $H$-elliptic bilinear form, and define the bilinear form $\tilde{a} : H \times H \to \mathbb{R}$ by

$$\tilde{a}(u,v) = a(u,v) + (u, \kappa v)_{L^2}.$$

If $H = H_0^1(0,1)$ and $\kappa(x)$ is continuous and satisfies $0 < \kappa_1 \le \kappa(x) \le \kappa_2$ for some constants $\kappa_1$ and $\kappa_2$, show that $\tilde{a}$ is continuous and $H$-elliptic.

# 6

# Orthonormal bases and Fourier series

In vector algebra it is often the case that computations are carried out using the components of vectors. A set of three mutually orthogonal unit vectors $\{\boldsymbol{i}, \boldsymbol{j}, \boldsymbol{k}\}$ is selected as a basis, and every vector $\boldsymbol{a}$ can then be written as $\boldsymbol{u} = \alpha\boldsymbol{i} + \beta\boldsymbol{j} + \gamma\boldsymbol{k}$, the coefficients $\alpha, \beta, \gamma$ being the components of $\boldsymbol{a}$ relative to the chosen basis. In Section 6.1 we start the process of extending this notion to vector spaces in general, by introducing finite-dimensional vector spaces. In Section 6.2 the vector space is endowed with an inner product or a norm, and this in turn permits the investigation of various properties that such inner product or normed spaces have by virtue of their being finite-dimensional. Section 6.3 is devoted to an examination of linear operators acting on finite-dimensional spaces; these are always continuous, and they also inherit in general the simple nature of their domains.

These concepts are extended to infinite-dimensional spaces in Section 6.4; if the space concerned is a Hilbert space, then the idea of an orthonormal basis carries over in a natural way from the finite-dimensional situation. The question of how one generates bases in infinite-dimensional spaces is partially answered by considering Sturm–Liouville problems, the topic of Section 6.5; these eigenvalue problems have a number of interesting properties, the most relevant of which is that their eigenfunctions form orthonormal bases in $L^2$.

## 6.1 Finite-dimensional spaces

In this section we discuss vector spaces that have the property that every member can be expressed as a *finite* sum of multiples (that is, a linear combination) of a selected subset of members of that space. The motivation for endowing vector spaces with this property once again comes from elementary vector algebra; every vector in three dimensions can be represented as a sum of multiples of three noncoplanar vectors.

**Linear combination.** Let $X$ be a linear space and $\mathbb{K}$ the set of real or complex numbers. Let $\{u_1, u_2, \ldots, u_n\}$ be a set of elements in $X$. The expression

$$\alpha_1 u_1 + \alpha_2 u_2 + \cdots + \alpha_n u_n,$$

where $\alpha_1, \ldots, \alpha_n \in \mathbb{K}$, is said to be a *linear combination* of the elements $u_1 \ldots, u_n$. Note that $\alpha_1 u_1 + \cdots + \alpha_n u_n \in X$.

In this section attention is restricted to *finite* linear combinations; as long as we do this, the theory that arises is purely algebraic. Infinite linear combinations of the form $\sum_{i=1}^{\infty} \alpha_i u_i$ require topological tools for their treatment and the discussion of this situation is postponed to Section 6.4.

**Linear dependence, independence.** Let $X$ be a linear space, and let $\{u_1, \ldots, u_n\}$ be a finite set of elements of $X$. Then this set is *linearly dependent* if there exist numbers $\alpha_1, \alpha_2, \ldots, \alpha_n$ in $\mathbb{K}$, not all of which are zero, such that

$$\alpha_1 u_1 + \cdots + \alpha_n u_n = 0. \tag{6.1}$$

The set $\{u_1, \ldots, u_n\}$ is *linearly independent* if (6.1) holds only when all of the $\alpha_i$ are zero. In other words, a set is linearly dependent if one of its elements can be written as a linear combination of the others; for if $\alpha_k$ is nonzero, then (6.1) may be rewritten in the form

$$u_k = -(1/\alpha_k)[\alpha_1 u_1 + \ldots + \alpha_{k-1} u_{k-1} + \alpha_{k+1} u_{k+1} + \ldots + \alpha_n u_n];$$

for a linearly independent set this is not possible.

### Examples

1. Let $X = \mathbb{R}^2$, and consider the vectors $\boldsymbol{a}_1 = (2, 1)$ and $\boldsymbol{a}_2 = (1, 2)$. To test for linear dependence, consider the linear combination

$$\alpha_1 \boldsymbol{a}_1 + \alpha_2 \boldsymbol{a}_2 = \mathbf{0} \quad \text{or} \quad (2\alpha_1 + \alpha_2)\boldsymbol{e}_1 + (\alpha + 2\alpha_2)\boldsymbol{e}_2 = \mathbf{0},$$

where $\boldsymbol{e}_1 = (1, 0)$ and $\boldsymbol{e}_2 = (0, 1)$. Accordingly, we must have

$$2\alpha_1 + \alpha_2 = 0 \quad \text{and} \quad \alpha_1 + 2\alpha_2 = 0.$$

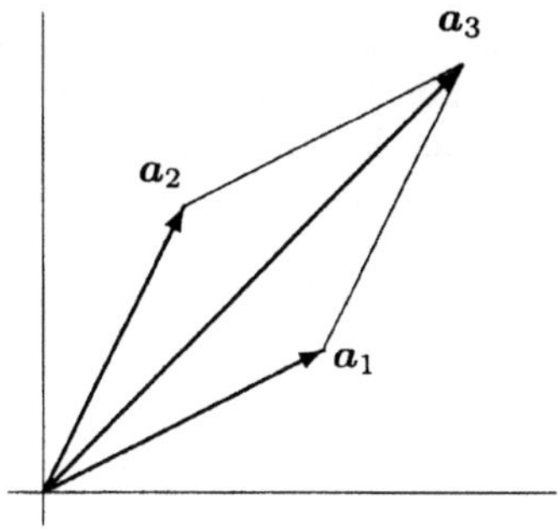


FIGURE 6.1. The vectors in Example 1

The only possible solution to these two equations is $\alpha_1 = \alpha_2 = 0$, and so $\boldsymbol{a}_1$ and $\boldsymbol{a}_2$ are *linearly independent*. Graphically this is easy to see (Figure 6.1), in that it is not possible to express $\boldsymbol{a}_2$ as a multiple of $\boldsymbol{a}_1$.

Now suppose that we also have the vector $\boldsymbol{a}_3$ as shown in Figure 6.1. This set is linearly dependent since, whatever the length and direction of $\boldsymbol{a}_3$, it is always possible to express it in the form $\boldsymbol{a}_3 = \beta_1\boldsymbol{a}_1 + \beta_2\boldsymbol{a}_2$ for some $\beta_1, \beta_2$. Hence there exist scalars $\beta_1, \beta_2$, and $\beta_3 = -1$ such that $\beta_1\boldsymbol{a}_1 + \beta_2\boldsymbol{a}_2 + \beta_3\boldsymbol{a}_3 = \boldsymbol{0}$.

2. Let $X = L^2(0,1)$ and consider the functions $u_k$ $(k = 1,2,3)$ defined by $u_1(x) = \cosh x$, $u_2(x) = \sinh x$, $u_3(x) = e^x$. Then the equation

$$\sum_{i=1}^{3} \alpha_i u_i = 0 \quad \text{or} \quad \alpha_1 \cosh x + \alpha_2 \sinh x + \alpha_3 e^x = 0$$

is satisfied for any nonzero $\alpha_i$ that are related to each other by $\alpha_1 = \alpha_2$, $\alpha_3 = -\alpha_1$, and so the set is linearly *dependent*.

**Basis, dimension**. A finite set $\{u_1, \ldots, u_n\}$ of elements of a vector space $X$ is said to *span* $X$ if *every* $u \in X$ can be written in the form $u = \alpha_1 u_1 + \cdots + \alpha_n u_n$ for some numbers $\alpha_i$, $i = 1, \ldots, n$ in $\mathbb{K}$. A set $\{u_1, \ldots, u_n\}$ of elements of $X$ is said to be a *basis* of $X$ if and only if

(i) the set spans $X$, and
(ii) the set $\{u_1, \ldots, u_n\}$ is linearly independent.

The number of elements that form a basis is called the *dimension* of $X$. We write $\dim X$ for the dimension of $X$. If $\{u_i\}_{i=1}^n$ is a basis for a vector space $X$ and

$$u = \sum_{i=1}^{n} \alpha_i u_i,$$

then $\alpha_i$ $(i = 1, \ldots, n)$ are called the *components* of $u$ relative to the basis $\{u_i\}$. Note that the components change with a change of basis. We stress the fact that the preceding definition applies only when $\{u_1, \ldots, u_n\}$ is a finite set; exactly what is meant by an infinite-dimensional space becomes clear later. We also note that, although the dimension of a space is fixed, it is possible to construct many different bases. These points should become clearer in the following examples.

## Examples

3. Consider the space $\mathbb{R}^3$: the set $\{\boldsymbol{e}_i\}_{i=1}^3 = \{(1,0,0),(0,1,0),(0,0,1)\}$ is linearly independent and also spans $\mathbb{R}^3$; hence $\{\boldsymbol{e}_i\}_{i=1}^3$ is a basis for $\mathbb{R}^3$ and dim $\mathbb{R}^3 = 3$. Consider the point $\boldsymbol{x} = 2\boldsymbol{e}_1 + 3\boldsymbol{e}_3$; this has components $(2,0,3)$ relative to the basis $\{\boldsymbol{e}_i\}$. But if we choose instead the basis $\{\boldsymbol{f}_i\}_{i=1}^3$ defined by $\boldsymbol{f}_1 = \boldsymbol{e}_1 + \boldsymbol{e}_2 + 2\boldsymbol{e}_3$, $\boldsymbol{f}_2 = \boldsymbol{e}_1 - \boldsymbol{e}_2 + \boldsymbol{e}_3$, $\boldsymbol{f}_3 = 2\boldsymbol{e}_1 + \boldsymbol{e}_2$, then the components of $\boldsymbol{x}$ relative to this basis are found from the fact that

$$\boldsymbol{x} = \boldsymbol{f}_1 + \boldsymbol{f}_2,$$

so that $\boldsymbol{x}$ has components $(1,1,0)$ with respect to the basis $\{\boldsymbol{f}_i\}$.

4. Consider the space $P_3[0,1]$ of polynomials of degree at most 3 defined on the interval $[0,1]$. Set $p_k(x) = x^k$, $k = 0, \ldots, 3$. Then $\{p_k\}_{k=0}^3$ is linearly independent since

$$\sum_{i=0}^{3} \alpha_i p_i = 0 \Rightarrow \alpha_0 + \alpha_1 x + \alpha_2 x^2 + \alpha_3 x^3 = 0$$

holds only if all the $\alpha_i$ are zero. Furthermore, every polynomial in $P_3[0,1]$ can be expressed in the form

$$p(x) = \sum_{i=0}^{3} \alpha_i p_i = \alpha_0 + \alpha_1 x + \alpha_2 x^2 + \alpha_3 x^3,$$

so that $\{p_k\}_{k=0}^3$ spans the space. Thus $E \equiv \{p_k\}_{k=0}^3$ forms a basis for $P_3[0,1]$ and $\dim P_3[0,1] = 4$. The components of the polynomial $p(x) = 2x - x^2 + x^3$ relative to the basis $E$ are $\{\alpha_i\} = \{0, 2, -1, 1\}$. But relative to the basis $F = \{(1-x),(1+x),x^2,x^3\}$ the components of $p$ are easily shown to be $\{-1, 1, -1, 1\}$ so that

$$p(x) = -1 \cdot (1-x) + 1 \cdot (1+x) - x^2 + x^3.$$

The following theorem describes an obvious but important property of finite-dimensional spaces.

THEOREM 1. *Let $X$ be a finite-dimensional linear space with dim $X = n$. Then any subset of $X$ containing more than $n$ members is linearly dependent.*

PROOF. Let $B = \{v_1, v_2, \ldots, v_n\}$ be a basis for $X$, and let $S = \{u_1, \ldots, u_n, u_{n+1}, \ldots, u_{n+k}\}$ be any set of $(n+k)$ elements in $X$. Then by definition there are scalars $A_{ij}$ such that

$$u_i = \sum_{j=1}^{n} A_{ij} v_j, \quad i = 1, \ldots, n+k.$$

For any set of scalars $\beta_1, \ldots, \beta_{n+k}$ we have

$$\begin{aligned} \beta_1 u_1 + \cdots + \beta_{n+k} u_{n+k} &= \sum_{i=1}^{n+k} \beta_i u_i = \sum_{i=1}^{n+k} \beta_i \sum_{j=1}^{n} A_{ij} v_j \\ &= \sum_{j=1}^{n} \left( \sum_{i=1}^{n+k} A_{ij} \beta_i \right) v_j. \end{aligned}$$

Thus if $\beta_1 u_1 + \cdots + \beta_{n+k} u_{n+k} = 0$, then we must have

$$\sum_{j=1}^{n} \gamma_j v_j = 0, \text{ where } \gamma_j = \sum_{i=1}^{n+k} A_{ij} \beta_i;$$

but since $\{v_j\}$ is linearly independent, this implies that $\gamma_j = 0$, or

$$\sum_{i=1}^{n+k} A_{ij} \beta_i = 0, \quad \text{or} \quad \boldsymbol{A}^t \boldsymbol{\beta} = \boldsymbol{0},$$

where $\boldsymbol{A}$ is an $(n+k) \times n$ matrix and $\boldsymbol{\beta}$ an $(n+k) \times 1$ column vector. From a standard result for sets of linear algebraic equations, every set of $n$ homogeneous (that is, right-hand side equal to 0) equations in $(n+k)$ unknowns has a nontrivial solution; hence there are scalars $\beta_1, \ldots, \beta_{n+k}$, not all zero, such that $\sum_{i=1}^{n+k} \beta_i u_i = 0$, so that $\{u_i\}_{i=1}^{n+k}$ is linearly dependent. □

## 6.2 Finite-dimensional inner product and normed spaces

Concepts such as linear dependence of a set and finite dimension of a space are algebraic: they require for their definition only the concept of a vector space. But if the vector space happens also to be an *inner product space*, it is possible to deduce a number of useful properties.

First, it is simple to check whether a set $\{u_i\}_{i=1}^k$ in an inner product space is linearly dependent. Confining attention to real inner product spaces, suppose that

$$\alpha_1 u_1 + \cdots + \alpha_k u_k = 0, \quad u_i \in X, \alpha_i \in \mathbb{R}, \quad i = 1, \ldots, k. \tag{6.2}$$

Take the inner product of both sides of this equation with $u_1$ to obtain

$$A_{11}\alpha_1 + A_{12}\alpha_2 + \cdots + A_{1k}\alpha_k = 0,$$

where $A_{ij} = (u_i, u_j) = (u_j, u_i) = A_{ji}$. By successively taking the inner product of (6.2) with each of the members $u_i$, we eventually find that

$$\sum_{j=1}^{k} A_{ij}\alpha_j = 0, \quad i = 1, \ldots, k, \text{ or } \boldsymbol{A}\boldsymbol{\alpha} = \boldsymbol{0}, \tag{6.3}$$

where $\boldsymbol{A}$ is the symmetric matrix with entries $A_{ij}$ and $\boldsymbol{\alpha}$ is the column vector $[\alpha_1, \ldots, \alpha_k]^t$. Now a necessary and sufficient condition for (6.3) to have a nontrivial solution is that $\det \boldsymbol{A} = 0$; hence, the set $\{u_1, \ldots, u_k\}$ *is linearly dependent if and only if* $\det \boldsymbol{A} = 0$.

**Examples**

5. Let $X = \mathbb{R}^2$ with $\boldsymbol{a}_1, \boldsymbol{a}_2$, and $\boldsymbol{a}_3$ as in Example 1. Then with $(\boldsymbol{a}, \boldsymbol{b}) \equiv \boldsymbol{a} \cdot \boldsymbol{b}$, $A_{ij} = \boldsymbol{a}_i \cdot \boldsymbol{a}_j$, and

$$\det \boldsymbol{A} = \det \begin{pmatrix} 5 & 4 & 2\beta_1 + \beta_2 \\ 4 & 5 & \beta_1 + 2\beta_2 \\ 2\beta_1 + \beta_2 & \beta_1 + 2\beta_2 & \beta_1^2 + \beta_2^2 \end{pmatrix}$$

which is easily shown to be identically zero for any values of $\beta_1$ and $\beta_2$. Hence the set $\{\boldsymbol{a}_1, \boldsymbol{a}_2, \boldsymbol{a}_3\}$ is linearly dependent.

6. The functions $u_1 = 1, u_2 = x, u_3 = x^2$ are linearly independent in $L^2(-1,1)$ since

$$\det \boldsymbol{A} = \det \left( \int_{-1}^{1} u_i(x) u_j(x)\, dx \right) = \det \begin{pmatrix} 2 & 0 & 2/3 \\ 0 & 2/3 & 0 \\ 2/3 & 0 & 2/5 \end{pmatrix} \neq 0.$$

**Orthonormal sets and bases.** If $X$ is an inner product space, a set $\{\phi_1, \ldots, \phi_k, \ldots\}$ of elements in $X$ is said to be an *orthonormal set* if the elements are mutually orthogonal and have unit length; that is,

$$(\phi_i, \phi_j) = \begin{cases} 1 & \text{if } i = j, \\ 0 & \text{otherwise.} \end{cases} \tag{6.4}$$

Any orthonormal set is linearly independent. To see this, consider

$$\alpha_1\phi_1 + \alpha_2\phi_2 + \cdots + \alpha_n\phi_n = 0;$$

now take the inner product with $\phi_1$ to obtain $\alpha_1 \cdot 1 + 0 + \ldots + 0 = 0$. Thus $\alpha_1 = 0$. In the same way, by taking the inner product with each $\phi_k$ in turn, we find that all the $\alpha_k$ are zero.

Now suppose that $X$ is a finite-dimensional inner product space with $\dim X = n$. Then a basis $\{\phi_1, \ldots, \phi_n\}$ of $X$ whose elements satisfy (6.4) is said to be an *orthonormal basis.*

**Examples**

7. The set $\{(1,0,0), (0,1,0), (0,0,1)\}$ forms an orthonormal basis for $\mathbb{R}^3$.

8. Consider the space $L^2(-1,1)$. The infinite set

$$\{\phi_k :\ \phi_k(x) = \sin k\pi x, \quad k = 1, 2, \ldots\}$$

is an orthonormal set since

$$(\phi_k, \phi_l)_{L^2} = \int_{-1}^{1} \sin k\pi x \sin l\pi x \, dx = \begin{cases} 1 & \text{if } k = l, \\ 0 & \text{otherwise.} \end{cases}$$

But $L^2(-1,1)$ is not finite-dimensional, so talk of an orthonormal basis is premature at this stage.

9. Consider again $L^2(-1,1)$, but this time as a space of complex-valued functions. The set $\{u_0, u_1, \ldots\}$ in which $u_k(x) = (1/\sqrt{2})e^{ik\pi x}$ is an orthonormal set since

$$(u_k, u_l) = \int_{-1}^{1} u_k(x)\overline{u_l(x)} \, dx = \tfrac{1}{2}\int_{-1}^{1} e^{i(k-l)\pi x} \, dx$$

which is 0 if $k \neq l$, and equal to 1 if $k = l$.

One of the main advantages of orthonormal bases over other bases is that computations involving the former are much simpler. For example, if $\{\phi_i\}_{i=1}^n$ is an orthonormal basis for $X$ and $u, v \in X$, then

$$\begin{aligned}(u,v) &= \left(\sum_{i=1}^n u_i\phi_i, \sum_{j=1}^n v_j\phi_j\right) \\ &= \sum_{i=1}^n\sum_{j=1}^n u_i\bar{v}_j(\phi_i, \phi_j) = \sum_{i=1}^n u_i\bar{v}_i.\end{aligned}$$

**Gram–Schmidt orthonormalization.** Given any nonorthonormal basis $S = \{\psi_i\}_{i=1}^n$, it is possible to construct from $S$ an orthonormal basis

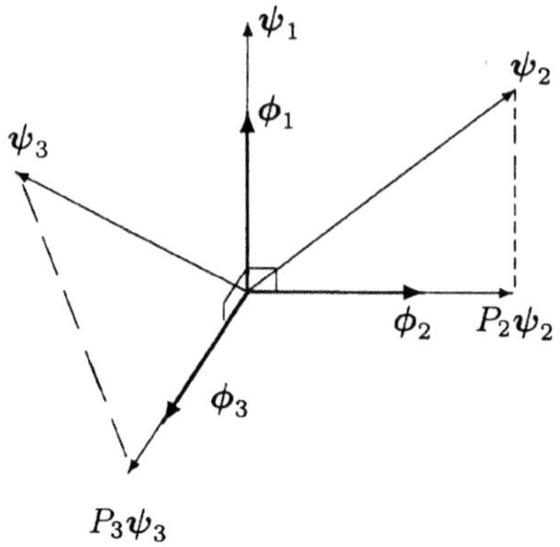


FIGURE 6.2. The Gram–Schmidt orthonormalization process

$E = \{\phi_i\}_{i=1}^n$, using the *Gram–Schmidt orthonormalization* procedure. We illustrate the procedure for vectors in three dimensions and then generalize from that.

Let $\{\boldsymbol{\psi}_1, \boldsymbol{\psi}_2, \boldsymbol{\psi}_3\}$ be any basis for $\mathbb{R}^3$, and construct $\boldsymbol{\phi}_1$ from

$$\boldsymbol{\phi}_1 = \frac{\boldsymbol{\psi}_1}{|\boldsymbol{\psi}_1|}.$$

Next, project $\boldsymbol{\psi}_2$ onto the plane orthogonal to $\boldsymbol{\phi}_1$ (Figure 6.2), using for this purpose the projection operator $P_2$ defined by

$$P_2\boldsymbol{a} = \boldsymbol{a} - (\boldsymbol{a}\cdot\boldsymbol{\phi}_1)\boldsymbol{\phi}_1.$$

Then set

$$\boldsymbol{\phi}_2 = \frac{P_2\boldsymbol{\psi}_2}{|P_2\boldsymbol{\psi}_2|}.$$

Finally, project $\boldsymbol{\psi}_3$ onto the line orthogonal to both $\boldsymbol{\phi}_1$ and $\boldsymbol{\phi}_2$, using the projection operator $P_3$ defined by

$$P_3\boldsymbol{a} = \boldsymbol{a} - (\boldsymbol{a}\cdot\boldsymbol{\phi}_1)\boldsymbol{\phi}_1 - (\boldsymbol{a}\cdot\boldsymbol{\phi}_2)\boldsymbol{\phi}_2.$$

Then set

$$\boldsymbol{\phi}_3 = \frac{P_3\boldsymbol{\psi}_3}{|P_3\boldsymbol{\psi}_3|}.$$

The resulting basis $\{\boldsymbol{\phi}_k\}_{k=1}^3$ is orthonormal.

Generally, the procedure may be summarized as follows. Given a basis $\{\psi_i\}_{i=1}^n$, form an orthonormal basis $\{\phi_i\}_{i=1}^n$ from

$$\phi_i = \frac{P_i\psi_i}{\|P_i\psi_i\|},$$

the projection operators $P_i$ being defined by

$$P_1 u = u, \quad P_i u = u - \sum_{k=1}^{i-1} (u, \phi_k)\phi_k \quad \text{for } i = 2, 3, \ldots, n.$$

We conclude this section with an important result on the completeness of finite-dimensional normed spaces. To prove this result the following lemma is required.

LEMMA 1. *Let $\{u_1, \ldots, u_n\}$ be a linearly independent set of members of a normed space $X$. Then there is a constant $c > 0$ such that for every choice of scalars $\alpha_1, \ldots, \alpha_n$,*

$$\|\alpha_1 u_1 + \cdots + \alpha_n u_n\| \geq c(|\alpha_1| + \cdots + |\alpha_n|).$$

THEOREM 2. *Every finite-dimensional normed space $X$ is complete.*

PROOF. Let $\{u_k\}_{k=1}^{\infty}$ be a Cauchy sequence in $X$; the aim then is to show that $u_k \to u$ in $X$. Suppose that $\dim X = n$, and let $\{e_1, \ldots, e_n\}$ be any basis for $X$. Then each $u_k$ can be expressed in the form

$$u_k = \alpha_{k1} e_1 + \alpha_{k2} e_2 + \ldots + \alpha_{kn} e_n,$$

where $\alpha_{ki}$ $(i = 1, \ldots, n)$ are the components of $u_k$. Using Lemma 1 and the fact that $\{u_k\}$ is Cauchy, it follows that for any given $\epsilon > 0$, there exists $N$ such that

$$\epsilon > \|u_k - u_l\| = \left\| \sum_{i=1}^{n} (\alpha_{ki} - \alpha_{li}) e_i \right\| \geq c \sum_{i=1}^{n} |\alpha_{ki} - \alpha_{li}|$$

for $k, l > N$. Since

$$|\alpha_{ki} - \alpha_{li}| \leq \sum_{i=1}^{n} |\alpha_{ki} - \alpha_{li}| < \frac{\epsilon}{c}$$

we see that $\{\alpha_{ki}\}$ is a Cauchy sequence in $\mathbb{K}$ for each fixed $i$. Hence $\alpha_{ki}$ converges to an element $\alpha_i$, say. Now define

$$u = \alpha_1 e_1 + \cdots + \alpha_n e_n.$$

Then

$$\|u_k - u\| = \left\| \sum_{i=1}^{n} (\alpha_{ki} - \alpha_i) e_i \right\| \leq \sum_{i=1}^{n} |\alpha_{ki} - \alpha_i| \, \|e_i\|;$$

but $\alpha_{ki} \to \alpha_i$ for each $i$, so that $u_k \to u$ in $X$; hence $X$ is complete. □

**Weak and strong convergence.** The notion of weak convergence was introduced in Section 5.4, and the point was also made there that strong convergence implies weak convergence of a sequence. The converse is generally not true, except for the particular case in which the space is finite-dimensional. This is the subject of the following theorem.

THEOREM 3. *Let $\{u_k\}$ be a weakly convergent sequence in a normed space $X$, with weak limit $u$. If $\dim X < \infty$, then $\{u_k\}$ converges strongly to $u$.*

PROOF. Suppose that $\dim X = n$ and let $\{e_i\}_{i=1}^n$ be a basis for $X$. Then we may express $u_k$ and $u$ in the form

$$u_k = \alpha_{k1}e_1 + \alpha_{k2}e_2 + \ldots + \alpha_{kn}e_n$$

and

$$u = \alpha_1 e_1 + \alpha_2 e_2 + \ldots + \alpha_n e_n.$$

Now

$$\langle \ell, u_k \rangle \to \langle \ell, u \rangle \tag{6.5}$$

by assumption, for every $\ell \in X'$. Take in particular the $n$ functionals $\ell_1, \ldots, \ell_n$ defined by

$$\langle \ell_i, e_j \rangle = \begin{cases} 1 & \text{for } i = j \\ 0 & \text{otherwise.} \end{cases}$$

Then it follows that $\langle \ell_i, u_k \rangle = \alpha_{ki}$ and $\langle \ell_i, u \rangle = \alpha_i$, and (6.5) implies that $\alpha_{ki} \to \alpha_i$ as $k \to \infty$, for each $i$. Thus

$$\begin{aligned} \|u_k - u\| &= \left\| \sum_{i=1}^n (\alpha_{ki} - \alpha_i) e_i \right\| \\ &\le \sum_{i=1}^n |\alpha_{ki} - \alpha_i| \, \|e_i\| \to 0 \end{aligned}$$

as $k \to \infty$. Thus $\lim_{k\to\infty} u_k = u$. □

## 6.3 Linear operators on finite-dimensional spaces

We turn now to the consideration of linear operators whose domains are finite-dimensional spaces. As might be expected, the nature of such operators is heavily influenced by the fact that their domains have finite dimension. For example, it turns out that if $T$ is a linear operator on a finite-dimensional normed space, then $T$ is always continuous, as the next theorem shows.

THEOREM 4. *Let $T : X \to Y$ be a linear operator, where $X$ and $Y$ are normed spaces, and $X$ has finite dimension. Then $T$ is bounded, and hence continuous.*

PROOF. Let $\{e_i, \dots, e_n\}$ be a basis for $X$; then any $u \in X$ has the representation $u = \alpha_1 e_1 + \cdots + \alpha_n e_n$ for certain scalars $\alpha_1, \dots, \alpha_n$, and so

$$\begin{aligned} \|Tu\| &= \|T(\alpha_1 e_1 + \cdots + \alpha_n e_n)\| = \|\alpha_1 Te_1 + \cdots + \alpha_n Te_n\| \\ &\leq |\alpha_1|\|Te_1\| + \cdots + |\alpha_n|\|Te_n\| \\ &\leq M(|\alpha_1| + \cdots + |\alpha_n|), \end{aligned}$$

where $M = \max\{\|Te_1\|, \dots, \|Te_n\|\}$. From Lemma 1 there is a constant $C > 0$ such that

$$C(|\alpha_1| + \cdots + |\alpha_n|) \leq \|u\|,$$

so that

$$\|Tu\| \leq \frac{M}{C}\|u\|.$$

Thus $T$ is bounded, hence continuous. □

There is a very simple relationship among the dimensions of the domain, null space, and range of a linear operator when the operator acts on a finite dimensional space, as we now show.

THEOREM 5. *Let $T : X \to Y$ be a linear operator with $\dim X = n$ and $\dim N(T) = k \leq n$, where $N(T)$ is the null space of $T$. Then*

*(a) if $\{e_1, \dots, e_k\}$ is a basis for $N(T)$ and $\{e_1, \dots, e_k, e_{k+1}, \dots, e_n\}$ is a basis for $X$, then $\{Te_{k+1}, \dots, Te_n\}$ is a basis for $R(T)$, the range of $T$;*

*(b) $\dim N(T) + \dim R(T) = \dim X$.*

PROOF. (a) The elements $Te_1, Te_2, \dots, Te_n$ certainly span $R(T)$ since any $v \in R(T)$ satisfies, for some $u \in X$,

$$v = Tu = T\left(\sum_{i=1}^{n} \alpha_i e_i\right) = \sum_{i=1}^{n} \alpha_i Te_i,$$

where $\alpha_i$ are the components of $u$ relative to the basis $e_i$. Since $e_1, \ldots, e_k$ are in $N(T)$ we have $Te_1 = \cdots = Te_k = 0$ so that $\{Te_{k+1}, \ldots, Te_n\}$ spans $R(T)$. We show next that this set is linearly independent. Suppose that there are scalars $\beta_{k+1}, \ldots, \beta_n$ such that

$$\sum_{i=k+1}^{n} \beta_i(Te_i) = 0;$$

by the linearity of $T$,

$$T\left(\sum_{i=k+1}^{n} \beta_i e_i\right) = 0$$

so that the sum $\sum_{i=k+1}^{n} \beta_i e_i$ belongs to $N(T)$, and may therefore be represented in the form

$$\sum_{i=k+1}^{n} \beta_i e_i = \sum_{j=1}^{k} \gamma_j e_j$$

for some scalars $\gamma_1, \ldots, \gamma_k$. It follows that, if we set $\beta_1 = -\gamma_1, \ldots, \beta_k = -\gamma_k$, then this expression may be rewritten in the form

$$\sum_{i=1}^{n} \beta_i e_i = 0.$$

But $\{e_1, \ldots, e_n\}$ is linearly independent, hence $\beta_1 = \cdots = \beta_n = 0$. So $\{Te_{k+1}, \ldots, Te_n\}$ is linearly independent and, since it spans $R(T)$, it forms a basis for $R(T)$. Part (b) is a trivial consequence of (a). □

For the special case in which $T : X \to Y$ with $\dim X = \dim Y = n$, we can deduce from Theorem 5 the following.

COROLLARY TO THEOREM 5. *Let $T : X \to Y$ be a linear operator with $\dim X = \dim Y = n$. Then $N(T) = \{0\}$ if and only if $R(T) = Y$, and when this is so, $T$ is one-to-one and surjective, with a unique inverse $T^{-1}$. Together with Theorem 4, it follows that $T$ is an isomorphism of $X$ onto $Y$.*

**Example**

10. Let $X = \mathbb{R}^3$, $Y = \mathbb{R}^2$, and let $\boldsymbol{T} : X \to Y$ be the matrix $\boldsymbol{T} = \begin{pmatrix} 1 & 0 & 2 \\ 3 & 4 & 2 \end{pmatrix}$. The null space of $\boldsymbol{T}$ consists of all vectors $\boldsymbol{x}$ satisfying

$$\boldsymbol{Tx} = \boldsymbol{0} \quad \text{or} \quad \begin{aligned} x_1 + 2x_3 &= 0, \\ 3x_1 + 4x_2 + 2x_3 &= 0. \end{aligned}$$

It is not difficult to see that $N(\boldsymbol{T})$ consists of all vectors of the form $\boldsymbol{x} = \alpha(-2,1,1)$ so that $\dim N(T) = 1$. We should then have $\dim R(T) = 3 - 1 = 2$; this is borne out by the fact that $\{\boldsymbol{x}, \boldsymbol{y}, \boldsymbol{z}\} = \{(-2,1,1),(1,2,0),(1,1,1)\}$ forms a basis for $X$. Now $\boldsymbol{x}$ spans $N(T)$ so that $\{\boldsymbol{Ty}, \boldsymbol{Tz}\}$ forms a basis for $Y$, as is readily verified.

**Isomorphisms.** We have seen that finite-dimensional spaces all "look" the same, in that their elements are uniquely described by the specification of the components relative to a given basis. The situation is even simpler, as it turns out: it is possible to set up a one-to-one correspondence between the elements of any $n$-dimensional real inner product space $X$ and the elements of $\mathbb{R}^n$ in such a way that the elements thus related have the same lengths. More precisely, let $T : X \to \mathbb{R}^n$ be a linear operator from $X$ to $\mathbb{R}^n$; we assert that it is possible to define an *isometric isomorphism* $T$ between these two spaces: that is, $T$ is bounded, bijective, and if $Tu = v$, then $\|u\| = \|Tu\| = \|v\|$ (see Section 5.2). So $X$ and $\mathbb{R}^n$ are, to all intents and purposes, one and the same thing.

THEOREM 6. *Let $X$ be any finite-dimensional inner product space with $\dim X = n$. Then $X \sim \mathbb{R}^n$; that is, there exists an isometric isomorphism from $X$ to $\mathbb{R}^n$.*

The proof of this theorem is the subject of Exercise 6.15.

**Representation of linear operators by matrices.** An $m \times n$ matrix is a linear operator from $\mathbb{R}^n$ to $\mathbb{R}^m$. It is natural to ask, then, whether there exists any way in which a linear operator from one arbitrary finite-dimensional space to another can be represented by a matrix. This is easily done, as we now show.

Let $T : X \to Y$, where $\dim X = n$ and $\dim Y = m$. Let $\{e_1, \ldots, e_n\}$ and $\{f_1, \ldots, f_m\}$ be bases for $X$ and $Y$, respectively. Then if $u$ is any member of $X$ with image $v \in Y$ under the mapping $T$, there are scalars $\alpha_1, \ldots, \alpha_n$ and $\beta_1, \ldots, \beta_m$ such that $u$ and $v$ have the representations

$$u = \alpha_1 e_1 + \cdots + \alpha_n e_n,$$
$$v = \beta_1 f_1 + \cdots + \beta_m f_m.$$

Since $Tu = v$ and $T$ is linear, it follows that

$$\alpha_1 Te_1 + \cdots + \alpha_n Te_n = \beta_1 f_1 + \cdots + \beta_m f_m. \tag{6.6}$$

Now $Te_j$ is in $Y$ for $j = 1, \ldots, n$, and so it is possible to express $Te_j$ in the form

$$Te_j = \sum_{i=1}^{m} T_{ij} f_i, \tag{6.7}$$

where $T_{ij}$ are scalars. We form a matrix $\boldsymbol{T}$ with components $T_{ij}$; then $\boldsymbol{T}$ is the *matrix of* $T$ relative to the bases $\{e_i\}$ and $\{f_j\}$, and (6.6) becomes

$$\sum_{i=1}^{m}\sum_{j=1}^{n} T_{ij}\alpha_j f_i = \sum_{i=1}^{m} \beta_i f_i$$

or

$$\sum_{i=1}^{m}\left(\sum_{j=1}^{n} T_{ij}\alpha_j - \beta_i\right) f_i = 0.$$

Since $f_1, \ldots, f_m$ form a linearly independent set we have

$$\sum_{j=1}^{n} T_{ij}\alpha_j = \beta_i \quad \text{or} \quad \boldsymbol{T\alpha} = \boldsymbol{\beta}, \tag{6.8}$$

where $\boldsymbol{\alpha} = (\alpha_1, \ldots, \alpha_n)$ and $\boldsymbol{\beta} = (\beta_1, \ldots, \beta_m)$. It follows that if the matrix corresponding to a linear operator is known, then (6.8) can be used to find the components of the image of any member of the domain of the operator.

## Example

11. Let $X = P_2[0,1]$ and $Y = P_1[0,1]$, where $P_k[0,1]$ is the set of polynomials of degree at most $k$ on $[0,1]$; $\dim P_k[0,1] = k+1$. Suppose that we choose as bases for $X$ and $Y$

$$\{e_1, e_2, e_3\} \quad \text{and} \quad \{f_1, f_2\},$$

where $e_1 = f_1 = 1, e_2 = f_2 = x, e_3 = x^2$, and let $T$ be the derivative operator $d/dx$; then

$$Te_1 = 0, \quad Te_2 = 1, \quad Te_3 = 2x.$$

The matrix $\boldsymbol{T}$ corresponding to $d/dx$ is found from (6.7):

$$\begin{aligned} 0 &= T_{11} + T_{21}x, \\ 1 &= T_{12} + T_{22}x, \\ 2x &= T_{13} + T_{23}x. \end{aligned}$$

The elements $T_{ij}$ are found by equating coefficients of $x^0$ and $x^1$, and are

$$\begin{aligned} &T_{11} = T_{21} = T_{22} = T_{13} = 0, \\ &T_{12} = 1, \quad T_{23} = 2, \end{aligned} \qquad \text{so that} \quad \boldsymbol{T} = \begin{pmatrix} 0 & 1 & 0 \\ 0 & 0 & 2 \end{pmatrix}.$$

Thus if we are given any polynomial $p$ in $P_2[0,1]$, the coefficients $\beta_i$ of its derivative may be found from (6.8); if $p(x) = 5 - x + 3x^2$, for example, then

$$\beta = \begin{pmatrix} 0 & 1 & 0 \\ 0 & 0 & 2 \end{pmatrix} \begin{pmatrix} 5 \\ -1 \\ 3 \end{pmatrix} = \begin{pmatrix} -1 \\ -6 \end{pmatrix},$$

or $dp/dx = Tp = -1 + 6x$.

**Linear functionals.** Linear functionals on finite-dimensional spaces have a particularly simple structure; in fact, they inherit the finite-dimensionality of their domain $X$, so that $\dim X' = \dim X$ is finite. To see this, let $\{e_1 \ldots, e_n\}$ be a basis for the $n$-dimensional normed space $X$, and define a total of $n$ linear functionals $\ell_1, \ldots, \ell_n$ on $X$ by

$$\ell_k : X \to \mathbb{R}, \quad \langle \ell_k, e_j \rangle = \begin{cases} 1 & \text{if } j = k, \\ 0 & \text{otherwise.} \end{cases} \tag{6.9}$$

We claim that the set $L = \{\ell_1, \ldots, \ell_n\}$ thus defined is a basis for $X'$; indeed, $L$ is linearly independent since, if

$$\alpha_1 \ell_1 + \alpha_2 \ell_2 + \cdots + \alpha_n \ell_n = 0, \tag{6.10}$$

then this implies that

$$\sum_{i=1}^{n} \alpha_i \langle \ell_i, u \rangle = 0 \quad \text{for all } \ u \in X,$$

which in turn gives

$$0 = \sum_{i=1}^{n} \alpha_i \langle \ell_i, e_j \rangle = \alpha_j, \quad j = 1, \ldots, n,$$

using (6.9). Thus (6.10) holds only if all $\alpha_j = 0$. Secondly, every $\ell \in X'$ has the unique representation

$$\ell = \beta_1 \ell_1 + \cdots + \beta_n \ell_n,$$

where $\beta_j = \langle \ell, e_j \rangle$. To see this, let $u = \alpha_1 e_1 + \cdots + \alpha_n e_n \in X$; then

$$\langle \ell, u \rangle = \langle \ell, \alpha_1 e_1 + \cdots + \alpha_n e_n \rangle = \sum_{i=1}^{n} \alpha_i \beta_i.$$

On the other hand,

$$\langle \ell_i, u \rangle = \langle \ell_i, \alpha_1 e_1 + \cdots + \alpha_n e_n \rangle = \alpha_i.$$

Hence $\langle \ell, u\rangle = \sum_{i=1}^{n} \beta_i \langle \ell_i, u\rangle$ or

$$\ell = \beta_1 \ell_1 + \beta_2 \ell_2 + \cdots + \beta_n \ell_n,$$

as asserted. Thus $\{\ell_1, \ldots, \ell_n\}$ spans $X'$, so that $\dim X' = n$. It can be shown, furthermore, that $X$ and $X'$ are isomorphic to each other. Of course, if $X$ is an inner product space, then by virtue of its finite-dimensionality it is a Hilbert space and thus $X$ and $X'$ are isometrically isomorphic according to the Riesz Representation Theorem.

## 6.4 Fourier series in Hilbert spaces

Our main aim in this section is to extend the idea of a basis to arbitrary Hilbert spaces, including spaces of functions. Now, generally speaking these spaces are *not finite-dimensional*; for example, it is not possible to find a finite set of functions in $L^2(\Omega)$ that spans $L^2(\Omega)$. The best that can be done is to construct an infinite sequence of functions with the property that any member of the space can be approximated arbitrarily closely by a finite linear combination of these functions, provided that a sufficiently large number of functions is used. This leads to the idea of a basis consisting of a countably infinite set. We work with such sets in inner product spaces and, although not necessary, the resulting theory is rendered more tidy if it is developed in the framework of *orthonormal sets*; these are sets of the form $\{\phi_1, \phi_2, \ldots, \phi_k, \ldots\}$ for which

$$(\phi_i, \phi_j) = \begin{cases} 1 & \text{if } i = j, \\ 0 & \text{otherwise.} \end{cases}$$

**Maximal orthonormal set, basis.** Let $X$ be any inner product space and let $\Phi = \{\phi_i\}_{i=1}^{\infty}$ be an orthonormal set in $X$. Then we say that $\Phi$ is a *maximal orthonormal set* in $X$ if there is no other non-zero member $\phi$ in $X$ that is orthogonal to all the $\phi_i$. That is, $\Phi$ is maximal if $(\phi, \phi_i) = 0$ for all $i$ implies that $\phi = 0$; it is not possible to add to $\Phi$ a further nonzero element that is orthogonal to all existing members of $\Phi$.

A maximal orthonormal set in a *Hilbert space H* is called an *orthonormal basis* for $H$.

It is clear that we are generalizing from the finite-dimensional case; indeed, since every finite-dimensional inner product space is complete and therefore a Hilbert space, the preceding definition of an orthonormal basis is equivalent to that given in Section 6.2.

**Example**

12. Let $H = L^2(-1,1)$ and consider the set $\Phi_1 = \{\sin \pi x, \sin 2\pi x, \dots\}$. $\Phi_1$ is an orthonormal set since

$$(\phi_k, \phi_l) = \int_{-1}^{1} \sin k\pi x \sin l\pi x \, dx = \begin{cases} 1 & \text{otherwise}, \\ 0 & \text{if } k \neq l. \end{cases}$$

But $\Phi_1$ is *not maximal*; in particular, any even function $u_e(x)$ is orthogonal to all $\phi_k$ since $\int_{-1}^{1} u_e(x) \sin k\pi x \, dx = 0$, $\sin k\pi x$ being an odd function. But it can be shown that the set

$$\Phi_2 = \{1/\sqrt{2}\} \cup \{\sin k\pi x, \cos k\pi x\}_{k=1}^{\infty}$$

is a maximal orthonormal set. Since $L^2(-1,1)$ is complete, $\Phi_2$ is an orthonormal basis.

Let $\{\phi_k\}$ be an orthonormal set in an inner product space $X$. Then, for any $u \in X$ the numbers $u_k = (u, \phi_k)$ are called the *Fourier coefficients* of $u$ with respect to $\{\phi_k\}$. These are the infinite-dimensional counterparts of the components $u_k$ of an element $u$ of a finite-dimensional space $X$. If $\{\phi_1, \dots, \phi_n\}$ is an orthonormal basis for an inner product space $X$ with dimension $n$, then for any $u \in X$,

$$u = \sum_{k=1}^{n} u_k \phi_k, \quad \text{where} \quad u_k = (u, \phi_k).$$

Precisely under what conditions the expression

$$u = \sum_{k=1}^{\infty} (u, \phi_k)\phi_k \tag{6.11}$$

is valid for an element $u$ of an *infinite-dimensional* inner product space $X$ is essentially the subject of this section. These conditions are discussed in a moment, but first we must digress and make clear exactly what is meant by an infinite sum of the form $\sum_{k=1}^{\infty} \alpha_k u_k$.

Suppose that $\{u_k\}$ is a sequence in an inner product space and $\{\alpha_k\}$ is a sequence of real or complex numbers, and define the corresponding $n$th *partial* sum $s_n$ of this sequence by

$$s_n = \sum_{k=1}^{n} \alpha_k u_k, \tag{6.12}$$

where $\{\alpha_1, \alpha_2, \dots\}$ is a set of real or complex numbers. Now suppose that we generate $s_1, s_2, \dots$ using (6.12); then the series $\sum_{k=1}^{\infty} \alpha_k u_k$ is said to *converge* to an element $u$ if the sequence $\{s_n\}$ of partial sums converges to

$u$. That is, we write $u = \sum_{k=1}^{\infty} \alpha_k u_k$ if, given any $\epsilon > 0$, it is possible to find a number $N$ such that

$$\|s_n - u\| < \epsilon \quad \text{whenever} \quad n > N$$

or, more briefly,

$$\lim_{n\to\infty} \|s_n - u\| = 0.$$

It is in this sense that the expression (6.11) must be interpreted: any partial sum $s_n$ is an approximation to $u$, and this approximation improves as $n$ increases.

**The Best Approximation Theorem.** We show next that the Fourier coefficients of a function are indeed special numbers in the following sense. Suppose that $v$ is an arbitrary member of the inner product space $X$, and we wish to find the best possible approximation to $v$ in the finite-dimensional subspace $\Phi$ spanned by $\{\phi_k\}_{k=1}^{n}$. The first question is, how does one measure such an approximation, and the second is, what is that best approximation?

To answer the first question, let $\tilde{v}$ be the element in $\Phi$ that is "closest" to $v$. A reasonable way of making this assertion mathematically is to interpret this to mean that

$$\|v - \tilde{v}\| \le \|v - w\| \quad \text{for any } w \in \Phi.$$

This is of course precisely the topic of Theorem 7, Chapter 4, according to which such an element $\tilde{v}$ exists, and is in fact unique. Here the task is one of characterizing $\tilde{v}$, given that $\Phi$ is finite-dimensional. Now since any member $w$ of $\Phi$ can be expressed in the form

$$w = \sum_{k=1}^{n} c_k \phi_k$$

it follows that the issue is one of determining what the coefficients $c_k$ must be; it turns out that the coefficients that provide the best approximation are precisely the *Fourier* coefficents.

THEOREM 7 (THE BEST APPROXIMATION THEOREM). *Let $X$ be an inner product space and $\{\phi_k\}_{k=1}^{\infty}$ an orthonormal set in $X$. Let $v$ be a member of $X$ and let $s_n$ and $t_n$ denote the partial sums*

$$s_n = \sum_{k=1}^{n} v_k \phi_k \quad \text{and} \quad t_n = \sum_{k=1}^{n} c_k \phi_k,$$

*where $v_k = (v, \phi_k)$ is the $k$th Fourier coefficient of $v$ and $c_k$ are arbitrary real or complex numbers. Then*

(a) (Best Approximation)

$$\|v - s_n\| \leq \|v - t_n\|; \tag{6.13}$$

(b) (Bessel's Inequality)

$$\sum_{k=1}^{\infty} |v_k|^2 \text{ converges, and } \sum_{k=1}^{\infty} |v_k|^2 \leq \|v\|^2; \tag{6.14}$$

(c) (Parseval's Formula)

$$\sum_{k=1}^{\infty} |v_k|^2 = \|v\|^2 \text{ if and only if } \|v - s_n\| \to 0 \text{ as } n \to \infty$$

(that is, $v = \sum_{k=1}^{\infty} v_k \phi_k$).

Proof. To prove (a), consider

$$\|v - t_n\|^2 = (v - t_n, v - t_n) = (v, v) - (t_n, v) - (v, t_n) + (t_n, t_n).$$

Now

$$\begin{aligned} (t_n, t_n) &= \left( \sum_{k=1}^{n} c_k \phi_k, \sum_{l=1}^{n} c_l \phi_l \right) = \sum_{k=1}^{n} \sum_{l=1}^{n} (c_k \phi_k, c_l \phi_l) \\ &= \sum_{k=1}^{n} \sum_{l=1}^{n} c_k \bar{c}_l (\phi_k, \phi_l) = \sum_{k=1}^{n} |c_k|^2; \end{aligned}$$

next,

$$(v, t_n) = \left( v, \sum_{l=1}^{n} c_l \phi_l \right) = \sum_{l=1}^{n} \bar{c}_l (v, \phi_l) = \sum_{l=1}^{n} \bar{c}_l v_l.$$

Likewise, $(t_n, v) = \sum_{l=1}^{n} c_l \bar{v}_l$. Assembling all these terms, we find that

$$\begin{aligned} \|v - t_n\|^2 &= \|v\|^2 + \sum_{k=1}^{n} \left( -\bar{c}_k v_k - c_k \bar{v}_k + |c_k|^2 \right) \\ &= \|v\|^2 + \sum_{k=1}^{n} |v_k - c_k|^2 - \sum_{k=1}^{n} |v_k|^2. \end{aligned} \tag{6.15}$$

The inequality (6.13) follows by comparing this equation with that obtained by setting $c_k = v_k$, for which case $t_n = s_n$ and the second term on the right-hand side is zero.

Parts (b) and (c) follow readily from (a), and are treated in Exercise 6.20. □

### Example

13. Let $X = L^2(-1,1)$ and let $\Phi$ be the subspace of $X$ spanned by the orthonormal set $\{1/\sqrt{2}, \cos\pi x, \sin\pi x\}$. Consider the function $v(x) = x^2$. Then the Fourier coefficients of $v$ are

$$\begin{aligned}
(v, 1/\sqrt{2}) &= \int_{-1}^{1} \frac{1}{\sqrt{2}} x^2 \, dx = \frac{\sqrt{2}}{3}, \\
(v, \cos\pi x) &= \int_{-1}^{1} x^2 \cos\pi x \, dx = -4/\pi^2, \\
(v, \sin\pi x) &= \int_{-1}^{1} x^2 \sin\pi x \, dx = 0,
\end{aligned}$$

and so

$$\tilde{v} = \frac{\sqrt{2}}{3} - \frac{4}{\pi^2} \cos\pi x.$$

The approximation $\tilde{v}$ of $v$ is shown in Figure 6.3.

The error in the approximation is found from

$$\|v - \tilde{v}\|_{L^2}^2 = \int_{-1}^{1} \left[ x^2 - \left( \frac{\sqrt{2}}{3} - \frac{4}{\pi^2} \cos\pi x \right) \right]^2 dx = 0.0516$$

so that the relative error is

$$\frac{\|v - \tilde{v}\|_{L^2}}{\|v\|_{L^2}} = 0.36.$$

The link between Theorem 7 and the Projection Theorem is obviously a close one, and in fact it can be shown that the partial sum $s_n$ is the *orthogonal projection* of $v$ onto $\Phi$. This point is taken further in Exercise 6.21.

The next thing we wish to do is to extend Theorem 7 to the case in which the orthonormal set is infinite, and to establish the conditions under which (6.11) is valid.

THEOREM 8 (THE FOURIER SERIES THEOREM). *Let $H$ be a Hilbert space, and let $\Phi = \{\phi_i\}_{i=1}^{\infty}$ be an orthonormal set in $H$. Then any $u \in H$ can be*

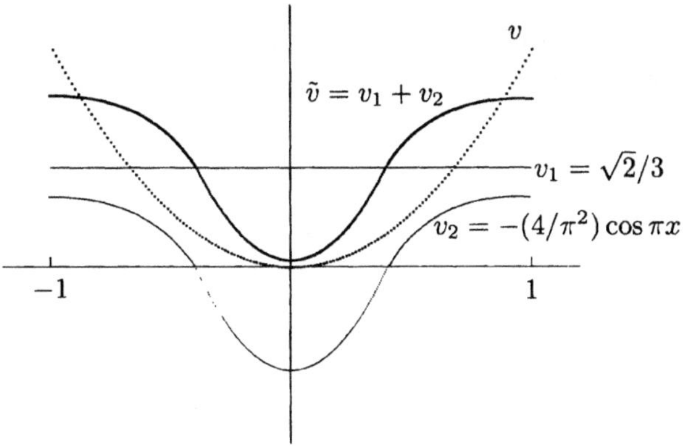


FIGURE 6.3. The function $v$ in Example 13 and its approximation

*expressed in the form*

$$u = \sum_{k=1}^{\infty} (u, \phi_k)\phi_k \tag{6.16}$$

*if and only if* $\Phi$ *is an orthonormal basis, that is, a maximal orthonormal set.*

PROOF. Assume first that $\Phi$ is an orthonormal basis. What we are required to show is that if $s_n$ denotes, as before, the partial sum $s_n = \sum_{k=1}^{n} u_k\phi_k$, where $u_k = (u, \phi_k)$, then $s_n \to u$ as $n \to \infty$, if and only if $\Phi$ is an orthonormal basis. We begin by showing that $\{s_n\}$ is a Cauchy sequence in $H$.

Let $n \geq m$, and consider

$$\|s_n - s_m\|^2 = (s_n - s_m, s_n - s_m) = \|s_n\|^2 - (s_n, s_m) - (s_m, s_n) + \|s_m\|^2.$$

Now recall that in the proof of the Best Approximation Theorem it was shown that $\|s_n\|^2 = \sum_{k=1}^{n} |u_k|^2$. Furthermore,

$$\begin{aligned}
(s_m, s_n) &= \left(\sum_{k=1}^{m} u_k\phi_k, \sum_{l=1}^{n} u_l\phi_l\right) \\
&= \left(\sum_{k=1}^{m} u_k\phi_k, \sum_{l=1}^{m} u_l\phi_l + \sum_{l=m+1}^{n} u_l\phi_l\right) \\
&= \left(\sum_{k=1}^{m} u_k\phi_k, \sum_{l=1}^{m} u_l\phi_l\right) + \left(\sum_{k=1}^{m} u_k\phi_k, \sum_{l=m+1}^{n} u_l\phi_l\right) \\
&= \sum_{k=1}^{m} |u_k|^2 + 0,
\end{aligned}$$

using the orthonormality of $\{\phi_k\}$. Likewise, $(s_n, s_m) = \sum_{k=1}^{m} |u_k|^2$. For convenience, set $\alpha_n = \sum_{k=1}^{n} |u_k|^2$. Then

$$\|s_n - s_m\|^2 = \alpha_n - \alpha_m = |\alpha_n - \alpha_m|;$$

the last term arises from the fact that $\alpha_m \leq \alpha_n$. Now from Bessel's inequality (6.14), $|\alpha_n| \leq \|u\|^2$, and since the right-hand side is independent of $n$, the sequence $\{\alpha_n\}$ is bounded; it is also monotone increasing, hence (see Exercise 1.14) it is convergent, and therefore also Cauchy (in $\mathbb{K}$). The sequence $\{s_n\}$ is thus also a Cauchy sequence (in $H$), and by the completeness of $H$ this sequence converges, to a member $u'$, say, of $H$. It remains to show that $u' = u$.

Consider

$$\begin{aligned}(u - u', \phi_l) &= \left(u - \lim_{n\to\infty} \sum_{k=1}^{n} u_k\phi_k, \phi_l\right) \\ &= \lim_{n\to\infty} \left(u - \sum_{k=1}^{n} u_k\phi_k, \phi_l\right) \quad \text{(using Exercise 4.2)} \\ &= \lim_{n\to\infty} [(u, \phi_l) - (u, \phi_l)] = 0.\end{aligned}$$

This holds for all $l$, and $\{\phi_l\}$ is a basis, so it follows that we must have $u' = u$. This proves the first part of the theorem.

Conversely, suppose that every $u \in H$ has the form (6.16). Then we have, for $u \in H$,

$$\begin{aligned}\|u\|^2 &= (u, u) = \left(\sum_{k=1}^{\infty} u_k\phi_k, \sum_{l=1}^{\infty} u_l\phi_l\right) \\ &= \sum_{k=1}^{\infty}\sum_{l=1}^{\infty} u_k\bar{u}_l(\phi_k, \phi_l) = \sum_{k=1}^{\infty} u_k\bar{u}_k = \sum_{k=1}^{\infty} |(u, \phi_k)|^2. \qquad (6.17)\end{aligned}$$

Now if $\Phi$ is not maximal, then there is a vector $\phi_0$, with $\|\phi_0\| = 1$, such that $(\phi_0, \phi_k) = 0$ for all $k$. But from (6.17),

$$1 = \|\phi_0\|^2 = \sum_i (\phi_0, \phi_i)^2 = 0,$$

a contradiction. Hence $\Phi$ is an orthonormal basis. □

The Fourier Series Theorem gives explicit conditions under which the elementary notion of representation of an element in terms of a basis is valid for infinite-dimensional spaces. The theorem does not guarantee, though, that it will always be possible to find such bases. Such a guarantee is provided by the next theorem.

THEOREM 9. *Every Hilbert space $H$ has an orthonormal basis.*

PROOF. The proof relies on an application of Zorn's Lemma, which was introduced in Chapter 1; the lemma is applied to the family of orthonormal sets in $H$. Let this family be denoted by $\mathcal{C}$; then it is a partially ordered set if as the partial ordering operation $\leq$ we use set inclusion, so that two sets $S_1$ and $S_2$ of orthonormal sets are ordered according to $S_1 \subseteq S_2$. The collection $\mathcal{C}$ is nonempty since the set $S = \{v/\|v\|\}$, where $v$ is any member of $H$, is trivially an orthonormal set.

Now let $\{S_n\}$ be a linearly ordered subset of $\mathcal{C}$; then $\cup_n S_n$ is itself an orthonormal set, and furthermore it contains each $S_n$, so that it is therefore an upper bound for $\{S_n\}$. Every linearly ordered subset of $\mathcal{C}$ has an upper bound, so by Zorn's Lemma $\mathcal{C}$ has a maximal element, that is, an orthonormal set not contained in any other orthonormal set. This is of course the definition of an orthonormal basis. □

If the Hilbert space $H$ is known to be *separable* (see Chapter 4, Section 4), then the proof of existence of an orthonormal basis is much more straightforward. Indeed, if $H$ is separable, then we can pick at least one countable set $U = \{u_n\}$ in $H$ that is dense in $H$. Now retain in $U$ only sufficient elements in order for the resulting set, $U'$, say, to consist of linearly independent elements. Then $U'$ is also dense in $H$. Finally, apply the Gram–Schmidt procedure to this set.

THEOREM 10. *Let $H$ be a separable Hilbert space. Then it has an orthonormal basis.*

We are now assured of the existence of orthonormal bases in Hilbert spaces; it remains only to find ways of *constructing such bases*. There are various procedures for doing this, and we explore one such approach, in the context of the space $L^2(\Omega)$, which is simple, and which has considerable relevance to physical problems. This is the subject of Sturm–Liouville problems.

## 6.5 Sturm–Liouville problems

**Separation of variables.** The method of separation of variables is a popular technique for solving simple linear initial boundary value problems. The use of this method leads to an eigenvalue problem whose solution is essential to the solution of the problem as a whole, and it is such eigenvalue problems that are examples of Sturm–Liouville problems, the *eigenfunctions* of which, it is shown, are candidates for orthonormal bases.

To review the method of separation of variables we return to the Introduction, and to the problem of heat conduction. This problem, which is formulated in Box 1 in that chapter, is considered in a somewhat simplified form, in that the following assumptions are made: the medium is homogeneous, heat sources are absent, and the problem is one-dimensional in space. We choose as the domain the interval $(-\ell, \ell)$, so that the problem becomes one of finding the temperature $u$ that satisfies the PDE

$$\frac{\partial u}{\partial t} - \kappa \frac{\partial^2 u}{\partial x^2} = 0 \text{ in } (-\ell, \ell), \tag{6.18}$$

in which $\kappa$, the thermal conductivity, is a positive constant. The temperature is assumed to be zero at each end, so that the boundary conditions are

$$u(-\ell, t) = 0 \quad \text{and} \quad u(\ell, t) = 0. \tag{6.19}$$

The initial temperature is specified by a given function $f$, so that

$$u(x, 0) = f(x). \tag{6.20}$$

The essence of the method of separation of variables is to seek a solution of the form

$$u(x, t) = M(x)N(t);$$

substitution in (6.18) and rearrangement leads to the equation

$$\frac{N'(t)}{N(t)} = \frac{\kappa M''(x)}{M(x)},$$

and since the left-hand side depends only on $t$ and the right-hand side only on $x$, it follows that each side of this equation is equal to a constant, which we denote by $-\lambda$, the minus sign being inserted for future convenience. The boundary conditions (6.19) become $M(-\ell) = M(\ell) = 0$ and so the first problem becomes one of finding $u$ and $\lambda$ that satisfy

$$\begin{aligned} -\kappa M''(x) &= \lambda M(x) \\ M(-\ell) = 0, \quad M(\ell) &= 0. \end{aligned} \tag{6.21}$$

The second problem involves finding $N$ that satisfies

$$N'(t) + \lambda N(t) = 0. \tag{6.22}$$

It is equation (6.21) which is the prototype of the Sturm–Liouville eigenvalue problem.

**Eigenvalue problems.** Let $L$ be a linear operator (the precise specification of its domain is not too important right now); then the eigenvalue problem for the operator $L$ is the problem of finding $u$ and $\lambda$ that satisfy

$$Lu = \lambda u, \tag{6.23}$$

where $\lambda$ is in general a complex number. If $L$ is a differential operator defined on a domain $\Omega$ with boundary $\Gamma$, then it is also necessary to specify *homogeneous* boundary conditions of the form

$$Bu = 0, \tag{6.24}$$

in which $B$ is also a linear operator. Alternatively, (6.23) could represent a matrix eigenvalue problem, of the kind that is encountered in elementary courses in linear algebra.

The defining features of an eigenvalue problem are, first, that $u = 0$ is a solution, known as the trivial solution; second, there are special nonzero values of $\lambda$, called *eigenvalues*, for which (6.23) and (6.24) have nontrivial solutions. In the context of matrix problems these solutions are known as eigenvectors whereas for differential equations they are known as *eigenfunctions.* In either case they are determined only up to a multiplicative constant; that is, if $u$ is an eigenfunction, then so is $\alpha u$ for any number $\alpha$, since (6.23) gives $\alpha Lu = \alpha \cdot \lambda u$ or $L(\alpha u) = \lambda(\alpha u)$. Because of this indeterminacy, it is customary to normalize eigenfunctions in some convenient manner.

Returning to problem (6.21), we now seek a solution to this eigenvalue problem in the form $M(x) = e^{ax}$. Substitution in $(6.21)_1$ leads to the equation $-\kappa a^2 = \lambda$, so that there are two possible solutions, viz. $a = \pm i\sqrt{\lambda/\kappa}$ (assuming that $\lambda > 0$). So the most general solution of $(6.21)_1$ is a linear combination of the solutions corresponding to the two values of $a$. Since $e^{iy} = \cos y + i \sin y$ for any real number $y$, this general solution may be expressed in the alternative form

$$M(x) = A\cos\alpha x + B\sin\alpha x$$

in which $\alpha = \sqrt{\lambda/\kappa}$. The boundary conditions $(6.21)_2$ are imposed next: these give

$$\left.\begin{array}{l} M(-\ell) = 0 \\ M(\ell) = 0 \end{array}\right\} \quad \Rightarrow \quad \begin{array}{l} A\cos\alpha\ell - B\sin\alpha\ell = 0 \\ A\cos\alpha\ell + B\sin\alpha\ell = 0 \end{array}$$

or, in matrix form,

$$\begin{pmatrix} \cos\alpha\ell & -\sin\alpha\ell \\ \cos\alpha\ell & \sin\alpha\ell \end{pmatrix} \begin{pmatrix} A \\ B \end{pmatrix} = \begin{pmatrix} 0 \\ 0 \end{pmatrix}.$$

In order for this set of equations to have a nontrivial solution it is necessary and sufficient that the determinant of the matrix be zero; that is, we require that

$$\cos\alpha\ell \sin\alpha\ell = 0 \quad \text{or} \quad \sin 2\alpha\ell = 0,$$

the solution of which is $2\alpha\ell = k\pi$ $(k = 0, 1, 2, \ldots)$, so that the problem (6.21) has an infinite sequence of eigenvalues $\lambda_k$, $k = 0, 1, 2, \ldots$, where

$$\lambda_k = \kappa\left(\frac{k\pi}{2\ell}\right)^2.$$

It also follows that there is an infinite sequence of eigenfunctions, denoted here for convenience by $M_k(x)$, where

$$M_k(x) = A_k \cos \alpha_k x + B_k \sin \alpha_k x$$

and $\alpha_k = k\pi/2\ell$. We are now able to return to the problem (6.22), which is considered in the form

$$N'(t) + \lambda_k N(t) = 0.$$

The solution of this equation, for each value of $k$, is

$$N_k(t) = \exp(-\lambda_k t).$$

The general solution of (6.18) and (6.19) may now be obtained by adding up the linear combinations of the possible solutions; we set aside for now the issue of convergence of the infinite sum that results, and express the general solution in the form

$$u(x,t) = \sum_{k=0}^{\infty} [A_k \cos \alpha_k x + B_k \sin \alpha_k x] \exp(-\lambda_k t).$$

All that remains is to obtain the constants $A_k$ and $B_k$. These may be found by using the last remaining condition to be satisfied, which is the initial condition; from (6.20), then,

$$f(x) = u(x,0) = M(x)N(0) = \sum_{k=0}^{\infty} [A_k \cos \alpha_k x + B_k \sin \alpha_k x]. \tag{6.25}$$

The representation (6.25) of the function $f$ is known as the *eigenfunction expansion* of $f$. The coefficients may be found by exploiting the orthogonality properties of the trigonometric functions; indeed, with the $L^2$-inner product on $(-\ell, \ell)$ denoted by $(\cdot, \cdot)$, recall that

$$\begin{aligned} &(\sin \alpha_k x, \sin \alpha_j x) = (\cos \alpha_k x, \cos \alpha_j x) = \begin{cases} \ell & \text{if } k = j, \\ 0 & \text{otherwise}, \end{cases} \\ &(\cos \alpha_k x, \sin \alpha_j x) = 0. \end{aligned} \tag{6.26}$$

So if we take the inner product of each side of (6.25) with $\sin \alpha_k x$ and with $\cos \alpha_k x$ in turn, we find that

$$(f, \sin \alpha_k x) = B_k.\ell \quad \text{and} \quad (f, \cos \alpha_k x) = A_k.\ell \tag{6.27}$$

and the coefficients $A_k$ and $B_k$ are all thus determined.

The relevance of the problem just discussed, and of Sturm–Liouville problems in general, may now be explored a little further. First, it has been seen that the eigenfunctions of the Sturm–Liouville problem (6.21) are the set of trigonometric functions $\Phi \equiv \{\cos \alpha_k x\}_{k=0}^{\infty} \cup \{\sin \alpha_k x\}_{k=1}^{\infty}$; second, in order to complete the solution of the problem it is required to expand the function $f(x)$ in terms of these trigonometric functions. The appearance of the $L^2$-inner product in (6.27) is not accidental; indeed, the question of whether the function $f$ may be represented in the form (6.25) is equivalent to asking whether the set $\Phi$ forms a *basis* for $L^2(-\ell, \ell)$ (this set is indeed orthogonal but not orthonormal, although easily orthonormalized). The answer is affirmative, and this is the relevance of Sturm–Liouville problems to the subject of orthonormal bases: *the eigenfunctions of Sturm–Liouville problems constitute orthonormal bases for $L^2$*. These considerations are precisely what motivate elementary Fourier analysis, and indeed (6.25) together with (6.27) simply gives the Fourier series repsentation of $f$.

We turn now to a more detailed study of Sturm–Liouville problems, the objective being to work towards this general result.

**Sturm–Liouville problems.** A Sturm–Liouville operator $L$ is a linear operator of the form

$$Lu = \frac{1}{\omega}[-(pu')' + qu],$$

defined on an interval $[a, b]$ of the real line. Here $p$, $p'$, $q$, and $\omega$ are continuous real-valued functions on $[a, b]$ that satisfy

$$\left.\begin{array}{l} p(x) > 0 \\ q(x) \geq 0 \\ \omega(x) > 0 \end{array}\right\} \quad \text{on } [a, b]. \tag{6.28}$$

Let $B_1$ and $B_2$ be linear operators that specify boundary values of a continuous function, and that are defined by

$$\begin{aligned} B_1 u &= \alpha_1 u(a) + \beta_1 u'(a), \\ B_2 u &= \alpha_2 u(b) + \beta_2 u'(b). \end{aligned} \tag{6.29}$$

The constants $\alpha$ and $\beta$ satisfy

$$\alpha_i \geq 0, \quad \beta_i \geq 0, \text{ and } \alpha_i + \beta_i > 0. \tag{6.30}$$

Then a *regular Sturm–Liouville problem* is an eigenvalue problem of the form

$$\begin{aligned} Lu = \lambda u \quad &\text{on } (a, b) \\ B_1 u = 0, \quad &B_2 u = 0. \end{aligned} \tag{6.31}$$

Equation $(6.31)_1$ is encountered in the form

$$-(pu')' + qu = \lambda \omega u,$$

rather than in the form in which $\omega$ is found on the left-hand side.

If any of the conditions in the definition differ from those given here, whether with respect to the boundedness of the interval, the requirements (6.28), or the conditions (6.30), the problem is then known as a *singular* Sturm–Liouville problem.

The problem (6.31) is considered in the space $L^2(a,b)$ endowed with the inner product $(\cdot,\cdot)$ defined by

$$(u,v) = \int_a^b u(x)\overline{v(x)}\,\omega(x)\;dx;$$

because of its role in the definition of the inner product, $\omega$ is called a *weighting function.*

Now the first issue that needs to be resolved is that concerning the domain $D(L)$ of the operator $L$. The problem is posed in $L^2(a,b)$, and of course not all members of this space have derivatives in the classical sense. It follows that $D(L)$ has to be a proper subspace of $L^2(a,b)$, and it suffices to take

$$D(L) = \{v \in L^2(a,b) \cap C^2[a,b],\; B_1 v = B_2 v = 0\}. \tag{6.32}$$

Since the space $C_0^\infty(a,b)$ is contained in $D(L)$, and since $C_0^\infty(a,b)$ is dense in $L^2(a,b)$ (see Chapter 4, Theorem 6 and the discussion that follows it), it follows that $D(L)$ is dense in $L^2(a,b)$.

## Examples

14. The problem (6.21) is a Sturm–Liouville problem with $[a,b] = [-\ell,\ell]$, $p(x) = \kappa$, $q(x) = 0$, and $\omega(x) = 1$. With regard to the boundary conditions, $\alpha_1 = \alpha_2 = 1$ and $\beta_1 = \beta_2 = 0$.

15. *Legendre's equation* arises when the method of separation of variables is applied to problems having spherical symmetry (see Exercise 6.23). This problem takes the form

$$\begin{gathered} -[(1-x^2)u']' = \lambda u \;\text{ on } (-1,1), \\ u(-1) \text{ and } u(1) \text{ are finite,} \end{gathered} \tag{6.33}$$

and is a singular Sturm–Liouville problem since the boundary conditions do not conform to the structure of (6.29) and $(6.31)_2$. Nevertheless, many of the properties of regular problems hold in this case as well.

**Symmetric operators.** It turns out that Sturm–Liouville operators are examples of what are known as symmetric operators, and symmetric operators have many of the nice properties that symmetric matrices possess in linear algebra.

Let $L$ be a linear operator defined on a Hilbert space $H$, with domain $D(L)$. Then $L$ is said to be a symmetric operator if

$$(Lu, v) = (u, Lv) \quad \text{for all } u, v \in D(L). \tag{6.34}$$

It is important to bear in mind that the definition applies to members of the domain of $L$, and given that boundary conditions play a role in the choice of the domain (as in (6.32)), these will be crucial in determining whether a given operator is symmetric.

The next two results concerning symmetric operators are direct generalizations of the situation that pertains for matrices.

LEMMA 2. *The eigenvalues of a symmetric linear operator are real.*

PROOF. Consider the eigenvalue problem $Lu = \lambda u$. Then

$$\begin{aligned}(\lambda - \overline{\lambda})(u, u) &= \lambda(u, u) - \overline{\lambda}(u, u) \\ &= (\lambda u, u) - (u, \lambda u) = (Lu, u) - (u, Lu) = 0,\end{aligned}$$

using (6.34). □

LEMMA 3. *Let $L$ be a symmetric linear operator defined on a Hilbert space $H$. Then the eigenfunctions corresponding to two distinct eigenvalues are orthogonal.*

PROOF. Let $\lambda_1$ and $\lambda_2$ be eigenvalues of $L$ with eigenfunctions $u_1$ and $u_2$, respectively. Then $Lu_i = \lambda_i u_i$ $(i = 1, 2)$ and so

$$\begin{aligned}(\lambda_1 - \lambda_2)(u_1, u_2) &= \lambda_1(u_1, u_2) - \lambda_2(u_1, u_2) \\ &= (\lambda_1 u_1, u_2) - (u_1, \lambda_2 u_2) = (Lu_1, u_2) - (u_1, Lu_2) = 0.\end{aligned}$$

Since $\lambda_2 \neq \lambda_1$ by assumption, it follows that $(u_1, u_2) = 0$. □

**Properties of Sturm–Liouville operators.** We begin by establishing that $L$ is symmetric; indeed, for any $u$ and $v$ in $D(L)$ we have

$$\begin{aligned}(Lu, v) - (u, Lv) &= \int_a^b [-(pu')'\overline{v} - qu\overline{v} + u(p\overline{v}')' + qu\overline{v}]\, dx \\ &= \int_a^b [-(pu')'\overline{v} + (p\overline{v}')'u]\, dx \\ &= \int_a^b [(p\overline{v}'u)' - (p\overline{v}u')']\, dx \\ &= [p\overline{v}'u - pu'\overline{v}]_a^b \\ &= p(b)[u(b)\overline{v}'(b) - u'(b)\overline{v}(b)] \\ &\quad -p(a)[u(a)\overline{v}'(a) - u'(a)\overline{v}(a)].\end{aligned} \tag{6.35}$$

Now since $v$ belongs to $D(L)$, so does $\overline{v}$ since the coefficients in the boundary terms are all real. It follows that $B_1u = B_1\overline{v} = 0$ or, recasting this in matrix form after using (6.29),

$$\begin{pmatrix} u(a) & u'(a) \\ \overline{v}(a) & \overline{v}'(a) \end{pmatrix} \begin{pmatrix} \alpha_1 \\ \beta_1 \end{pmatrix} = \begin{pmatrix} 0 \\ 0 \end{pmatrix}.$$

From the set of conditions (6.30) at least one of $\alpha_1$ and $\beta_1$ must be nonzero, and this is only possible if the matrix is singular, that is, if

$$u(a)\overline{v}'(a) - u'(a)\overline{v}(a) = 0.$$

Repeating the exercise for the boundary condition $B_2u = B_2\overline{v} = 0$, we obtain for that case

$$u(b)\overline{v}'(b) - u'(b)\overline{v}(b) = 0.$$

From these two equations it follows that the right-hand of side of (6.35) is zero, as desired. This result, together with a related result, is summarized in the following theorem.

THEOREM 11.

(a) *The Sturm–Liouville operator is symmetric;*

(b) *The Sturm–Liouville operator $L$ is positive; that is, $(Lu, u) \geq 0$ for all $u \in D(L)$.*

The proof of part (b) is deferred to Exercise 6.25, as is the proof of the following corollary.

COROLLARY TO THEOREM 11. *The eigenvalues of $L$ are all nonnegative, and form a countable set.*

Thus we have established that the eigenvalues of $L$ may be arranged in the sequence $0 \leq \lambda_1 \leq \lambda_2 \leq \ldots$. It can further be shown that $\lambda_n \to \infty$ as $n \to \infty$, although we do not pursue this result here.

We come now to the main result of this section.

THEOREM 12. *The eigenfunctions of a regular Sturm–Liouville problem form an orthonormal basis for $L^2(a, b)$.*

By way of preparing for the proof of this theorem, we introduce the *Rayleigh quotient* $R$, a functional defined on $D(L)$ by

$$R(v) = \frac{(Lv, v)}{\|v\|^2} \quad \text{for all } v \in D(L).$$

Note that $R(v) \geq 0$ by the positivity of $L$.

LEMMA 4. *The minimum of $R(v)$ over all functions $v \in D(L)$ that are orthogonal to the first $n$ eigenfunctions is $\lambda_{n+1}$. That is,*

$$\min\{R(v): \ v \in D(L), \ (v,\phi_1) = (v,\phi_2) = \ldots = (v,\phi_n) = 0\} = \lambda_{n+1}.$$

*Furthermore, the minimizing function is $\phi_{n+1}$.*

The proof of this lemma is treated in Exercise 6.26.

PROOF OF THEOREM 12. We have already established from Lemma 3 and the Corollary to Theorem 11 that there is a countable infinity of eigenfunctions of the Sturm–Liouville problem, and that these are mutually orthogonal. Since the eigenfunctions $\{\phi_1, \phi_2, \ldots\}$ are determined up to a multiplicative constant, we assume once and for all that these are normalized, so that $\|\phi_k\| = 1, \quad k = 1, 2, \ldots$.

It therefore remains to show that, given any member $u$ of $L^2(a,b)$ and a Sturm–Liouville problem defined on $D(L)$, the eigenfunctions $\{\phi_k\}_{k=1}^{\infty}$ form an orthonormal basis for $L^2(a,b)$ in the sense that for every $u \in H$

$$u = \sum_{k=1}^{\infty} u_k \phi_k \quad \text{or} \quad \lim_{n\to\infty} \|u - s_n\| = 0,$$

where $u_k = (u, \phi_k)$ and $s_n = \sum_{k=1}^{n} u_k \phi_k$ (recall Theorem 8 and its proof).

It is convenient to introduce the remainder $r_n = u - s_n$, and we now estimate $R(r_n)$. For $k = 1, \ldots, n$ we have

$$\begin{aligned} (r_n, \phi_k) &= (u - s_n, \phi_k) = (u, \phi_k) - (s_n, \phi_k) \\ &= u_k - u_k = 0. \end{aligned}$$

In other words, the remainder is orthogonal to the first $n$ eigenfunctions. Now according to Lemma 4, the smallest possible value of $R(v)$ over $v$ in the subspace of functions that are orthogonal to the first $n$ eigenfunctions, is $\lambda_{n+1}$. It follows, therefore, that since $r_n$ belongs to this subspace, $R(r_n)$ cannot be less than $\lambda_{n+1}$:

$$R(r_n) = \frac{(Lr_n, r_n)}{\|r_n\|^2} \geq \lambda_{n+1}. \tag{6.36}$$

Suppose now that we are able to show that the numerator $(Lr_n, r_n)$ is bounded independently of $n$; then as $n \to \infty$ the right-hand side of (6.36) goes to infinity, and hence $r_n \to 0$. The theorem is thus proved if it can be shown that $(Lr_n, r_n)$ is bounded. This is achieved by showing that $(Lr_n, r_n)$

can be bounded above by $(Lu, u)$; indeed, we have

$$\begin{aligned}
(Lu,u) &= (L(r_n+s_n), r_n+s_n) \\
&= (Lr_n,r_n)+(Ls_n,r_n)+(Lr_n,s_n)+(Ls_n,s_n) \\
&= (Lr_n,r_n)+(Ls_n,r_n)+(r_n,Ls_n)+(Ls_n,s_n) \\
&\quad \text{(using the symmetry of } L) \\
&= (Lr_n,r_n)+(Ls_n,s_n)+2\text{Re}\,(Ls_n,r_n) \\
&\geq (Lr_n,r_n)+2\text{Re}\,(Ls_n,r_n) \\
&\quad \text{since } (Ls_n,s_n)\geq 0.
\end{aligned}$$

Now it can be shown that in fact $(Ls_n, r_n) = 0$ (see Exercise 6.27), and so we are left with the inequality

$$(Lr_n, r_n) \leq (Lu, u),$$

which shows that $(Lr_n, r_n)$ is bounded. This concludes the proof of the theorem. □

## Examples

16. Theorem 12 confirms that the set of eigenfunctions $\{\cos\alpha_k x\}_{k=0}^{\infty} \cup \{\sin\alpha_k x\}_{k=1}^{\infty}$ of the Sturm–Liouville problem (6.21) forms a basis for $L^2(a,b)$. In this form the basis is orthogonal, but not orthonormal; however, it is easily converted to an orthonormal basis by using the relations (6.26), according to which $\|\cos\alpha x\| = \|\sin\alpha x\| = \sqrt{\ell}$ and, for the case $k = 0$, $\|1\| = \sqrt{2\ell}$. The orthonormal basis is thus

$$\{1/\sqrt{2\ell}\} \cup \{1/\sqrt{\ell}\cos\alpha_k x\}_{k=1}^{\infty} \cup \{1/\sqrt{\ell}\sin\alpha_k x\}_{k=1}^{\infty}.$$

17. We return to the problem (6.21), but this time express the solution in the form

$$M(x) = Ce^{i\alpha x} + De^{-i\alpha x},$$

rather than rewriting it in terms of trigonometric functions. By following the same procedure as before (that is, substituting in the boundary conditions and seeking a nontrivial solution) we find that the eigenvalues are $\lambda_n = (n\pi/\ell)^2$, $n = 0, \pm1, \pm2, \ldots$ so that the set of eigenfunctions is now

$$\{e^{in\pi/\ell}\}_{n=-\infty}^{\infty},$$

and this forms a basis for the space $L^2(-\ell,\ell)$ of complex-valued functions (see also Example 9). Normalization is easy, since

$$\|e^{in\pi/\ell}\|^2 = \int_{-\ell}^{\ell} e^{in\pi/\ell}\overline{e^{in\pi/\ell}}\,dx = \int_{-\ell}^{\ell} e^{in\pi/\ell}e^{-in\pi/\ell}\,dx = 2\ell.$$

18. The problem associated with Legendre's equation (6.33) is a singular Sturm–Liouville problem. Nevertheless, by a minor modification of the theory it can be shown that the properties of the regular problem carry over. In particular, the eigenfunctions of Legendre's equation, the Legendre polynomials $P_n$, form an orthogonal basis for $L^2(-1,1)$: these are defined by

$$P_n(x) = \frac{1}{2^n n!}\frac{d^n}{dx^n}(x^2-1)^n, \quad n = 0, 1, 2, \ldots .$$

Since

$$(P_n, P_n) = \int_{-1}^{1} P_n^2(x)dx = \frac{2}{2n+1},$$

the corresponding *orthonormal* set is $\{\phi_n(x),\ n = 0, 1, \ldots\}$ where $\phi_n = [(2n+1)/2]^{1/2} P_n$. It is also worth noting that the Legendre polynomials can be obtained by applying the Gram-Schmidt procedure to the set $\{1, x, x^2, \ldots\}$ of monomials (cf. Exercise 6.8).

## 6.6 Bibliographical remarks

Finite-dimensional spaces are given a detailed treatment in the texts by Halmos [18], Hoffman and Künze, [20] and Lang [28]. The texts by Noble [35] and by Strang [50] also give good, applications-oriented expositions of linear algebra. The subject of orthonormal bases in Hilbert spaces is developed in a systematic fashion in Naylor and Sell [33] and Oden [36], as are the other topics in this chapter. Other good treatments of the topic of orthonormal bases are to be found in Reed and Simon [40] and Apostol [2], the latter devoting attention to the situation in $L^2$. The proof of Lemma 1 may be found in Naylor and Sell. The text by Zauderer [52] is a good source for applications-oriented material on Sturm–Liouville problems, whereas Naylor and Sell treat the functional analytic aspects.

## 6.7 Exercises

### Finite-dimensional spaces

6.1. Which of the following subsets of $\mathbb{R}^3$ is linearly independent?
(a) $\{(1,\ 4,\ 9), (1,\ 0,\ 9), (-1,\ 4,\ -9)\}$; (b) $\{(1,\ 4,\ 9), (1,\ 0,\ 9), (2,\ 4,\ 8)\}$.

6.2. Let $p_k(x) = e^{ikx}$. Is $\{p_1, p_2, \ldots, p_n\}$ linearly independent in $L^2(0, 2\pi)$?

6.3. Let $X$ be the set of solutions to the ordinary differential equation

$$u'' - 2u' + u = 0, \quad u \in C^2[0,1].$$

Show that $X$ is a vector space, find $\dim X$, and display a basis for $X$.

6.4. Let $M$ be the vector space of all real $3 \times 3$ matrices and $K$ the subset of matrices of the form

$$\begin{pmatrix} \alpha & \alpha & -\alpha \\ \beta & \gamma & \delta \\ \beta & -\gamma & \delta \end{pmatrix}$$

for all nonzero real numbers $\alpha$, $\beta$, $\gamma$, $\delta$. What are $\dim M$ and $\dim K$? Display a basis for $K$.

6.5. Let $V$ and $W$ be proper subspaces of a finite-dimensional vector space $X$ such that $X = V \oplus W$. Show that

$$\dim(V \oplus W) = \dim V + \dim W.$$

6.6. Let $V$ be a subspace of a linear space $X$, with $\dim X = n$. Show (a) that every linearly independent subset of $V$ is part of a basis for $X$; and (b) that if $V$ is a proper subspace of $X$, then $\dim V < \dim X$.

**Finite-dimensional inner product and normed spaces**

6.7. Apply the Gram-Schmidt procedure to the vectors $\{(1,\, 0,\, 1),\ (1,\, 0,\, -1),\ (0,\, 3,\, 4)\}$ to obtain an orthonormal basis for $\mathbb{R}^3$.

6.8. Let $V$ be the four-dimensional subspace of $L^2(-1,1)$ spanned by $\{1,\, x,\, x^2,\, x^3\}$. Use the Gram-Schmidt procedure to construct an orthonormal basis for $V$.

6.9. Test the set $\{u_1, u_2\} = \{e^x, e^{-3x}\}$ for linear dependence in $L^2(0,1)$ by evaluating $\det \boldsymbol{A}$ where $A_{ij} = (u_i, u_j)_{L^2}$.

6.10. Show that any norm $\|\cdot\|_1$ on a finite-dimensional space $X$ is equivalent to any other norm $\|\cdot\|_2$ on $X$ (recall the definition (3.10) of equivalent norms). [Lemma 1 may be useful.]

**Linear operators on finite-dimensional spaces**

6.11. Let $X$ and $Y$ be the spaces of polynomials of degree 3 and 1, respectively, and let $\{1+t,\, t(1+t),\, 1+t^3\}$ and $\{1,\, t\}$ be bases for $X$ and $Y$, respectively. Let $T : X \to Y$ be the linear operator defined by

$$Tp = d^2p/dx^2.$$

Find the matrix corresponding to $T$.

6.12. Let $X$ be the linear space of all functions of the form $u(x) = \alpha + \beta\cos x + \gamma\sin x$, $0 \le x \le 2\pi$, and define $T : X \to X$ by

$$Tu = \int_0^{2\pi} [1 + \cos(x - \xi)]u(\xi)\, d\xi.$$

Find the matrix corresponding to $T$.

6.13. Let $\boldsymbol{T}$ be an $n \times m$ matrix with transpose $\boldsymbol{T}^t$, and consider the equation $\boldsymbol{T}\boldsymbol{a} = \boldsymbol{b}$, where $\boldsymbol{a} \in \mathbb{R}^m$ and $\boldsymbol{b} \in \mathbb{R}^n$. Suppose that we wish to solve for $\boldsymbol{a}$: show that a necessary condition for such a solution to exist is

$$(\boldsymbol{c}, \boldsymbol{b}) = 0 \quad \text{for all} \quad \boldsymbol{c} \in N(\boldsymbol{T}^t);$$

that is, $\boldsymbol{b} \in N(\boldsymbol{T}^t)^{\perp}$. (Note that $(\boldsymbol{x}, \boldsymbol{T}\boldsymbol{y}) = (\boldsymbol{T}^t\boldsymbol{x}, \boldsymbol{y})$.) This shows that $R(\boldsymbol{T}) \subset N(\boldsymbol{T}^t)^{\perp}$. Show that $R(\boldsymbol{T}) = N(\boldsymbol{T}^t)^{\perp}$.
Determine $N(\boldsymbol{T}^t)$ for the matrix

$$\boldsymbol{T} = \begin{bmatrix} 3 & -1 & 1 \\ 1 & 2 & 1 \\ 4 & 1 & 2 \end{bmatrix}$$

and hence find the general form of the vector $\boldsymbol{b}$ such that $\boldsymbol{T}\boldsymbol{a} = \boldsymbol{b}$.

6.14. Find a basis for the null space of the functional $\ell : \mathbb{R}^3 \to \mathbb{R}$, $\langle \ell, \boldsymbol{x} \rangle = \alpha_1 x_1 + \alpha_2 x_2 + \alpha_3 x_3$, where $\alpha_1 \ne 0$.

6.15. Prove Theorem 6, which states that there exists an isometric isomorphism from any $n$-dimensional inner product space to $\mathbb{R}^n$. Show that this does not hold in general for finite-dimensional normed spaces by verifying that $(\mathbb{R}^2, \|\cdot\|_1)$ and $(\mathbb{R}^2, \|\cdot\|_2)$ are not isometrically isomorphic.

6.16. Let $X = \mathbb{R}^3$ with the norm $\|\cdot\|_1$. If $\ell$ is as defined in Exercise 6.14, find $\|\ell\|$.

### Fourier series in Hilbert spaces

6.17. The set $\{1/\sqrt{2\pi}, \cos kx, \sin kx, k = 1, 2, \ldots\}$ is an orthonormal basis for $L^2(-\pi, \pi)$. Find the Fourier coefficients $u_i$ if (i) $u(x) = 1$; (ii)

$$u(x) = \begin{cases} -1, & -\pi \le x \le 0 \\ 1, & 0 < x \le \pi. \end{cases}$$

6.18. Determine the first three terms of the expansion $u = \sum_{k=0}^{\infty} u_k e_k$ on $[-1, 1]$ when $e_k$ are the normalized Legendre polynomials and

$$u(x) = \begin{cases} -1, & -1 \le x \le 0 \\ x, & 0 < x \le 1. \end{cases}$$

6.19. The trigonometric Fourier series representation

$$u(x) = \frac{u_0}{\sqrt{2}} + \sum_{k=1}^{\infty}(u_{2k}\cos k\pi x + u_{2k-1}\sin k\pi x)$$

can be written in complex form as $u(x) = \sum_{k=-\infty}^{\infty} c_k e^{ikx}$. Express the coefficients $c_k$ in terms of $u_k$.

6.20. (a) If $\Phi$ is an orthonormal *set* in an inner product space $X$, derive the *Bessel inequality* (part (b) of Theorem 7). [Use (6.15) to show that the inequality in (6.14) holds for finite sums. Then consider why it should hold for an infinite sum.]

(b) Derive *Parseval's formula* (part (c) of Theorem 7).

6.21. Let $\{\phi_1, \phi_2, \ldots, \phi_n\}$ be an orthonormal set in a Hilbert space $H$, and let $V = \text{span}\,\{\phi_1, \ldots, \phi_n\}$. Show that the orthogonal projection $P$ of $H$ onto $V$ is given by

$$Pu = \sum_{k=1}^{n}(u, \phi_k)\phi_k.$$

6.22. Let $V$ be the subspace of $L^2(-1,1)$ spanned by the first four orthonormal Legendre polynomials $\phi_n = (n + \frac{1}{2})^{1/2}P_n(x)$, $n = 0, 1, \ldots, 3$. Write out $\phi_n$ explicitly, find the orthogonal projection of $u(x) = x^4$ onto $V$, and verify the inequality $\|u - Pu\| \le \|u - v\|$, $v \in V$, for any suitable choice of $v$.

## Sturm–Liouville problems

6.23. Spherical coordinates $(r, \theta, \phi)$ are related to Cartesian coordinates $(x, y, z)$ through

$$x = r\sin\theta\cos\phi, \quad y = r\sin\theta\sin\phi, \quad z = r\cos\theta,$$

and the Laplacian operator in spherical coordinates is given by

$$\nabla^2 u = \frac{1}{r^2}\frac{\partial}{\partial r}\left(r^2\frac{\partial u}{\partial r}\right) + \frac{1}{r^2\sin\theta}\frac{\partial}{\partial\theta}\left(\sin\theta\frac{\partial u}{\partial\theta}\right) + \frac{1}{r^2\sin^2\theta}\frac{\partial^2 u}{\partial\phi^2}.$$

Consider the problem of steady heat conduction in a body with spherical symmetry.

(a) Assume that heat sources are absent, and that there is no dependence on $\phi$; then use the method of separation of variables to obtain two problems with independent variables $r$ and $\theta$, respectively. By making the change of variable $\xi = \cos\theta$, show that the second of these equations reduces to Legendre's equation (Example 15). Hence obtain the general solution to this problem.

(b) A known temperature distribution $f(\theta)$ is maintained over the surface of a sphere of unit radius. Find the steady-state temperature at any point in the region $\{(r, \theta, \phi) : r < 1\}$.

6.24. Determine the eigenvalues (approximately) and the eigenfunctions of the problem

$$v'' + \lambda v = 0, \quad v(0) = 0, \; v'(l) + \beta v(l) = 0,$$

where $\beta > 0$. Verify that the properties of Sturm–Liouville eigenpairs are satisfied. Which boundary conditions in the IBVP for the heat equation would give rise to this problem?

6.25. Show that the Sturm–Liouville operator is positive, that its eigenvalues are nonnegative, and that they form a countable set. [For the last part of this exercise use the fact that $L^2$ is separable, together with the properties of the set of eigenfunctions.]

6.26. If $R$ represents the Rayleigh quotient, show that $\min\{R(u) : (u, \phi_1) = \cdots = (u, \phi_{n-1}) = 0\} = \lambda_n$, where $(\cdot, \cdot)$ is the usual weighted inner product. Show also that $R$ is minimized by the $n$th eigenfunction $\phi_n$.

6.27. Show that $(Ls_n, r_n) = 0$ in the proof of Theorem 12.

6.28. Modify, where necessary, the theory for regular Sturm–Liouville problems to accommodate the following singular problems.

(a) $-(pu')' + qu = \lambda \omega u, \quad x \in (-\infty, \infty)$;

(b) $-(pu')' = \lambda \omega u \quad$ with $\; u(-L) = u(L), \; u'(-L) = u'(L)$.

6.29. The Schrödinger operator $S$ defined by

$$Su = -\frac{d^2 u}{dx^2} + a^2 x^2 u, \quad x \in (-\infty, \infty),$$

arises in the study of the harmonic oscillator, and is a singular Sturm–Liouville operator.

(a) Show that $S$ is symmetric on the space $C_0^2(-\infty, \infty)$ of twice continuously differentiable functions with compact support.

(b) Consider the Sturm–Liouville problem $Su = \lambda u$; make the changes of variables $y = \sqrt{a}x$ and $\mu = \lambda / a$ to obtain the equation

$$-v'' + x^2 u = \lambda u; \tag{6.37}$$

then make the further change of variable $v = u \exp(-x^2/2)$ to obtain the *Hermite differential equation*

$$v'' - 2yv' + \lambda v = 0. \tag{6.38}$$

(c) The Hermite polynomials $H_n$ may be defined by the Rodrigues formula

$$H_n(x) = (-1)^n e^{x^2} \frac{d^n}{dx^n}\left(e^{-x^2}\right), \quad n = 0, 1, 2, \ldots .$$

Show that $H_n' = 2nH_{n-1}$ and that $H_{n+1} - 2xH_n + H_n' = 0$, and that $H_n$ is a solution of (6.37) with $\lambda = 2n$. The functions $\phi_n$ $(n = 0, 1, \ldots)$ defined by

$$\phi_n(x) = (\sqrt{\pi}2^n n!)^{-1/2} H_n(x) \exp(-x^2/2)$$

are an orthonormal collection of eigenfunctions for the problem (6.38), and form a basis for $L^2(-\infty, \infty)$.

# 7
# Distributions and Sobolev spaces

In the Introduction the relevance of qualitative information about boundary value problems and their solutions was discussed, by way of motivating the introduction to functional analysis. Given the problem of finding a function $u$ that satisfies a partial differential equation and one or more boundary conditions, it is clearly of great value to know beforehand whether such a solution exists and, if so, whether it is unique, and finally how smooth this function is.

When approaching such issues one is essentially dealing with the properties of an operator (in this case a partial differential operator) $A$ from one space of functions $X$ to another space $Y$, so to start with it is necessary to choose suitable spaces for $X$ and $Y$. The spaces $C^m(\Omega)$ would appear at first sight to be appropriate since they are spaces of $m$-times continuously differentiable functions and we are, after all, dealing with differential equations. However, these spaces suffer from the drawback that, although $C(\overline{\Omega})$ with the sup-norm, and $C^m(\overline{\Omega})$ with a corresponding norm, are Banach spaces, these are not *Hilbert spaces*. Also, the spaces $C^m(\Omega)$ are not the natural spaces in which to consider variational formulations of problems, and such formulations lie at the heart of our treatment of boundary value problems and their approximation by finite elements.

The *Sobolev spaces* provide, as we show, a very natural setting for boundary value problems. First of all, there is a category of Sobolev spaces that are Hilbert spaces; second, it is possible to obtain quite general results regarding existence and uniqueness of solutions in a variational setting, using these spaces. A third advantage is that, like the spaces $C^m(\Omega)$, Sobolev spaces provide a means of characterizing the degree of smoothness of functions. Finally, and perhaps of most importance is the fact that approximate solution methods such as the Galerkin and finite element methods are most conveniently and correctly formulated in finite-dimensional subspaces of Sobolev spaces.

In order to discuss Sobolev spaces it is necessary first of all to learn a little about distributions. This necessary background is provided in Sections 7.1 and 7.2. Then in Section 7.3 we introduce Sobolev spaces, and discuss some of their more important properties. The apparently innocuous question of how one obtains the value of a function on the boundary of a domain, given the function on the domain, is shown in Section 7.4 to be a nontrivial problem. We show that a function has to satisfy certain requirements in order for its values on the boundary to be defined unambiguously. Finally, in Section 7.5 we discuss the Sobolev spaces $H_0^m(\Omega)$ of functions that, together with their derivatives of order less than $m$, vanish on the boundary. We also introduce in this section the space $H^{-m}(\Omega)$ (for $m > 0$) which is defined to be the dual space of $H_0^m(\Omega)$.

## 7.1 Distributions

In this section and in those that follow it is often necessary to deal with partial derivatives of all orders, and when discussing general ideas the notation can sometimes become very clumsy. As a prelude to the main topic of this chapter the very useful multi-index notation for partial derivatives is introduced.

**Multi-index notation**. Let $\mathbb{Z}_+^n$ denote the set of all ordered $n$-tuples of nonnegative integers: a member of $\mathbb{Z}_+^n$ is usually denoted by $\alpha$ or $\beta$, where, for example,

$$\alpha = (\alpha_1, \alpha_2, \ldots, \alpha_n),$$

each component $\alpha_i$ being a nonnegative integer.

We denote by $|\alpha|$ the sum $|\alpha| = \alpha_1 + \alpha_2 + \cdots + \alpha_n$ and by $D^\alpha u$ the partial derivative

$$D^\alpha u = \frac{\partial^{|\alpha|} u}{\partial x_1^{\alpha_1}\, \partial x_2^{\alpha_2} \cdots \partial x_n^{\alpha_n}}.$$

Thus if $|\alpha| = m$, then $D^\alpha u$ denotes one of the $m$th partial derivatives of $u$.

**Examples**

1. If $n = 3$, then a multi-index $\alpha \in Z_+^3$ is an ordered triple of nonnegative integers. For example, $\alpha = (1, 0, 3)$ belongs to $Z_+^3$, with $|\alpha| = \alpha_1 +$

$\alpha_2 + \alpha_3 = 1 + 0 + 3 = 4$. Furthermore, in this case the partial derivative $D^\alpha u$ is the fourth derivative defined by

$$D^\alpha u = \frac{\partial^4 u}{\partial x^{\alpha_1}\,\partial y^{\alpha_2}\,\partial z^{\alpha_3}} = \frac{\partial^4 u}{\partial x^1\,\partial y^0\,\partial z^3} = \frac{\partial^4 u}{\partial x\,\partial z^3}.$$

2. Let $n = 2$, and consider the expression

$$I = \sum_{|\alpha|\le 2} a_\alpha D^\alpha u,$$

where $a_\alpha$ are given functions of $x$ and $y$. Thus

$$I = \sum_{|\alpha|=0} a_\alpha D^\alpha u + \sum_{|\alpha|=1} a_\alpha D^\alpha u + \sum_{|\alpha|=2} a_\alpha D^\alpha u.$$

When $|\alpha| = 0$ the only possibility is $\alpha = (0,0)$ (remember that $n = 2$ here, so we are dealing with ordered pairs); the other values are

$$\begin{aligned} |\alpha| = 1: &\quad \alpha = (0,1) \text{ and } (1,0), \\ |\alpha| = 2: &\quad \alpha = (2,0) \text{ and } (1,1) \text{ and } (0,2). \end{aligned}$$

Suppose now that the functions $a_\alpha$ are given as

$$a_{00} = a_{20} = a_{02} = 1, \quad a_{10} = a_{01} = 2x, \quad a_{11} = x^2,$$

where we have written, for example, $a_{10}$ for $a_{(1,0)}$. Then

$$\begin{aligned} \sum_{|\alpha|=0} a_\alpha D^\alpha u &= a_{00}\frac{\partial^0 u}{\partial x^0\,\partial y^0} = 1\cdot u, \\ \sum_{|\alpha|=1} a_\alpha D^\alpha u &= a_{10}\frac{\partial^1 u}{\partial x^1\,\partial y^0} + a_{01}\frac{\partial^1 u}{\partial x^0\,\partial y^1} = 2x\left(\frac{\partial u}{\partial x} + \frac{\partial u}{\partial y}\right), \\ \sum_{|\alpha|=2} a_\alpha D^\alpha u &= a_{20}\frac{\partial^2 u}{\partial x^2\,\partial y^0} + a_{11}\frac{\partial^2 u}{\partial x^1\,\partial y^1} + a_{02}\frac{\partial^2 u}{\partial x^0\,\partial y^2} \\ &= 1\cdot\frac{\partial^2 u}{\partial x^2} + x^2\frac{\partial^2 u}{\partial x\,\partial y} + 1\cdot\frac{\partial^2 u}{\partial y^2}. \end{aligned}$$

Collecting all terms, it turns out that

$$\sum_{|\alpha|\le 2} a_\alpha D^\alpha u = \frac{\partial^2 u}{\partial x^2} + x^2\frac{\partial^2 u}{\partial x\,\partial y} + \frac{\partial^2 u}{\partial y^2} + 2x\left(\frac{\partial u}{\partial x} + \frac{\partial u}{\partial y}\right) + u.$$

Hence $\sum_{|\alpha|\le k} a_\alpha D^\alpha u$ is, in general, shorthand for a linear combination of partial derivatives of $u$, up to and including those of order $k$. The advantages of using multi-index notation should be evident from this simple example.

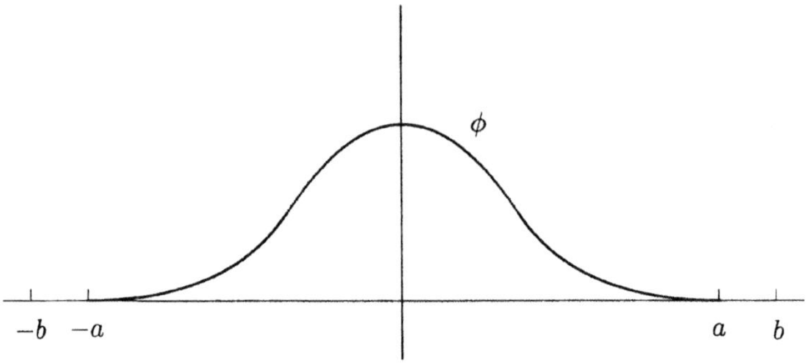


FIGURE 7.1. An example of a member of $\mathcal{D}(\Omega)$

In Section 5.4 we discussed an example that showed that the Dirac delta $\delta$ is not a function at all, but is more correctly viewed as a continuous linear *functional*, in that it operates on a continuous function $u$ to produce a real number, namely, $u(0)$:

$$\delta : C[-1,1] \to \mathbb{R}, \quad \langle \delta, u \rangle = u(0).$$

The Dirac delta belongs to a rather special space of functionals called *distributions*, and these in turn play a central role in the definition of Sobolev spaces. In order to introduce distributions formally, we first set up a space of very smooth functions on which these distributions can operate.

**The space $\mathcal{D}(\Omega)$.** For reasons that become evident later, it is desirable to consider the action of distributions not on all of $C(\Omega)$, but on only the small subset $C_0^\infty(\Omega)$ of infinitely differentiable functions with compact support; the notion of functions having compact support was, of course, introduced in Section 4.4. In the context of distribution theory it is conventional to use the notation $\mathcal{D}(\Omega)$ for $C_0^\infty(\Omega)$, and to refer to $\mathcal{D}(\Omega)$ as the space of *test functions*, because it is against functions in this space that distributions are tested, in a sense to be made precise.

## Example

3. A canonical example of a member of $\mathcal{D}(\Omega)$ is the function

$$\phi(x) = \begin{cases} 0, & |x| \geq a, \\ \exp[1/(x^2 - a^2)], & |x| < a, \end{cases}$$

defined on $\Omega = (-b, b)$, where $b > a > 0$, as shown in Figure 7.1. It is not difficult to show that $\phi$ is infinitely differentiable, and that the support of $\phi$ and all of its derivatives is the set $[-a, a]$.

It is possible to provide the space $\mathcal{D}(\Omega)$ with a topology known as an inductive limit topology, but such considerations are rather complicated.

Fortunately, the only topological concept we require is the notion of convergence of sequences in $\mathcal{D}(\Omega)$, for which the following definition suffices.

**Convergence in $\mathcal{D}(\Omega)$.** Let $\{\phi_n\}$ be a sequence of functions in $\mathcal{D}(\Omega)$. Then this sequence is said to converge to $\phi \in \mathcal{D}(\Omega)$ if

(a) there is a fixed compact set $K$ in $\Omega$ that contains the supports of all $\phi_n$; and
(b) the sequence $\{D^\alpha \phi_n\}$ converges uniformly on $K$ to $D^\alpha \phi$ for any $\alpha$.

**Distributions.** We define a distribution on a domain $\Omega$ in $\mathbb{R}^n$ to be a *continuous linear functional* on $\mathcal{D}(\Omega)$. That is, a distribution is a continuous linear map from $\mathcal{D}(\Omega)$ to $\mathbb{R}$. Thus the space of distributions is the *dual space* of $\mathcal{D}(\Omega)$ and, in keeping with the notation introduced in Section 5.4 for dual spaces, we denote the space of distributions by $\mathcal{D}'(\Omega)$.

Again, the topological notions that are required are best defined through the actions of sequences. Thus, to say that $f$ is continuous on $\mathcal{D}(\Omega)$ means that for every convergent sequence $\{\phi_n\}$ in $\mathcal{D}(\Omega)$, with limit $\phi$,

$$\langle f, \phi_n \rangle \to \langle f, \phi \rangle \quad \text{as } n \to \infty.$$

**Example**

4. By now the idea of the Dirac delta as a distribution should be a familiar one. In fact, $\delta$ belongs to $\mathcal{D}(-a, a)$ for any $a > 0$ since it is more generally defined by

$$\delta : \mathcal{D}(-a, a), \to \mathbb{R}, \quad \langle \delta, \phi \rangle = \phi(0), \quad \phi \in \mathcal{D}(\Omega),$$

and is therefore a continuous linear functional on $\mathcal{D}(\Omega)$.

**Regular distributions**. It is not only highly irregular objects such as the Dirac delta that are distributions. In fact, there are many ordinary functions that can be identified with distributions. All we require of a function $f$ is that the integral $\int_K |f(x)| dx$ be finite on every compact subset $K$ of $\Omega$. When this is so, $f$ is said to be *locally integrable* on $\Omega$, and a distribution $F$ associated with $f$ can then be defined in a very natural way by

$$F : \mathcal{D}(\Omega) \to \mathbb{R}, \quad \langle F, \phi \rangle = \int_\Omega f\phi \, dx, \quad \phi \in \mathcal{D}(\Omega).$$

If the support of $\phi$ is $K \subset \Omega$, then

$$|\langle F, \phi \rangle| = \left| \int_\Omega f\phi \, dx \right| = \left| \int_K f\phi \, dx \right| \le \sup_{x \in K} |\phi(x)| \int_K |f(x)| \, dx,$$

which is finite, and so we are assured that $\langle F, \phi \rangle$ has meaning. Under these circumstances $F$ is said to be the distribution *generated* by $f$. In future the different notation for a function $(f)$ and its associated distribution $(F)$ is dispensed with, and we simply write $f$ for both quantities. Whether $f$ is the function or the distribution that it generates is clear from the context; for example, $f$ in the expression $\int_\Omega f\phi \, dx$ is clearly a function whereas $f$ in $\langle f, \phi \rangle$ is a distribution.

## Examples

5. Every bounded continuous function is locally integrable and hence generates a distribution, but there are also many irregular and discontinuous functions that are locally integrable. One such example is the function $f(x) = |x|^{-1/2}$ on $[-1, 1]$. This function has a singularity at the origin, but is locally integrable since

$$\int_a^b |f(x)| \, dx = \int_a^b |x|^{-1/2} \, dx = 2(\operatorname{sgn} b|b|^{1/2} - \operatorname{sgn} a|a|^{1/2}),$$

and is bounded for every closed interval $[a, b]$ in $[-1, 1]$. Thus $f$ generates the distribution, also denoted by $f$, and defined by

$$\langle f, \phi \rangle = \int_{-1}^{1} |x|^{-1/2} \phi \, dx.$$

6. The *step function* $H(x)$, defined on $[-1, 1]$ by

$$H(x) = \begin{cases} 0, & -1 \le x < 0, \\ 1 & 0 \le x \le 1, \end{cases}$$

is locally integrable and generates the distribution $H$ that satisfies

$$\langle H, \phi \rangle = \int_{-1}^{1} H(x)\phi(x) \, dx = \int_0^1 \phi(x) \, dx.$$

A distribution that is generated by a locally integrable function is called a *regular distribution.* If a distribution cannot be generated by a locally integrable function, it is then said to be a *singular* distribution. An important example of a singular distribution is the Dirac delta; it is not difficult to show (see Exercise 7.2) that there does not exist a locally integrable function that generates $\delta$.

It is possible to define in a very natural way the *product* of a function and a distribution. Specifically, if $\Omega \subset \mathbb{R}^n$, $u$ belongs to $C^\infty(\Omega)$, and $f$ is a distribution on $\Omega$, then by $uf$ we understand the distribution satisfying

$$\langle (uf), \phi \rangle = \langle f, u\phi \rangle \quad \text{for all } \phi \in \mathcal{D}(\Omega). \tag{7.1}$$

Note that (7.1) is a generalization of the trivial identity

$$\int_\Omega [u(\boldsymbol{x})f(\boldsymbol{x})]\phi(\boldsymbol{x})\,dx = \int_\Omega f(\boldsymbol{x})[u(\boldsymbol{x})\phi(\boldsymbol{x})]\,dx$$

that holds when $f$ is locally integrable.

**Example**

7. The distribution $u\delta$ on $(-1,1)$, where $u(x) = x$, satisfies

$$\begin{aligned}\langle x\delta, \phi\rangle &= \langle \delta, x\phi\rangle \\ &= (x\phi)|_{x=0} = 0 \quad \text{for all } \phi \in \mathcal{D}(-1,1).\end{aligned}$$

Hence $x\delta = 0$, the zero distribution.

## 7.2 Derivatives of distributions

Quantities such as the Dirac delta and the Heaviside step function do not have derivatives in the ordinary sense. However, if they are treated as distributions it is possible to extend the concept of a derivative in such a way that any number of derivatives can be defined for these quantities and, indeed, for any distribution. Furthermore, we can define the derivative of a distribution in such a way that, should the distribution be (generated by) a continuously differentiable function, then the classical notion of a derivative is recovered. To this end we appeal to *Green's theorem*; the classical version of this well-known theorem states that the identity

$$\int_\Omega u\frac{\partial v}{\partial x_i}\,dx = \int_\Gamma uv\nu_i\,ds - \int_\Omega v\frac{\partial u}{\partial x_i}\,dx \tag{7.2}$$

holds for all functions $u, v$ in $C^1(\overline{\Omega})$, where $\nu_i$ is the $i$th component of the outward unit normal vector $\boldsymbol{\nu}$ to the boundary $\Gamma$ of a domain $\Omega$, the boundary being assumed to be sufficiently smooth (Figure 7.2). The one-dimensional version of (7.2), the integration-by-parts formula, states that

$$\int_a^b uv'\,dx = [uv]_a^b - \int_a^b vu'\,dx, \quad u, v \in C^1[a,b]. \tag{7.3}$$

Indeed, (7.3) is just a special case of (7.2) with $\Omega = (a,b), \Gamma = \{a,b\}$, and $\boldsymbol{\nu} = \pm 1$ at $x = b$ and $a$, respectively.

The theorem is easily generalized to a result involving partial derivatives of order $m$ of functions $u$, $v \in C^m(\overline{\Omega})$ (see Exercise 7.4): by replacing $u$ by $D^\alpha u$ in (7.2), and with $|\alpha| = m$, one can show that

$$\int_\Omega (D^\alpha u)v\,dx = (-1)^{|\alpha|}\int_\Omega uD^\alpha v\,dx + \int_\Gamma h(u,v)\,ds, \tag{7.4}$$

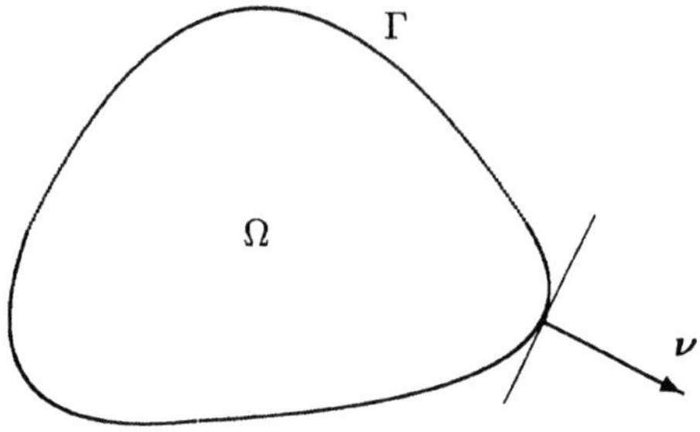


FIGURE 7.2. A domain $\Omega$ with boundary $\Gamma$ and outward unit normal $\boldsymbol{\nu}$

where $h(u, v)$ is an expression involving a sum of products of derivatives of $u$ and $v$ of order less than $m$.

Now replace $v$ in (7.4) by a function $\phi$ belonging to $\mathcal{D}(\Omega)$. Then since $\phi = 0$ on the boundary, (7.4) becomes

$$\int_\Omega (D^\alpha u)\phi \, dx = (-1)^{|\alpha|} \int_\Omega u D^\alpha \phi \, dx. \tag{7.5}$$

Since $u$ is $m$-times continuously differentiable it generates a distribution, also denoted by $u$, so that

$$\langle u, \phi \rangle = \int_\Omega u\phi \, dx$$

or, since $D^\alpha \phi$ also belongs to $\mathcal{D}(\Omega)$,

$$\langle u, D^\alpha \phi \rangle = \int_\Omega u D^\alpha \phi \, dx.$$

Furthermore, $D^\alpha u$ is continuous so it is able to generate a regular distribution (denoted by $D^\alpha u$) satisfying

$$\langle D^\alpha u, \phi \rangle = \int_\Omega (D^\alpha u)\phi \, dx.$$

Hence (7.5) can be written as

$$\langle D^\alpha u, \phi \rangle = (-1)^{|\alpha|} \langle u, D^\alpha \phi \rangle \quad \text{for all } \phi \in \mathcal{D}(\Omega). \tag{7.6}$$

*We take (7.6) as the basis for defining the derivative of any distribution* $f$, as follows. The $\alpha$th *distributional* or *generalized partial derivative* of a distribution $f$ is *defined* to be a distribution, denoted by $D^\alpha f$, that satisfies

$$\langle D^\alpha f, \phi \rangle = (-1)^{|\alpha|} \langle f, D^\alpha \phi \rangle \quad \text{for all } \phi \in \mathcal{D}(\Omega). \tag{7.7}$$

Thus we use the same notation for the generalized derivative of a distribution as that used for the conventional derivative of a function. Of course, if the function belongs to $C^m(\overline{\Omega})$, then the generalized derivative coincides with the conventional $\alpha$th partial derivative for $|\alpha| \leq m$, as can be seen immediately from (7.5) and (7.6).

For the special case of first derivatives the multi-index notation can be dispensed with, in which case (7.7) becomes

$$\left\langle \frac{\partial f}{\partial x_i}, \phi \right\rangle = -\left\langle f, \frac{\partial \phi}{\partial x_i} \right\rangle, \quad i = 1, \ldots, n.$$

Furthermore, for the case $\Omega = (a, b) \subset \mathbb{R}$ all derivatives are with respect to $x$ only, and so (7.7) becomes

$$\left\langle \frac{d^k f}{dx^k}, \phi \right\rangle = (-1)^k \left\langle f, \frac{\partial^k \phi}{\partial x^k} \right\rangle. \tag{7.8}$$

**Examples**

8. The first generalized derivative of the Heaviside step function $H(x)$ is the distribution $H'$ satisfying, for all test functions $\phi$,

$$\begin{aligned} \langle H', \phi \rangle &= (-1)^1 \left\langle H, \frac{d\phi}{dx} \right\rangle \\ &= -\int_{-1}^{1} H(x) \frac{d\phi}{dx}\, dx \quad (H \text{ is locally integrable}) \\ &= -\int_{0}^{1} \frac{d\phi}{dx}\, dx = -[\phi]_0^1 = \langle \phi, 0 \rangle = \langle \delta, \phi \rangle \end{aligned}$$

so that, symbolically, $H' = \delta$; that is, the derivative $H'$ of the step function is the *Dirac delta*.

9. The ramp function $R(\boldsymbol{x})$ on $\Omega = (-1, 1) \times (-1, 1) \subset \mathbb{R}^2$ is defined by

$$R(\boldsymbol{x}) = \begin{cases} xy & \text{if} \quad x \geq 0,\ y \geq 0, \\ 0 & \text{if} \quad x < 0 \text{ or } y < 0. \end{cases}$$

The generalized derivative $D^{(1,0)}R = \partial R/\partial x$ is found from

$$\begin{aligned} \left\langle \frac{\partial R}{\partial x}, \phi \right\rangle &= -\left\langle R, \frac{\partial \phi}{\partial x} \right\rangle = -\int_{-1}^{1}\int_{-1}^{1} R(\boldsymbol{x}) \frac{\partial \phi}{\partial x}\, dxdy \\ &\qquad (R \text{ is locally integrable}) \\ &= -\int_{0}^{1}\int_{0}^{1} xy \frac{\partial \phi}{\partial x}\, dxdy = \int_{0}^{1}\int_{0}^{1} y\phi\, dxdy \end{aligned}$$

after using Green's theorem (7.2). Furthermore, $D^{(1,1)}R = \partial^2 R/\partial x \partial y$ is found from

$$\begin{aligned}\left\langle \frac{\partial^2 R}{\partial x \partial y}, \phi \right\rangle &= (-1)^2 \left\langle R, \frac{\partial^2 \phi}{\partial x \partial y} \right\rangle = \int_0^1 \int_0^1 xy \frac{\partial^2 \phi}{\partial x\, \partial y}\, dxdy \\ &= \int_0^1 \int_0^1 \phi\, dxdy \ \text{(applying Green's theorem twice)} \\ &= \int_\Omega H(\boldsymbol{x})\phi(x)\, dxdy,\end{aligned}$$

where $H$ is the two-dimensional step function:

$$H(\boldsymbol{x}) = \begin{cases} 1 & \text{if} \quad x \geq 0,\ y \geq 0, \\ 0 & \text{if} \quad x < 0 \text{ or } y < 0. \end{cases}$$

Hence

$$\left\langle \frac{\partial^2 R}{\partial x \partial y}, \phi \right\rangle = \langle H, \phi \rangle \text{ so that } D^{(1,1)}R = H.$$

**Weak derivatives.** Suppose that a function $u$ is locally integrable so that it generates a distribution, also denoted by $u$, that satisfies

$$\langle u, \phi \rangle = \int_\Omega u\phi\, dx \quad \text{for all } \phi \in \mathcal{D}(\Omega).$$

Furthermore, the distribution $u$ possesses distributional derivatives of all orders: in particular, the derivative $D^\alpha u$ is defined by (7.7). Of course $D^\alpha u$ may or may not be a regular distribution; if it is a regular distribution, then naturally it is generated by a locally integrable function so that

$$\langle D^\alpha u, \phi \rangle = \int_\Omega D^\alpha u(\boldsymbol{x})\phi(\boldsymbol{x})\, dx. \tag{7.9}$$

It follows in this case from (7.7) and (7.9) that the functions $u$ and $D^\alpha u$ are related by

$$\int_\Omega D^\alpha u(\boldsymbol{x})\phi(\boldsymbol{x})\, dx = (-1)^m \int_\Omega u(\boldsymbol{x}) D^\alpha \phi(\boldsymbol{x})\, dx$$

for $|\alpha| = m$. We call the *function* (more precisely, the equivalence class of functions; see the discussion following Example 10) $D^\alpha u$ obtained in this way the $\alpha$th *weak derivative* of the function $u$. Of course, if $u$ is sufficiently smooth to belong to $C^m(\overline{\Omega})$, then its weak derivatives $D^\alpha u$ coincide with its classical derivatives for $|\alpha| \leq m$.

A remark concerning notation is in order here. We have reached the stage where $D^\alpha u$ may represent the classical partial derivative of a function, or

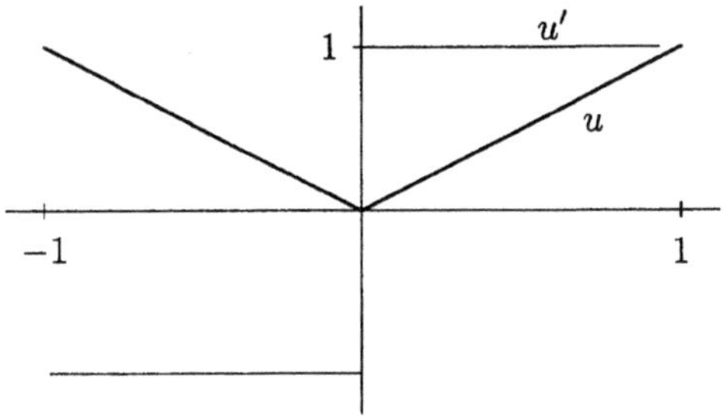


FIGURE 7.3. The function $u(x) = |x|$ and its weak derivative

the weak partial derivative of a function, or the generalized derivative of a distribution (possibly generated by a function). For the most part it should be clear from the context exactly which derivative is being used, but should there be any danger of ambiguity it is made quite clear exactly what $D^\alpha u$ stands for. The same applies to the notation $\partial u/\partial x$, and the like; this may represent any one of the various types of derivatives.

## Example

10. The function $u(x) = |x|$ belongs to $C[-1,1]$, but the classical derivative $u'$ does not exist, in that it is not defined at the origin. However, the *weak* derivative of $u$ is the function

$$u' = \begin{cases} -1 & \text{for} \quad -1 \le x < 0, \\ +1 & \text{for} \quad 0 \le x \le 1 \end{cases}$$

(see Figure 7.3), since the identity $\int_{-1}^{1} u'\phi\, dx = -\int_{-1}^{1} u\phi'\, dx$ is easily shown to hold. Note furthermore that $u' \in L^2(-1,1)$, and is therefore of course locally integrable.

The preceding example illustrates one fundamental difference between classical and weak derivatives. The classical derivative, if it exists, is a function defined *pointwise* on an interval, so it must be at least continuous. A weak derivative, on the other hand, need only be locally integrable. Thus any function $v$ differing from a weak derivative $u'$ on a set of measure zero (for example, at a finite number of points in the real line) is itself a weak derivative of $u$.

**Distributional differential equations.** Since we now have at our disposal the concept of the derivative of a distribution, it is natural to consider next differential equations involving distributions. For example, suppose that we are required to find the distribution $g$ that satisfies

$$g' = f \tag{7.10}$$

for a given distribution $f$, on some interval of the real line. If $f$ and $g$ were ordinary functions (for example, $f \in C[a,b]$ and $g \in C^1[a,b]$), then

(7.10) would be a simple first order differential equation. Since $f$ and $g$ are actually distributions, we go back to the definition (7.8) of a generalized derivative; then (7.10) really reads

$$\langle g', \phi\rangle = \langle f, \phi\rangle \quad \text{or} \quad -\langle g, \phi'\rangle = \langle f, \phi\rangle \quad \text{for all } \phi \in \mathcal{D}(a,b). \tag{7.11}$$

If by (7.10) we understand (7.11), then (7.10) is said to be a *distributional differential equation.*

The same procedure applies in the case of more general differential equations. For example, suppose that we are required to find the distribution $g$ satisfying

$$Ag = f, \tag{7.12}$$

where $A$ is the (generalized) differential operator given by

$$A = a_0(x)\frac{d^k}{dx^k} + a_1(x)\frac{d^{k-1}}{dx^{k-1}} + \ldots + a_k(x).$$

We interpret (7.12) as a differential equation involving *generalized* derivatives of $g$, and seek $g$ such that

$$\langle Ag, \phi\rangle = \langle f, \phi\rangle \quad \text{for all} \quad \phi \in \mathcal{D}(a,b),$$

which is equivalent to

$$\langle g, A^*\phi\rangle = \langle f, \phi\rangle, \qquad \phi \in \mathcal{D}(a,b),$$

the operator $A^*$ resulting from successive applications of (7.1) and (7.7); thus

$$A^*\phi = (-1)^k\frac{d^k}{dx^k}(a_0\phi) + \ldots + a_k\phi. \tag{7.13}$$

Generally, for partial differential equations involving distributions the same procedure is adopted. The problem of finding a distribution $g$ satisfying

$$Ag = f, \tag{7.14}$$

where $Ag = \sum_{|\alpha|\le k} a_\alpha D^\alpha g$, is equivalent to the problem of finding $g$ such that

$$\langle Ag, \phi\rangle = \langle f, \phi\rangle \quad \text{or} \quad \langle g, A^*\phi\rangle = \langle f, \phi\rangle, \tag{7.15}$$

where $A^*$ is obtained, as in (7.13), by repeated application of (7.1) and (7.7).

Naturally one would expect that if $f$ is continuous (that is, a distribution generated by a continuous function), then the solution $g$ should be a function that is $k$ times continuously differentiable. This is indeed so; in other words, when the distributions involved are generated by sufficiently differentiable functions, we recover the classical concept of a differential equation. In this case $g$ is called a *classical* solution. More generally, though, if $f$ is a regular distribution generated by a function that is locally integrable but not continuous, or indeed if it is a singular distribution, then equation (7.14) cannot be expected to have any meaning in the classical sense. The solution in this case is called a *weak* or *generalized* solution.

When $g$ satisfies an equation of the form (7.14) or (7.15) we say that $Ag = f$ *in the sense of distributions*, or that $g$ satisfies (7.14) *distributionally.*

**Examples**

11. The equation

$$xg' = 0 \quad \text{on} \quad \Omega = (-1, 1) \tag{7.16}$$

has the classical solution $g =$ constant. But if (7.16) is regarded as a *distributional* differential equation, then the weak solution is

$$g(x) = c_1 H(x) + c_2,$$

where $c_1$ and $c_2$ are constants. We check as follows: $g' = c_1\delta$ so that

$$\langle xg', \phi\rangle = \langle g', x\phi\rangle = c_1\langle \delta, x\phi\rangle = c_1[(x\phi)(0)] = 0;$$

hence $xg' = 0$ in the sense of distributions.

12. The equation $g'' = \delta'$ has no classical solution on $(-1, 1)$ but its weak solution is

$$g = H + c_1 x + c_2.$$

This is verified by considering the fact that

$$\langle g'', \phi\rangle = \langle H'', \phi\rangle = \langle H, \phi''\rangle = \int_0^1 \phi''\, dx = -\phi'(0) = -\langle \delta, \phi'\rangle = \langle \delta', \phi\rangle.$$

## 7.3 The Sobolev spaces $H^m(\Omega)$

Before we actually get down to discussing Sobolev spaces, it is appropriate at this stage to elaborate on the degree of smoothness that we expect the boundary $\Gamma$ of a domain $\Omega$ in $\mathbb{R}^n$ $(n \geq 2)$ to have, since some results concerning Sobolev spaces hold only when the boundary is sufficiently smooth. Let $\Omega$ be a domain in $\mathbb{R}^n$ $(n \geq 2)$ with boundary $\Gamma$. Let $\boldsymbol{x}_0$ be an arbitrary point on $\Gamma$ and construct $B(\boldsymbol{x}_0, \epsilon)$, the open ball of radius $\epsilon$, center $\boldsymbol{x}_0$, for some $\epsilon > 0$; that is, $B(\boldsymbol{x}_0, \epsilon) = \{\boldsymbol{x} \in \mathbb{R}^n : |\boldsymbol{x} - \boldsymbol{x}_0| < \epsilon\}$.

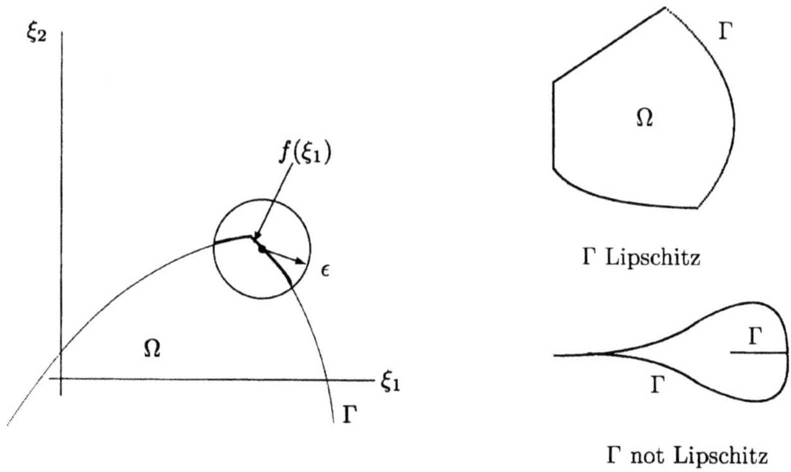


FIGURE 7.4. A local coordinate system for classifying the boundary of a domain, and examples of Lipschitz and non-Lipschitz domains

Next, set up a coordinate system $(\xi_1, \ldots, \xi_n)$ such that the segment $\Gamma \cap B(\boldsymbol{x}_0, \epsilon)$ can be expressed as the function

$$\xi_n = f(\xi_1, \ldots, \xi_{n-1}).$$

If the function $f$ is $m$-times continuously differentiable for every $\boldsymbol{x}_0 \in \Gamma$, we say that $\Gamma$ *is of class* $C^m$; $\Gamma$ is said to be *Lipschitz* if $f$ is Lipschitz-continuous, that is, if there is a constant $k$ such that

$$|f(\hat{\boldsymbol{\xi}}) - f(\hat{\boldsymbol{\eta}})| \leq k|\hat{\boldsymbol{\xi}} - \hat{\boldsymbol{\eta}}|,$$

where $\hat{\boldsymbol{\xi}} = (\xi_1, \ldots, \xi_{n-1})$ and $\hat{\boldsymbol{\eta}} = (\eta_1, \ldots, \eta_{n-1})$ (recall that a Lipschitz-continuous function is uniformly continuous). The situation is illustrated in Figure 7.4 for $n = 2$. Unless otherwise stated $\Gamma$ is assumed to be Lipschitz; this includes, in $\mathbb{R}^2$, boundaries that are triangular, rectangular, and annular, whereas in $\mathbb{R}^3$ tetrahedra and cubes are Lipschitz. Boundaries that are not Lipschitz include those with cusps and those that have the domain $\Omega$ on both sides, as shown in Figure 7.4.

**The Sobolev spaces $H^m(\Omega)$.** The Sobolev space of order $m$, denoted by $H^m(\Omega)$, is defined to be the space consisting of those functions in $L^2(\Omega)$ that, together with all their *weak* partial derivatives up to and including those of order $m$, belong to $L^2(\Omega)$:

$$H^m(\Omega) = \{u : D^\alpha u \in L^2(\Omega) \text{ for all } \alpha \text{ such that } |\alpha| \leq m\}.$$

We consider real-valued functions only, and make $H^m(\Omega)$ an *inner product space* by introducing the Sobolev inner product $(\cdot,\cdot)_{H^m}$ defined by

$$(u,v)_{H^m} = \int_\Omega \sum_{|\alpha|\le m} (D^\alpha u)(D^\alpha v)\, dx \quad \text{for} \quad u,v \in H^m(\Omega).$$

This inner product in turn generates the *Sobolev norm* $\|\cdot\|_{H^m}$ defined by

$$\|u\|_{H^m}^2 = (u,u)_{H^m} = \int_\Omega \sum_{|\alpha|\le m} (D^\alpha u)^2\, dx.$$

Note that $H^0(\Omega) = L^2(\Omega)$, and furthermore that we may write $(u,v)_{H^m}$ as

$$(u,v)_{H^m} = \sum_{|\alpha|\le m} (D^\alpha u, D^\alpha v)_{L^2};$$

in other words, the Sobolev inner product $(u,v)_{H^m}$ is equal to the sum of the $L^2$-inner products of $D^\alpha u$ and $D^\alpha v$ over all $\alpha$ such that $|\alpha| \le m$. Of course, we could also write

$$\|u\|_{H^m}^2 = \sum_{|\alpha|\le m} \|D^\alpha u\|_{L^2}^2;$$

when written out in full for the case $m = 2$, and for a function $u$ defined on $\Omega \subset \mathbb{R}^2$, this becomes

$$\|u\|_{H^2}^2 = \int_\Omega \left[u^2 + \left(\frac{\partial u}{\partial x}\right)^2 + \left(\frac{\partial u}{\partial y}\right)^2 + \left(\frac{\partial^2 u}{\partial x^2}\right)^2 + 2\left(\frac{\partial^2 u}{\partial x \partial y}\right)^2 + \left(\frac{\partial u}{\partial y^2}\right)^2\right] dx.$$

**Examples**

13. Consider the function $u(x)$ defined on $\Omega = (0,2)$ by

$$u(x) = \begin{cases} x^2, & 0 < x \le 1, \\ 2x^2 - 2x + 1, & 1 < x < 2. \end{cases}$$

Then

$$v(x) \equiv u'(x) = \begin{cases} 2x, & 0 < x \le 1, \\ 4x - 2, & 1 < x < 2, \end{cases}$$

which is a continuous function. The (weak) derivative of this function is

$$w(x) \equiv u''(x) = \begin{cases} 2, & 0 < x \le 1, \\ 4, & 1 < x < 2. \end{cases}$$

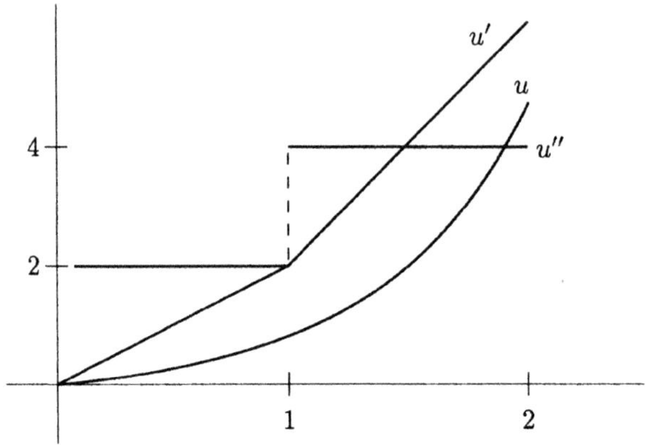


FIGURE 7.5. The function $u$ of Example 13, and its derivatives

By inspection $u, u'$, and $u''$ all belong to $L^2(0,2)$; however, the (generalized) derivative of $u''$ is $u''' = 2\delta(x-1) \notin L^2(0,2)$. Hence $u$ is a member of $H^2(0,2)$, the function $v$ belongs to $H^1(0,2)$, and $w$ belongs to $L^2(0,2) = H^0(0,2)$ (see Figure 7.5). The respective norms of these functions are

$$\begin{aligned}
\|u\|_{H^2}^2 &= \int_0^2 [u^2 + (u')^2 + (u'')^2]\,dx = 71.37, \\
\|v\|_{H^1}^2 &= \int_0^2 [v^2 + (v')^2]\,dx = 39, \\
\|w\|_{L^2}^2 &= \int_0^2 w^2\,dx = 20.
\end{aligned}$$

14. The function $u$ defined on $\Omega = (-1,1) \times (-1,1)$ by

$$u(\boldsymbol{x}) = \begin{cases} x & \text{for} \quad x > 0, \\ 0 & \text{for} \quad x \le 0, \end{cases}$$

is shown in Figure 7.6; this function belongs to $H^1(\Omega)$. To see this, we start by evaluating its first derivatives: for $\phi \in \mathcal{D}(\Omega)$,

$$\begin{aligned}
\int_\Omega \frac{\partial u}{\partial y}\phi\,dxdy &= -\int_\Omega u\frac{\partial \phi}{\partial y}\,dxdy = -\int_0^1 x\left[\int_{-1}^1 \frac{\partial \phi}{\partial y}dy\right]dx \\
&= -\int_0^1 x\,[\phi(x,1) - \phi(x,-1)]\,dy = 0.
\end{aligned}$$

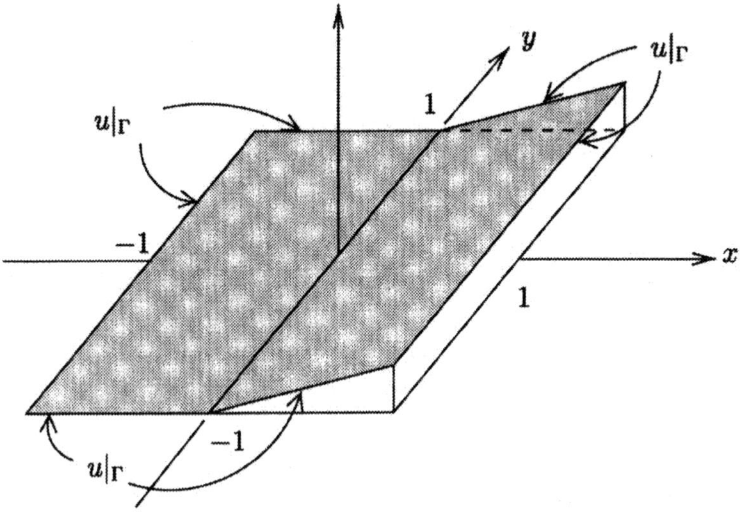


FIGURE 7.6. The function in $u$ in Example 14

Hence $\partial u/\partial y = 0$. Secondly,

$$\begin{aligned}
\int_\Omega \frac{\partial u}{\partial x}\phi \, dxdy &= -\int_\Omega u\frac{\partial \phi}{\partial x}\, dxdy = -\int_{-1}^{1}\left(\int_0^1 x\frac{\partial \phi}{\partial x}dx\right)dy \\
&= -\int_1^1\left([x\phi]_0^1 - \int_0^1 \phi \, dx\right)dy \\
&= \int_{-1}^{1}\int_0^1 \phi \, dxdy = \int_\Omega H(x)\phi(x,y)\, dxdy.
\end{aligned}$$

Hence $\partial u/\partial x = H(x)$, the Heaviside step function in the $x$ direction. We can show next that $\partial^2 u/\partial x^2 = \delta_x$, the two-dimensional Dirac delta defined by $\delta_x(\phi) = \phi(0,y)$, so that $\partial^2 u/\partial x^2 \notin L^2(\Omega)$. Hence $u \in H^1(\Omega)$.

The picture that emerges is that the spaces $H^m(\Omega)$ provide a very logical means for characterizing the degree of smoothness of a function. When dealing with the spaces $C^m(\overline{\Omega})$, by "degree of smoothness" is understood "how many times can the function be differentiated?" In the case of Sobolev spaces "degree of smoothness" is understood to mean "how many times can the function be differentiated (weakly) before it ceases to belong to $L^2(\Omega)$?" The following theorem summarizes the most important properties of the space $H^m(\Omega)$.

THEOREM 1. *Let $H^m(\Omega)$ be the Sobolev space of order $m$, and $\Omega$ a bounded domain with Lipschitz boundary. Then* *(i)* $H^r(\Omega) \subseteq H^m(\Omega)$ *if* $r \geq m$;

*(ii)* $H^m(\Omega)$ *is a Hilbert space with respect to the norm* $\|\cdot\|_{H^m}$;

*(iii)* $H^m(\Omega)$ *is the completion or closure, with respect to the norm* $\|\cdot\|_{H^m}$, *of the space* $C^\infty(\overline{\Omega})$.

PROOF. Only Parts (i) and (ii) are proved; the proof of (iii) is rather long and technical, and its details may be found in the references at the end of this chapter.

(i) If $u \in H^r(\Omega)$, then $D^\alpha u$ belongs to $L^2(\Omega)$ for all $\alpha$ such that $|\alpha| \leq r$, and thus for all $\alpha$ such that $|\alpha| \leq m$. So $u \in H^m(\Omega)$, and $H^r(\Omega) \subseteq H^m(\Omega)$.

(ii) We know that $H^m(\Omega)$ is an inner product space, so what remains to be shown is that $H^m(\Omega)$ is complete. Let $\{u_k\}$ be a Cauchy sequence in $H^m(\Omega)$. We have to show that $u_k$ converges to a function $u$ in $H^m(\Omega)$. First, by definition

$$\lim_{k,l\to\infty} \|u_k - u_l\|_{H^m} = 0$$

or, using the definition of the $H^m$-norm,

$$\lim_{k,l\to\infty} \sum_{|\alpha|\leq m} \|D^\alpha u_k - D^\alpha u_l\|_{L^2}^2 = 0.$$

Since each term in this sum is positive, it follows that

$$\lim_{k,l\to\infty} \|D^\alpha u_k - D^\alpha u_l\|_{L^2} = 0 \quad \text{for all } \alpha,\ |\alpha| \leq m.$$

Hence $\{D^\alpha u_k\}$ is a Cauchy sequence in $L^2(\Omega)$ for each $\alpha$ such that $|\alpha| \leq m$. Since $L^2$ is complete, it follows that $D^\alpha u_k$ converges to a function $u^{(\alpha)}$, say, that belongs to $L^2$. In particular, for $|\alpha| = 0$, $u_k$ converges to a function $u$, say, in $L^2$.

We show next that $u$ is in $H^m(\Omega)$. Consider

$$\begin{aligned}
\int_\Omega u^{(\alpha)}\phi\, dx &= \int_\Omega (\lim_{k\to\infty} D^\alpha u_k)\phi\, dx = (\lim_{k\to\infty} D^\alpha u_k, \phi)_{L^2} \\
&= \lim_{k\to\alpha} (D^\alpha u_k, \phi)_{L^2} = \lim_{k\to\infty} (-1)^{|\alpha|}(u_k, D^\alpha\phi)_{L^2} \\
&= (-1)^{|\alpha|}(\lim_{k\to\infty} u_k, D^\alpha\phi)_{L^2} = (-1)^{|\alpha|}\int_\Omega u D^\alpha\phi\, dx,
\end{aligned}$$

where we have used the result of Exercise 2 of Chapter 4, (7.7), and the fact that $D^\alpha u_k$ is a regular distribution. Thus $u^{(\alpha)}$ is the $\alpha$th weak derivative of $u$ and since $u$, as well as all of its weak derivatives of order $\leq m$, is in $L^2(\Omega), u$ belongs to $H^m(\Omega)$. Hence $H^m(\Omega)$ is complete. □

Part (iii) of the theorem has an important interpretation: from the definition of the completion of a space (Section 4.3) we know that $C^\infty(\overline{\Omega})$ is *dense* in $H^m(\Omega)$; hence, for any $u \in H^m(\Omega)$ it is always possible to find an infinitely differentiable function $f(x)$, say, that is arbitrarily close to $u$ in the sense that

$$\|u - f\|_{H^m} < \epsilon$$

for any given $\epsilon > 0$. In other words, *every member of $H^m(\Omega)$ is either a member of $C^\infty(\overline{\Omega})$, or may be approximated arbitrarily closely by a function from this space.*

**Example**

15. Refer to Example 13: from what was said there we conclude that, given any $\epsilon > 0$, it is possible to find functions $f, g$, and $h$ in $C^\infty(\overline{\Omega})$ that satisfy

$$\begin{aligned}
\|u - f\|_{H^2} &= \left[\int_0^1 (u-f)^2 + (u'-f')^2 + (u''-f'')^2 dx\right]^{1/2} < \epsilon, \\
\|v - g\|_{H^1} &= \left[\int_0^1 (v-g)^2 + (v'-g')^2 dx\right]^{1/2} < \epsilon, \\
\|w - h\|_{L^2} &= \left[\int_0^1 (w-h)^2 dx\right]^{1/2} < \epsilon.
\end{aligned}$$

When $m = 0$ we can deduce the following property of $H^0(\Omega) = L^2(\Omega)$ from Theorem 1.

COROLLARY TO THEOREM 1. *$L^2(\Omega)$ is the completion, with respect to the $L^2$-norm, of the space $C^\infty(\overline{\Omega})$.*

It is worth recalling that this result is contained also in Theorem 6 of Chapter 4.

**The Sobolev Embedding Theorem.** A glance at the examples discussed earlier in this section may lead one to wonder whether it is true that members of $H^m(\Omega)$ are simply functions that, together with their derivatives of order $\leq m - 1$, are continuous. After all it is not easy, for example, to conceive of a function in $H^1(\Omega)$ that is not continuous. A famous theorem due to Sobolev asserts that, as we would expect, all members of $H^1(a, b)$ are indeed continuous functions, *but that this does not hold for higher-dimensional domains.*

Before stating the theorem we give a simple example to show that intuition would be misleading. Let $\Omega$ be the disc of radius $\frac{1}{2}$ in $\mathbb{R}^2$, and let $u = \ln(\ln(1/r))$, where $r^2 = x^2 + y^2$. Then, using polar coordinates $(r, \theta)$,

$$\iint_\Omega u^2 \, dxdy = \int_0^{1/2} \int_0^{2\pi} [\ln(\ln(1/r))]^2 \, rdrd\theta = \int_0^{2\pi} \int_{\ln 2}^{\infty} (e^{-t} \ln \, t)^2 \, dtd\theta$$

(making the change in coordinates $t = -\ln r$) which is easily shown to be bounded. Furthermore,

$$\iint_\Omega \left[(\partial u/\partial x)^2 + (\partial u/\partial y)^2\right] \, rdrd\theta = \iint_\Omega (\ln r)^{-2} d(\ln r) \, d\theta = 2\pi/\ln 2.$$

Hence $\|u\|_{H^1}$ is finite and so $u$ belongs to $H^1(\Omega)$. But $u$ is *not* continuous at the origin.

Let $X$ and $Y$ be two Banach spaces, with $X \subseteq Y$. The notion of $X$ being *embedded* in $Y$ goes a little further than the fact that $X$ is a subset of $Y$, in the following sense. Let $u$ be any member of $X$; then of course $u$ is also a member of $Y$, and this may be represented by making use of the identity operator $\iota : X \to Y$, that simply takes a member of $X$ to the same member, viewed as an element in $Y$: that is, $\iota(u) = u$. This exercise is of more than trivial interest because the two spaces $X$ and $Y$ are, in general, endowed with different norms $\|\cdot\|_X$ and $\|\cdot\|_Y$, so we may enquire as to whether the operator $\iota$ is *bounded*, that is, whether it is the case that $\|\iota(u)\|_Y = \|u\|_Y \leq K\|u\|_X$, for some constant $K > 0$. When this is the case, then $X$ is said to be *continuously embedded in* $Y$. The following theorem gives conditions under which Sobolev spaces are embedded in spaces of continuous functions.

THEOREM 2 (THE SOBOLEV EMBEDDING THEOREM). *Let $\Omega$ be a bounded domain in $\mathbb{R}^n$ with a Lipschitz boundary $\Gamma$. If $m - k > n/2$, then every function in $H^m(\Omega)$ belongs to $C^k(\overline{\Omega})$. Furthermore, the embedding*

$$H^m(\Omega) \subseteq C^k(\overline{\Omega}) \tag{7.17}$$

*is continuous.*

REMARK. Some care has to be exercised in the interpretation of Theorem 2. Recall that members of $H^m(\Omega)$ are equivalence classes of functions, given that they are members of $L^2$, whereas continuous functions, by contrast, are defined unambiguously. The embedding (7.17) has therefore to be interpreted in the sense that each member of $H^m(\Omega)$ may be identified with a function in $C^k(\overline{\Omega})$, possibly after changing its values on a set of measure zero.

According to the Sobolev Embedding Theorem, if $n = 1$ so that $\Omega$ is a subset of the real line, then the functions in $H^1(\Omega)$ are continuous. For domains that are subsets of the plane, though, $n = 2$ and we require that a function be a member of $H^2(\Omega)$ in order to guarantee its continuity.

**An alternative definition of Sobolev spaces.** The definition of Sobolev spaces used here is one that is phrased in terms of generalized derivatives, and whether these belong to $L^2$. An alternative definition takes as a starting point the spaces of $m$-times continuously differentiable functions; these are *not* complete with respect to the norm $\|\cdot\|_{H^m}$, and the Sobolev space $H^m$ is defined as precisely the *completion* of $C^m(\Omega)$ in this norm, for $m \geq 1$. That the two definitions are in fact equivalent is a well-known result, that is contained in the following theorem.

THEOREM 3. *Let $\Omega$ be a bounded domain. Then $H^m(\Omega)$ is the completion or closure, with respect to the norm $\|\cdot\|_{H^m}$, of the space $\widehat{C}^m(\Omega)$ of $m$-times continuously differentiable functions that have a finite norm $\|\cdot\|_{H^m}$.*

The main point about Theorem 3 is that every function in $H^m(\Omega)$ can be approximated arbitrarily closely by a member of $C^m(\Omega)$.

We conclude this section with an important and frequently useful inequality.

THEOREM 4. (THE POINCARÉ INEQUALITY). *Let $\Omega$ be a domain in $\mathbb{R}^n$ with a Lipschitz boundary. Then for any $u \in H^1(\Omega)$ there exist constants $c_1$ and $c_2$ such that*

$$\|u\|_{L^2}^2 \leq c_1 \int_\Omega \sum_{|\alpha|=1} |D^\alpha u|^2 \, dx + c_2 \left[\int_\Omega u(x) \, dx\right]^2. \tag{7.18}$$

*More generally, for any $u \in H^k(\Omega)$ there exist constants $c_3$ and $c_4$ such that*

$$\|u\|_{H^k}^2 \leq c_1 \int_\Omega \sum_{|\alpha|=k} (D^\alpha u)^2 dx + c_2 \sum_{|\alpha|<k} \left(\int_\Omega D^\alpha u \, dx\right)^2. \tag{7.19}$$

PROOF. We prove (7.19) for the case $\Omega = (a, b) \subset \mathbb{R}$; the higher-dimensional results follow in a similar way. Thus for $n = 1$ we have to derive the inequality

$$\|u\|_{H^k}^2 \leq c_1 \int_a^b \left(\frac{d^k u}{dx^k}\right)^2 dx + c_2 \sum_{j<k} \left(\int_a^b \frac{d^j u}{dx^j}\right)^2 dx. \tag{7.20}$$

Let $\xi$ and $\eta$ be two points in $(a, b)$ with $\eta < \xi$, so that

$$u(\xi) - u(\eta) = \int_\eta^\xi u' \, dx$$

and so, using the Cauchy–Schwarz inequality,

$$[u(\xi) - u(\eta)]^2 = \left[\int_\eta^\xi u'\ dx\right]^2 \leq \left[\int_\eta^\xi 1^2 dx\right]\left[\int_\eta^\xi (u')^2 dx\right]$$
$$\leq (b-a)\int_a^b (u')^2 dx.$$

Now we integrate with respect to $\xi$, keeping $\eta$ fixed, and then with respect to $\eta$, to get

$$(b-a)\left[\int_a^b u^2(\xi)d\xi + \int_a^b u^2(\eta)d\eta\right] - 2\int_a^b u(\xi)d\xi \int_a^b u(\eta)d\eta$$
$$\leq (b-a)^3 \int_a^b (u')^2 dx$$

or

$$\int_a^b u^2 dx \leq C_1 \int_a^b (u')^2 dx + C_2\left(\int_a^b u\ dx\right)^2, \tag{7.21}$$

where $C_1 = \frac{1}{2}(b-a)^2$ and $C_2 = 2/(b-a)$. Since $u \in H^k(a,b)$, (7.21) is still valid if we replace $u$ by $u'$, or by $u''$, and so on, up to $d^{k-1}u/dx^{k-1}$. That is, we also have

$$\int_a^b (u')^2 dx \leq C_1 \int_a^b (u'')^2 dx + C_2\left(\int_a^b u'\ dx\right)^2,$$
$$\vdots$$
$$\int_a^b \left(\frac{d^{k-1}u}{dx^{k-1}}\right)^2 dx \leq C_1 \int_a^b \left(\frac{d^k u}{dx^k}\right)^2 dx + C_2\left(\int_a^b \frac{d^{k-1}u}{dx^{k-1}} dx\right)^2.$$

To obtain (7.20) for $k=1$ we add $\int_a^b (u')^2 dx$ to both sides of (7.21) to get

$$\|u\|_{H^1}^2 \leq (1+C_1)\int_a^b (u')^2 dx + C_2\left(\int_a^b u\ dx\right)^2. \tag{7.22}$$

Next, to get (7.20) for $k=2$ we add $\int_a^b (u'')^2 dx$ to both sides of (7.22) and use (7.21), to obtain

$$\|u\|_{H^2}^2 \leq (1 + C_1(1+C_1))\int_a^b (u'')^2 dx + C_2\left(\int_a^b u\ dx\right)^2$$
$$+ C_2(1+C_1)\left(\int_a^b u'\ dx\right)^2.$$

Continuing in this manner we can derive (7.20) for any value of $k$. □

**The Sobolev spaces $W^{m,p}(\Omega)$.** The Sobolev spaces $H^m(\Omega)$ have been defined by taking as a point of departure the Hilbert space $L^2(\Omega)$; in this way it has been possible to introduce a family of Hilbert spaces, of which $L^2$ is a special case, viz. the case $m = 0$. In much the same way it is possible to take as a point of departure the spaces $L^p(\Omega)$ for $1 \leq p \leq \infty$, and in this way to introduce, for each $p$, a Sobolev space that is a *Banach space*. Thus, for $m = 0, 1, \ldots$, the Sobolev space $W^{m,p}(\Omega)$ is defined to be the space of functions that, together with all their weak derivatives up to and including those of order $m$, belong to $L^p(\Omega)$:

$$W^{m,p}(\Omega) = \{u \in L^p(\Omega) : \; D^\alpha u \in L^p(\Omega), \quad |\alpha| \leq m\}. \tag{7.23}$$

Clearly we have $H^m(\Omega) = W^{m,2}(\Omega)$. The space $W^{m,p}(\Omega)$ is a *normed space* when endowed with the Sobolev norm $\|\cdot\|_{W^{m,p}}$, or simply $\|\cdot\|_{m,p}$, defined by

$$\|u\|_{m,p} = \left( \sum_{|\alpha| \leq m} \int_\Omega |D^\alpha u|^p \, dx \right)^{1/p}$$

for $1 \leq p < \infty$, and by

$$\|u\|_{m,\infty} = \sum_{|\alpha| \leq m} \text{ess sup} \, |D^\alpha u|$$

for $p = \infty$. Recalling the definition (3.6) of the $L^p$-norm, the Sobolev norm may be expressed in the alternative form

$$\|u\|_{m,p} = \left( \sum_{|\alpha| \leq m} \|D^\alpha u\|^p_{L^p} \right)^{1/p}.$$

That is, the Sobolev norm $\|u\|^p_{m,p}$ is equal to the sum of the $p$th powers of the $L^p$-norms of $D^\alpha u$ over all $\alpha$ such that $|\alpha| \leq m$.

Theorems 1 to 3 all have counterparts for the spaces $W^{m,p}(\Omega)$, and some of these extensions are given in the following theorem.

THEOREM 5. *Let $W^{m,p}(\Omega)$ be the Sobolev space defined by (7.23), and $\Omega$ a bounded domain with Lipschitz boundary. Then*

*(i) $W^{m,p}(\Omega)$ is a Banach space with respect to the norm $\|\cdot\|_{m,p}$;*

*(ii) $W^{m,p}(\Omega)$ is the completion, with respect to the norm $\|\cdot\|_{W^{m,p}}$, of the space $\widehat{C}^m(\Omega)$ of $m$-times continuously differentiable functions that have a finite norm $\|\cdot\|_{W^{m,p}}$;*

(iii) *$W^{m,p}(\Omega)$ is the completion, with respect to the norm $\|\cdot\|_{W^{m,p}}$, of the space $C^\infty(\overline{\Omega})$;*

(iv) *for nonnegative integers $m$, $k$, and $p$ such that $1 \le p \le \infty$, $W^{m,p}(\Omega)$ is continuously embedded in $C^k(\overline{\Omega})$ if $m - k > n/p$.*

All of the discussions later on involving boundary value problems and their approximation by finite elements take place in a Hilbert space context, and the spaces $H^m(\Omega)$ suffice for these purposes. The focus in this chapter therefore remains on this special case of $W^{m,p}(\Omega)$.

## 7.4 Boundary values of functions and trace theorems

**Traces of functions in $H^m(\Omega)$.** Later on, when dealing with boundary value problems, we are concerned not only with values of functions on an open domain $\Omega$, but also with their values on the boundary $\Gamma$. Now, in the case of continuous functions defined on $\overline{\Omega} = \Omega \cup \Gamma$, one simply finds the values of a function $u$ on the boundary by evaluating $u$ on $\Gamma$; we then write this as $u|_\Gamma$. For example, if $\overline{\Omega} = [0,1]$ and $u = x + 2$, then $u|_\Gamma$ consists of $u$ evaluated at $x = 0$ and $x = 1$ : $u\,|_\Gamma = \{u(0), u(1)\} = \{2, 3\}$. Similarly, if $\Omega$ is the square $\{(x,y) : \ |x| < 1, \ |y| < 1\}$ with boundary $\Gamma = \{(x,y) : \ |x| = 1, \ |y| \le 1\} \cup \{(x,y) : \ |y| = 1, \ |x| \le 1\}$, and we choose $u \in C(\overline{\Omega})$ as in Example 14, then the restriction of $u$ to the boundary is the function $u|_\Gamma$ shown in Figure 7.6. This procedure may be formalized by introducing an operator $\gamma$ called the *trace operator*: $\gamma$ is a *linear* operator that acts on a continuous function $u \in C(\overline{\Omega})$ to produce its restriction to the boundary $\Gamma$; that is,

$$\gamma : \ C(\overline{\Omega}) \to C(\Gamma), \quad \gamma(u) = u|_\Gamma. \tag{7.24}$$

Note that since $u$ is continuous, its restriction to $\Gamma$ is a continuous function on $\Gamma$, so that $u|_\Gamma \in C(\Gamma)$. Thus if $\Omega \subset \mathbb{R}^2$, the graph of $u|_\Gamma$ can be drawn as a *continuous* curve above $\Gamma$ (as in Figure 7.6). The case of $\Omega \subset \mathbb{R}$ is of course a degenerate special case.

We are interested in the problem of how to define $u|_\Gamma$ when $u$ belongs to $L^2(\Omega)$ or, more generally, to one of the Sobolev spaces $H^m(\Omega)$. Note that functions belonging to $H^m(\Omega)$ are defined on $\Omega$ and not on $\overline{\Omega}$ since $\Gamma$ is a set of measure zero and members of $H^m(\Omega)$ are in fact equivalence classes of functions, two functions being equivalent if they differ on a set of measure zero. One way of defining $u|_\Gamma$ (or $\gamma(u)$) for a function $u$ in $L^2(\Omega)$ would be to set up a sequence $\{u_k\}$ of *continuous* functions on $\overline{\Omega}$ that converges to $u$ in the $L^2$-norm (recall from the remarks following Theorem 6 of Chapter 4, or indeed, from Theorem 3, that $C(\overline{\Omega})$ is dense in $L^2(\Omega)$). Since $u_k$ is in

$C(\overline{\Omega})$ we can unambiguously define, as in (7.24),

$$\gamma(u_k) = u_k|_\Gamma,$$

so we hope to be able to define $\gamma(u)$ according to

$$\gamma(u) = \lim_{k\to\infty} \gamma(u_k) \quad \text{or} \quad \gamma(\lim_{k\to\infty} u_k) = \lim_{k\to\infty} \gamma(u_k).$$

However, a glance back to Chapter 5, Theorem 4 shows that this is tantamount to requiring that $\gamma$ be a *continuous* (or bounded) linear operator from $C(\overline{\Omega})$ to $C(\Gamma)$, and we hope to extend $\gamma$ to a bounded operator from $L^2(\Omega)$ to $L^2(\Gamma)$. But this is not possible. It can be shown that there is no continuous mapping of the kind we are looking for. In other words, it is not possible to define *unambiguously* $u|_\Gamma$ when $u$ is in $L^2(\Omega)$, as is further illustrated by the following example.

**Example**

16. Let $u(x) = 1$ on $\Omega = (0,1)$; $u$ is actually continuous and it would seem logical to define $u|_\Gamma$ by $u(0) = u(1) = 1$. But if we set up the sequence $\{u_k\}_{k=3}^{\infty}$ defined by (Figure 7.7)

$$u_k(x) = \begin{cases} (-1)^{k+1}(1-kx), & 0 \le x < 1/k, \\ 0, & 1/k \le x < 1 - 1/k, \\ (-1)^{k+1}(1+k(x-1)), & 1 - 1/k \le x \le 1, \end{cases}$$

that converges to $u$ in $L^2(0,1)$, we find that $\gamma(u_k) = u_k|_\Gamma = \{u_k(0), u_k(1)\} = \{(-1)^{k+1}, (-1)^{k+1}\}$; thus $u_k|_\Gamma$ oscillates between $-1$ and $+1$ and $\lim_{k\to\infty} u_k$ does not exist.

As the preceding example shows, even though an apparently obvious value for $u|_\Gamma$ may be assigned ($u(0) = u(1) = 1$), there is still ambiguity that results from the fact that $\gamma$ is not a continuous operator. This has far-reaching consequences, as can well be appreciated by the following considerations. If $\gamma : L^2(\Omega) \to L^2(\Gamma)$ were a continuous operator, then we would have

$$\|\gamma(u)\|_{L^2(\Gamma)} \le C\|u\|_{L_2(\Omega)} \tag{7.25}$$

for some constant $C > 0$. Thus if $u$ and $v$ are two functions in $L^2(\Omega)$ (they could be continuous functions) that are close in the sense that $\|u - v\|_{L^2(\Omega)} < \epsilon$ for some small $\epsilon > 0$, then (7.25) gives immediately

$$\|\gamma(u) - \gamma(v)\|_{L^2(\Gamma)} < C\epsilon$$

so that $u|_\Gamma$ and $v|_\Gamma$ are correspondingly close. However, if $\gamma$ is not continuous there is no guarantee that this situation would obtain. This is obviously untenable if we are to develop a coherent theory of boundary value problems.

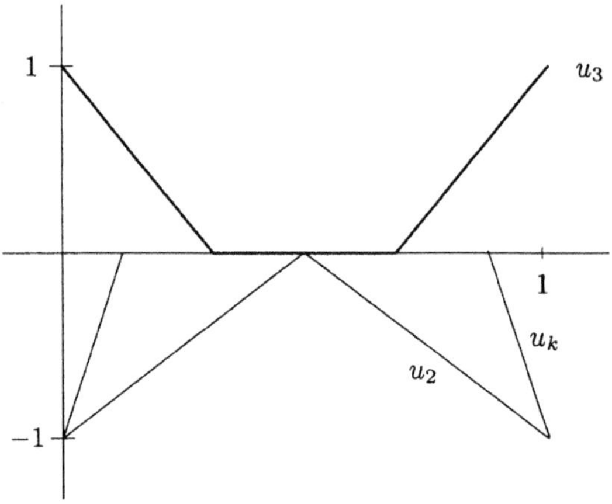


FIGURE 7.7. A sequence of continuous functions with nonconvergent boundary values

All is not lost, however; if a function $u$ belongs to $C^1(\overline{\Omega})$, then it can be shown that the operator $\gamma$ mapping $u$ to its value on $\Gamma$ is a *continuous* operator from $C^1(\overline{\Omega})$ to $C(\Gamma)$, with respect to the norms $\|\cdot\|_{H^1(\Omega)}$ and $\|\cdot\|_{L^2(\Gamma)}$. That is,

$$\gamma : C^1(\overline{\Omega}) \to C(\Gamma), \quad \gamma(u) = u|_\Gamma$$

satisfies

$$\|\gamma(u)\|_{L^2(\Gamma)} \leq C\|u\|_{H^1(\Omega)} \tag{7.26}$$

for some constant $C > 0$ (note the norms used). The proof of this result is contained in the next lemma.

LEMMA 1. *Let $\Omega$ be a domain with Lipschitz boundary. Then the estimate (7.26) holds for all functions $u \in C^1(\overline{\Omega})$.*

PROOF. We prove the result for the case $n = 2$; the proof for the more general case follows in a similar way. We consider a local piece of the boundary and set up coordinates $(\xi, \eta)$ so that this can be represented in the form

$$\eta = f(\xi), \quad \xi \in [-a, a],$$

where $f$ is a Lipschitz function. It follows that there exists a number $b > 0$ such that the set

$$S = \{-a \leq \xi \leq a, \quad f(\xi) - b \leq \eta \leq f(\xi)\}$$

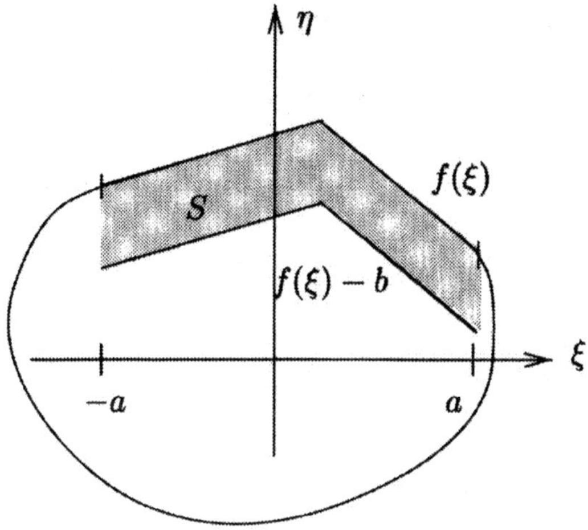


FIGURE 7.8. The subset $S$ of $\overline{\Omega}$

belongs to $\overline{\Omega}$ (Figure 7.8). Now let $u \in C^1(\overline{\Omega})$. Then

$$u(\xi, f(\xi)) = \int_s^{f(\xi)} \frac{\partial u}{\partial \eta}(\xi, \eta)\, d\eta + u(\xi, s),$$

where $f(\xi) - b \leq s \leq f(\xi)$. We use the elementary inequality $(\alpha + \beta)^2 \leq 2\alpha^2 + 2\beta^2$ to obtain

$$u(\xi, f(\xi))^2 \leq 2\left(\int_s^{f(\xi)} \frac{\partial u}{\partial \eta}(\xi, \eta)\, d\eta\right)^2 + 2u(\xi, s)^2. \qquad (7.27)$$

Now the integral in (7.27) may be simplified by applying the Cauchy–Schwarz inequality, to give

$$\begin{aligned} \left(\int_s^{f(\xi)} 1 \cdot \frac{\partial u}{\partial \eta}(\xi, \eta)\, d\eta\right)^2 &\leq \left(\int_s^{f(\xi)} 1^2 d\eta\right)\left(\int_s^{f(\xi)} \left(\frac{\partial u}{\partial \eta}\right)^2 d\eta\right)^2 \\ &= (f(\xi) - s)^2 \left(\int_s^{f(\xi)} \left(\frac{\partial u}{\partial \eta}\right)^2 d\eta\right)^2 \\ &\leq b^2 \left(\int_{f(\xi)-b}^{f(\xi)} \left(\frac{\partial u}{\partial \eta}\right)^2 d\eta\right)^2 \end{aligned}$$

using the fact that $s \geq f(\xi) - b$. After substitution in (7.27) we next integrate with respect to $s$ to obtain

$$bu(\xi, f(\xi))^2 \leq 2\int_{f-b}^{f} \left[b^2 \left(\frac{\partial u}{\partial \eta}\right)^2 + u^2\right] d\eta,$$

and integrate again, this time with respect to $\xi$; this gives

$$\int_{-a}^{a} bu(\xi, f(\xi))^2 \, d\xi \le 2 \int_S \left[ b^2 \left( \frac{\partial u}{\partial \eta} \right)^2 + u^2 \right] dx. \tag{7.28}$$

If $\Gamma$ is a $C^1$ boundary, then $f \in C^1$ and the differential of arc length is given by

$$ds = [1 + (f')^2]^{1/2} d\xi.$$

Furthermore, $f'$ is bounded so that $1 \le [1 + (f')^2]^{1/2} \le C$, where $C$ is a constant independent of $f$. Substitution in the left-hand side of (7.28) yields

$$\int_{-a}^{a} b \frac{u(\xi, f(\xi))^2}{[1 + (f')^2]^{1/2}} \, d\xi \ge C_1 \int_{\Gamma_S} u(\xi, f(\xi))^2 \, ds,$$

for some constant $C_1$, where $\Gamma_S$ is the portion of the boundary corresponding to the interval $\xi \in [-a, a]$. The right-hand side is easily estimated, and (7.28) becomes

$$\int_{\Gamma_S} u(\xi, f(\xi))^2 \, ds \le C_2 \int_S \left[ \left( \frac{\partial u}{\partial \eta} \right)^2 + u^2 \right] dx = C_2 \int_S \sum_{|\alpha| \le 1} (D^\alpha u)^2 \, dx.$$

So the inequality is established for the domain $S$; in order to obtain (7.26) we simply sum over all such patches.

In the event that $\Gamma$ is merely Lipschitzian, it is still the case that $f'$ is bounded and the proof carries over virtually unchanged. □

THEOREM 6 (THE TRACE THEOREM). *Let $\Omega$ be a bounded domain in $\mathbb{R}^n$ with a Lipschitz boundary $\Gamma$. Then*

*(i) there exists a unique bounded linear operator $\gamma$ that maps $H^1(\Omega)$ into $L^2(\Gamma)$; that is,*

$$\gamma : H^1(\Omega) \to L^2(\Gamma), \quad \|\gamma(u)\|_{L^2(\Gamma)} \le C \|u\|_{H^1(\Omega)},$$

*with the property that if $u \in C^1(\overline{\Omega})$, then $\gamma(u) = u|_\Gamma$ in the conventional sense;*

*(ii) the range of $\gamma$ is dense in $L^2(\Gamma)$.*

PROOF. We prove (i). The proof follows immediately from (7.26) and the fact that $H^1(\Omega)$ and $L^2(\Gamma)$ are the completions of $C^1(\Omega)$ and $C(\Gamma)$, respectively, in the appropriate norms. Indeed, for any $u \in H^1(\Omega)$ we can set up a sequence $\{u_k\}$ in $C^1(\Omega)$ that converges to $u$ in the $H^1$-norm. Thus

$$\lim_{k \to \infty} \|u_k - u\|_{H^1} = 0,$$

and so, using (7.26), one sees that $\{\gamma_k\}$ is a Cauchy sequence in $L^2(\Gamma)$ and therefore converges to $v$, say, in $L^2(\Gamma)$. Define $\gamma(u) = v$; then

$$\begin{aligned}\|\gamma(u)\|_{L^2(\Gamma)} &= \|\gamma(\lim_{k\to\infty} u_k)\| = \lim_{k\to\infty} \|\gamma(u_k)\| \\ &\leq C \lim_{k\to\infty} \|u_k\|_{H^1(\Omega)} = C\|u\|_{H^1(\Omega)}.\end{aligned}$$

Part (ii) of the theorem implies that, although the range of $\gamma$ is not all of $L^2(\Gamma)$, any member of $L^2(\Gamma)$ can be approximated arbitrarily closely by a function lying in the range of $\gamma$. □

The trace theorem enables us to define unambiguously $\gamma(u)$ or $u|_\Gamma$, provided that $u$ is smooth enough to be in $H^1(\Omega)$. Now suppose that $u$ is even smoother, so that $u$ belongs to $H^2(\Omega)$. Then $u$ is a member of $H^1(\Omega)$ and so in fact is $D^\alpha u$ for $|\alpha| = 1$:

$$u, \frac{\partial u}{\partial x_1}, \ldots, \frac{\partial u}{\partial x_n} \in H^1(\Omega).$$

This means that the boundary values of the first derivatives of $u$ can also be defined unambiguously, using the trace theorem.

The argument can be generalized to the space $H^m(\Omega)$; indeed, when $m > 1$ then for any $u \in H^m(\Omega)$ we have $D^\alpha u \in H^1(\Omega)$ for $|\alpha| \leq m-1$. By the trace theorem the value of $D^\alpha u$ on the boundary is well-defined and belongs to $L^2(\Gamma)$ :

$$\gamma(D^\alpha u) \in L^2(\Gamma), \quad |\alpha| \leq m-1.$$

Furthermore, if $u$ is in fact $m$-times continuously differentiable, then $D^\alpha u$ is at least continuously differentiable for $|\alpha| \leq m-1$ and

$$\gamma(D^\alpha u) = (D^\alpha u)|_\Gamma.$$

We introduce the notation $\gamma_\alpha$ to denote the operator that, when applied to a member $u$ of $H^m(\Omega)$, gives the trace or boundary value of $D^\alpha u$ for $|\alpha| \leq m-1$ :

$$\gamma_\alpha : H^m(\Omega) \to L^2(\Gamma), \quad \gamma_\alpha(u) = \gamma(D^\alpha u), \quad |\alpha| \leq m-1. \tag{7.29}$$

If $u \in C^m(\overline{\Omega})$, then $\gamma_\alpha(u) = \gamma(D^\alpha u) = D^\alpha u|_\Gamma$. Clearly $\gamma_\alpha$ is a bounded operator.

A word about notation is in order at this point. Henceforth we deal with boundary values of a function only if these boundary values can be defined unambiguously, in the sense of Theorem 6 (or its extension to (7.29)); when referring to the value of a function $u$ or that of its derivatives on the boundary we simply write $u, \partial u/\partial x, \ldots$, instead of $\gamma(u), \gamma_{(1,0\ldots)}u$, *it being understood that the boundary values are to be interpreted in the sense of the trace theorem.* So if we see, for example,

$$u = u_0 \quad \text{on} \quad \Gamma,$$

this means that $\gamma(u)$ takes on the value $u_0$ a.e. on $\Gamma$. Sometimes, in order to make this clearer, we may write "$u = u_0$ in the sense of traces".

Now that the issue of boundary values of functions in $H^m(\Omega)$ has been clarified, it is fairly straightforward to extend Green's theorem, equation (7.2), to functions in $H^1(\Omega)$ (see Exercise 7.16): given functions $u, v \in H^1(\Omega)$, the identity

$$\int_\Omega u \frac{\partial v}{\partial x_i}\, dx = \int_\Gamma uv\nu_i\, ds - \int_\Omega \frac{\partial u}{\partial x_i} v\, dx \tag{7.30}$$

holds for $i = 1, 2, \ldots, n$. From this identity we can deduce higher-order identities; for example, if $u$ is replaced by $\partial u/\partial x_i$ (assuming now that $u \in H^2(\Omega)$) and the resulting equation is summed over $i$ from 1 to $n$, then we find that

$$\int_\Omega \nabla u \cdot \nabla v\, dx = \int_\Gamma \frac{\partial u}{\partial \nu} v\, ds - \int_\Omega (\nabla^2 u) v\, dx$$

for $u \in H^2(\Omega)$, $v \in H^1(\Omega)$, where $\nabla^2$ is the Laplacian (see (8)).

We conclude this section with a set of inequalities that are useful later.

THEOREM 7. *Let $\Omega$ be a bounded domain in $\mathbb{R}^n$ with Lipschitz boundary $\Gamma$ if $n \geq 2$. Then*

*(i) for any $u \in H^1(\Omega)$ there are positive constants $c_1$ and $c_2$ such that*

$$\int_a^b u^2\, dx \leq c_1 \int_a^b (u')^2\, dx + c_2[u^2(a) + u^2(b)] \tag{7.31}$$

*for $n = 1$, and*

$$\int_\Omega u^2\, dx \leq c_1 \int_\Omega |\nabla u|^2\, dx + c_2 \int_\Gamma u^2\, ds \tag{7.32}$$

*for $n \geq 2$;*

*(ii) for any $u \in H^2(\Omega)$ there exists a constant $c_3$ such that that*

$$\|u\|_{H^2}^2 \leq c_3 \left( \sum_{|\alpha|=2} \int_\Omega |D^\alpha u|^2\, dx + \int_\Gamma u^2\, ds \right). \tag{7.33}$$

## 7.5 The spaces $H_0^m(\Omega)$ and $H^{-m}(\Omega)$

The space $H_0^m(\Omega)$ is a subspace of $H^m(\Omega)$ that arises frequently in boundary value problems because members of $H_0^m(\Omega)$ are distinguished by the fact that certain of their derivatives vanish on the boundary. We define $H_0^m(\Omega)$ to be the completion, in the Sobolev norm $\|\cdot\|_{H^m}$, of the space $C_0^m(\Omega)$ of functions with continuous derivatives of order $\leq m$, all of which have compact support in $\Omega$. In other words, $H_0^m(\Omega)$ is formed by taking the union of $C_0^m(\Omega)$ and all those limits of Cauchy sequences in $C_0^m(\Omega)$ that are not in $C_0^m(\Omega)$.

Since $D^\alpha u_k = 0$ on $\Gamma$ $(|\alpha| \leq m)$ for each member of a Cauchy sequence $\{u_k\}$ in $C_0^m(\Omega)$, it suggests that the limit of such a Cauchy sequence, that of course belongs to $H_0^m(\Omega)$, also satisfies $D^\alpha u = 0$ on the boundary. This is borne out by the following theorem, which also gives other properties of $H_0^m(\Omega)$.

THEOREM 8. *Let $\Omega$ be a bounded domain in $\mathbb{R}^n$ with a sufficiently smooth boundary $\Gamma$ and let $H_0^m(\Omega)$ be the completion of $C_0^m(\Omega)$ in the norm $\|\cdot\|_{H^m}$. Then*

*(a) $H_0^m(\Omega)$ is also the completion of $C_0^\infty(\Omega)$ in the norm $\|\cdot\|_{H^m}$;*

*(b) $H_0^m(\Omega) \subset H^m(\Omega)$;*

*(c) if $u \in H^m(\Omega)$ belongs to $H_0^m(\Omega)$, then*

$$D^\alpha u = 0 \quad on \quad \Gamma, \quad |\alpha| \leq m-1.$$

PROOF. The proof of (a) is similar to that of Theorem 1 (iii). Part (b) is obvious. To prove (c), we use the *continuity* of the trace operator: let $\{u_k\}$ be a Cauchy sequence in $C_0^m(\Omega)$ with limit $u$ in $H_0^m(\Omega)$. Then from the definition of $\gamma_\alpha$ we have

$$\gamma_\alpha(u_k) = 0 \quad \text{for} \quad |\alpha| \leq m-1.$$

Hence $\lim_{k\to\infty} \gamma_\alpha(u_k) = \gamma_\alpha \lim_{k\to\infty} u_k = \gamma_\alpha(u) = D^\alpha u = 0.$ □

Part (c) of Theorem 8 is particularly useful in characterizing members of $H_0^m(\Omega)$, as the following example shows.

**Example**

17. The function $u$ defined by

$$u(x) = \begin{cases} x^2, & 0 \leq x < \frac{1}{2} \\ -x^2 + 2x - \frac{1}{2}, & \frac{1}{2} \leq x \leq \frac{3}{2} \\ (2-x)^2, & \frac{3}{2} < x \leq 2 \end{cases}$$

is a member of $H^2(0,2)$, as Figure 7.9 shows. Also, $u$ and $du/dx$ are equal to zero on the boundary $x = 0$, $x = 2$. Hence $u \in H_0^2(0,2)$.

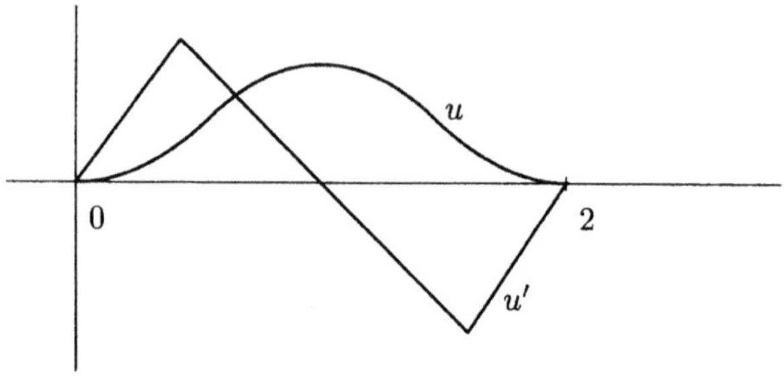


FIGURE 7.9. The function in Example 17

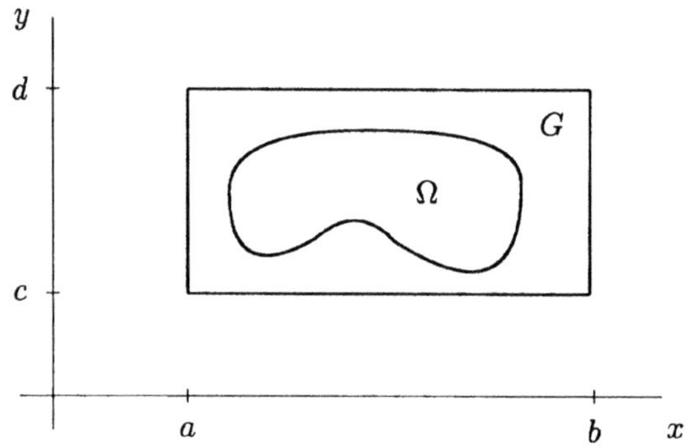


FIGURE 7.10. The construction used in the derivation of (7.34)

**Equivalent norms on $H_0^m(\Omega)$.** We begin with a famous inequality that serves as a basis for defining a norm on $H_0^1(\Omega)$ that is equivalent to the standard $H^1$-norm.

THEOREM 9 (THE POINCARÉ–FRIEDRICHS INEQUALITY). *Let $\Omega$ be a bounded domain in $\mathbb{R}^n$. Then there exists a constant $C > 0$ such that*

$$\int_\Omega |u|^2 \, dx \le C \int_\Omega |\nabla u|^2 \, dx \quad \text{for all } u \in H_0^1(\Omega). \tag{7.34}$$

PROOF. The inequality is first established for the case $u \in C_0^\infty(\Omega)$, after which the density of this space in $H_0^1(\Omega)$ may be used to obtain (7.34). We focus on the situation in which $n = 2$, for convenience. Let $G = [a, b] \times [c, d]$ be a rectangle that includes $\Omega$ as a proper subset, as in Figure 7.10, and note that

$$u(x,y) = \int_c^y \frac{\partial u}{\partial t}(x,t)\, dt \quad \text{for all } (x,y) \in G$$

because $u(x,c) = 0$. From the Cauchy–Schwarz inequality we have

$$\begin{aligned} u^2(x,y) &= \left(\int_c^y 1 \cdot \frac{\partial u}{\partial t}(x,t)\, dy\right)^2 \leq \int_c^y dt \int_c^y \left(\frac{\partial u}{\partial t}(x,t)\right)^2 dt \\ &\leq (d-c)\int_c^d \left(\frac{\partial u}{\partial t}(x,t)\right)^2 dt. \end{aligned}$$

Integrating over $G$, and bearing in mind that $u = 0$ outside of $\Omega$, we obtain

$$\int_\Omega |u|^2\, dx \leq C \int_\Omega \left(\frac{\partial u}{\partial y}\right)^2 dx\, dy.$$

The inequality (7.34) may now be obtained by repeating the argument, this time integrating in the $x$-direction, and then adding.

The extension to functions in $H_0^1(\Omega)$ is left as an exercise (Exercise 7.16). □

At this stage it is convenient to introduce a family of *seminorms* on $H^m(\Omega)$. A seminorm $|\cdot|$ satisfies all the norm axioms except that of positive-definiteness (Axiom N2 in Section 3.3), in that $|u| \geq 0$, but $|u| = 0$ does not imply that $u = 0$. The quantity $|\cdot|_m$ defined on $H^m(\Omega)$ by

$$|u|_m^2 = \sum_{|\alpha|=m} \int_\Omega |D^\alpha u|^2\, dx \tag{7.35}$$

is a seminorm; indeed, $|u|_m = 0$ implies that $D^\alpha u = 0$ for $|\alpha| = m$, which of course does not imply that $u$ itself is zero.

The relevance of the semi-norm to the present discussion is that, with the aid of the Poincaré–Friedrichs inequality, it is possible to show that $|\cdot|_1$ is in fact a *norm* on $H_0^1(\Omega)$.

COROLLARY TO THEOREM 9. *The quantity $|\cdot|_1$ is a norm on $H_0^1(\Omega)$, equivalent to the standard $H^1$-norm.*

This result is treated in Exercise 7.17; note in particular that (7.34) can be expressed in the form

$$\|u\|_{L^2}^2 \leq C|u|_1^2.$$

It is possible to extend Theorem 9 and its Corollary to the spaces $H_0^m(\Omega)$ for any $m \geq 1$; this is also discussed in Exercise 7.17, and the result is summarized in the following.

THEOREM 10. *Let $\Omega$ be a bounded domain in $\mathbb{R}^n$. Then there exists a constant $C > 0$ such that*

$$\|u\|_{L^2}^2 \leq C|u|_m^2 \quad \textit{for all } u \in H_0^m(\Omega). \tag{7.36}$$

*Furthermore, $|\cdot|_m$ is a norm on $H_0^m(\Omega)$, equivalent to the standard $H^m$-norm.*

**The Space $H^{-m}(\Omega)$.** In Section 5.4 we discovered that the space $L^2(\Omega)$ is self-dual. The question now arises as to how we can characterize $[H^m(\Omega)]'$, the space of bounded linear functionals on $H^m(\Omega)$. Now we would hope to find out about $[H^m(\Omega)]'$ by considering functionals $\ell$ on $\mathcal{D}(\Omega)$ (that is, distributions), and by looking at the limits of $\langle \ell, \phi_k \rangle$ as $k \to \infty$, where $\{\phi_k\}$ is a Cauchy sequence in $\mathcal{D}(\Omega)$. There is a complication here, however, in that $\mathcal{D}(\Omega)$ is *not* dense in $H^m(\Omega)$, so that not every $u \in H^m(\Omega)$ is the limit of a Cauchy sequence $\{\phi_k\}$ in $\mathcal{D}(\Omega)$. This dilemma is resolved by restricting attention instead to the dual of $H_0^m(\Omega)$; $\mathcal{D}(\Omega)$ is dense in $H_0^m(\Omega)$, by Theorem 8(a), and this property is used to define $H_0^m(\Omega)'$ in the following theorem. Before stating the theorem we introduce the convention whereby the dual of $H_0^m(\Omega)$ is denoted by $H^{-m}(\Omega)$:

$$[H_0^m(\Omega)]' \equiv H^{-m}(\Omega).$$

As shown in the following, this notation makes complete sense.

THEOREM 11. *A distribution $q$ is in the dual space $H^{-m}(\Omega)$ of $H_0^m(\Omega)$ if and only if it can be expressed in the form*

$$q = \sum_{|\alpha| \leq m} D^\alpha q_\alpha, \tag{7.37}$$

*where $q_\alpha$ are functions in $L^2(\Omega)$.*

PROOF. Let $f$ be any function in $L^2(\Omega)$ $(= [L^2(\Omega)]')$; then, for any $\phi \in \mathcal{D}(\Omega)$

$$\begin{aligned} |\langle D^\alpha f, \phi \rangle| &= |\langle f, D^\alpha \phi \rangle| \\ &= \left| \int_\Omega f(D^\alpha \phi)\, dx \right| \\ &\leq \|f\|_{L^2} \|D^\alpha \phi\|_{L^2} \leq \|f\|_{L^2} \|\phi\|_{H^m} \end{aligned} \tag{7.38}$$

using the Cauchy–Schwarz inequality. If $\{\phi_k\}$ is any Cauchy sequence in $\mathcal{D}(\Omega)$ with limit $u$ in $H_0^m(\Omega)$, then by replacing $\phi$ with $\phi_k$ in (7.38) and taking the limit as $k \to \infty$, we see that $D^\alpha f$ is a bounded linear functional on $H_0^m(\Omega)$ for $f \in L^2(\Omega)$ and $|\alpha| \leq m$. That is, $D^\alpha f$ belongs to $[H_0^m(\Omega)]' = H^{-m}(\Omega)$.

Conversely, if $q$ belongs to $H^{-m}(\Omega)$, then by the Riesz Representation Theorem there is a $u \in H_0^m(\Omega)$ such that

$$\langle q, \phi \rangle = (u, \phi)_{H^m} \quad \text{for all} \quad \phi \in \mathcal{D}(\Omega).$$

Now, for any $v \in H_0^m(\Omega)$, let $\{\phi_k\} \subset \mathcal{D}(\Omega)$ with $\lim_{k\to\infty} \phi_k = v$. Then

$$\begin{aligned}\langle q, \phi_k \rangle &= (u, \phi_k)_{H^m} \\ &= \int_\Omega \sum_{|\alpha|\le m} (D^\alpha u)(D^\alpha \phi_k)\, dx \\ &= \sum_{|\alpha|\le m} (-1)^{|\alpha|} \langle D^\alpha(D^\alpha u), \phi_k \rangle.\end{aligned}$$

Hence, as $k \to \infty$ we have

$$\langle q, v \rangle = \left\langle \sum_{|\alpha|\le m} (-1)^{|\alpha|} D^\alpha(D^\alpha u), v \right\rangle$$

so that $q$ is of the form

$$q = \sum_{|\alpha|\le m} (-1)^{|\alpha|} D^\alpha(D^\alpha u)$$

which gives the desired result since $D^\alpha u \in L^2(\Omega)$. □

**Example**

18. Theorem 11 gives a useful way of characterizing the negative Sobolev spaces $H^{-m}(\Omega)$; indeed, (7.37) indicates that if we differentiate a member of $L^2(\Omega)$ up to $m$ times, we get a functional $q$ on $H_0^m$. For example, take

$$H(x) = \begin{cases} 0, & -1 < x < 0, \\ 1, & 0 < x < 1, \end{cases}$$

which belongs to $L^2(\Omega)$. We know that $H' = \delta$, the Dirac delta, so we conclude that

$$\delta \in H^{-1}(-1, 1).$$

This should give some idea of the nature of members of $H^{-m}(\Omega)$; as $m$ gets larger, we find progressively more irregular distributions in $H^{-m}(\Omega)$. We note here that

$$H^m(\Omega) \subset H^{m-1}(\Omega) \subset \cdots \subset H^0(\Omega) = L^2(\Omega) \subset H^{-1}(\Omega) \subset H^{-m}(\Omega)$$

(recall that $[H^0(\Omega)]' = H^0(\Omega)$).

## 7.6 Bibliographical remarks

The definition of a distribution given in Section 7.1 is a simplified one that avoids a number of complicated topological considerations, but which is adequate for our purposes. The classical reference to distributions is Schwartz [46], who was responsible for developing much of the theory. The book [47] by Schwartz on mathematical methods also contains an account of distributions. Other references worth consulting are those by Dautray and Lions [13], Oden and Reddy [38], Roman [42], Showalter [48], and Zeidler [54, 53].

We have restricted attention to the Hilbert space $W^{m,2}(\Omega) \equiv H^m(\Omega)$. A comprehensive treatment of the Sobolev spaces $W^{m,p}(\Omega)$ may be found in the books by Adams [1], Dautray and Lions [13], Oden and Reddy [38], Showalter [48], and Zeidler [54, 53]. Very accessible treatments may also be found in the works by Rektorys [41] and, at a more advanced level, Nečas [34].

It is possible to define Sobolev spaces $W^{s,p}(\Omega)$ for *real* values of $s$. These spaces are important in certain classes of boundary value problems, particularly nonlinear problems. Of some relevance to the treatment given in this chapter is the boundary space $W^{1/2,2}(\Gamma)$ or $H^{1/2}(\Gamma)$, the chief property of this space being that it is the *range* of the trace operator (recall that the range of $\gamma$ is merely dense in $L^2(\Gamma)$). A proper treatment of this space, and of fractional Sobolev spaces in general, is necessarily a lengthy process, and is omitted. Certainly it suffices in later chapters to work with $L^2(\Gamma)$.

## 7.7 Exercises

### Distributions

7.1. If $\alpha$ is a multi-index in $\mathbb{Z}_+^n$, define $\alpha!$ and $\boldsymbol{x}^\alpha$ by

$$\alpha! = \alpha_1!\,\alpha_2!\,\ldots\,\alpha_n!, \quad \boldsymbol{x}^\alpha = x_1^{\alpha_1} x_2^{\alpha_2} \ldots x_n^{\alpha_n} \quad \text{for } \boldsymbol{x} \in \mathbb{R}^n.$$

Verify that the conventional Taylor expansion of a function $f$ about the origin takes the form

$$f(\boldsymbol{x}) = \sum_{|\alpha|=0}^{\infty} \frac{\boldsymbol{x}^\alpha}{\alpha!} D^\alpha f(\boldsymbol{0}).$$

[Expand the right-hand side for the case $n = 2$, by working out the first few terms.]

7.2. Show that the Dirac delta $\delta$ is not generated by a locally integrable function $\delta(x)$, as follows. Let $\phi_a(x)$ be the test function defined by

$$\phi_a(x) = \begin{cases} \exp\left[\frac{a^2}{x^2-a^2}\right], & |x| < a, \\ 0, & b > |x| \geq a, \end{cases}$$

for $b > a > 0$. Assume that a function $\delta(x)$ exists, and show that

$$\left| \int_{-b}^{b} \delta(x)\phi(x)\, dx \right| \leq \frac{1}{e} \int_{-a}^{a} \delta(x)\, dx.$$

Consider the limit as $a \to 0$ and obtain a contradiction.

7.3. Prove that the only continuous function $f$ for which $\langle f, \phi \rangle = \int f(x)\phi(x)\, dx = 0$ for all $\phi \in \mathcal{D}(\Omega)$, is the zero function. [Use the result in Exercise 2.6.]

## Derivatives of distributions

7.4. Use the standard form of Green's theorem to show that

$$\int_\Omega (D^\alpha u) v\, dx = (-1)^{|\alpha|} \int_\Omega u(D^\alpha v)\, dx + \int_\Gamma h(u, v)\, ds$$

for $|\alpha| = m$, where $u, v \in C^m(\overline{\Omega})$, and $h(u, v)$ is a function of $u$ and $v$.

7.5. Show that $(\mathrm{sgn})' = 2\delta$ on $(-1, 1)$, where

$$\mathrm{sgn}\; x = \begin{cases} x/|x| & \text{for} \quad x \neq 0, \\ 0 & \text{for} \quad x = 0. \end{cases}$$

7.6. Show that $f'' = a\delta - a^2(\sin ax)H$ on $(-1, 1)$, where $f$ is (the distribution generated by) the function $f(x) = (\sin ax)H(x)$, and $a$ is constant.

7.7. Show that the function

$$f(x) = \begin{cases} x, & -1 < x \leq 0, \\ x + c, & 0 < x \leq 1, \end{cases}$$

has a generalized derivative given by $f' = df/dx + c\delta = 1 + c\delta$.

7.8. Let $f$ be defined on $\Omega = (-1, 1) \times (-1, 1) \subset \mathbb{R}^2$ by

$$f(\boldsymbol{x}) = \begin{cases} xy, & \text{if } xy \geq 0, \\ 0, & \text{if } xy < 0. \end{cases}$$

Show that $D^{(1,1)} f \equiv \partial^2 f / \partial x \partial y = g(\boldsymbol{x}) = \begin{cases} +1, & x \geq 0,\ y \geq 0, \\ -1, & x \leq 0,\ y \leq 0, \\ 0, & \text{otherwise.} \end{cases}$

7.9. Find the solution to the distributional differential equation

$$u' + u = e^{-x}\delta.$$

[Try $u$ in the form $u = Hf$, where $H$ is the step function and $f \in C^2(-1,1)$.]

**The Sobolev spaces $H^m(\Omega)$**

7.10. To which spaces $H^m(\Omega)$ do the following functions $u$ belong?

(a) $$u'(x) = \begin{cases} 0, & 0 < x < 1, \\ x-1, & 1 \le x < 2, \\ x^3 - x^2 - 3, & 2 \le x < 3; \end{cases}$$

(b) $$u(x,y) = \begin{cases} xy, & 0 \le x \le 1,\ 0 \le y < 1, \\ x(2-y), & 0 \le x \le 1,\ 1 < y \le 2, \\ x & 0 \le x < \frac{1}{2},\ y = 1, \\ 3\pi, & x = \frac{1}{2},\ y = 1, \\ x, & \frac{1}{2} < x \le 1,\ y = 1. \end{cases}$$

7.11. Show that the functions

$$u(x) = \begin{cases} x, & 0 \le x \le 1, \\ 2-x, & 1 \le x \le 2, \end{cases} \quad \text{and} \quad v(x) = \sin \pi x,$$

are orthogonal in $L^2(0,2)$. Investigate whether they are orthogonal in $H^1(0,2)$. Find the distance between $u$ and $v$ in $L^2(0,2)$ and in $H^1(0,2)$.

7.12. For the functions $u$ and $u - v$ in Exercise 7.11, verify the Cauchy–Schwarz inequality for $L^2(0,2)$ and $H^1(0,2)$.

7.13. Use the Sobolev Embedding Theorem to show that the function

$$u(\boldsymbol{x}) = \begin{cases} x^2y^2, & x > 0,\ y > 0, \\ 0, & \text{otherwise}, \end{cases}$$

is continuous on $\Omega = (-1,1) \times (-1,1)$.

**Boundary values of functions in $H^m(\Omega)$**

7.14. Starting with (7.2), derive Green's theorem (7.30) for functions in $H^1(\Omega)$. [Apply (7.2) to sequences $u_n$, $v_n$ in $C^1(\overline{\Omega})$ and use the fact that $D^\alpha u_n \to D^\alpha u$ in $L^2$, together with the continuity of the inner product and of the trace operator.]

7.15. Derive the Green's formula

$$\int_\Omega \nabla^2 u \nabla^2 v \, dx = \int_\Omega (\nabla^4 u) v \, dx + \int_\Gamma \left[ (\nabla^2 u) \frac{\partial v}{\partial \nu} - \frac{\partial}{\partial \nu}(\nabla^2 u) v \right] dx$$

for $u \in H^4(\Omega), v \in H^2(\Omega)$.

**The spaces $H_0^m(\Omega)$ and $H^{-m}(\Omega)$**

7.16. Complete the proof of Theorem 9 by extending the inequality from $C_0^\infty(\Omega)$ to $H_0^1(\Omega)$. Show also that $|\cdot|_1$ is a norm on $H_0^1(\Omega)$.

7.17. Show that the seminorm (7.35) is a norm on $H_0^m(\Omega)$, equivalent to the standard $H^m$-norm.

7.18. Use Green's theorem to show that

$$\|\nabla^2 v\|_{L^2} = |v|_{H^2} \quad \text{for all} \quad v \in H_0^2(\Omega).$$

7.19. Since $H^{-m}(\Omega)$ consists of continuous linear functionals on $H_0^m(\Omega)$, the norm on $H^{-m}(\Omega)$ is defined by (see Section 5.4)

$$\|f\|_{H^{-m}} = \sup \frac{|\langle f, v\rangle|}{\|v\|_{H^m}}, \quad v \in H_0^m(\Omega).$$

Under what conditions is the Dirac delta a member of $H^{-m}(\Omega)$?

7.20. In the space $H^1(\Omega)$ show that the orthogonal complement of $H_0^1(\Omega)$ is the subspace of functions $u \in H^1(\Omega)$ for which $\nabla^2 u = u$ (distributionally). Find a basis for $H_0^1(\Omega)^\perp$ for the case $\Omega = (0,1) \subset \mathbb{R}$.

7.21. Show that $u(x) = \ln x$ is a member of $L^2(0,1)$, and hence that $v(x) = 1/x$ belongs to $H^{-1}(0,1)$.

# Part II

# Elliptic Boundary Value Problems

# 8
# Elliptic boundary value problems

In this chapter we return to the topic of the Introduction, and set about the process of developing a mathematically coherent framework for boundary value problems. Section 8.1 sets the stage by introducing a range of problems involving differential equations; we saw some examples in the Introduction, and here the opportunity is taken to introduce a few more.

In the remaining four sections we build up towards a general theory for the existence, uniqueness, and regularity of solutions to elliptic boundary value problems. The problem is posed as one involving an elliptic operator from one Sobolev space to another. To the uninitiated, the ideas discussed here may seem esoteric at times; rather than discuss techniques for *solving* boundary value problems, the results obtained are of a *qualitative* nature. This is precisely the program of investigation that was proposed in the Introduction, and the intention is that the motivating ideas of that chapter together with the theory developed here, convey the relevance of these qualitative results to a proper understanding of the problem.

## 8.1 Differential equations, boundary conditions, and initial conditions

The main ideas of this section have in fact already been stated in the Introduction, albeit rather succinctly. Here we expand on many of those notions, and introduce a few more definitions relevant to the study of boundary value problems.

**Differential equations.** Differential equations are the lifeblood of any mathematical modeling process in which real-life situations are translated into mathematical language through the use of *functions* of position or time, or both. The assumption that such functions are differentiable to some extent, together with sets of equations that represent natural laws (conservation or balance laws, for example) and that capture in mathematical form the behavior of particular media, lead to mathematical models in the form of differential equations.

At the heart of a differential equation is an unknown function $u$, say, that could be a function of one or more independent variables $x_1, x_2, \dots, x_n, t$. The variables $x_1, x_2, \dots, x_n$, of which there are invariably three or less, usually refer to coordinates of a point in space. As before we use $\boldsymbol{x} = (x, y, z)$ rather than $\boldsymbol{x} = (x_1, x_2, x_3)$ for a point in $\mathbb{R}^3$, whenever this is more convenient. The variable $t$ refers in a physical context to time.

A differential equation (DE) is any equation involving the independent variables $x_1, x_2, \dots, x_n, t$, a function $u$ of these variables, and some of the derivatives of $u$ with respect to these variables. If there is only one independent variable, then the DE is called an *ordinary differential equation* or ODE; on the other hand, if there are two or more independent variables, it is a *partial differential equation* (PDE).

In addition to the variables mentioned there may be other given functions that appear in the DE; these, together with any other information that is given beforehand, constitute the *data* of the problem.

The *order* of a DE is defined to be the order of the highest derivative appearing in the equation.

It may be the case that the unknown function is vector- rather than scalar-valued. If, for example, the unknown function $\boldsymbol{u}$ has components $u_i$ $(i = 1, 2, 3)$, each of which is a function of $\boldsymbol{x} \in \mathbb{R}^3$ and $t$, then there will be not one but a *system* of three PDEs defining the problem, one equation for each unknown.

A DE (or a system of DEs) is *linear* if it can be written in the form $Au = f$, where $A$ is a *linear operator*. Otherwise it is a *nonlinear* DE.

## Examples

1. Biological population growth. Suppose that we wish to model the change in a biological population with time. The population at time $t$ is denoted by $u(t)$, and since there is only one independent variable $t$, it is an ODE that will model this process.

   A simple example of such a model is one in which it is assumed that the rate of change of population depends on the current population, and on the difference between the birth rate per capita $b(u)$, and the death rate per capita $d(u)$. The functions $b$ and $d$ constitute part of the data of the problem, and the ODE corresponding to this model is given by

$$\frac{du}{dt} = [b(u) - d(u)]u. \tag{8.1}$$

This is a first-order nonlinear ODE, since the operator $Au \equiv du/dt - [b(u) - d(u)]u$ is nonlinear.

2. Heat conduction or diffusion. The unsteady heat or diffusion equation

$$\frac{\partial u}{\partial t} - \frac{1}{c\rho}\text{div}\,(K\nabla u) = Q \tag{8.2}$$

derived in the Introduction (see equation (2) there) is an example of a second-order PDE. The functions $c$, $\rho$, $K$, and $Q$ constitute part of the data of the problem, as does the domain on which the problem is posed. Note that this is a *linear* PDE since the operator $A$ defined by $Au = \partial u/\partial t - (1/c\rho)\text{div}\,(K\nabla u)$ is linear.

3. The Poisson equation. The assumption that heat conduction is steady (that is, time-independent), and that the medium is homogeneous (so that $c$, $\rho$, and $K$ in (8.2) are constant) leads to the second-order PDE known as the Poisson equation; this is given by

$$-k\nabla^2 u = Q$$

in which

$$\nabla^2 = \partial^2/\partial x^2 + \partial^2/\partial y^2 + \partial^2/\partial z^2$$

is the Laplacian operator in $\mathbb{R}^3$. Recall from the Introduction that this equation also arises, on a domain in two dimensions, in the problem of the deflection of an elastic membrane.

4. One-dimensional heat conduction. An example of a spatial ODE may be obtained by specializing Example 2 to a situation in which, first, the conduction is steady (so that time disappears as a variable) and, second, all the data depend on one variable, $x$, say. Then the problem of steady one-dimensional heat conduction corresponds to

$$-\frac{1}{c\rho}\frac{d}{dx}\left(K\frac{du}{dx}\right) = Q. \tag{8.3}$$

Note that the left-hand side of (8.3) has the form of a Sturm–Liouville operator (recall Section 6.5).

5. Linear elasticity. This next example is new, and yields a *system* of PDEs in which the unknown is a vector-valued function.

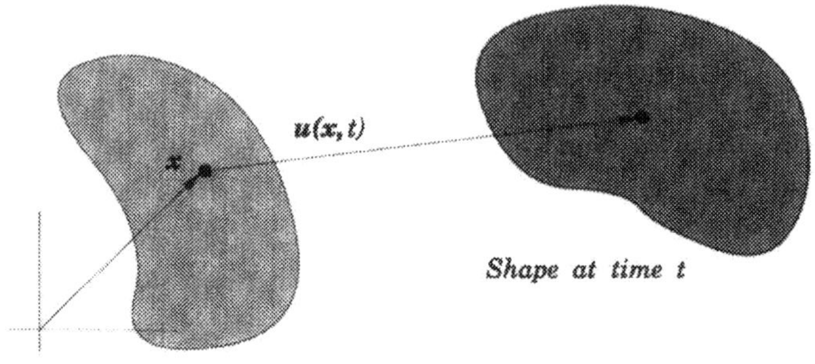


FIGURE 8.1. The deformation of an elastic body

An elastic body is defined to be a solid which, once deformed, will revert to its original shape if the forces causing the deformation are removed. As in the derivation of the heat equation in the Introduction, the equations of elasticity are obtained from a balance law, viz. balance of momentum, together with a constitutive law, viz. Hooke's law.

The analogue of the temperature $u$ is the *displacement* $\boldsymbol{u}$ whereas the function analogous to the heat flux $\boldsymbol{q}$ is the *stress* $\boldsymbol{\sigma}$. The displacement is a vector with components $(u_1, u_2, u_3)$, and $\boldsymbol{x} + \boldsymbol{u}(\boldsymbol{x}, t)$ gives the position at time $t$ occupied by a material particle originally located at $\boldsymbol{x}$ (Figure 8.1). The stress is a second-order tensor, but can be regarded as a symmetric $3 \times 3$ matrix with components $\sigma_{ij}$ for the purposes of this discussion. The stress characterizes the internal forces at any point in a body in a very simple way according to Cauchy's law, which states that the (vector) force per unit area $\boldsymbol{t}$ acting on a surface with unit normal $\boldsymbol{\nu}$ is given by

$$\boldsymbol{t} = \boldsymbol{\sigma\nu}. \tag{8.4}$$

Proceeding in a manner completely analogous to that for the case of the heat equation (cf. (2) through (7) in the Introduction, and see also Exercise 8.2), we obtain *Cauchy's equation of motion*

$$\rho \frac{\partial^2 \boldsymbol{u}}{\partial t^2} - \operatorname{div} \boldsymbol{\sigma} = \boldsymbol{Q} \tag{8.5}$$

in which $\rho$ is the mass density and $\boldsymbol{Q}$ is a prescribed body force per unit volume. Just as the operator div maps a vector to a scalar, when applied to a matrix it produces a *vector*, according to the formula

$$\operatorname{div} \boldsymbol{\sigma} = \sum_{i,j=1}^{3} \frac{\partial \sigma_{ij}}{\partial x_j} \boldsymbol{e}_i.$$

In component form, (8.5) is therefore the set of equations

$$\rho\frac{\partial^2 u_i}{\partial t^2} - \sum_{j=1}^{3}\frac{\partial \sigma_{ij}}{\partial x_j} = Q_i \quad (i = 1, 2, 3). \tag{8.6}$$

The analogue of Fourier's law is the elasticity law, sometimes known as the generalized Hooke's law. Just as Fourier's law relates the heat flux $\boldsymbol{q}$ to derivatives of the temperature, in the same way the elasticity law relates the stress to certain derivatives of the displacement. These derivatives are contained in the symmetric *strain* tensor or matrix $\boldsymbol{\epsilon}$, which measures deformation in the body, and whose components are given by

$$\epsilon_{ij}(\boldsymbol{u}) = \tfrac{1}{2}\left(\frac{\partial u_i}{\partial x_j} + \frac{\partial u_j}{\partial x_i}\right). \tag{8.7}$$

The constitutive law for linear elastic materials then states that the stress depends linearly on the strain at every point of the body; that is,

$$\boldsymbol{\sigma} = \boldsymbol{C}\boldsymbol{\epsilon}(\boldsymbol{u}), \tag{8.8}$$

so that $\boldsymbol{C}$ is a linear operator that takes strains to stresses, and is known as the *elasticity tensor*. When written out in component form (8.8) becomes

$$\sigma_{ij} = \sum_{k,l=1}^{3} C_{ijkl}\epsilon_{kl}(\boldsymbol{u}),$$

so that in general each component of $\boldsymbol{\sigma}$ depends on every component of $\boldsymbol{\epsilon}$. In practice this dependence can be narrowed down quite considerably, and we in fact focus on one special but very important case, viz. that corresponding to *isotropic* elasticity. For this case (8.8) reads

$$\boldsymbol{\sigma} = \lambda(\operatorname{tr}\boldsymbol{\epsilon})\boldsymbol{I} + 2\mu\boldsymbol{\epsilon} \quad \text{or} \quad \sigma_{ij} = \lambda\left(\sum_{k=1}^{3}\epsilon_{kk}\right)I_{ij} + 2\mu\epsilon_{ij}$$

in which $\lambda$ and $\mu$ are material coefficients known as Lamé's constants, and the trace $\operatorname{tr}\boldsymbol{M}$ of a matrix $\boldsymbol{M}$ is defined by $\operatorname{tr}\boldsymbol{M} = \sum_{i=1}^{n} M_{ii}$. Thus the components of the elasticity tensor are given by

$$C_{ijkl} = \lambda I_{ij}I_{kl} + \mu(I_{ik}I_{jl} + I_{il}I_{jk}). \tag{8.9}$$

It is of course possible to express the stress directly as a function of the displacement, by writing

$$\boldsymbol{\sigma} = \Diamond\boldsymbol{u},$$

where the *elasticity operator* $\diamond$ is defined by

$$\diamond \boldsymbol{u} = \lambda[\operatorname{tr} \boldsymbol{\epsilon}(\boldsymbol{u})]\boldsymbol{I} + 2\mu\boldsymbol{\epsilon}(\boldsymbol{u}).$$

When written out in full, this reads

$$\begin{aligned} \sigma_{11} &= (\lambda + 2\mu)\frac{\partial u_1}{\partial x_1} + \lambda\left(\frac{\partial u_2}{\partial x_2} + \frac{\partial u_3}{\partial x_3}\right), \\ \sigma_{12} = \sigma_{21} &= \mu\left(\frac{\partial u_1}{\partial x_2} + \frac{\partial u_2}{\partial x_1}\right), \end{aligned}$$

and so on. Thus although the constitutive equation is a little more complex than that for the case of heat conduction, the structure is exactly the same; for heat conduction the operator in question is $\nabla$ whereas for elasticity it is $\diamond$.

Continuing the analogy, we eliminate the flux (in this case the stress) from (8.5) to obtain a system of equations in the components of $\boldsymbol{u}$; these are known as *Navier's equations*, and are found by straightforward substitution to be

$$\rho\frac{\partial^2 \boldsymbol{u}}{\partial t^2} - \operatorname{div}[\boldsymbol{C}\boldsymbol{\epsilon}(\boldsymbol{u})] = \boldsymbol{Q} \tag{8.10}$$

or, after substituting for $\boldsymbol{C}$,

$$\rho\frac{\partial^2 \boldsymbol{u}}{\partial t^2} - (\lambda + \mu)\nabla(\operatorname{div} \boldsymbol{u}) - \mu\nabla^2\boldsymbol{u} = \boldsymbol{Q}. \tag{8.11}$$

This represents a set of three second-order linear PDEs in the three components of $\boldsymbol{u}$; their structure should be compared with that of the heat equation (Box 1 in the Introduction).

6. **Deflection of a plate.** The next example also comes from linear elasticity, and concerns the special case in which the body is a thin plate. That is, one of its dimensions, in the $z$ direction, say, is very much smaller than the other two, and the body occupies the region $\Omega \times (-h/2, h/2)$, where $\Omega$ is a domain in $\mathbb{R}^2$, so that geometrically the plate is flat (Figure 8.2). It is assumed that external forces act only in the $z$ direction. This set of circumstances allows various assumptions to be made about the deformation of the plate. First, the midsurface $\Omega$ is assumed to undergo a displacement with components $u_1(x, y, 0) = u_2(x, y, 0) = 0$ and $u_3(x, y, 0) \equiv w(x, y)$. Second, we invoke a key geometrical assumption known as the *Kirchhoff-Love hypothesis*: this states that sections of the plate that are straight, and normal to the midplane $\Omega$, remain straight and normal after deformation. The Kirchhoff–Love hypothesis has an immediate consequence,

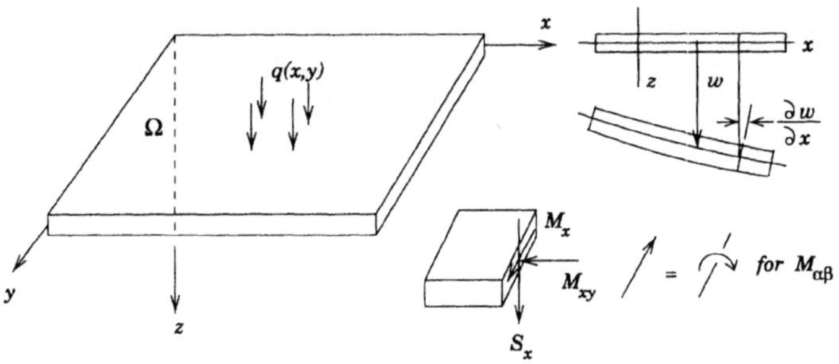


FIGURE 8.2. A thin elastic plate

which is that the inplane displacements can be expressed in terms of the transverse displacement; indeed, from Figure 8.2 we see that

$$u_1(x,y,z) = -z\frac{\partial w}{\partial x} \quad \text{and} \quad u_2(x,y,z) = -z\frac{\partial w}{\partial y}, \tag{8.12}$$

to which is added

$$u_3(x,y,z) = w(x,y). \tag{8.13}$$

The governing equation for an elastic plate is obtained by imposing these assumptions on the elasticity equations. First, we adopt the convention that Greek suffixes range over 1 and 2. Next, we define the components $S_\alpha$ and $M_{\alpha\beta}$ of the *shear force* vector $\boldsymbol{S}$ and *bending moment* matrix $\boldsymbol{M}$ by

$$S_\alpha = \int_{-h/2}^{h/2} \sigma_{3\alpha}\, dz \quad \text{and} \quad M_{\alpha\beta} = \int_{-h/2}^{h/2} z\sigma_{\alpha\beta}\, dz.$$

These are quantities that are averaged over the thickness of the plate; their interpretations are illustrated in Figure 8.2.

The shear force is eventually eliminated, but a constitutive equation is required for $\boldsymbol{M}$. This may be derived from the generalized Hooke's law, which together with (8.12) becomes

$$M_{\alpha\beta} = -D\left[\nu(\nabla^2 w)I_{\alpha\beta} + (1-\nu)\frac{\partial^2 w}{\partial x_\alpha x_\beta}\right]. \tag{8.14}$$

Here $\nabla^2$ is the two-dimensional Laplace operator, $I_{\alpha\beta}$ are the components of the $2 \times 2$ identity matrix, and $D$ is called the bending stiffness; it depends on the material and the geometry, and is defined by $D = Eh^3/12(1-\nu^2)$, in which $E$ and $\nu$ are material constants known, respectively, as Young's modulus and Poisson's ratio (nothing to do with the Poisson equation!). These two constants may be expressed in terms of the Lamé moduli if desired.

Assuming static (time-independent) behavior and an external force per unit area $q$ acting only in the vertical direction, the use of (8.7) together with the definitions of shear force and bending moment can be shown (see Exercise 8.4) to lead to the pair of equations

$$\begin{aligned} \sum_{\beta=1}^{2} \partial M_{\alpha\beta}/\partial x_\beta &= S_\alpha, \quad \alpha = 1,2, \\ \sum_{\alpha=1}^{2} \partial S_\alpha/\partial x_\alpha + q &= 0. \end{aligned} \tag{8.15}$$

Finally, elimination of $S_\alpha$ from these equations, and use of the constitutive equation (8.14) leads to the linear *fourth-order PDE*

$$D\nabla^4 w = q,$$

in which $\nabla^4$, the *biharmonic operator*, is defined by

$$\nabla^4 w = \nabla^2(\nabla^2 w) = \frac{\partial^4 w}{\partial x^4} + 2\frac{\partial^4 w}{\partial x^2 \partial y^2} + \frac{\partial^4 w}{\partial y^4}. \tag{8.16}$$

7. Deflection of a beam. A body that is rectilinear in shape, and whose length is considerably greater than its two other dimensions, is known as a *beam*. The equations of elasticity, when applied to beams, simplify in much the same way as they do for plates, the difference being that the theory for beams is one-dimensional.

   Consider then inplane deflection of the beam shown in Figure 8.3; it has length $L$, breadth $b$, and depth $d$, and its breadth and depth are assumed to be much smaller than its length. The beam is subjected to a force of intensity $q$ per unit length. The assumptions underlying beam theory are very similar to those for plates, so these are discussed only briefly. First, the midplane $z = 0$ of the beam is identified, and it is assumed that the midplane displacements are of the form $u_1(x, y, 0) = u_2(x, y, 0) = 0$, and $u_3(x, y, 0) \equiv w(x)$. The analogue of the Kirchhoff–Love assumption is the *Euler–Bernoulli hypothesis*, according to which plane sections that are normal to the midplane

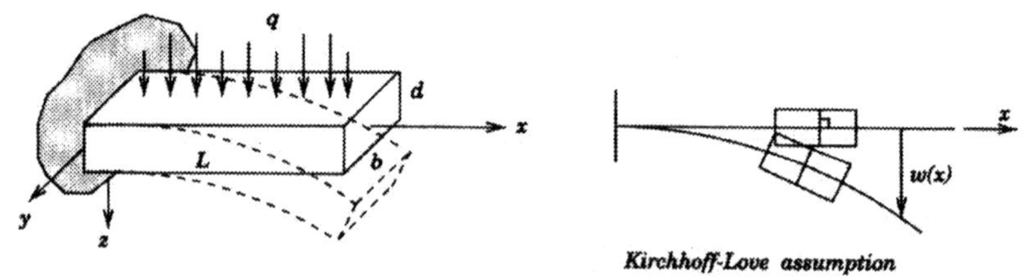


FIGURE 8.3. Inplane deformation of a beam

before deformation remain plane and normal; by analogy with (8.12) we thus obtain

$$u_1(x,y,z) = -z\frac{dw}{dx}, \quad u_2(x,y,z) = 0, \quad u_3(x,y,z) = w(x).$$

Thus in particular, $\epsilon_{11} = -zw''$.

Next, we define the bending moment $M$ and shear force $S$ according to

$$M(x) = \int_A \sigma_{11} z \; dydz, \quad S(x) = \int_A \sigma_{31} \; dydz;$$

by carrying out a series of manipulations similar to those that lead to (8.15) for plates (see Exercise 8.4), we find that two of the equilibrium equations give

$$\begin{aligned} M' - S &= 0, \\ S' + f &= 0. \end{aligned} \tag{8.17}$$

The constitutive equation for the bending moment comes from the assumption that Poisson's ratio $\nu$ is very nearly zero; thus from (8.14), for example, $\sigma_{11} = E\epsilon_{11}$ and so, after substitution for $\epsilon_{11}$, multiplication by $z$ and integration with respect to $y$ and $z$, we find that

$$M = -EIw'', \tag{8.18}$$

in which $I$ is a property of the cross-sectional area known as the second moment of area, and is defined by $I = \int_A z^2 \, dy \, dz = bd^3/12$. The constitutive equation for the shear stress may be found from $(8.17)_1$, and is

$$S = -EIw'''. \tag{8.19}$$

Elimination of $S$ from (8.17) thus leads to the governing equation

$$EI\frac{d^4w}{dx^4} = f \tag{8.20}$$

for the deflection of a beam.

**Specification of the domain of interest.** Physical conditions invariably dictate that a DE is required to be satisfied only on an open subset $\Omega$ of $\mathbb{R}^n$ and, if time is present as a variable, over a prescribed length of time. It follows that a proper description of the physical system must include, in addition to the DE, a statement indicating the spatial and temporal ranges of interest. For example, (8.1) needs to be supplemented by a statement to the effect that we require $u(t)$ for $t$ lying in the range $0 < t \leq T$ or $(0, T]$, where $t = 0$ represents some datum and $T$ is the longest time of interest. If $t = 0$ is taken to be the present, and a solution is required for all time in the future, then the range of $t$ is $(0, \infty)$. Similarly, if for example, the problem has to do with heat conduction in a slab occupying the region $(0,1) \times (0,1) \times (0,1)$, and if we require a solution for all time, then (8.2) has to be supplemented by the statements

$$\boldsymbol{x} \in \Omega = (0,1)^3 \quad \text{and} \quad t \in (0, \infty).$$

**Boundary conditions and initial conditions.** Once the domain of interest has been specified, the next stage in the formulation of the problem involves the specification of the unknown function and possibly some of its derivatives on the boundary $\Gamma$ and at the initial time $t = 0$ (if time is present as a variable). The former are known as *boundary conditions* (BCs) and the latter are called *initial conditions* (ICs). Once again, these are normally dictated by physical considerations. These ideas were of course stated in the Introduction, in the context of the heat equation, but they are reiterated here in this more general context.

If the domain of specification of a DE is purely spatial and denoted by $\Omega$, then only boundary conditions need to be specified, and the DE together with the set of BCs is called a *boundary value problem* (BVP). A special kind of BVP is one defined on an interval $[a, b]$ of the real line; then $\Omega = (0, 1)$, $\Gamma = \{a, b\}$, and boundary conditions are given at $x = a$ and $x = b$. This kind of problem is called, for obvious reasons, a *two-point boundary value problem.*

When the domain is purely temporal, the problem consists of an ODE defined for $t \in (0, T)$ – $T$ may be infinity – and one or more initial conditions that specify the unknown function and possibly some of its derivatives at $t = 0$. This kind of problem is known as an *initial value problem* (IVP).

Finally, when the domain is both spatial and temporal, the problem comprises a PDE (or a set of PDEs) together with boundary conditions and initial conditions. This problem is called an *initial boundary value problem* (IBVP).

**Examples**

8. Population dynamics. Returning to Example 1, the complete specification of the problem becomes: find $u(t)$ satisfying (8.1), with $u(0) = u_0$. Thus the initial population is prescribed, and this is an initial value problem, which is summarized in Box 1.

| Box 1: The IVP for Population Growth | |
|---|---|
| ODE: | $du/dt = [b(u) - d(u)]u, \quad t \in (0, \infty)$ |
| IC: | $u(0) = u_0$ |

9. **Heat conduction.** Suppose for example that the domain $\Omega$ is the cylindrical region $r < a$ and $0 < z < L$, where $r^2 = x^2 + y^2$. Suppose further that the ends $z = 0$ and $z = L$ are insulated and the temperature is a prescribed constant on the curved part of the boundary (Figure 8.4). In this case it is more convenient to use cylindrical coordinates $(r, \theta, z)$; then if the initial temperature is known, and is given by the function $f(r, \theta, z)$, the initial boundary value problem corresponding to heat conduction is summarized as in Box 2.

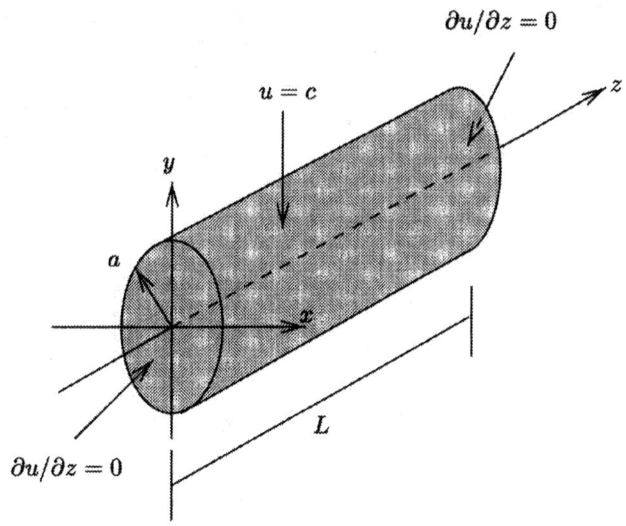


FIGURE 8.4. Heat conduction in a cylindrical domain

**Box 2: The IBVP for Heat Conduction**

PDE: $$\frac{\partial u}{\partial t} - \frac{1}{c\rho}\operatorname{div}(K\nabla u) = Q$$

BCs: $$\frac{\partial u}{\partial z}(r,\theta,0,t) = \frac{\partial u}{\partial z}(r,\theta,L,t) = 0$$
$$u(a,\theta,z,t) = c$$

IC: $$u(r,\theta,z,0) = f(r,\theta,z)$$

10. **One-dimensional steady heat conduction.** Suppose that Example 4 applies to heat conduction in a bar such as that shown in Figure 8.5, and that the circumstances along the longitudinal sides of the bar are consistent with the assumption that all variables depend only on $x$ (for example, the conditions on the surfaces $x = 0$ and $x = l$ are independent of $y$ and $z$).

    We give an example of the kinds of boundary conditions that may be specified at the ends $x = 0$ and $x = l$ of the slab. Suppose then that the end $x = 0$ is held at a prescribed temperature, and that at the other end $x = l$ the heat flux is proportional to the difference between the ambient temperature $u_a$ and the temperature $u(l)$ at that end of the bar; this condition is known as Newton's law of cooling. The full two-point BVP is then as summarized in Box 3.

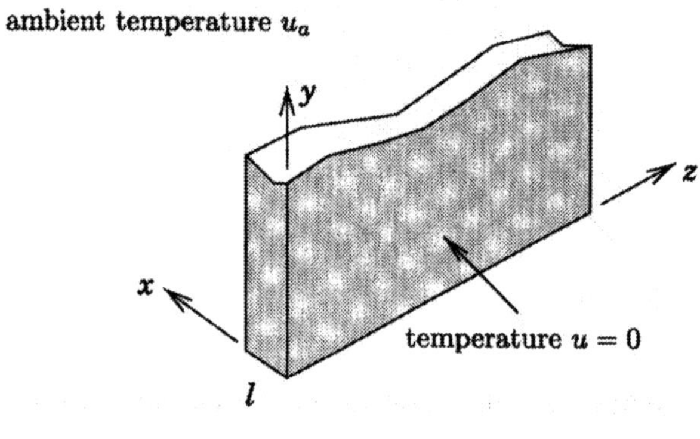


FIGURE 8.5. One-dimensional heat conduction in a bar

**Box 3: The Two-Point BVP for Steady 1D Heat Conduction**

ODE: $$-\frac{1}{c\rho}\frac{d}{dx}\left(K\frac{du}{dx}\right) = Q$$

BCs: $u(0) = 0$ and $-Ku'(l) = \alpha(u(l) - u_a)$

The constant $\alpha$ is assumed positive; this makes physical sense, since heat then flows from a high to a low temperature.

11. Elasticity. Suppose that the elastic body under consideration is the bar shown in Figure 8.6; this bar is fixed at the end $x = 0$, it is subjected to a time-independent (vectorial) force per unit area $\boldsymbol{f}(y, z)$ at the end $x = L$, and on the remainder of its surface there are no forces acting. To specify the force boundary conditions we make use of (8.4) and (8.8), with the appropriate choice of $\boldsymbol{\nu}$. In this way we arrive at the boundary value problem in Box 4.

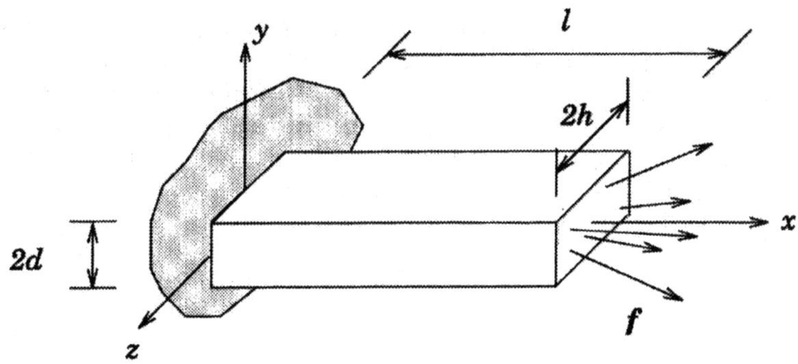


FIGURE 8.6. Deformation of an elastic bar

**Box 4: The BVP for Linear Elasticity**

| | |
|---|---|
| PDE: | $-(\lambda+\mu)\nabla(\operatorname{div}\boldsymbol{u}) - \mu\nabla^2\boldsymbol{u} = \mathbf{0}$ |
| BCs: | $\boldsymbol{u}(0,y,z) = \mathbf{0}$ |
| | $(\Diamond\boldsymbol{u})(l,y,z)\boldsymbol{e}_x = \boldsymbol{f}(y,z)$ |
| | $(\Diamond\boldsymbol{u})(x,y,z)\boldsymbol{e}_y = \mathbf{0}$ for $y = \pm d$ |
| | $(\Diamond\boldsymbol{u})(x,y,z)\boldsymbol{e}_z = \mathbf{0}$ for $z = \pm h$ |

12. **Elastic plate.** The fourth-order plate problem requires *two boundary conditions* at each point on the boundary, as we show in the theory that follows. These are of two kinds: those in which the displacement or its first derivatives are prescribed, and those in which the shear force or bending moment along the boundary are prescribed. We take a concrete example to show what form some of these boundary conditions can take.

    Consider then the rectangular plate shown in Figure 8.7. It is constrained against motion along the ends $x = \pm h$, whereas the other two ends $y = \pm l$ rest on supports that permit rotation, but not vertical displacement. The boundary conditions along $x = \pm h$ therefore stipulate that the displacement and slope are both zero; in other words, $w = 0$ and $\partial w/\partial x = 0$.

    In order to write down the boundary conditions at the other two ends we must first be clear about what it is that they stipulate. One of the conditions is straightforward: $w = 0$ there. But the condition that these ends are free to rotate is equivalent to stating that the plate experiences no restraining moment or couple there. Referring to Figure 8.2, we see that it is the moment $M_{11}$ that is required to be zero. From (8.14) this is

$$M_{11} = -D\left(\frac{\partial^2 w}{\partial y^2} + \nu\frac{\partial^2 w}{\partial x^2}\right).$$

    But since $w = 0$ along the edge $y = \pm l$, it follows that $\partial^2 w/\partial x^2 = 0$ there. So the condition $M_{xx} = 0$ becomes, along that edge, $\partial^2 w/\partial y^2 = 0$.

    The boundary value problem for the rectangular plate is summarized in Box 5.

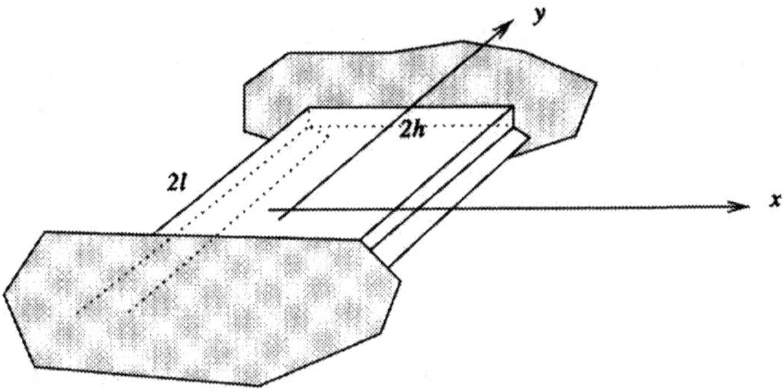


FIGURE 8.7. Boundary conditions for a rectangular plate

**Box 5: The BVP for a Plate**

PDE: $$D\nabla^4 w = Q \quad \text{in } \Omega = (-h,h)\times(-l,l)$$

BCs: $$w(\pm h, y) = 0, \quad (\partial w/\partial x)(\pm h, y) = 0, \quad y \in [-l, l]$$

$$w(x, \pm l) = 0, \quad (\partial^2 w/\partial y^2)(x, \pm l) = 0, \quad x \in [-h, h]$$

Although all classes of problems introduced here are important in their own right, subsequent discussions are limited to boundary value problems in order to keep the scope of this work within reasonable limits. Certainly BVPs provide the ideal vehicle with which to introduce and motivate the finite element method later on; more generally, the study of BVPs presented here may be regarded as a suitable prerequisite to the study of time-dependent problems.

We begin in earnest the study of BVPs in the following section, which is devoted to a study of an important class of (ordinary or partial) differential operators called *elliptic operators*. The corresponding DE together with an appropriate set of boundary conditions is referred to as an *elliptic boundary value problem*.

## 8.2 Linear elliptic operators

Let $A$ be a partial differential operator of even order $2m$ in $n$ variables, and of the form

$$Au = \sum_{|\alpha|,|\beta|\le m} (-1)^{|\alpha|} D^\alpha \left(a_{\alpha\beta}(\boldsymbol{x}) D^\beta u\right), \quad \boldsymbol{x} \in \Omega \subset \mathbb{R}^n, \tag{8.21}$$

where $\Omega$ is an open bounded set in $\mathbb{R}^n$ (recall the discussion of multi-index notation in Section 7.1). The coefficients $a_{\alpha\beta}$ are real-valued functions of position, and $D^\alpha$ represents a partial differential operator of order $|\alpha|$; that is,

$$D^\alpha = \frac{\partial^{|\alpha|}}{\partial x_1^{\alpha_1} \dots \partial x_n^{\alpha_n}}.$$

The term $(-1)^{|\alpha|}$ is not essential, but is included here for future convenience.

The operator $A$ is assumed to occur in a PDE (or system of PDEs) of the form

$$Au = f,$$

where $f$ lies in the range of $A$. For now we restrict attention to scalar-valued functions $u$, and make the extension to vector-valued functions (that occur in elasticity, for example) later.

The classification of $A$ depends only on the coefficients of the highest-order derivatives, that is, the derivatives of order $2m$, and the terms involving these derivatives are said to constitute the *principal part of* $A$, denoted by $A_0$, and which for the operator (8.21) is given by

$$A_0 \equiv \sum_{|\alpha|,|\beta| \leq m} a_{\alpha\beta} D^{\alpha+\beta} u.$$

Let $\boldsymbol{\xi}$ be a vector in $\mathbb{R}^n$, and let

$$\boldsymbol{\xi}^\alpha = \xi_1^{\alpha_1} \cdots \xi_n^{\alpha_n}, \quad \alpha \in \mathbb{Z}_n^+.$$

Then

(i) $A$ is *elliptic* at $\boldsymbol{x}_0 \in \Omega$ if

$$\sum_{|\alpha|,|\beta|=m} a_{\alpha\beta}(\boldsymbol{x}_0)\boldsymbol{\xi}^{\alpha+\beta} \neq 0 \quad \text{for all} \quad \boldsymbol{\xi} \neq \mathbf{0}; \tag{8.22}$$

(ii) $A$ is *elliptic* if it is elliptic at all points in $\Omega$;

(iii) $A$ is *strongly elliptic* if there exists a number $\mu > 0$ such that

$$\left| \sum_{|\alpha|,|\beta|=m} a_{\alpha\beta}(\boldsymbol{x}_0)\boldsymbol{\xi}^{\alpha+\beta} \right| \geq \mu |\boldsymbol{\xi}|^{2m} \tag{8.23}$$

holds at every point $\boldsymbol{x}_0$ in $\Omega$, and for all $\boldsymbol{\xi} \in \mathbb{R}^n$. Here $|\boldsymbol{\xi}| = (\xi_1^2 + \dots + \xi_n^2)^{1/2}$ is the length of the vector $\boldsymbol{\xi}$.

For the case in which $A$ is a *second-order* operator (that is, $m = 1$), the notation can be simplified. Indeed, suppose that the problem is posed in $\mathbb{R}^n$; then (8.21) takes the form

$$Au = -\sum_{i,j=1}^{n} \frac{\partial}{\partial x_i}\left(a_{ij}(\boldsymbol{x})\frac{\partial u}{\partial x_j}\right) + \sum_{j=1}^{n} a_j \frac{\partial u}{\partial x_j} + a_0 u = f \text{ in } \Omega \qquad (8.24)$$

for suitable coefficients $a_{ij}$, $a_j$, and $a_0$, and the condition of ellipticity is examined by considering, instead of (8.22) and (8.23), the conditions

$$\sum_{i,j=1}^{n} a_{ij}(\boldsymbol{x}_0)\xi_i\xi_j \neq 0 \text{ for all } \boldsymbol{\xi} \neq \mathbf{0}, \qquad (8.25)$$

for ellipticity, and

$$\sum_{i,j=1}^{n} a_{ij}(\boldsymbol{x}_0)\xi_i\xi_j \geq \mu|\boldsymbol{\xi}|^2 \qquad (8.26)$$

for *strong ellipticity*.

These ideas are best appreciated by looking at a few examples.

## Examples

13. Consider the operator that appears in the steady, nonhomogeneous heat equation (that is, the steady version of Example 9), and assume that the problem is plane, so that $n = 2$. The operator $A$ is thus (ignoring the coefficient $1/(c\rho)$) given by

$$\begin{aligned} Au &= -\operatorname{div}(K\nabla) \\ &= -\frac{\partial}{\partial x}\left(K\frac{\partial u}{\partial x}\right) - \frac{\partial}{\partial y}\left(K\frac{\partial u}{\partial y}\right) \end{aligned}$$

so that, in the notation of (8.24), $a_{11} = a_{22} = K$ and $a_{12} = a_{21} = 0$. The principal part of this operator is

$$K\left(\frac{\partial^2 u}{\partial x^2} + \frac{\partial^2 u}{\partial y^2}\right) \quad \text{or} \quad K\nabla^2 u.$$

The left-hand side of (8.25) is equal to $K(\xi_1^2+\xi_2^2)$, and so this operator is strongly elliptic, with $\mu = K$ in (8.26).

14. The biharmonic operator given by (8.16) is strongly elliptic: $a_{\alpha\beta} = 1$ only when $\alpha = \beta = (2,0)$ or $(0,2)$ or $(1,1)$; so, for $\Omega \subset \mathbb{R}^2$ and writing $\boldsymbol{\xi} = (\xi, \eta)$,

$$\sum_{|\alpha|,|\beta|=2} a_{\alpha\beta}\boldsymbol{\xi}^{\alpha+\beta} = \xi^4 + 2\xi^2\eta^2 + \eta^4 = |\boldsymbol{\xi}|^4.$$

15. The operator

$$A = (1-x)\frac{\partial^2}{\partial x^2} + 3\frac{\partial^2}{\partial y^2} - y\frac{\partial}{\partial x}$$

is elliptic only in the half plane $x < 1$; to see this, we evaluate

$$\sum_{|\alpha|,|\beta|=1} a_{\alpha\beta}\boldsymbol{\xi}^\alpha = (1-x)\xi^2 + 3\eta^2;$$

this expression is nonzero for all nonzero vectors $\boldsymbol{\xi} = (\xi, \eta)$ provided that $x < 1$. However, for any point $(x_0, y_0)$ in the half plane $x \geq 1$ this expression is zero for all vectors of the form $\boldsymbol{\xi} = (\sqrt{3}, \sqrt{x_0 - 1})$.

The definition of elliptic operators has deliberately been confined to operators of *even* order, since it is possible to show that *all elliptic operators in* $\mathbb{R}^n$ *are of even order when* $n \geq 2$. It is also worth noting that the operators that occur in physically realistic problems such as those discussed in Section 8.1 are always of even order.

Though the definitions (8.22) and (8.23) are given in the context of PDEs involving a single scalar-valued function, the extension to systems of PDEs is immediate. Exercise 8.8 addresses this point in the context of the elasticity problem.

## 8.3 Normal boundary conditions

Boundary conditions cannot be specified arbitrarily; there must be restrictions on their number, the order of the differential operators appearing in them, and so on, if the boundary value problem is to admit a solution. For example, if two boundary conditions are identical or, in any case, not independent of each other, the formulation is defective. Similar considerations apply if two boundary conditions are contradictory; for example, suppose we have a domain $\Omega \subset \mathbb{R}^2$ with boundary $\Gamma$, and let the two boundary conditions be

$$u = g, \tag{8.27}$$

$$\nabla u \cdot \boldsymbol{s} \equiv du/ds = h, \tag{8.28}$$

where $du/ds$ is the tangential derivative, $\boldsymbol{s}$ being the unit tangent vector to the curve defining the boundary. The condition (8.27) implies that $du/ds = dg/ds$, which contradicts (8.28), unless $dg/ds = h$ (Figure 8.8). Hence these two equations are inadmissible as boundary conditions when specified together. In order to avoid situations such as these, we restrict the manner in which boundary conditions might be specified. First, recall that we restrict attention to boundary value problems involving differential

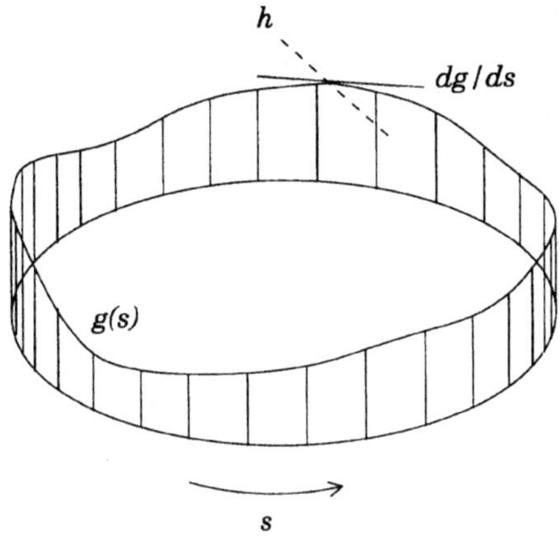


FIGURE 8.8. A pair of contradictory boundary conditions

equations of *even* order $2m$ ($m = 1, 2, \dots$), say, and the boundary is assumed to be smooth (that is, of class $C^\infty$). Then the following restrictions are imposed on the boundary conditions.

(i) A total of $m$ conditions must be specified at each point of the boundary. These are written in the form

$$\begin{aligned} B_0 u &= g_0, \\ B_1 u &= g_1, \\ &\vdots \\ B_{m-1} u &= g_{m-1}, \end{aligned} \tag{8.29}$$

where $g_0, g_1, \dots, g_{m-1}$ are given functions and $B_0, B_1, \dots, B_{m-1}$ are a set of linear differential operators called *boundary operators.* (The boundary conditions are numbered $0, 1, 2, \dots$ rather than $1, 2, \dots$ for reasons of convenience, as becomes apparent). The $j$th boundary operator is of the form

$$B_j u = \sum_{|\alpha| \le q_j} b_\alpha^{(j)} D^\alpha u;$$

that is, it is a linear operator of order $q_j$. The coefficients $b_\alpha^{(j)}$ are given functions of $\boldsymbol{x}$ for $\boldsymbol{x} \in \Gamma$. We assume that $b_\alpha^{(j)}$ and $g_j$ are smooth functions;

(ii) the order of the highest derivative appearing in each boundary condition must be less than the order of the PDE: in other words,

$$0 \le q_j \le 2m - 1 \quad \text{for } j = 0, 1, \dots, m-1;$$

(iii) $q_i \neq q_j$ for $i \neq j$; that is, no two boundary conditions should have differential operators of the same order;

(iv) the final requirement is a restriction on the coefficients of the highest order derivatives, the *principal part* of $B_j$. We require that

$$\sum_{|\alpha|=q_j} b_\alpha^{(j)} \boldsymbol{\nu}^\alpha \neq 0 \quad \text{for all } \boldsymbol{x} \in \Gamma, \tag{8.30}$$

where $\boldsymbol{\nu}^\alpha = \nu_1^{\alpha_1} \nu_2^{\alpha_2} \dots \nu_n^{\alpha_n}$.

For second-order problems these conditions may once again be simplified. First, from (i) and (ii) we have a single boundary condition, which is of order at most equal to one. This condition may therefore be expressed in the form

$$Bu = \sum_{j=1}^{n} b_j \frac{\partial u}{\partial x_j} + cu, \tag{8.31}$$

in which $b_j$ ($j = 1, \dots, n$) and $c$ are real-valued functions. Finally, requirement (iv), when recast using the notation in (8.31), becomes

$$\sum_{j=1}^{n} b_j \nu_j \neq 0 \quad \text{or} \quad \boldsymbol{b} \cdot \boldsymbol{\nu} \neq 0. \tag{8.32}$$

Requirements (i) through (iii) are self-explanatory but the fourth requirement needs some explanation, which is best done by means of a simple example. Suppose that we have a second-order problem with the boundary condition

$$\nabla u \cdot \boldsymbol{a} = h$$

specified on $\Gamma \subset \mathbb{R}^2$, where $\boldsymbol{a}$ is an arbitrary unit vector; $\nabla u \cdot \boldsymbol{a}$ is the directional derivative in the direction of $\boldsymbol{a}$, and is equal to $a_x \partial u/\partial x + a_y \partial u/\partial y$. Clearly $b_j = a_j$ in (8.31), and (8.32) yields the condition

$$\boldsymbol{b} \cdot \boldsymbol{\nu} \neq 0.$$

Thus (8.30) or (8.32) requires that the vector $\boldsymbol{a}$ should not be orthogonal to $\boldsymbol{\nu}$; this condition ensures that we do not have a situation such as that which occurred with the pair of boundary conditions (8.27) and (8.28) discussed earlier. There, $\boldsymbol{a} = \boldsymbol{s}$ and the two conditions are contradictory.

When Conditions (i) to (iv) are satisfied, the set $\{B_0, B_1, \dots, B_{m-1}\}$ is said to be a set of *normal boundary conditions*. An important special case of a set of normal boundary conditions arises when the order $q_j$ of the highest derivative in the $j$th boundary condition is equal to $j$, for $j = 1, \dots, m-1$; such a set of boundary conditions is called a *Dirichlet system of order m*.

**Examples**

16. As observed in Example 12, the PDE corresponding to the plate problem requires two boundary conditions to be specified at each point on the boundary. One possibility is to specify that the displacement and the slope are both zero along $\Gamma$; in other words, the plate is *clamped* along its edge. In this case the boundary conditions are

$$B_0 u \equiv u = 0,$$
$$B_1 u \equiv \nabla u \cdot \boldsymbol{\nu} = \partial u/\partial \nu = 0,$$

which is a Dirichlet system of order 2 since $q_0 = 0$ and $q_1 = 1$. The system

$$\partial u/\partial x = g_0,$$
$$\partial u/\partial y = g_1,$$

on the other hand, violates requirement (iii) since $q_0 = q_1 = 1$.

17. It is not necessary that the total of $m$ boundary conditions has to be in the form of $m$ equations, each of which applies to the whole of $\Gamma$. The requirement is that $m$ conditions be specified *at each point* in $\Gamma$. We have already seen in Example 12 how it is possible – and indeed often dictated by the physical description of the problem – that different boundary conditions may be prescribed on different parts of $\Gamma$. The boundary conditions in that example are specified on two complementary parts $\Gamma_1$ and $\Gamma_2$ of the boundary:

$$\Gamma_1 = \{(x,y) : \ x = \pm h, \ y \in [-l, l]\} \ \text{ and}$$
$$\Gamma_2 = \{(x,y) : \ y = \pm l, \ x \in [-h, h]\}.$$

These are known as *mixed* boundary conditions.

18. Consider the two-point BVP

$$\frac{d^4u}{dx^4} + 2\frac{d^2u}{dx^2} + 3\frac{du}{dx} + u = f \quad \text{on } \Omega = (0,1),$$

$$u(0) = 0, \quad u(1) = 0,$$
$$u'(0) = 1, \quad u'(1) = 2.$$

The boundary conditions form a normal set (note that requirement (iv) is trivial in the case of two-point BVPs); in fact, the boundary conditions constitute a Dirichlet system of order 2. We observe also that the BCs can be written in the format (8.29) if $B_0$ and $B_1$ are regarded as maps from $C^2(\overline{\Omega})$, say, to $\mathbb{R}^2$, and are defined by

$$B_0 u = (u(0), u(1)), \quad B_1 u = (u'(0), u'(1));$$

then we have

$$B_0 u = (0,0), \quad B_1 u = (1,2).$$

The conditions for a set of boundary conditions to be normal in the case of vector-valued functions may be extended from the scalar case, although the end result is less straightforward. We carry out this extension for the case of elasticity, but rather than make allowance for the most general set of conditions possible, we confine attention to those cases that are likely to occur in practice.

Boundary conditions for problems of elasticity are almost always expressed as conditions involving the displacement $\boldsymbol{u}$ or the surface traction $\boldsymbol{t}$ (equation (8.4)). Now recall that the elasticity operator is one of second order, so that a single boundary condition is required at each point along the boundary. However, because we are dealing with a vector-valued unknown variable, it follows that it is a single *vector-valued* boundary condition that is required. In other words, we require a total of $n$ conditions, corresponding to the $n$ components of the vector.

For convenience we assume that the $n$ components of the boundary conditions are referred to a local basis made up of the unit outward normal, and either one or two unit tangent vectors, accordingly as the domain is in $\mathbb{R}^2$ or $\mathbb{R}^3$. These bases are denoted by $\{\boldsymbol{\nu}_k\}_{k=1}^2 \equiv \{\boldsymbol{\nu}, \boldsymbol{s}\}$ and $\{\boldsymbol{\nu}_k\}_{k=1}^3 = \{\boldsymbol{\nu}, \boldsymbol{s}_1, \boldsymbol{s}_2\}$, respectively (Figure 8.9). The case of nonsmooth but otherwise Lipschitz boundaries may be treated as shown in Figure 8.9.

The vectors $\boldsymbol{u}$ and $\boldsymbol{t}$ are resolved relative to this basis, and the boundary conditions are assumed to be, most generally, linear combinations of the normal and tangential components of $\boldsymbol{u}$ and $\boldsymbol{t}$; that is, for a domain in $\mathbb{R}^n$,

$$\sum_{l=1}^{n} (b_{kl}\boldsymbol{\nu}_l \cdot \boldsymbol{u} + c_{kl}\boldsymbol{\nu}_l \cdot \boldsymbol{t}) = g_k, \quad k = 1, \dots, n, \tag{8.33}$$

in which $b_{kl}$ and $c_{kl}$ are most generally sets of functions. Note that these two matrices do *not* contain derivative operators, and that $\boldsymbol{t}$ is a function of the displacement through (8.4) and (8.8).

As in the case of scalar problems, the functions $b_{kl}$ and $c_{kl}$ cannot be specified arbitrarily. For example, it is necessary that conditions be specified for the normal and each of the tangential $n$ components. Such a requirement is met by specifying that the functions appearing in the boundary conditions (8.33) satisfy the condition:

*for boundary condition $k$, the coefficients $b_{kk}$ and $c_{kk}$ are not both zero.* (8.34)

This requirement rules out the possibility of a set of boundary conditions in which not all $n$ components of $\boldsymbol{u}$ appear in the boundary condition.

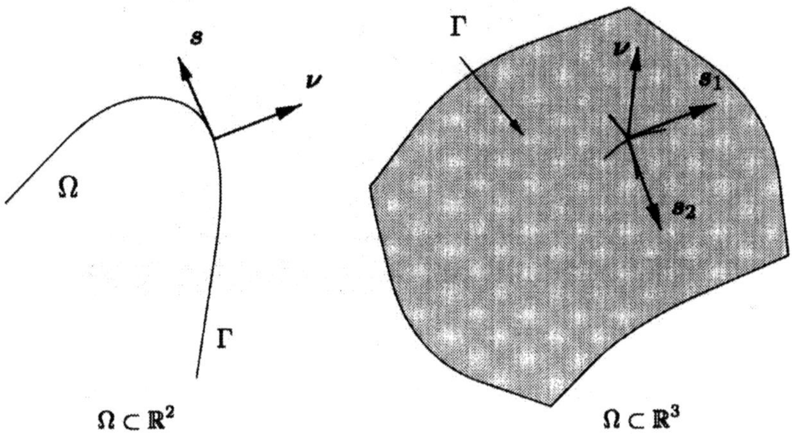


FIGURE 8.9. Local bases for the formulation of boundary conditions

**Example**

19. A very common boundary condition encountered in problems of elasticity is that in which the displacement is specified at every point on the boundary, so that $\boldsymbol{u} = \boldsymbol{g}$. For this case $\boldsymbol{b} = \boldsymbol{I}$ and $\boldsymbol{c} = \boldsymbol{0}$. A second common condition is one in which the surface traction $\boldsymbol{t}$ is specified, so that $\boldsymbol{t} \equiv \boldsymbol{\sigma\nu} = \boldsymbol{g}$. In this case $\boldsymbol{c} = \boldsymbol{I}$ and $\boldsymbol{b} = \boldsymbol{0}$.

    Consider a domain in $\mathbb{R}^2$; then the pair of boundary conditions

$$\begin{aligned} \boldsymbol{u} \cdot \boldsymbol{\nu} &= 0, \\ \boldsymbol{t} \cdot \boldsymbol{s} &= 0 \end{aligned}$$

    corresponds to a situation such as that shown in Figure 8.10, in which frictionless sliding is possible along the boundary; for this case $b_{11} = c_{22} = 1$ and all other components are zero. The pair of conditions

$$\begin{aligned} \boldsymbol{u} \cdot \boldsymbol{\nu} &= 0, \\ \boldsymbol{t} \cdot \boldsymbol{\nu} &= 0, \end{aligned}$$

    on the other hand, is not acceptable since no conditions are specified in respect to tangential components.

We return to scalar problems. Having ensured that the boundary conditions are consistent and contain no ambiguities, we must now ensure also that they are compatible with the partial differential equation of the problem. Intuitively it should be clear that one cannot expect an arbitrary

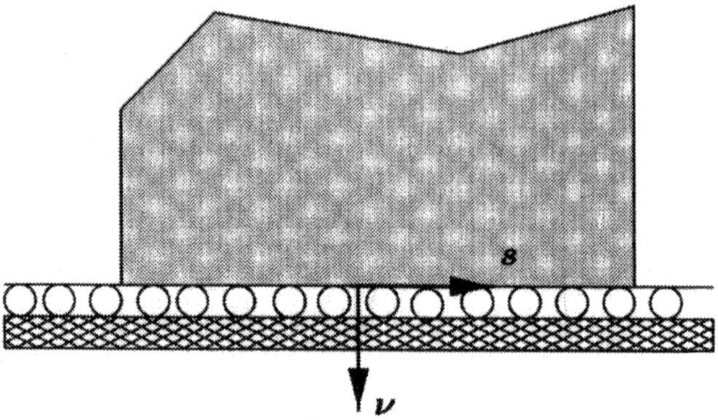


FIGURE 8.10. A typical mixed boundary condition in elasticity

set of boundary conditions to be compatible with the PDE, and that it is therefore necessary that further restrictions be placed on them in order to ensure that the problem as a whole is well-posed.

Let $\boldsymbol{s}$ be a unit tangent vector to $\Gamma$ at a point $\boldsymbol{x}$, and let $\boldsymbol{\nu}$ be the outward unit normal at this point. Now consider the pair of equations

$$\sum_{|\alpha|,|\beta|=m} a_{\alpha\beta}(\boldsymbol{x})\left[\boldsymbol{s} - i\boldsymbol{\nu}\frac{d}{ds}\right]^{\alpha+\beta} u(s) = 0, \quad s > 0, \tag{8.35}$$

$$\sum_{|\alpha|=q_j} b_\alpha^{(j)}(\boldsymbol{x})\left[\boldsymbol{s} - i\boldsymbol{\nu}\frac{d}{ds}\right]^{\alpha} u(s)\bigg|_{s=0} = 0, \quad j = 0,\ldots,m-1, \tag{8.36}$$

that involve only the principal parts of $A$ and of $B_j$ (recall that $\boldsymbol{a}^\alpha = a_1^{\alpha_1}a_2^{\alpha_2}\ldots a_n^{\alpha_n}$ for any vector $\boldsymbol{a}$ in $\mathbb{R}^n$). The set $\{B_0, B_1, \ldots, B_{m-1}\}$ of boundary operators is compatible with $A$, and is said to *cover $A$ at $\boldsymbol{x}$*, if the only solution of (8.35), (8.36) is $u(s) = 0$. We require that $\{B_j\}$ cover $A$ at every point $\boldsymbol{x}$ in $\Gamma$.

Precisely why a requirement such as the covering condition should ensure compatibility between $B_j$ and $A$ is not an obvious matter; the details are lengthy, and may be pursued in the references given at the end of this chapter.

**Example**

20. Consider the Poisson equation

$$-\nabla^2 u = f \quad \text{in } \Omega \subset \mathbb{R}^2;$$

the most general normal boundary condition is of the form

$$B_0 u = a\frac{\partial u}{\partial x} + b\frac{\partial u}{\partial y} + cu = g \quad \text{on } \Gamma, \tag{8.37}$$

and so we must investigate the restrictions placed on $a$, $b$, and $c$ by the covering condition. At a point $\boldsymbol{x}$ on the boundary with tangent $\boldsymbol{s} = (\sigma, 0)$, equation (8.35) gives (with $a_{20} = a_{02} = -1$)

$$-\sigma^2 u + \frac{d^2 u}{ds^2} = 0 \tag{8.38}$$

whereas (8.36) gives

$$a\sigma u(0) - ibu'(0) = 0. \tag{8.39}$$

A general solution of (8.38) is $u(s) = c_1 e^{\sigma s} + c_2 e^{-\sigma s}$, and since we require $u(s)$ to be finite as $s \to \infty$, we must have $c_1 = 0$. The use of (8.39) now gives

$$(a + ib)\sigma c_2 = 0,$$

so that $c_2 = 0$ and hence $u(s) = 0$ provided that $a \neq 0$ or $b \neq 0$, so that (8.37) covers $A$ at $\boldsymbol{x}$ for any values of $a$, $b$, and $c$. In order to investigate the covering condition at other points on the boundary, we simply introduce new axes $\overline{x}, \overline{y}$ so that $\boldsymbol{\nu} = (0, 1)$ relative to these axes, at the point under consideration.

## 8.4 Green's formulas and adjoint problems

In this and the following sections we concern ourselves with boundary value problems of the form

$$\begin{aligned} Au &= f \qquad \text{in } \Omega \subset \mathbb{R}^n, \\ \left.\begin{aligned} B_0 u &= g_0 \\ B_1 u &= g_1 \\ &\vdots \\ B_{m-1} u &= g_{m-1} \end{aligned}\right\} &\quad \text{on } \Gamma, \end{aligned} \tag{8.40}$$

where $A$ is a linear elliptic partial differential operator of order $2m$, of the form

$$Au = \sum_{|\alpha| \le m} (-1)^{|\alpha|} D^\alpha \left( \sum_{|\beta| \le m} a_{\alpha\beta}(\boldsymbol{x}) D^\beta u \right), \quad \boldsymbol{x} \in \Omega \subset \mathbb{R}^n; \tag{8.41}$$

the coefficients $a_{\alpha\beta}$ are functions of $\boldsymbol{x}$, are smooth, and satisfy the condition for ellipticity. The set $B_0, B_1, \ldots, B_{m-1}$ of boundary operators is of the form

$$B_j u = \sum_{|\alpha| \le q_j} b_\alpha^{(j)} D^\alpha u \tag{8.42}$$

and constitutes a set of normal boundary conditions that cover $A$. The coefficients $b_\alpha^{(j)}$ are also assumed to be smooth functions. We refer to (8.40) through (8.42) as a *regularly elliptic boundary value problem of order* $2m$.

In the case of second-order problems, (8.41) and (8.42) can be expressed in the form (8.24) with the single boundary condition

$$Bu = \sum_{j=1}^{n} b_j \frac{\partial u}{\partial x_j} + cu \quad = \quad g \text{ on } \Gamma,$$

in which the ellipticity of $A$ and the normality of the boundary operator $B$ are defined through (8.25), (8.26), and (8.32).

A central question of the theory of elliptic boundary value problems relates to the conditions under which one may expect a *unique* solution of (8.40) to *exist*. In other words, given data in the form of the functions $f$, $a_{\alpha\beta}$, $b_\alpha^{(j)}$, $g_j$, as well as the geometry of the domain $\Omega$, under what conditions can we expect to find a unique solution? Furthermore, if such a solution exists, then it is equally important to know something about the *regularity* or *smoothness* of this solution. If, for example, $f$ belongs to $H^r(\Omega)$ and the functions $g_j$ are members of the boundary spaces $H^{s_j}(\Gamma)$, we would like to know the largest integer $\sigma$ for which the solution $u$ belongs to $H^\sigma(\Omega)$, since this conveys information about the degree of smoothness of $u$. As one would expect, the regularity of $u$ depends very much on that of the data: the smoother the data, the smoother $u$ can be expected to be.

Before we can discuss questions of existence and uniqueness in any detail it is necessary to introduce the concept of a Green's formula associated with the operator $A$.

**Green's formula and the formal adjoint operator.** With the operator $A$ given by (8.41), we denote by $A^*$ the operator defined by

$$A^*u = \sum_{|\alpha|\le m} (-1)^{|\alpha|} D^\alpha \left( \sum_{|\beta|\le m} a_{\beta\alpha}(x) D^\beta u \right);$$

$A^*$ is referred to as the *formal adjoint* of $A$. The relevance of the formal adjoint is that if Green's theorem (7.4) is applied to the integral $\int_\Omega vAu\, dx$, then we obtain

$$\int_\Omega vAu\ dx = \int_\Omega uA^*v\ dx + \int_\Gamma F(u,v)\ ds \tag{8.43}$$

in which $F(u, v)$ represents boundary terms that arise from the application of the theorem. If $A^* = A$, that is, $a_{\alpha\beta} = a_{\beta\alpha}$, the operator $A$ is then said to be *formally self-adjoint*.

In the case of second-order problems, two successive applications of Green's theorem (7.2) yield, for fixed $i$ and $j$,

$$\begin{aligned}
-\int_\Omega v\frac{\partial}{\partial x_i}\left(a_{ij}\frac{\partial u}{\partial x_j}\right)\,dx &= -\int_\Gamma va_{ij}\frac{\partial u}{\partial x_j}\nu_i\,ds + \int_\Omega a_{ij}\frac{\partial u}{\partial x_j}\frac{\partial v}{\partial x_i}\,dx \\
&= -\int_\Gamma\left[va_{ij}\frac{\partial u}{\partial x_j}\nu_i - ua_{ij}\frac{\partial v}{\partial x_i}\nu_j\right]\,ds \\
&\quad -\int_\Omega u\frac{\partial}{\partial x_j}\left(a_{ij}\frac{\partial v}{\partial x_i}\right)\,dx.
\end{aligned}$$

By summing over $i$ and $j$ we therefore find that (8.43) holds with

$$A^*v = -\sum_{i,j=1}^{n}\frac{\partial}{\partial x_i}\left(a_{ji}(\boldsymbol{x})\frac{\partial v}{\partial x_j}\right), \tag{8.44}$$

and

$$F(u,v) = -\sum_{i,j=1}^{n} a_{ij}\left(v\frac{\partial u}{\partial x_j}\nu_i - u\frac{\partial v}{\partial x_i}\nu_j\right), \tag{8.45}$$

so that $A$ is formally self-adjoint if $a_{ji} = a_{ij}$.

## Examples

21. Consider the second-order ordinary differential operator

$$A = -\frac{d^2}{dx^2} + 1;$$

using integration by parts we have, for sufficiently smooth $u$ and $v$ and for $\Omega = (0,1)$,

$$\begin{aligned}
\int_0^1 vAu\,dx &= \int_0^1\left(-v\frac{d^2u}{dx^2} + vu\right)\,dx \\
&= -\left[v\frac{du}{dx}\right]_0^1 + \int_0^1\left(\frac{dv}{dx}\frac{du}{dx} + vu\right)\,dx \\
&= -\left[v\frac{du}{dx}\right]_0^1 + \left[\frac{dv}{dx}u\right]_0^1 - \int_0^1\left(\frac{d^2v}{dx^2} + v\right)u\,dx.
\end{aligned}$$

The Green's formula is thus

$$\int_0^1 v\left(-\frac{d^2u}{dx^2} + u\right)\,dx = \underbrace{\left[-v\frac{du}{dx} + \frac{dv}{dx}u\right]_0^1}_{F(u,v)} + \int_0^1\underbrace{\left(-\frac{d^2v}{dx^2} + v\right)}_{A^*v}u\,dx, \tag{8.46}$$

and since $A^* = A$, $A$ is formally self-adjoint.

22. Consider next the operator defined by

$$Au = -\frac{\partial^2 u}{\partial x^2} - 2\frac{\partial^2 u}{\partial x \partial y} - \frac{\partial^2 u}{\partial y^2}. \tag{8.47}$$

Since $A$ is a second-order operator and this problem is posed on $\mathbb{R}^2$, (8.44) and (8.45) can be used; thus

$$Au = -\sum_{i,j=1}^{2} \frac{\partial}{\partial x_i}\left(a_{ij}\frac{\partial u}{\partial x_j}\right),$$

where $a_{11} = a_{12} = a_{21} = a_{22} = 1$, so that $A$ is formally self-adjoint. Furthermore,

$$F(u,v) = -v\frac{\partial u}{\partial \nu} + u\frac{\partial v}{\partial \nu} - v\left(\frac{\partial u}{\partial x}\nu_y - \frac{\partial u}{\partial y}\nu_x\right) + u\left(\frac{\partial v}{\partial x}\nu_y - \frac{\partial v}{\partial y}\nu_x\right).$$

23. The analogue of (8.43) is readily derived for the elasticity problem. We disregard dependence on time as before, and write the system of PDEs (8.10) or (8.11) corresponding to the elasticity problem in the form

$$\boldsymbol{A}\boldsymbol{u} = \boldsymbol{Q}; \tag{8.48}$$

the elasticity operator is denoted here by $A$ in keeping with the notation of this section, and is defined by the composition $A(\cdot) = -\operatorname{div} \boldsymbol{C}\epsilon(\cdot)$. To obtain (8.43) we take the scalar product of $A\boldsymbol{u}$ with an arbitrary smooth vector function $\boldsymbol{v}$, integrate, and use Green's theorem to obtain

$$\begin{aligned}\int_\Omega A\boldsymbol{u}\cdot\boldsymbol{v}\,dx &= -\int_\Omega \operatorname{div}[\boldsymbol{C}\epsilon(\boldsymbol{u})]\cdot\boldsymbol{v}\,dx \\ &= -\int_\Gamma [\boldsymbol{C}\epsilon(\boldsymbol{u})]\boldsymbol{\nu}\cdot\boldsymbol{v}\,ds + \int_\Omega [\boldsymbol{C}\epsilon(\boldsymbol{u})]\cdot\epsilon(\boldsymbol{v})\,dx \end{aligned}\tag{8.49}$$

in which the scalar product of two matrices $\boldsymbol{\sigma}$ and $\boldsymbol{\tau}$ has been written as $\boldsymbol{\sigma}\cdot\boldsymbol{\tau} = \sum_{i,j=1}^{n} \sigma_{ij}\tau_{ij}$. The details of the derivation of (8.49) are discussed in Exercise 8.16. Now another application of Green's theorem, this time to the volume integral on the right-hand side of (8.49), yields

$$\int_\Omega [\boldsymbol{C}\epsilon(\boldsymbol{u})]\cdot\epsilon(\boldsymbol{v})\,dx = \int_\Gamma \boldsymbol{u}\cdot[\boldsymbol{C}\epsilon(\boldsymbol{v})]\boldsymbol{\nu}\,ds - \int_\Omega \operatorname{div}[\boldsymbol{C}\epsilon(\boldsymbol{v})]\cdot\boldsymbol{u}\,dx, \tag{8.50}$$

in which symmetry properties of the components $C_{ijkl}$ are exploited (see Exercise 8.15). Putting together (8.49) and (8.50) we have, finally,

$$\int_\Omega A\boldsymbol{u}\cdot\boldsymbol{v}\,dx = \int_\Gamma \underbrace{-[\boldsymbol{C}\epsilon(\boldsymbol{u})]\boldsymbol{\nu}\cdot\boldsymbol{v} + [\boldsymbol{C}\epsilon(\boldsymbol{v})]\boldsymbol{\nu}\cdot\boldsymbol{v}}_{F(\boldsymbol{u},\boldsymbol{v})}\,ds + \int_\Omega \underbrace{-\operatorname{div}[\boldsymbol{C}\epsilon(\boldsymbol{v})]}_{A^*\boldsymbol{v}}\cdot\boldsymbol{u}\,dx. \tag{8.51}$$

Comparison with the definition reveals that the elasticity operator $A$ is formally self-adjoint. The boundary term in (8.51) may be rewritten in a more readily recognizable form if we recall that the dependence of the surface traction $\boldsymbol{t}$ on displacement is, after combining (8.4) and (8.8),

$$\boldsymbol{t}(\boldsymbol{u}) = [\boldsymbol{C}\epsilon(\boldsymbol{u})]\boldsymbol{\nu}.$$

It therefore follows that $F$ may be written in the more compact form

$$F(\boldsymbol{u},\boldsymbol{v}) = -\boldsymbol{t}(\boldsymbol{u})\cdot\boldsymbol{v} + \boldsymbol{t}(\boldsymbol{v})\cdot\boldsymbol{u}. \tag{8.52}$$

It turns out that the boundary integral appearing in a Green's formula can be expressed very concisely in terms of four sets of boundary operators. One of these sets is $B_j$, that forms part of the description of the original BVP. The second set of boundary operators is denoted by $S_j$ $(j = 0, \ldots, m-1)$ and has the property that the $2m$ operators

$$B_0, B_1, \ldots, B_{m-1}, S_0, S_1, \ldots, S_{m-1} \tag{8.53}$$

form a *Dirichlet system of order* $2m$. Given these two sets of operators, it is possible to write the Green's formula in the form

$$\int_\Omega vAu\,dx = \int_\Omega uA^*v\,dx + \sum_{j=0}^{m-1}\int_\Gamma \left(S_juB_j^*v - B_juS_j^*v\right)\,ds, \tag{8.54}$$

where $S_j$ and $B_j$ are as previously defined, and the operators $B_j^*$ and $S_j^*$ $(j = 0, 1, \ldots, m-1)$, which are uniquely defined, have the properties:

$$\begin{aligned}&B_j^* \text{ is of order } 2m-1-p_j, \text{ where } p_j \text{ is the order of } S_j;\\ &S_j^* \text{ is of order } 2m-1-q_j, \text{ where } q_j \text{ is the order of } B_j;\\ &\text{the system } B_0^*, B_1^*, \ldots, B_{m-1}^*, S_0^*, S_1^*, \ldots, S_{m-1}^*\\ &\quad\text{is a Dirichlet system of order } 2m.\end{aligned} \tag{8.55}$$

We return to the previous examples to illustrate these ideas.

## Examples

24. In the Green's formula (8.46) we wish to express the boundary term in the form

$$[S_0 u B_0^* v - B_0 u S_0^* v]_0^1$$

(remember that $m = 1$ here). Exactly what form this integral takes depends of course on the boundary condition. Suppose that this problem has the boundary condition

$$u(0) = u(1) = 0 \quad \text{or} \quad B_0 u \equiv (u(0), u(1)) = (0, 0).$$

Thus $q_0 = 0$, and so $S_0$ must be of order 1 for $\{B_0, S_0\}$ to be a Dirichlet system of order 2. Furthermore, $S_0^*$ must be of order $2m - 1 - q_0 = 1$ and $B_0^*$ must be of order $2m - 1 - p_0 = 0$. By inspection of (8.46) we have the correspondence

$$B_0 u = B_0^* u = (u(0), u(1)), \quad S_0 u = S_0^* u = (-u'(0), -u'(1)).$$

25. In the Green's formula corresponding to the operator in Example 22, the function $F(u, v)$ ought to be expressible as

$$S_0 u B_0^* v - B_0 u S_0^* v.$$

Suppose that we are given the boundary condition $\partial u / \partial \nu = g$, so that $B_0 = \partial / \partial \nu$. This is a first-order operator ($q_0 = 1$) so in order that $\{B_0, S_0\}$ form a Dirichlet system of order 2 the operator $S_0$ must be of order 0; that is,

$$S_0 u = \beta u \quad (p_0 = 0)$$

for some function $\beta$. Next, $S_0^*$ must be of order $2m - 1 - q_0 = 2 - 1 - 1 = 0$ and $B_0^*$ must be of order $2m - 1 - p_0 = 2 - 1 - 0 = 1$. Thus $B_0^*$ and $S_0^*$ must be of the form

$$\begin{aligned} S_0^* v &= \gamma v, \\ B_0^* v &= \rho v + \sigma \frac{\partial v}{\partial x} + \tau \frac{\partial v}{\partial y}, \end{aligned}$$

for some functions $\gamma$, $\rho$, $\sigma$, and $\tau$, from which it follows that

$$\begin{aligned} &(\beta u) \left( \rho v + \sigma \frac{\partial v}{\partial x} + \tau \frac{\partial v}{\partial y} \right) + \frac{\partial u}{\partial \nu} (\gamma v) = \left( v \frac{\partial u}{\partial \nu} - u \frac{\partial v}{\partial \nu} \right) \\ &+ v \left( \frac{\partial u}{\partial x} \nu_y + \frac{\partial u}{\partial y} \nu_x \right) - u \left( \frac{\partial v}{\partial x} \nu_y + \frac{\partial v}{\partial y} \nu_x \right). \end{aligned}$$

Since $\partial u/\partial\nu = \nu_x \partial u/\partial x + \nu_y \partial u/\partial y$ we obtain, by equating coefficients,

$$\begin{aligned}
u\,\partial v/\partial x: &\quad \beta\sigma = \nu_x + \nu_y \\
u\,\partial v/\partial y: &\quad \beta\tau = \nu_x + \nu_y \\
uv: &\quad \beta\rho = 0 \\
v\,\partial u/\partial x: &\quad \gamma\nu_x = -\nu_x - \nu_y \\
v\,\partial u/\partial y: &\quad \gamma\nu_y = -\nu_x - \nu_y.
\end{aligned}$$

The last two of these equations give, after using the fact that $\nu_x^2 + \nu_y^2 = 1$,

$$\gamma = -1 - 2\nu_x\nu_y.$$

This leaves three equations with the four unknowns $\beta$, $\rho$, $\sigma$, $\tau$. One of these may be chosen arbitrarily, so we set $\beta = 1$. Then

$$\rho = 0, \quad \sigma = \tau = \nu_x + \nu_y.$$

Hence the boundary integral can be written in the form

$$\int_\Gamma \underbrace{u}_{S_0 u} \underbrace{\left[(\nu_x + \nu_y)\left(\frac{\partial v}{\partial x} + \frac{\partial v}{\partial y}\right)\right]}_{B_0^* v} - \underbrace{\left(\frac{\partial u}{\partial \nu}\right)}_{B_0 u} \underbrace{(-1 - 2\nu_x\nu_y)v}_{S_0^* v}\, ds.$$

26. The analogue of (8.54) in the case of the elasticity problem may be formulated by considering the specific form (8.52) taken by the boundary integrand $F(\boldsymbol{u}, \boldsymbol{v})$. First we denote the left-hand side of the boundary conditions (8.33) by $B_i$ $(i = 1, \ldots, n)$, so that this set of equations reads $B_i\boldsymbol{u} = g_i$, (remember that $\boldsymbol{t}$ also depends on $\boldsymbol{u}$.) Then (8.52) is expressible in the form

$$F(\boldsymbol{u}, \boldsymbol{v}) = \sum_{i=1}^{n} S_i\boldsymbol{u}\, B_i^*\boldsymbol{v} - B_i\boldsymbol{u}\, S_i^*\boldsymbol{v}, \tag{8.56}$$

in which the new operators $S_i$, $B_i^*$, $S_i^*$ are defined in exactly the same way as in (8.54) and (8.55), with $m = 1$; thus for $i = 1, \ldots, n$, $\{B_i, S_i\}$ forms a Dirichlet system of order 2, $B_i^*$ is of order $1 - p_j$, where $p_i$ is the order of $S_i$, $S_i^*$ is of order $1 - q_i$, where $q_i$ is the order of $B_i$, and $\{B_i^*, S_i^*\}$ forms a Dirichlet system of order 2.

Suppose then that for a problem posed in $\mathbb{R}^2$ $(n = 2)$ the pair of boundary conditions is

$$\begin{aligned}
\boldsymbol{u} \cdot \boldsymbol{\nu} &= 0, \\
\boldsymbol{t} \cdot \boldsymbol{s} &= 0.
\end{aligned}$$

Thus $B_1\boldsymbol{u} = \boldsymbol{u}\cdot\boldsymbol{\nu}$ and $B_2\boldsymbol{u} = \boldsymbol{t}(\boldsymbol{u})\cdot\boldsymbol{s} = [C\epsilon(\boldsymbol{u})\boldsymbol{\nu}]\cdot\boldsymbol{s}$. By resolving the vectors into their tangential and normal components, denoted, respectively, by subscripts $s$ and $\nu$, (8.52) can be recast in the following form, in which the forms of the various operators are plain.

$$\begin{array}{rcccc} F(\boldsymbol{u},\boldsymbol{v}) = & -t_\nu(\boldsymbol{u})v_\nu & -t_s(\boldsymbol{u})v_s & +t_\nu(\boldsymbol{v})u_\nu & +t_s(\boldsymbol{v})u_s; \\ & | & | & | & | \\ & -S_1\boldsymbol{u}B_1^*\boldsymbol{v} & +B_2\boldsymbol{u}S_2^*\boldsymbol{v} & -S_1^*\boldsymbol{v}B_1\boldsymbol{u} & +B_2^*\boldsymbol{v}S_2\boldsymbol{u}. \end{array}$$

Note that the boundary operators $S_j$ have to be partial differential operators of such orders as to make (8.53) a Dirichlet system, but further than that they are not unique. Indeed, in the last example we saw that the function $\beta$ could be chosen arbitrarily. However, once $S_j$ are fixed then so are the forms that the sets of operators $B_j^*$ and $S_j^*$ take.

With each regularly elliptic problem of the form (8.40) may be associated an *adjoint problem*

$$\begin{aligned} A^*u &= f^* \qquad \text{in } \Omega \subset \mathbb{R}^n, \\ \left.\begin{aligned} B_0^*u &= g_0^* \\ B_1^*u &= g_1^* \\ &\vdots \\ B_{m-1}^*u &= g_{m-1}^* \end{aligned}\right\} &\quad \text{on } \Gamma, \end{aligned}$$

where $f^*$, $g_0^*$, $g_1^*$, ..., $g_{m-1}^*$ are given functions.

Like the original problem, the adjoint problem is also a regularly elliptic boundary value problem of order $2m$. We soon show that the adjoint problem plays a key role in determining whether the original problem has solutions, and whether these are unique.

**Example**

27. Returning to Example 22, $A^* = A$, and the operator $B_0^*$ is given in Example 25. The adjoint problem is thus

$$\begin{aligned} -\frac{\partial^2 u}{\partial x^2} - 2\frac{\partial^2 u}{\partial x \partial y} - \frac{\partial^2 u}{\partial y^2} &= f^* \quad \text{in } \Omega, \\ \frac{\partial u}{\partial \nu} + \nu_x\frac{\partial u}{\partial y} - n_y\frac{\partial u}{\partial x} &= g_0^* \quad \text{on } \Gamma. \end{aligned}$$

## 8.5 Existence, uniqueness, and regularity of solutions

We come now to the main topic of this chapter, namely, the discussion of well-posedness of solutions to problems of the form (8.40). In order to keep the discussion as simple as possible we confine attention to problems having *homogeneous boundary conditions*, that is, problems for which $g_0, g_1, \ldots, g_{m-1}$ are all zero. This is no real restriction, since it is not difficult to show (see Exercise 8.19) that any problem with nonhomogeneous boundary conditions can be converted to one with homogeneous boundary conditions in a fairly straightforward manner.

We also assume that the domain $\Omega$ is *bounded*, and is *smooth* (in the language of Section 7.3, the boundary is assumed to be $C^\infty$). This assumption, although rather restrictive, permits the development of a fairly general existence theory.

Thus we consider the problem

$$Au = f \quad \text{in } \Omega \subset \mathbb{R}^n, \qquad \left.\begin{aligned} B_0 u &= 0 \\ B_1 u &= 0 \\ &\vdots \\ B_{m-1} u &= 0 \end{aligned}\right\} \text{ on } \Gamma, \tag{8.57}$$

where $A$ and $B_j$ are given by (8.41) and (8.42). Our aim is to settle the questions of

(a) *existence:* under what circumstances (8.57) has a solution $u$ that belongs to $H^s(\Omega)$, $s$ being an integer greater than or equal to $2m$;

(b) *uniqueness:* whether there is only one such solution;

(c) *continuous dependence on the data*: whether the solution depends on the data in the sense that the estimate

$$\|u\|_{H^s} \leq C\|f\|_{H^{s-2m}} \tag{8.58}$$

holds for some constant $C > 0$, independent of the solution; and

(d) *regularity:* to establish the largest value of $s$ for which $u \in H^s(\Omega)$.

If the problem (8.57) has a unique solution that depends continuously on the data, then the problem is said to be *well-posed*. Regularity is a supplementary issue, the goal of which is to establish the maximum smoothness of the solution consistent with the data. Note that if $u$ belongs to $H^s(\Omega)$, then $Au \in H^{s-2m}(\Omega)$ since $A$ is a differential operator of order $2m$.

The inequality (8.58) has the following implication. Suppose that problem (8.57) is considered with two different sets of data $f_1$ and $f_2$, and that the solutions corresponding to these two sets of data are, respectively, $u_1$ and $u_2$. Since $A$ is linear it follows that

$$Au_1 = f_1 \quad \text{and} \quad Au_2 = f_2$$

imply that

$$A(u_2 - u_1) \equiv A\Delta u = \Delta f,$$

where $\Delta u = u_2 - u_1$ and $\Delta f = f_2 - f_1$. So $\Delta u$ is a solution to the problem with data $\Delta f$, and the inequality (8.58) then gives

$$\|\Delta u\|_{H^s} \leq C\|\Delta f\|_{H^{s-2m}},$$

from which it can be concluded that if $f_1$ and $f_2$ are close to each other in the sense that $\|\Delta f\|$ is small, $\|\Delta f\| < \epsilon$, say, where $\epsilon$ is a small number, then $\|\Delta u\| < C\epsilon$ so that $u_1$ and $u_2$ are correspondingly close.

The question of existence and uniqueness of a solution is best approached by adopting the language of linear operator theory (Chapter 5). First, we denote by $N(B_j)$ the *null space* of the boundary operator $B_j$; that is, if $B_j$ is regarded as an operator from $H^s(\Omega)$ to $L^2(\Gamma)$, then

$$N(B_j) = \{u \in H^s(\Omega) : \ B_j u = 0 \ \text{ on } \Gamma\}, \quad j = 0, 1, \ldots, m-1.$$

It now follows that a solution of (8.57), if it exists, will belong to the subspace of $H^s(\Omega)$ consisting of all functions that are also in $N(B_j)$. We consequently take the domain of $A$ to be the space $D(A)$ defined by

$$\begin{aligned} D(A) &= H^s(\Omega) \cap N(B_0) \cap \cdots \cap N(B_{m-1}) \\ &= \{u \in H^s(\Omega) : \ B_j u = 0 \ \text{ on } \Gamma\}, \end{aligned} \tag{8.59}$$

so that problem (8.57) now reads: find $u$ that satisfies

$$A : D(A) \to H^{s-2m}(\Omega), \quad Au = f \ \text{ in } \Omega. \tag{8.60}$$

Our first task is to determine the set of functions $f$ in $H^{s-2m}(\Omega)$ for which (8.60) admits a solution. That is, we must identify $R(A)$, the *range* of $A$. This enables us to solve the problem of the *existence* of a solution. We find that $R(A)$ is not all of $H^{s-2m}(\Omega)$; there are functions $f$ in $H^{s-2m}(\Omega)$ that do not lie in $R(A)$, and for which no solution exists. The situation is shown diagrammatically in Figure 8.11. The second task is to ascertain the conditions under which the solution is *unique*; in other words, we wish to know the conditions under which $A$ is one-to-one. For this purpose we define the null space $N(A)$ of $A$ by

$$\begin{aligned} N(A) &= \{u \in D(A) : \ Au = 0\} \\ &= \{u \in H^s(\Omega) : \ Au = 0 \text{ in } \Omega, \ B_j u = 0 \ \text{ on } \Gamma\}. \end{aligned}$$

Clearly if $N(A) \neq \{0\}$, then we cannot expect to have a unique solution since, if $u_0$ is a solution, so is $u_0 + w$ for any $w \in N(A)$ because $A(u_0 + w) = Au_0 + Aw = Au_0 = f$. So elements of $N(A)$ have to be excluded from the domain of $A$ in order to ensure uniqueness. This is no problem, since we

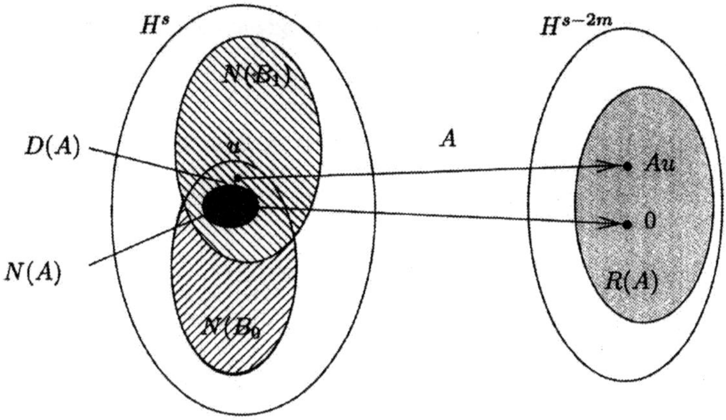


FIGURE 8.11. The various spaces occurring in the problem (8.60)

have simply to introduce the *orthogonal complement* $N(A)^{\perp}$ of $N(A)$ with respect to the $L^2$-inner product, which is defined by

$$N(A)^{\perp} = \{v \in D(A) : \ (v, w) = 0 \ \text{ for all } w \in N(A)\}.$$

Now it can be shown that $N(A)$ is finite-dimensional, and hence complete, so that by the Projection Theorem (Theorem 8 of Chapter 4) we have

$$D(A) = N(A) \oplus N(A)^{\perp};$$

in other words, every $u \in D(A)$ is of the form $u = v + w$ for $v \in N(A)^{\perp}$ and $w \in N(A)$, and furthermore $N(A) \cap N(A)^{\perp} = \{0\}$. Since $N(A)$ and $N(A)^{\perp}$ have in common only the zero element, we simply restrict the domain of $A$ to $N(A)^{\perp}$ to ensure uniqueness.

Similar remarks apply of course to the adjoint problem

$$\begin{aligned} A^*u = f \quad & \text{in } \Omega \subset \mathbb{R}^n, \\ \left.\begin{aligned} B_0^* u &= 0 \\ B_1^* u &= 0 \\ &\vdots \\ B_{m-1}^* u &= 0 \end{aligned}\right\} & \text{ on } \Gamma; \end{aligned} \tag{8.61}$$

we define

$$D(A^*) = \{u \in H^s(\Omega) : \ B_0^* u = B_1^* u = \ldots = B_{m-1}^* u = 0 \text{ on } \Gamma\}$$

by analogy with (8.59), and rephrase (8.61) as the problem of finding $u$ that satisfies

$$A^* : D(A^*) \to H^{s-2m}(\Omega), \quad A^* u = f^* \ \text{ in } \Omega.$$

The null space $N(A^*)$ of $A^*$ and its orthogonal complement $N(A^*)^\perp$ are then

$$\begin{aligned} N(A^*) &= \{w \in D(A^*) : \ A^*w = 0\}, \\ N(A^*)^\perp &= \{v \in D(A^*) : \ (v,w)_{L^2} = 0 \text{ for all } w \in N(A^*)\}. \end{aligned}$$

Like $N(A)$, the space $N(A^*)$ is finite-dimensional. Indeed, for most problems of practical interest

$$\dim N(A) = \dim N(A^*).$$

We are not particularly concerned with solutions to the adjoint problem, but when discussing the existence of solutions to (8.60) it is necessary to call on properties of the space $N(A^*)^\perp$. We now give a few examples.

## Examples

28. Consider the problem

$$\begin{aligned} Au &= -u'' = f \ \text{ in } \ \Omega = (0,1), \\ B_0u &= (u(0), u(1)) = (0,0). \end{aligned}$$

Assume that $f \in L^2(0,1)$, so that a solution $u \in H^2(0,1)$ is sought. Also,

$$N(B_0) = \{u \in H^2(\Omega) : \ u(0) = u(1) = 0\} = D(A).$$

The null space of $A$ is the set of solutions to the problem

$$w'' = 0 \ \text{ in } (0,1), \quad w(0) = w(1) = 0;$$

the only solution to this problem is $w = 0$ so that $N(A) = \{0\}$ and a solution, if it exists, will be unique. Alternatively, suppose that the boundary condition is

$$B_0u = (u'(0), u'(1)) = (0,0);$$

then $N(A) = \{w : w(x) = \text{const.}\}$ so that

$$N(A)^\perp = \left\{v : \ (v,w)_{L^2} = 0 \text{ or } \int_0^1 v(x)\, dx = 0\right\}.$$

The operator $-d^2/dx^2$ is formally self-adjoint and $B_0^* = B_0$, so that $N(A^*) = N(A)$.

29. Consider the problem

$$\begin{aligned} Au &= -\frac{\partial^2 u}{\partial x^2} - 2\frac{\partial^2 u}{\partial x \partial y} - \frac{\partial^2 u}{\partial y^2} = f \quad \Omega \subset \mathbb{R}^2, \\ B_0u &= \frac{\partial u}{\partial \nu} = 0 \ \text{ on } \Gamma. \end{aligned}$$

Clearly

$$N(A) = \{w : \ w(\boldsymbol{x}) = \text{const.}\}$$

from which it follows that

$$N(A)^\perp = \left\{v : \ \int_\Omega v(\boldsymbol{x}) \ dx = 0\right\}.$$

The self-adjointness of $A$ has been established in Example 22, and $B_0$ and $B_0^*$ are given in Example 25, as

$$B_0 = \frac{\partial}{\partial \nu} \quad \text{and} \quad B_0^* = (\nu_x + \nu_y)\left(\frac{\partial}{\partial x} + \frac{\partial}{\partial y}\right),$$

so that the condition $B_0^* w = 0$ is the same as

$$\frac{\partial w}{\partial x} + \frac{\partial w}{\partial y} = 0.$$

The null space of $A^* = A$ is the set of solutions to

$$A^* w = 0 \ \text{ in } \Omega, \quad B_0^* w = 0 \ \text{ on } \Gamma,$$

and this is given by

$$N(A^*) = \{w : \ w(\boldsymbol{x}) = \alpha + \beta(x - y), \quad \alpha, \beta \in \mathbb{R}\}$$

and

$$N(A^*)^\perp = \left\{v : \ \int_\Omega v(\boldsymbol{x})[\alpha + \beta(x - y)] \ dxdy = 0\right\}$$

or, since $\alpha$ and $\beta$ are arbitrary,

$$N(A^*)^\perp = \left\{v : \ \int_\Omega v(\boldsymbol{x}) \ dx = 0, \quad \int_\Omega v(\boldsymbol{x})(x - y) \ dxdy = 0\right\}.$$

We are now in a position to state the main result of this section.

THEOREM 1. *Consider the regularly elliptic boundary value problem (8.57), with $s \geq 2m$, and posed on a bounded domain $\Omega$ with smooth boundary $\Gamma$. Then*

*(i)* (uniqueness) *assuming that the solution $u$ exists, it is unique if $u \in N(A)^\perp$, that is, if*

$$(u, w)_{L^2} = 0 \quad \textit{for all } w \in N(A); \tag{8.62}$$

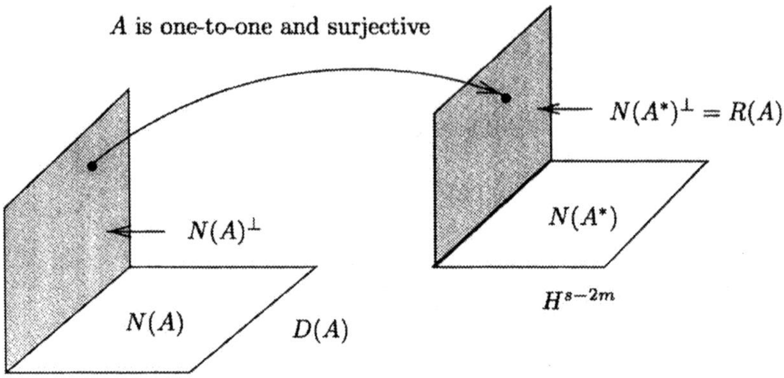


FIGURE 8.12. The domain and range of the operator $A$ in Theorem 1

*(ii)* (existence) *there exists at least one solution if and only if* $f \in N(A^*)^\perp$, *that is, if*

$$(f, v)_{L^2} = 0 \quad \textit{for all } v \in N(A^*); \tag{8.63}$$

*(iii)* (continuous dependence on data) *if a unique solution exists, then there is a constant* $C > 0$, *independent of* $u$, *such that*

$$\|u\|_{H^s} \leq C\|f\|_{H^{s-2m}} \,.$$

REMARKS. 1. The theorem states that $A$ is a *surjective operator* from $D(A)$ onto the subspace of functions in $H^{s-2m}$ that satisfy (8.63). Furthermore, $A$ is *one-to-one* if its domain is restricted to the subspace of functions that satisfy (8.62) (Figure 8.12).

2. Theorem 1, in a slightly modified form, is referred to as the *Closed Range Theorem*, since Part (ii) of the theorem is equivalent to the requirement that $R(A)$ be closed. This equivalence is made apparent in the proof.

3. Part (ii) of the theorem expresses the fact that the data cannot be specified arbitrarily: they have to satisfy (8.63) if a solution is to exist. This is known as a *compatibility condition*, and when (8.63) is satisfied we say that the data are compatible with the operator $A$. The condition is, however, trivial in the event that $N(A^*) = \{0\}$.

4. Part (iii) may be interpreted also as a regularity result, in the sense that it shows that $u \in H^{s+2m}(\Omega)$ if $f \in H^s(\Omega)$.

PROOF. (i) Take any $w \in N(A)$ and assume that there are two solutions $u_1, u_2$ satisfying

$$(u_1, w) = (u_2, w) = 0;$$

that is, $u_1$ and $u_2$ belong to $N(A)^\perp$. Since $Au_1 = Au_2 = f$, we have $A(u_1 - u_2) = 0$ so that $u_1 - u_2 \in N(A)$. But $D(A) = N(A) \oplus N(A)^\perp$ from Chapter 4, Theorem 8, and since $N(A) \cap N(A)^\perp = \{0\}$ it follows that $u_1 - u_2 = 0$, or $u_1 = u_2$. Hence the solution is unique.

(ii) First assume that (8.57) has a solution. Then for any $v \in N(A^*)$ we have, using Green's formula (8.54),

$$\begin{aligned}(f, v)_{L^2} = (Au, v)_{L^2} &= (u, A^*v)_{L^2} + \sum_{j=0}^{m-1} \int_\Gamma \left(S_j u B_j^* v - B_j u S_j^* v\right) \, ds \\ &= (u, 0)_{L^2} + \sum_{j=0}^{m-1} \int_\Gamma \left(S_j u \cdot 0 - 0 \cdot S_j^* v\right) \, ds = 0.\end{aligned}$$

Hence $f \in N(A^*)^\perp$.

We sketch the proof of the converse and leave some of the details to the exercises. The aim is to prove that if $f \in N(A^*)^\perp$, then $f \in R(A)$; that is, $N(A^*)^\perp \subset R(A)$. First we note from (i) that, since $A$ is one-to-one from $N(A)^\perp$ onto $R(A)$, it is possible to define the inverse operator $A^{-1} : R(A) \to N(A)^\perp$. Second, it can be shown (see Exercise 8.22) that both $A$ and $A^{-1}$ are bounded operators, and furthermore that $R(A)$ is closed. It follows then from Chapter 4, Lemma 1 that $R(A)^{\perp\perp} = \overline{R(A)} = R(A)$.

Next, if $v \in R(A)^\perp$ and $u \in D(A)$, then

$$(v, Au)_{L^2} = 0 = (u, A^*v)_{L^2} + \sum_{j=0}^{m-1} \int_\Gamma S_j u B_j^* v \, ds,$$

so that $v \in N(A^*)$ (since $u$ is arbitrary we must have $A^*v = 0$ and $B_j^* v = 0$). Hence $R(A)^\perp \subset N(A^*)$, which implies that $N(A^*)^\perp \subset R(A)^{\perp\perp} = R(A)$ (see Exercise 4.26 and Lemma 1, Chapter 4), which completes the proof.

(iii) Once again we use the fact that $A$ is a bounded, one-to-one linear operator from $N(A)^\perp$ onto $R(A)$; then (Exercise 8.23) there is a constant $C > 0$ such that

$$\|u\|_{H^s} \le C\|Au\|_{H^{s-2m}} = C\|f\|_{H^{s-2m}}. \qquad \square$$

## Examples

30. Consider the Poisson problem

$$\begin{aligned} -k\nabla^2 u &= f \ \text{ in } \Omega, \\ u &= 0 \ \text{ on } \Gamma. \end{aligned}$$

In this case $A = A^* = -k\nabla^2$, so $A$ is formally self-adjoint. Assume that $f \in L^2(\Omega)$, and take $s = 2$ ($m = 1$ here). Thus

$$N(A) = N(A^*) = \{u \in H^2(\Omega) : -k\nabla^2 u = 0 \text{ on } \Omega,\ u = 0 \text{ on } \Gamma\} = \{0\}$$

which should come as no surprise if one considers the various physical problems for which the Poisson equation is a model; whether it is the membrane problem or that of steady heat conduction, clearly one will expect that the solution in the absence of any forcing function $f$, with $u$ prescribed to be zero along the boundary, is going to be zero.

Returning to Theorem 1, (8.62) and (8.63) are satisfied identically. It follows that $-k\nabla^2$ is one-to-one from $D(A)$ onto $L^2(\Omega)$. Furthermore, from Part (iii) of the theorem there is a constant $C > 0$ such that

$$\|u\|_{H^2} \leq C\|f\|_{L^2}.$$

31. Consider now the problem

$$\begin{aligned} -k\nabla^2 u &= f \text{ in } \Omega, \\ \partial u/\partial \nu &= 0 \text{ on } \Gamma. \end{aligned}$$

This would correspond physically to the problem of a membrane constrained around its boundary in such a way that the slope there is zero, or in the case of heat conduction, to a medium that is perfectly insulated along its boundary.

In this case $N(A) = N(A^*) = \{c\}$, $c$ being a constant function. From Part (ii) of the theorem we thus deduce that there exists a solution if and only if

$$(f, c) = 0, \quad \text{or } c\int_\Omega f\ dx = 0, \quad \text{or} \quad \int_\Omega f\ dx = 0,$$

since $c$ is arbitrary. Physically, this compatibility condition means that the *net force* on the membrane must be zero, or in the case of heat conduction, the net heat source must be zero. Again this condition makes physical sense: in the case of the membrane, there is no constraint against vertical motion along the boundary, so that the membrane would fly off unless the forces acting on it were in equilibrium.

From (i) the solution is unique if we prescribe the condition

$$(u, c) = 0 \quad \text{or} \quad \int_\Omega u\ dx = 0.$$

Such a condition would serve to determine the value of any arbitrary constant in the solution.

32. We return to the problem of elasticity, and show that Theorem 1 is applicable to this problem as well. Suppose that the boundary condition is

$$u_\nu = u_s = 0 \quad \text{or, equivalently,} \quad \boldsymbol{u} = \boldsymbol{0} \ \text{ on } \Gamma.$$

We first investigate the structure of $N(A) = \{\boldsymbol{u} : A\boldsymbol{u} = \boldsymbol{0} \text{ on } \Omega,\ \boldsymbol{u} = \boldsymbol{0} \text{ on } \Gamma\}$ (recall (8.48)). Now let $(\cdot,\cdot)$ denote the inner product on $[L^2(\Omega)]^n$, with $(\boldsymbol{u},\boldsymbol{v}) \equiv \int_\Omega \boldsymbol{u}\cdot\boldsymbol{v}\,dx$, and consider the inner product $(A\boldsymbol{u},\boldsymbol{u})$, where $\boldsymbol{u} \in N(A)$; this inner product is of course zero since $A\boldsymbol{u} = \boldsymbol{0}$, and so (8.9) and (8.49) give

$$0 = (A\boldsymbol{u},\boldsymbol{u}) = \int_\Omega \sum_{i,j,k,l=1}^{n} C_{ijkl}\epsilon_{ij}(\boldsymbol{u})\epsilon_{kl}(\boldsymbol{u})\,dx. \tag{8.64}$$

Now in order that various features of realistic elastic materials be encapsulated in the specification of $\boldsymbol{C}$, it is necessary that this tensor possess a property akin to that of positive-definiteness in the case of matrices. In the present context this is known as *pointwise stability*; the elasticity tensor is said to be pointwise stable if there exists a constant $C_0 > 0$ such that

$$\sum_{i,j,k,l=1}^{n} C_{ijkl}M_{ij}M_{kl} \geq C_0 \sum_{i,j=1}^{n} M_{ij}M_{ij} \quad \text{for all matrices } \boldsymbol{M}. \tag{8.65}$$

For an isotropic elastic material, pointwise stability is equivalent to the requirement that (Exercise 8.8)

$$\mu > 0 \quad \text{and} \quad \lambda + 2\mu > 0.$$

Returning to (8.64), and assuming that the elasticity tensor is pointwise stable, we now have

$$0 = \int_\Omega \sum_{i,j,k,l=1}^{n} C_{ijkl}\epsilon_{ij}(\boldsymbol{u})\epsilon_{kl}(\boldsymbol{u})\,dx \geq C_0 \int_\Omega |\epsilon(\boldsymbol{u})|^2\,dx, \tag{8.66}$$

where $|\cdot|$ represents the norm of a matrix; that is, $|\epsilon|^2 = \sum_{i,j=1}^{n} \epsilon_{ij}\epsilon_{ij}$. Now define the norm on $[H^1(\Omega)]^n$ in an obvious way, according to

$$\|\boldsymbol{u}\|_{H^1}^2 = \|u_1\|_{H^1}^2 + \ldots + \|u_n\|_{H^1}^2$$

and define also the norm on the space $[L^2(\Omega)]^{n\times n}$ of matrix-valued functions whose components are in $L^2$, by

$$\|\epsilon\|_{L^2}^2 = \int_\Omega |\epsilon(\boldsymbol{u})|^2\,dx.$$

We would like next to bound (8.66) from below in terms of a Sobolev norm, in order to conclude that $\boldsymbol{u} = \boldsymbol{0}$, and herein lies a problem: the right-hand side of (8.66) contains only specific first derivatives of the displacement. This impasse can fortunately be resolved by appealing to a result known as *Korn's inequality*, which plays a vital role in

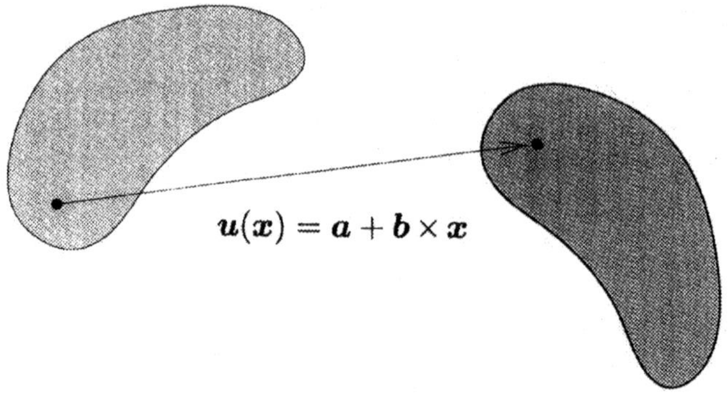


FIGURE 8.13. An elastic body subject to a traction boundary condition and a rigid body displacement

analyses of problems in elasticity, and according to which there is a constant $C_2 > 0$ such that

$$\|\epsilon(\boldsymbol{v})\|_{L^2}^2 \geq C_2 \|\boldsymbol{v}\|_{H^1}^2 \quad \text{for all } \boldsymbol{v} \in [H^1(\Omega)]^n, \tag{8.67}$$

whenever $\boldsymbol{v} = \mathbf{0}$ on a part $\Gamma_v$ of the boundary $\Gamma$, with $\mu(\Gamma_v) \neq 0$. Putting (8.66) and (8.67) together we therefore find that $\|\boldsymbol{u}\|_{H^1} = 0$, so that $\boldsymbol{u} = \mathbf{0}$.

Thus the only member of the null space $N(A)$ is the zero element, and so according to Theorem 1, the problem (8.48) together with the boundary condition $\boldsymbol{u} = \mathbf{0}$ has a unique solution, and furthermore there is a constant $C > 0$ such that

$$\|\boldsymbol{u}\|_{H^1} \leq C \|\boldsymbol{Q}\|_{L^2}.$$

33. A more interesting situation arises when the boundary condition is given by

$$\boldsymbol{t}(\boldsymbol{u}) = \mathbf{0} \quad \text{on } \Gamma.$$

Physically, the body is not constrained against movement anywhere on its boundary, so we would expect an element of nonuniqueness in the solution, inasmuch as the body could be translated and rotated from whatever its current position is, without affecting its state at all (Figure 8.13). Such a motion, which takes place without adding any deformation to the body, is known as a *rigid body displacement.* Its most general form is

$$\boldsymbol{u}(\boldsymbol{x}) = \boldsymbol{a} + \boldsymbol{b} \times \boldsymbol{x},$$

and it is easy to verify that $\epsilon(\boldsymbol{u}) = \mathbf{0}$ for such a displacement field.

For the problem with a traction boundary condition, the most general solution of the problem $A\boldsymbol{u} = \boldsymbol{0}$ in $\Omega$ and $\boldsymbol{t}(\boldsymbol{u}) = \boldsymbol{0}$ on $\Gamma$ is $\boldsymbol{\epsilon} = \boldsymbol{0}$, in other words, a rigid body displacement, and so

$$N(A) = \{\boldsymbol{u} : \ \boldsymbol{u}(\boldsymbol{x}) = \boldsymbol{a} + \boldsymbol{b} \times \boldsymbol{x}, \quad \boldsymbol{a}, \boldsymbol{b} \in \mathbb{R}^n\}.$$

A solution therefore exists, according to Part (ii) of Theorem 1, if and only if the force $\boldsymbol{Q}$ satisfies the condition

$$\int_\Omega \boldsymbol{Q} \cdot [\boldsymbol{a} + \boldsymbol{b} \times \boldsymbol{x}] \, dx = 0 \quad \text{for all } \boldsymbol{a}, \boldsymbol{b} \in \mathbb{R}^n$$

or, equivalently, if

$$\int_\Omega \boldsymbol{Q} \, dx = \boldsymbol{0} \quad \text{and} \quad \int_\Omega \boldsymbol{Q} \times \boldsymbol{x} \, dx = \boldsymbol{0}.$$

These conditions stipulate that $\boldsymbol{Q}$ may not be specified arbitrarily, but rather that the net total force and total couple acting on the body be zero (Figure 8.13), a condition that makes physical sense.

Uniqueness is also subject to a condition: the solution is unique only if it is in $N(A)^\perp$, that is, if it satisfies

$$\int_\Omega \boldsymbol{u} \cdot [\boldsymbol{a} + \boldsymbol{b} \times \boldsymbol{x}] \, dx = 0 \quad \text{for all } \boldsymbol{a}, \boldsymbol{b} \in \mathbb{R}^n$$

or, equivalently, if

$$\int_\Omega \boldsymbol{u} \, dx = \boldsymbol{0} \quad \text{and} \quad \int_\Omega \boldsymbol{u} \times \boldsymbol{x} \, dx = \boldsymbol{0}.$$

These two conditions suffice to ensure that $\boldsymbol{u}$ contains no rigid body displacement.

## 8.6 Bibliographical remarks

The concepts in Section 8.1 are elementary, and are normally encountered in beginning courses on differential equations. Further details, including various techniques for finding solutions, may be found in texts such as that of Zauderer [52], for example.

The theory of elliptic boundary value problems developed in Sections 8.2 and 8.3 draws heavily on the account given in the extended survey by Babuška and Aziz ([3], Chapter 3, which was written by B. Kellogg). This account is based in turn on the treatment of Lions and Magenes [30]; indeed, the presentation given in Sections 8.2 and 8.3 avoids some rather delicate technical issues, full details of which may be found in [30], and concentrates on the aspects that are most accessible, and most relevant, to readers of this text. Accessible treatments of an alternative approach to regularity, using what is known as the method of differentials, may be found in the monographs by Zeidler [53] and by Dautray and Lions [13]. The latter text may also be consulted for further details of Korn's inequalities. The article by Horgan [21] summarizes the major results concerning Korn's inequalities for bounded domains, and discusses bounds on the constants appearing in the inequalities.

Attention has been focused deliberately on those aspects of the theory of elliptic boundary value problems that are relevant to the primary objective, viz. that of presenting the theory of variational boundary value problems and their approximation by finite elements. Some of the more complex topics that have been omitted include the question of well-posedness in the presence of nonhomogeneous boundary conditions, and in the presence of data in $H^{-r}(\Omega)$ for $r > 0$. The latter would cover problems such as $-\nabla^2 u = f$ in $\Omega$ where, for example, $f$ is a Dirac delta. Naturally the solution $u$ is correspondingly irregular. These topics require some knowledge of Sobolev spaces $H^s(\Omega)$ and $H^s(\Gamma)$ for which $s$ is real; the theory of such spaces is covered in the references to Sobolev spaces given at the end of Chapter 7.

We have assumed the boundary to be of class $C^\infty$; when the boundary is less smooth (for example, Lipschitz or polygonal) then the theory on regularity becomes more complicated, although in many cases the results look similar to those given here. For a comprehensive treatment of problems in nonsmooth domains the monograph by Grisvard [17] is recommended.

## 8.7 Exercises

### Differential equations, boundary conditions, and initial conditions

8.1. For each of the following differential equations specify the order of the equation, state whether it is linear, and sketch the spatial domain $\Omega$.

(a) $\dfrac{\partial^2 u}{\partial x^2} + \dfrac{\partial u}{\partial x}\dfrac{\partial u}{\partial y} = y \quad \text{in } \Omega = \{\boldsymbol{x} \in \mathbb{R}^2 : x^2 + y^2 < 1, y > 0\};$

(b) $\dfrac{\partial^4 u}{\partial x^4} + 2\dfrac{\partial^4 u}{\partial x^2 \partial y^2} + \dfrac{\partial^4 u}{\partial y^4} - a\dfrac{\partial u}{\partial t} = 0 \quad \text{in } \Omega = \{\boldsymbol{x} \in \mathbb{R}^2 : x > 0,\ y > 0,\ x + y < 1,\ t \in (0, \infty)\}.$

8.2. The purpose of this exercise is to derive Navier's equation (8.11) for elastic bodies, by retracing the steps employed in the Introduction (equations (0.1) through (0.7)) in the derivation of the heat equation.

(a) The balance law in this case is balance of linear momentum, which states that the rate of change of total momentum equals the total force acting on the body. Express this balance law in mathematical form, and obtain Cauchy's equation of motion (8.5).

(b) Eliminate the stress from Cauchy's equation, using the constitutive equation (8.8) and (8.9), to obtain Navier's equation.

8.3. Using the general approach of the Introduction, obtain the *Helmholtz equation*

$$-\nabla^2 u + ku = f$$

for the behavior of a membrane that is connected to a foundation with stiffness $k$; that is, the foundation exerts a resisting force that is proportional to the displacement of the membrane, the coefficient of proportionality being the stiffness $k$.

8.4. The purpose of this exercise is to fill in some of the missing details in the derivation of the plate equation (Example 12, Box 5).

(a) Assuming static (time-independent) behavior and an external force acting only transversely, consider the first two of equations (8.6), that is, $\sum_{j=1}^{3} \partial\sigma_{\alpha j}/\partial x_j = 0 \quad (\alpha = 1, 2)$; multiply by $z$ and integrate to obtain $(8.15)_1$.

(b) Consider next the third of equations (8.7), that is, $-\sum_{j=1}^{3} \partial\sigma_{3j}/\partial x_j = Q_3$, and integrate to obtain $(8.15)_2$.

(c) Use Parts (a) and (b) together with the constitutive equation (8.14) to derive the biharmonic equation in Box 5.

8.5. The problem of an elastic beam (Example 7), being a fourth-order differential equation, requires two boundary conditions at each end. Sketch and formulate the boundary conditions corresponding to the situations in which

(a) the end of the beam is unable to rotate, but may displace vertically;

(b) the end of the beam is unable to displace vertically, but is free to rotate.

**Linear elliptic operators**

8.6. Find the regions in the $xy$ plane in which the operator

$$A = (1-x)^2 \frac{\partial^4}{\partial x^4} + 2(1-x)(1-y)\frac{\partial^4}{\partial x^2 \partial y^2} + (1-y)^2 \frac{\partial^4}{\partial y^4}$$

is (i) elliptic; (ii) strongly elliptic.

8.7. Show that the operator $A$ defined by

$$A = -\frac{\partial}{\partial x}\left[(1+x^2)\frac{\partial}{\partial x}\right] + 3\frac{\partial^2}{\partial y^2} + 2(1+z^2)\frac{\partial^2}{\partial z^2}$$

is not elliptic anywhere in $\mathbb{R}^3$.

8.8. In the context of elasticity the definition of ellipticity given in Section 8.2 is extended in a very natural way to systems of PDEs involving the displacement vector as unknown variable. Suppose we consider only time-independent second-order problems in $\mathbb{R}^3$. Then clearly the principal part of Navier's equation can be written in the form

$$\sum_{j,k,l=1}^{3} C_{ijkl}\frac{\partial^2 u_k}{\partial x_j \partial x_l},$$

where the coefficients $C_{ijkl}$ are defined by (8.9). The elasticity operator is then said to be *elliptic* if for all vectors $\boldsymbol{\xi}$ and $\boldsymbol{\eta}$,

$$\sum_{i,j,k,l=1}^{3} C_{ijkl}\xi_i\eta_j\xi_k\eta_l \geq 0.$$

Furthermore, it is said to be *strongly elliptic* if this inequality holds *strictly* for all nonzero vectors. Show that the operator in Navier's equation is strongly elliptic, and that it is also pointwise stable (see (8.65)) if and only if the Lamé constants satisfy $\mu \geq \mu_0 > 0$ and $3\lambda + 2\mu \geq k_0 > 0$, for constants $\mu_0$ and $k_0$.

## Normal boundary conditions

8.9. Express the boundary condition

$$\frac{\partial^2 u}{\partial \nu^2} = g \qquad \text{on } \Gamma$$

in the form $\sum_{|\alpha|\leq 2} b_\alpha D^\alpha u = g$. Is it normal?

8.10. Show that in the theory of elastic plates, the boundary condition $S_1 = 0$ along the edge $x = L$ of the plate can be expressed in the form

$$-\frac{\partial}{\partial x}(\nabla^2 w) + (1-\nu)\frac{\partial^3 w}{\partial x \partial y^2} = 0.$$

Write the equation in the form $\sum b_\alpha D^\alpha u = 0$ and investigate whether it fails to be a normal boundary condition for any values of $\nu$.

8.11. Determine the conditions under which the pair of boundary conditions

$$\begin{aligned} B_0 u &= u, \\ B_1 u &= \alpha \frac{\partial^3 u}{\partial x^3} + \beta \frac{\partial^3 u}{\partial x^2 \partial y} + \gamma \frac{\partial^3 u}{\partial x \partial y^2} + \frac{\partial^3 u}{\partial y^3}, \end{aligned}$$

cover the biharmonic operator $A = \nabla^4$, at a point on the boundary with normal $\boldsymbol{\nu} = (0, 1)$.

8.12. An elastic body occupies the domain $\Omega = (0,1) \times (0,1)$. The sides $x = 0$, $x = 1$, and $y = 1$ are traction-free, whereas the side $y = 0$ is constrained by a flexible foundation, in the sense that the normal component of the surface traction acting on the boundary is proportional to the normal component of displacement; the tangential component of displacement is zero along this side. Do the boundary conditions along $y = 0$ satisfy (8.34)?

8.13. Consider again the elastic body discussed in Exercise 8.12, but this time suppose that the boundary condition along $y = 0$ is that corresponding to Coulomb friction: the normal component of displacement is zero, whereas the tangential component of traction is proportional to the normal component of traction. Formulate this boundary condition.

### Green's formulas and adjoint problems

8.14. Show that the Green's formula for the operator $A$ defined by

$$\frac{d^4 u}{dx^4} = f \qquad \text{in } \Omega = (0,1)$$

is $\displaystyle\int_0^1 vu'''' dx = \int_0^1 uv'''' dx + [u'''v - u''v' + u'v'' - uv''']_0^1.$

Given that $B_0 u = (u''(0), u''(1))$ and $B_1 u = (u'''(0), u'''(1))$, find the operators $B_j^*$, $S_j$, and $S_j^*$ $(j = 0, 1)$.

8.15. Show that the Green's formula for the Laplacian operator $Au = \nabla^2 u$ can be expressed in the form

$$\int_\Omega (\nabla^2 u) v \, dx = \int_\Omega u(\nabla^2 v) \, dx + \int_\Gamma (v \nabla u \cdot \boldsymbol{\nu} - u \nabla v \cdot \boldsymbol{\nu}) \, ds.$$

Given that $B_0 = \partial/\partial\nu$, identify the boundary operators $B_0^*, S_0, S_0^*$.

8.16. The purpose of this exercise is to derive the identities (8.49) and (8.50). First, use (7.2) to show that

$$\sum_{i,j=1}^n \int_\Omega \frac{\partial \sigma_{ij}}{\partial x_j} v_i \, dx = \sum_{i,j=1}^n \int_\Gamma \sigma_{ij} \nu_j v_i \, ds - \sum_{i,j=1}^n \int_\Omega \sigma_{ij} \frac{\partial v_i}{\partial x_j} \, dx, \tag{8.68}$$

where $\sigma_{ij}$ are the elements of a matrix $\boldsymbol{\sigma}$. Show furthermore that if $\boldsymbol{\sigma}$ is symmetric, then the integrand over $\Omega$ on the right-hand side of (8.68) is in fact equal to $\boldsymbol{\sigma} \cdot \boldsymbol{\epsilon}(\boldsymbol{v}) = \sum_{i,j=1} \sigma_{ij}\epsilon_{ij}(\boldsymbol{v})$. Next, use (8.8) to obtain (8.49). Apply Green's theorem again to find (8.50).

8.17. Derive the Green's formula for the operator $A$ given by

$$Au = \nabla^4 u = f \quad \text{in } \Omega,$$

$$\left.\begin{array}{l} u = g_0 \\ \partial u/\partial \nu = g_1 \end{array}\right\} \text{ on } \Gamma.$$

**Existence, uniqueness, and regularity of solutions**

8.18. Consider the BVP

$$\begin{array}{lll} Au = f & \text{in } \Omega & \\ B_j u = g_j & \text{on } \Gamma & (j = 0, 1, \ldots, m - 1), \end{array}$$

where $A$ is a $2m$th order operator. Let $\phi$ be a known function in $C^{2m}(\overline{\Omega})$ such that $B_j\phi = g_j$ on $\Gamma$. Show that the BVP can be transformed to the problem

$$\begin{array}{rcll} Aw & = & \hat{f} & \text{in } \Omega, \\ B_j w & = & 0 & \text{on } \Gamma, \end{array}$$

where $w = u - \phi$ and $\hat{f} = f - A\phi$.

8.19. Investigate the existence, uniqueness, and regularity of solutions to the problem of an elastic beam, which is described by

$$\begin{aligned} &\frac{d^4u}{dx^4} = f \text{ in } (0,1), \\ &u''(0) = u''(1) = 0, \\ &u'''(0) = u'''(1) = 0. \end{aligned}$$

In particular, determine the conditions that must be placed on the loading $f$.

8.20. Investigate the existence, uniqueness, and regularity of solutions to the problem

$$\begin{aligned} &\frac{\partial^2 u}{\partial x^2} + 2\frac{\partial^2 u}{\partial x \partial y} + \frac{\partial^2 u}{\partial y^2} = f \text{ in } \Omega, \\ &\frac{\partial u}{\partial \nu} = 0 \text{ on } \Gamma. \end{aligned}$$

If $\Omega = (-1,1) \times (-1,1)$, show that any loading $f$ satisfying $f(x,y) = f(y,x)$ with $f$ odd in $x$ or $y$ is compatible.

8.21. Verify that $\boldsymbol{\epsilon}(\boldsymbol{u}) = \mathbf{0}$ for the rigid body displacement $\boldsymbol{u}(\boldsymbol{x}) = \boldsymbol{a} + \boldsymbol{b} \times \boldsymbol{x}$.

8.22. The purpose of this exercise is to fill in some of the details of the proof of Theorem 1.

(a) Show that $A : H^s(\Omega) \to H^{s-2m}(\Omega)$, $A$ as in (8.41), is a bounded operator if the coefficients have bounded derivatives of all orders.

(b) Use the fact that $A$ is one-to-one from $N(A)^{\perp}$ onto its range, so that $A$ has an inverse $A^{-1} : R(A) \to N(A) \perp$. Now use the *Banach Theorem*, Theorem 6 of Chapter 5, to conclude that $A^{-1}$ is bounded. Use the boundedness of $A^{-1}$ to show that $R(A)$ is closed.

8.23. Investigate the conditions under which unique solutions exist to the elasticity problem with boundary conditions given in Exercises 8.12 and 8.13.

# 9

# Variational boundary value problems

In the preceding few sections we have built up a theory of regularly elliptic BVPs, in which the typical problem involves finding a function $u$ that satisfies

$$\text{PDE:}\quad Au = f \text{ in } \Omega,$$

$$\text{BCs:}\quad \left.\begin{aligned} B_0 u &= g_0 \\ &\vdots \\ B_{m-1} u &= g_{m-1} \end{aligned}\right\} \text{ on } \Gamma,$$

where $A$ is an elliptic PDE of order $2m$ in a domain $\Omega$, whereas the boundary conditions are normal, and cover $A$. The question of well-posedness of solutions to elliptic BVPs has been settled, at least for the case of a smooth domain and homogeneous BCs; provided that certain conditions are met, a unique solution exists. Furthermore, if $f \in H^s(\Omega)$, then $u$ is smooth enough to belong to $H^{s+2m}(\Omega)$.

In this chapter we broaden the concept of a boundary value problem by introducing what is known as a *variational boundary value problem* (VBVP). The variational formulation is a weaker one than the conventional formulation, since it demands less smoothness of the solution $u$. Nevertheless, there is a VBVP corresponding to every BVP, and vice versa, so that we have the option of formulating a problem in either of these two settings.

We start by examining a typical VBVP in Section 9.1; we take a simple example and show explicitly the relationship between the variational and conventional formulations. Then in Section 9.2 the general features of VBVPs are examined: how they are formulated and how they are related to BVPs. In Section 9.3 we consider the questions of existence and uniqueness of solutions to VBVPs. Finally, we show in Section 9.4 that certain VBVPs can be formulated alternatively as minimization problems, in which it is required to find the function that mimimizes a given functional.

## 9.1 A simple variational boundary value problem

In the present context we understand a variational boundary value problem to be one of the form: find a function $u$ that belongs to a Hilbert space $V$, and that satisfies the equation

$$a(u,v) = \langle \ell, v\rangle$$

for all functions $v$ in $V$, where $a$ is a *bilinear form* and $\ell$ a *linear functional.*

Before discussing general ideas, we consider the following simple, concrete example of a VBVP.

Find $u \in H_0^1(\Omega)$, $\Omega \subset \mathbb{R}^2$, that satisfies

$$\int_\Omega \nabla u \cdot \nabla v \; dxdy = \int_\Omega fv \; dxdy \quad \text{for all } v \in H_0^1(\Omega). \tag{9.1}$$

Here $V = H_0^1(\Omega)$,

$$a(u,v) = \int_\Omega \nabla u \cdot \nabla v \; dx = \int_\Omega \left( \frac{\partial u}{\partial x}\frac{\partial v}{\partial x} + \frac{\partial u}{\partial y}\frac{\partial v}{\partial y} \right) dxdy$$

and

$$\langle \ell, v\rangle = \int_\Omega fv \; dxdy.$$

The first question we ask is: in what sense is (9.1) equivalent to a BVP, and what does this BVP look like? This is resolved by observing first that since $v$ in (9.1) is arbitrary, we can set $v = \phi \in \mathcal{D}(\Omega)$ (note that $\mathcal{D}(\Omega) \subset H_0^1(\Omega)$), to give

$$a(u,\phi) = \int_\Omega \left( \frac{\partial u}{\partial x}\frac{\partial \phi}{\partial x} + \frac{\partial u}{\partial y}\frac{\partial \phi}{\partial y} \right) dxdy = \langle \ell, \phi\rangle. \tag{9.2}$$

Suppose for definiteness that $f$ is in $L^2(\Omega)$; then $f$ is locally integrable and generates a regular distribution, also denoted $f$, so that

$$\langle \ell, \phi\rangle = \langle f, \phi\rangle = \int_\Omega f\phi \; dxdy. \tag{9.3}$$

Now the functions $\partial u/\partial x$ and $\partial u/\partial y$ appearing in (9.2) belong to $L^2(\Omega)$ (since $u \in H_0^1(\Omega)$) and also generate regular distributions $\partial u/\partial x$ and $\partial u/\partial y$, from which it follows that

$$a(u,v) = \left\langle \frac{\partial u}{\partial x}, \frac{\partial \phi}{\partial x} \right\rangle + \left\langle \frac{\partial u}{\partial y}, \frac{\partial \phi}{\partial y} \right\rangle, \tag{9.4}$$

the right-hand side indicating the action of the distributions $\partial u/\partial x_i$ on $\partial \phi/\partial x_i$. Furthermore, from the definition of the generalized derivative of a distribution,

$$\left\langle \frac{\partial u}{\partial x_i}, \frac{\partial \phi}{\partial x_i} \right\rangle = -\left\langle \frac{\partial^2 u}{\partial x_i^2}, \phi \right\rangle, \tag{9.5}$$

$\partial^2 u/\partial x_i^2$ being a distribution, although not necessarily regular. Bringing together (9.2) through (9.5) we thus obtain

$$\langle \nabla^2 u - f, \phi \rangle = 0 \quad \text{for all } \phi \in \mathcal{D}(\Omega); \tag{9.6}$$

in other words (9.1) implies the problem of finding $u \in H_0^1(\Omega)$ that satisfies the Poisson equation

$$-\nabla^2 u = f \tag{9.7}$$

*in the sense of distributions* (see Section 7.2). We could even go one step further, and make use of the fact that $\mathcal{D}(\Omega)$ is dense in $H_0^1(\Omega)$ to argue, using (9.6), that (9.7) makes sense in $H^{-1}(\Omega)$, the dual space of $H_0^1(\Omega)$. Furthermore, since $u \in H_0^1(\Omega)$ it vanishes on the boundary, and we have

$$u = 0 \quad \text{on } \Gamma, \tag{9.8}$$

in the sense of traces.

It is important to remember that by (9.7) we mean (9.6). That is, the PDE (9.7) may only make sense when viewed as a distributional differential equation. For example, suppose that we consider the physical context of the membrane problem, in which case $f$ represents the force acting on the membrane, and suppose further that this force is a point load of intensity $P$ (Figure 9.1) acting at $\boldsymbol{x} = \mathbf{0}$. Then instead of (9.3) we have

$$\langle \ell, v \rangle = P\langle \delta, v \rangle = Pv(0). \tag{9.9}$$

where $\delta$ is the Dirac singular distribution, and the same procedure leads to the equation

$$-\nabla^2 u = \delta$$

which, as we know from Section 7.2, only has meaning in the distributional sense.

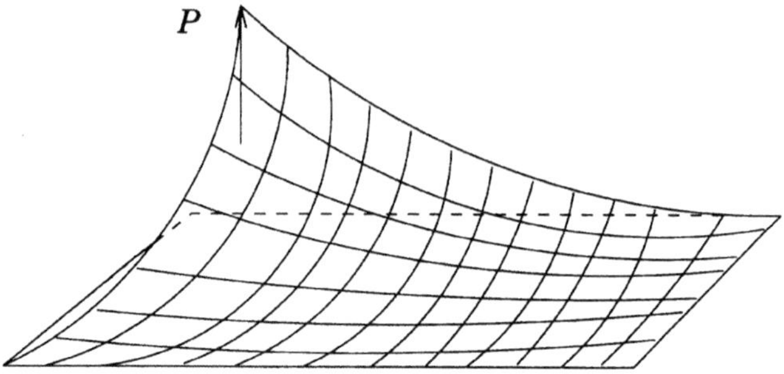


FIGURE 9.1. A membrane subjected to a point force

As (9.7) and (9.8) stand, a solution is sought in the space $H_0^1(\Omega)$. Whether this solution coincides with a "classical" solution of the kind discussed in Chapter 8, depends on the smoothness of $f$. If $f \in H^s(\Omega)$ with $s \geq 0$, then $u \in H^{s+2}(\Omega)$, and so the solution to the VBVP is the same as that of the classical BVP.

So far we have shown that the VBVP (9.1) implies (9.7) (in the sense of distributions) and (9.8). What of the converse? Suppose that we start with the Dirichlet problem for the Poisson equation, that is,

$$-\nabla^2 u = f \text{ in } \Omega, \tag{9.10}$$

$$u = 0 \text{ on } \Gamma, \tag{9.11}$$

with $f \in L^2(\Omega)$, and we wish to derive the corresponding VBVP. First we select $V$ to be $H_0^1(\Omega)$ (the general procedure for selecting this space is discussed in detail in the following section); next, we multiply (9.10) by an arbitrary function $v$ from $H_0^1(\Omega)$ and integrate over $\Omega$, to obtain

$$-\int_\Omega (\nabla^2 u)v \, dx = \int_\Omega fv \, dx. \tag{9.12}$$

Green's theorem in the form (7.30) is now applied to the left-hand side of (9.12), to reduce this to

$$-\int_\Omega (\nabla^2 u)v \, dx = -\int_\Gamma \left(\frac{\partial u}{\partial \nu}\right) v \, ds + \int_\Omega \nabla u \cdot \nabla v \, dx.$$

Since $v \in H_0^1(\Omega)$ the boundary integral vanishes and so (9.1) is seen to hold.

To summarize, then: the solution to the Dirichlet problem (9.10) and (9.11) satisfies the VBVP (9.1). Conversely, the VBVP (9.1) implies the problem (9.7) and (9.8) or (9.10) and (9.11) *provided that this problem is interpreted in the broader sense of seeking $u \in H_0^1(\Omega)$ that satisfies (9.6).*

Thus the variational formulation contains all the information found in the classical formulation *and more*, since we are able, when dealing with VBVPs, to work in a larger space and also to consider very irregular data such as that given by (9.9). This is an important consideration since physical problems may well require that data be modeled using distributions such as the Dirac delta: the case discussed earlier of the membrane subjected to a point force is one such example, and there are other similar examples, such as in heat conduction, in which one might want to consider a point heat source of the form $P\delta$. Whereas the classical formulation does not permit a treatment of such problems, the variational formulation offers a natural setting.

## 9.2 Formulation of variational boundary value problems

The ideas developed in the previous section are readily applicable to BVPs of arbitrary order. We confine attention to regularly elliptic BVPs of order $2m$, and go on now to discuss details of the general procedure for formulating the corresponding variational boundary value problems.

In anticipation of the fact that each boundary condition plays a role that depends on the order of the condition, we partition the set of boundary conditions into two subsets:

(i) *essential boundary conditions*, which are those of order $< m$;

(ii) *natural boundary conditions*, which are those of order $\geq m$.

The reason for making the distinction is this: the aim is to formulate a VBVP in which the solution is required only to be in a subspace of $H^m(\Omega)$. If this is so, then by the trace theorem it is only the derivatives of order less than $m$ that make sense as boundary values; thus the set of essential boundary conditions may be included in the description of the space in which a solution is sought (as with the inclusion of (9.11) in the problem description by choosing $V = H_0^1(\Omega)$). The natural boundary conditions, on the other hand, have to be accommodated in a different way.

As in the case of the theory in Section 8.5, we restrict our considerations to *bounded domains* $\Omega$ having *smooth boundaries* $\Gamma$.

In order to simplify matters, attention is confined to problems with *homogeneous* essential boundary conditions. This assumption does not imply any restriction on the class of problems that may be considered, since it is a straightforward matter to convert any problem with nonhomogeneous boundary conditions to one whose boundary conditions are homogeneous, as has already been discussed in Exercise 8.18.

So if the BCs are written down in the order of the highest derivatives appearing in each one, so that the first $p$ BCs are essential, then the BVP to be considered has the form

$$\text{PDE:} \quad Au \equiv \sum_{|\alpha|,|\beta|\le m}(-1)^{|\alpha|}D^\alpha(a_{\alpha\beta}(\boldsymbol{x})D^\beta u) = f \text{ in } \Omega \tag{9.13}$$

$$\text{BCs:} \quad \left.\begin{array}{rcl} B_0 u &=& 0 \\ &\vdots& \\ B_{p-1}u &=& 0 \end{array}\right\} (\textit{essential}) \tag{9.14}$$

$$\left.\begin{array}{rcl} B_p u &=& g_p \\ &\vdots& \\ B_{m-1}u &=& g_{m-1} \end{array}\right\} (\textit{natural}). \tag{9.15}$$

The first step is to define a space $V$ in which the solution to the VBVP is to be sought. This corresponds to the space $H_0^1(\Omega)$ in problem (9.1). The space $V$ is known as the *space of admissible functions*, and is defined by

$$V = \{v \in H^m(\Omega) : \ v \text{ satisfies all essential boundary conditions}\}$$

or

$$V = \{v \in H^m(\Omega) : \ B_j v = 0 \text{ on } \Gamma, \ j = 1, \dots, p-1\}.$$

As with the simple example worked through earlier, the next step is to multiply both sides of (9.13) by an arbitrary function $v$ from $V$, integrate, and use Green's theorem to reduce the expression so obtained to one of the form

$$a(u, v) = \langle \ell, v \rangle \tag{9.16}$$

in which the bilinear form $a$ is given by

$$a(u, v) = \int_\Omega \sum_{|\alpha|,|\beta|\le m} a_{\alpha\beta}(\boldsymbol{x})D^\beta u D^\alpha v \, dx + \text{boundary terms}.$$

Although the essential BCs are taken care of by the requirement that $u \in V$, the natural BCs are substituted into (9.16) directly. Once the formulation (9.16) is arrived at we may disregard any smoothness initially assumed of $u$, and pose the VBVP: find $u \in V$ that satisfies (9.16) for all $v \in V$. Since the VBVP is derived from the setting (9.13) through (9.15), every solution of (9.13) through (9.15) is a solution of the VBVP. Conversely, it can be shown that every solution of (9.16) solves the classical problem, possibly in a weak or distributional sense.

**Examples**

1. Consider the problem

$$\begin{aligned} -\nabla^2 u + au &= f \quad \text{in } \Omega, \\ \partial u/\partial \nu + bu &= g \quad \text{on } \Gamma, \end{aligned} \tag{9.17}$$

where $a$ and $b$ are continuous functions and it is assumed that $f \in L^2(\Omega)$ and $g \in L^2(\Gamma)$. This problem arises in steady heat conduction, in which the heat source is temperature-dependent, and of the form $f - au$, and there is Newton cooling on the boundary. In this problem $m = 1$, so that the boundary condition is a natural one. The space of admissible functions is thus $V = H^1(\Omega)$. Multiplying both sides of the PDE by $v \in H^1(\Omega)$, integrating, and using Green's theorem, we get

$$\int_\Omega (\nabla u \cdot \nabla v + auv)\, dx - \int_\Gamma \left(\frac{\partial u}{\partial \nu}\right) v\, ds = \int_\Omega fv\, dx.$$

The introduction of the natural boundary condition into the boundary term reduces this equation to the VBVP of finding $u \in H^1(\Omega)$ that satisfies

$$\underbrace{\int_\Omega (\nabla u \cdot \nabla v + auv)\, dx + \int_\Gamma buv\, ds}_{a(u,v)} = \underbrace{\int_\Omega fv\, dx + \int_\Gamma gv\, ds}_{\langle l,v\rangle} \tag{9.18}$$

for all $v \in H^1(\Omega)$. Thus the solution to problem (9.17) for $f \in L^2(\Omega)$ also solves the VBVP (9.18).

Conversely, if $u$ is a solution of (9.18), then upon setting $v = \phi \in \mathcal{D}(\Omega)$ we get

$$\langle -\nabla^2 u + au - f, \phi \rangle = 0, \tag{9.19}$$

so that $(9.17)_1$ is satisfied distributionally.

The interpretation of the boundary integrals in (9.18) is less straightforward, though, unless we assume that $u \in H^2(\Omega)$, in which case Green's theorem may be used to obtain

$$\begin{aligned} 0 &= \int_\Gamma \left(bu - g + \frac{\partial u}{\partial \nu}\right) v\, ds - \int_\Omega (\nabla^2 u - au + f)\, v\, dx \\ &= \int_\Gamma \left(bu - g + \frac{\partial u}{\partial \nu}\right) v\, ds \end{aligned}$$

using (9.19). The boundary value $\partial u/\partial \nu$ is, of course, well-defined since $u \in H^2(\Omega)$ by assumption, and so $\partial u/\partial \nu \in L^2(\Gamma)$. By choosing a function $v \in V$ that has a trace $\phi \in C^\infty(\Gamma)$ (this is always possible), and by exploiting the density of test functions in $L^2(\Gamma)$, the boundary condition $(9.17)_2$ can be shown to be valid in $L^2(\Gamma)$, and hence holds almost everywhere on $\Gamma$.

2. We consider next an example involving the biharmonic equation

$$\nabla^4 w = f,$$

where

$$\nabla^4 w = \frac{\partial^4 u}{\partial x^4} + 2\frac{\partial^4 u}{\partial x^2 \partial y^2} + \frac{\partial^4 u}{\partial y^4}.$$

Recall from Section 8.1 (Box 5) that physically this equation represents the behavior of a flat plate with stiffness $D$ subject to a transverse force $Q$ per unit area, with $f = Q/D$. For simplicity we confine attention here to a rectangular plate such as that shown in Figure 8.6.

Suppose that the plate is supported on its entire boundary in such a way that rotation is permitted, but the boundary is constrained against displacement (as in the second boundary in Section 8.1, Box 5). Then there are two boundary conditions, the first of which is $w = 0$ on $\Gamma$. To formulate the second boundary condition we must consider the edges $x = \pm h$ and $y = \pm l$ separately. For the edges $y = \pm l$ we have, as in Box 5, the condition $\partial^2 w/\partial y^2 = 0$. By a similar argument, that essentially entails reversing the roles of $x$ and $y$, we arrive at the condition $\partial^2 w/\partial x^2 = 0$ along the edge $x = \pm h$. In summary, we require that

$$\begin{aligned} w &= 0 \quad \text{on } \Gamma, \\ \partial^2 w/\partial x^2 &= 0 \quad \text{for } x = \pm h,\ y \in [-l, l], \\ \partial^2 w/\partial y^2 &= 0 \quad \text{for } y = \pm l,\ x \in [-h, h]. \end{aligned} \tag{9.20}$$

In this problem $m = 2$, of course, which accounts for the two boundary conditions. The condition $w = 0$ is an essential BC whereas the remaining two are natural conditions. Hence

$$V = \{v \in H^2(\Omega) :\ v = 0 \text{ on } \Gamma\} = H^2(\Omega) \cap H_0^1(\Omega).$$

To obtain the bilinear form corresponding to this problem, we first observe that

$$\begin{aligned} \int_\Omega v \frac{\partial^4 w}{\partial x^4}\, dx &= \int_\Gamma v \frac{\partial^3 w}{\partial x^3} \nu_x\, ds - \int_\Omega \frac{\partial v}{\partial x}\frac{\partial^3 w}{\partial x^3}\, dx \\ &= \int_\Gamma \left( v\frac{\partial^3 w}{\partial x^3} - \frac{\partial v}{\partial x}\frac{\partial^2 w}{\partial x^2}\right) \nu_x\, ds + \int_\Omega \frac{\partial^2 v}{\partial x^2}\frac{\partial^2 w}{\partial x^2}\, dx \end{aligned} \tag{9.21}$$

after two applications of Green's theorem. Similarly,

$$\int_\Omega v\frac{\partial^4 w}{\partial y^4}\,dx = \int_\Gamma\left(v\frac{\partial^3 w}{\partial y^3} - \frac{\partial v}{\partial y}\frac{\partial^2 w}{\partial y^2}\right)\nu_y\,ds + \int_\Omega \frac{\partial^2 v}{\partial y^2}\frac{\partial^2 w}{\partial y^2}\,dx. \tag{9.22}$$

Finally,

$$2\int_\Omega v\frac{\partial^4 u}{\partial x^2\partial y^2}\,dx = \int_\Omega v\left(\frac{\partial^4 w}{\partial x\partial y\partial x\partial y} + \frac{\partial^4 w}{\partial y\partial x\partial y\partial x}\right)\,dx$$

(this decomposition is carried out in order to preserve the symmetry inherent in the biharmonic operator)

$$= \int_\Gamma\left[v\left(\nu_y\frac{\partial^3 w}{\partial x^2\partial y} + \nu_x\frac{\partial^3 w}{\partial y^2\partial x}\right) - \frac{\partial^2 w}{\partial x\partial y}\left(\frac{\partial v}{\partial x}\nu_y + \frac{\partial v}{\partial y}\nu_x\right)\right]\,ds$$
$$+2\int_\Omega \frac{\partial^2 w}{\partial x\partial y}\frac{\partial^2 v}{\partial x\partial y}\,dx. \tag{9.23}$$

Now $v$ is assumed to be in $V$ so, in particular, $v = 0$ on $\Gamma$. Thus it follows that the first terms on the right-hand sides of (9.21) through (9.23) vanish. This leaves boundary terms involving second derivatives of $w$. The terms involving $\partial^2 w/\partial x^2$ and $\partial^2 w/\partial y^2$ all vanish, either due to $(9.20)_{2-3}$ or because $w = 0$ along $x = \pm h$ implies that $\partial^2 w/\partial y^2 = 0$ there, with a similar argument along $y = \pm l$. To see that the term involving the mixed derivative $\partial^2 w/\partial x\partial y = 0$ also vanishes, note that this can be written as

$$\frac{\partial}{\partial x}\left(\frac{\partial w}{\partial y}\right),$$

and $\partial w/\partial y$ vanishes along $x = \pm h$. The other two sides are treated in the same way, by swapping $x$ and $y$.

Thus all the boundary terms vanish, and we finally obtain the VBVP: find $w \in V$ such that

$$\underbrace{\int_\Omega\left(\frac{\partial^2 w}{\partial x^2}\frac{\partial^2 v}{\partial x^2} + 2\frac{\partial^2 w}{\partial x\partial y}\frac{\partial^2 v}{\partial x\partial y} + \frac{\partial^2 w}{\partial y^2}\frac{\partial^2 v}{\partial y^2}\right)\,dx}_{a(w,v)} = \underbrace{\int_\Omega fv\,dx}_{\langle l,v\rangle}. \tag{9.24}$$

In Exercise 9.3 we show how the VBVP may be arrived at in a more direct way, which also allows the boundary conditions to be applied more easily.

3. We return to the problem of the deformed elastic bar, summarized in Box 4, Chapter 8, and discussed further in Example 11 of that chapter. The same procedure applies as that adopted for scalar-valued

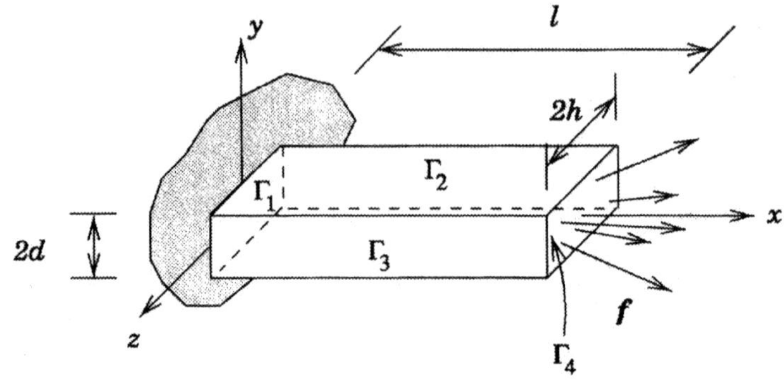


FIGURE 9.2. The domain of Example 3, and its boundary

functions, so we begin by identifying the essential boundary conditions: there is only one, that is,

$$\boldsymbol{u} = \boldsymbol{0} \quad \text{on } \Gamma_1,$$

where $\Gamma_1 = \{\boldsymbol{x} : \ x = 0, \ \ y \in (-d, d), \ \ z \in (-h, h)\}$ (Figure 9.2). So the space of admissible functions is

$$V = \{\boldsymbol{v} : \ v_i \in H^1(\Omega), \ \ \boldsymbol{v} = \boldsymbol{0} \text{ on } \Gamma_1\}.$$

Now fortunately much of the work entailed in deriving the VBVP appropriate to this problem has already been done in Chapter 8, in the course of arriving at the adjoint problem. Indeed, by taking the inner product of the left-hand side of (8.10) (without the time derivative) with an arbitrary function $\boldsymbol{v} \in V$, integrating, and using Green's theorem, we arrive at (8.50). The boundary term

$$-\int_\Gamma [\boldsymbol{C}\epsilon(\boldsymbol{u})]\boldsymbol{\nu} \cdot \boldsymbol{v} \ ds$$

may be written as a sum of integrals over the parts $\Gamma_1, \ldots, \Gamma_4$ making up $\Gamma$; now the integrals over $\Gamma_1$, $\Gamma_2$, and $\Gamma_3$ vanish, either because $\boldsymbol{v} = \boldsymbol{0}$ (on $\Gamma_1$) or because the surface traction vanishes (on $\Gamma_2$ and $\Gamma_3$). The integral over $\Gamma_4$ becomes simply $\int_{\Gamma_4} \boldsymbol{f} \cdot \boldsymbol{v} \ ds$, after substitution of the natural boundary conditions, and the desired VBVP is: find $\boldsymbol{u} \in V$ such that

$$\underbrace{\int_\Omega [\boldsymbol{C}\epsilon(\boldsymbol{u})] \cdot \epsilon(\boldsymbol{v}) \ dx}_{a(\boldsymbol{u},\boldsymbol{v})} = \underbrace{\int_{\Gamma_4} \boldsymbol{f} \cdot \boldsymbol{v} \ ds}_{\langle \ell, \boldsymbol{v} \rangle} \tag{9.25}$$

4. Consider the problem of deflection of a linear elastic beam; the differential equation for this problem has been derived in Chapter 8,

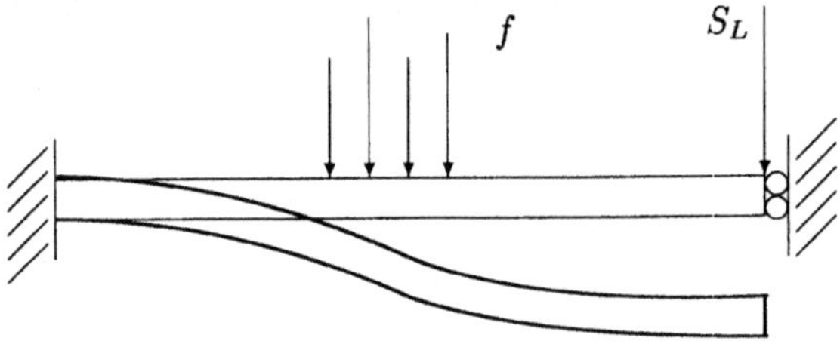


FIGURE 9.3. The beam corresponding to Example 4

Example 7, and various boundary conditions have been discussed in Exercise 8.5. Take, for example, the case in which the beam is constrained against displacement and rotation at one end, whereas at the other it is constrained merely against rotation, and is subjected to a shear force of magnitude $S_L$ at that end, as shown in Figure 9.3. The boundary conditions are thus

$$\begin{aligned} u(0) &= 0 \qquad & u'(L) &= 0 \\ u'(0) &= 0 \qquad & u'''(L) &= -S_L/EI. \end{aligned}$$

Of these, all except the condition $u'''(L) = S/EI$ are essential conditions, so it follows that the space of admissible functions is

$$V = \{v \in H^2(0, L) : \; u(0) = u'(0) = u'(L) = 0\}.$$

In order to obtain the VBVP we multiply the left-hand side of the differential equation (8.20) by an arbitrary function $v$ and integrate twice by parts; this gives

$$\begin{aligned} \int_0^L w^{(iv)} v \, dx &= [w'''v]_0^L - \int_0^L w'''v' \, dx \\ &= [w'''v - w''v']_0^L + \int_0^L w''v'' \, dx. \end{aligned}$$

Now with the assumption that the function $v$ belongs to $V$, the boundary condition reduces to the single term $w'''(L)v(L) = -(S_L/EI)v(L)$, after imposition of the natural boundary condition. After rearranging terms we therefore arrive at the VBVP: find $w \in V$ that satisfies

$$\underbrace{\int_0^L w''v'' \, dx}_{a(w,v)} = \underbrace{\int_0^L qv \, dx + (S_L/EI)v(L)}_{\langle \ell, v \rangle},$$

where $q = f/EI$.

## 9.3 Existence, uniqueness, and regularity of solutions

**Existence and uniqueness of solutions to VBVPs.** Earlier, in Chapter 8, we discussed the conditions under which solutions to regularly elliptic BVPs exist and are unique. The results there apply, of course, to what is referred to as the classical formulation, that consists of a PDE and a collection of homogeneous boundary conditions.

Now in much the same way we wish to know the conditions under which a unique solution to the corresponding variational boundary value problem may be found. Just as the issues of existence and uniqueness of the solution to (8.57) depend on various properties of the differential operators $A$ and $B_0, \ldots, B_{m-1}$, in the case of VBVPs these issues can be expected to be tied closely to properties of the bilinear form $a(\cdot, \cdot)$ and the linear functional $\ell$.

It turns out that there is exactly one solution to a VBVP of the form (9.16) provided that $\ell$ is continuous and provided that $a$ is *continuous* and $V$*-elliptic*: recall from Section 5.5 that a bilinear operator $a$ is continuous if there is a constant $M > 0$ such that

$$|a(u, v)| \leq M\|u\|_V \,\|v\|_V \quad \text{for all } u,\, v \in V,$$

and $V$*-elliptic* if there is a constant $\alpha > 0$ such that

$$a(v, v) \geq \alpha\|v\|_V^2 \quad \text{for all} \quad v \in V, \tag{9.26}$$

$V$ being the space of admissible functions and $\|\cdot\|_V$ the norm on this space. Without further ado we present the basic existence and uniqueness theorem for VBVPs, after which a few specific examples are considered.

THEOREM 1. *Let $V$ be a Hilbert space and let $a(\cdot, \cdot) : V \times V \to \mathbb{R}$ be a continuous, $V$-elliptic bilinear form on $V$. Furthermore, let $\ell : V \to \mathbb{R}$ be a continuous linear functional on $V$. Then*

*(i) the VBVP of finding $u \in V$ that satisfies*

$$a(u, v) = \langle \ell, v \rangle \quad \textit{for all } v \in V, \tag{9.27}$$

*has one and only one solution;*

*(ii) the solution depends continuously on the data, in the sense that*

$$\|u\|_V \leq \frac{1}{\alpha}\|\ell\|_{V'}, \tag{9.28}$$

*where $\|\cdot\|_{V'}$ is the norm in the dual space $V'$ of $V$ and $\alpha$ is the constant in (9.26).*

PROOF. The proof of this theorem follows from the Lax–Milgram theorem (Theorem 5.13). Since $a$ is continuous and $V$-elliptic, every bounded linear functional, and in particular the functional $\ell$, can be expressed in the form

$$\langle \ell, v\rangle = a(u, v),$$

where $u$ is unique. This proves Part (i); Part (ii) follows by setting $v = u$ in (9.26) and using (9.27). This gives

$$\alpha\|u\|_V^2 \le a(u,u) = \langle \ell, u\rangle \le \|\ell\|_{V'}\|u\|_V,$$

the last inequality coming from the fact that $\ell$ is bounded. Dividing throughout by $\|u\|$, we obtain (9.28). □

Recall from the discussion in Section 8.5 the significance of a result such as (9.28). This inequality assures us that a small change in the functional $\ell$ leads to a correspondingly small change in the solution.

The inequality (9.28) may be expressed in an alternative form if $\ell$ is given by

$$\langle \ell, v\rangle = \int_\Omega fv\; dx + \int_\Gamma gv\; ds,$$

where $f$ is in $L^2(\Omega)$ and $g \in L^2(\Gamma)$ (as in (9.18)); for then we have

$$\begin{aligned}\alpha\|u\|_V^2 \le a(u,u) = \langle \ell, u\rangle &= (f,u)_{L^2(\Omega)} + (g,u)_{L^2(\Gamma)}\\ &\le \|f\|_{L^2(\Omega)}\|u\|_{L^2(\Omega)} + \|g\|_{L^2(\Gamma)}\|u\|_{L^2(\Gamma)}\\ &\le c\|u\|_{L^2(\Omega)}(\|f\|_{L^2(\Omega)} + \|g\|_{L^2(\Gamma)})\end{aligned}$$

using the Cauchy–Schwarz inequality and the trace theorem. Since $V$ is a subspace of a Sobolev space $H^m(\Omega)$, the norm $\|\cdot\|_V$ is the $H^m$-norm, and of course $\|u\|_{L^2} \le \|u\|_V$. Hence

$$\|u\|_V \le \frac{c}{\alpha}(\|f\|_{L^2(\Omega)} + \|g\|_{L^2(\Gamma)}). \tag{9.29}$$

We now show how Theorem 1 is applied to actual problems.

## Examples

5. Consider the BVP

$$\begin{aligned}-\nabla^2 u + ku = f &\quad \text{in } \Omega,\\ u = 0 &\quad \text{on } \Gamma,\end{aligned} \tag{9.30}$$

where $k(\boldsymbol{x})$ is continuous on $\Omega$. We assume first of all that

$$M_2 \ge k(\boldsymbol{x}) \ge M_1 > 0 \tag{9.31}$$

for some constants $M_1, M_2$. The VBVP corresponding to (9.30) is: find $u \in H_0^1(\Omega)$ that satisfies

$$\underbrace{\int_\Omega (\nabla u \cdot \nabla v + kuv)\, dx}_{a(u,v)} = \underbrace{\int_\Omega fv\, dx}_{\langle \ell, v\rangle}, \quad v \in H_0^1(\Omega). \tag{9.32}$$

If it can be shown that $\ell$ is continuous and that $a$ is both continuous and $H_0^1$ - elliptic, then we are guaranteed the existence of a unique solution to (9.32). First, $\ell$ is continuous since

$$\begin{aligned} |\langle \ell, v\rangle| &= \left|\int_\Omega fv\, dx\right| \leq \|f\|_{L^2}\|v\|_{L^2} \quad \text{(Cauchy–Schwarz inequality)} \\ &\leq \|f\|_{L^2}\|v\|_{H^1}; \end{aligned}$$

if we set $\|f\|_{L^2} = K$, then $|\langle \ell, v\rangle| \leq K\|v\|_{H^1}$ and so $\ell$ is bounded, and hence continuous.

Next, $a$ is continuous since

$$\begin{aligned} |a(u,v)| &= \left|\int_\Omega (\nabla u \cdot \nabla v + kuv)\, dx\right| \\ &\leq \left|\int_\Omega \nabla u \cdot \nabla v\, dx\right| + \left|\int_\Omega kuv\, dx\right| \quad \text{(triangle inequality)} \\ &\leq \sum_{i=1}^n \left|\left(\frac{\partial u}{\partial x_i}, \frac{\partial v}{\partial x_i}\right)_{L^2}\right| + M_2|(|u|,|v|)_{L^2}| \quad \text{(from (9.39))} \\ &\leq \sum_{i=1}^n \left\|\frac{\partial u}{\partial x_i}\right\|_{L^2}\left\|\frac{\partial v}{\partial x_i}\right\|_{L^2} + M_2\|u\|_{L^2}\|v\|_{L^2} \\ &\leq M\left[\sum_{i=1}^n \left\|\frac{\partial u}{\partial x_i}\right\|_{L^2}\left\|\frac{\partial v}{\partial x_i}\right\|_{L^2} + \|u\|_{L^2}\|v\|_{L^2}\right] \\ &= M\|u\|_{H^1}\|v\|_{H^1} \end{aligned}$$

in which $M = \max\{1, M_2\}$. The last line is arrived at by defining the vector $\boldsymbol{u} = (\|\partial u/\partial x_1\|_{L^2}, \ldots, \|\partial u/\partial x_n\|_{L^2}, \|u\|_{L^2})$ in $\mathbb{R}^{n+1}$, and by defining similarly a vector $\boldsymbol{v}$. Then the penultimate line is simply $M\boldsymbol{u} \cdot \boldsymbol{v}$, and from the Cauchy–Schwarz inequality, this is bounded above by $M|\boldsymbol{u}|\,|\boldsymbol{v}|$; the observation that $|\boldsymbol{u}| = \|u\|_{H^1}$ (similarly for $\boldsymbol{v}$) yields the last line. Thus $a$ is continuous.

Finally, to show that $a$ is $H_0^1$-elliptic, consider

$$\begin{aligned} a(v,v) &= \int_\Omega (\nabla v \cdot \nabla v + kv^2)\,dx \\ &\geq \int_\Omega (\nabla v \cdot \nabla v + M_1 v^2)\,dx \quad \text{(from (9.31))} \\ &\geq \begin{cases} \int_\Omega (\nabla v \cdot \nabla v + v^2)\,dx & \text{if } M_1 \geq 1, \\ M_1 \left[ \int_\Omega (\nabla v \cdot \nabla v + v^2)\,dx \right] & \text{if } M_1 \leq 1. \end{cases} \end{aligned}$$

In other words,

$$a(v,v) \geq \alpha \int_\Omega (\nabla v \cdot \nabla v + v^2)\,dx = \alpha \|v\|_{H^1}^2,$$

where $\alpha = \min(1, M_1)$. Thus $a$ is $H_0^1$-elliptic. All requirements are met, and so a unique solution to (9.32) exists.

6. Consider once again the BVP (9.30), but this time assume only that $k(\boldsymbol{x})$ is nonnegative and bounded above, so that

$$0 \leq k(\boldsymbol{x}) \leq M_2$$

(note that the Poisson equation $-\nabla^2 u = f$ is a particular example of this situation). Continuity of $\ell$ and $a$ follow from the same arguments as those used in Example 4, but the proof of $H_0^1$-ellipticity does not apply here since we made use in Example 4 of the condition $a(\boldsymbol{x}) \geq M_1$. In order to show that $a$ is $H_0^1$-elliptic, we exploit the *Poincaré–Friedrichs inequality* (7.34); from (9.32) we have

$$a(v,v) = \int_\Omega (\nabla v \cdot \nabla v + kv^2)\,dx \geq \int_\Omega \nabla v \cdot \nabla v\,dx$$

and (7.34) gives

$$(C+1) \int_\Omega \nabla v \cdot \nabla v\,dx \geq \int_\Omega (v^2 + \nabla v \cdot \nabla v)\,dx = \|v\|_{H^1}^2$$

so that

$$a(v,v) \geq \frac{1}{C+1} \|v\|_{H^1}^2,$$

and $a$ is $H_0^1$-elliptic.

7. Returning to Example 3, that concerns the deformation of an elastic bar, we consider the case of an isotropic material for which $\boldsymbol{C}$ is given

by (8.9). The bilinear form $a(\cdot,\cdot)$ is defined in (9.25); first consider the question of continuity of $a$, that is defined in (9.25). Using (8.9) we have

$$\begin{aligned}
|a(\boldsymbol{u},\boldsymbol{v})| &= \left|\int_\Omega \sum_{i,j,k,l=1}^{n} C_{ijkl}\epsilon_{ij}(\boldsymbol{u})\epsilon_{kl}(\boldsymbol{v})\,dx\right| \\
&= \left|\int_\Omega [\lambda(\operatorname{div}\boldsymbol{u})(\operatorname{div}\boldsymbol{v}) + 2\mu\boldsymbol{\epsilon}(\boldsymbol{u})\cdot\boldsymbol{\epsilon}(\boldsymbol{v})]\,dx\right| \\
&= \lambda(\operatorname{div}\boldsymbol{u},\operatorname{div}\boldsymbol{v})_{L^2} + 2\mu(\boldsymbol{\epsilon}(\boldsymbol{u}),\boldsymbol{\epsilon}(\boldsymbol{v}))_{L^2} \\
&\le \lambda\|\operatorname{div}\boldsymbol{u}\|_{L^2}\,\|\operatorname{div}\boldsymbol{v}\|_{L^2} + 2\mu\|\boldsymbol{\epsilon}(\boldsymbol{u})\|_{L^2}\,\|\boldsymbol{\epsilon}(\boldsymbol{v})\|_{L^2}\,.
\end{aligned}$$

Given that all the norms are of various combinations of first derivatives of $\boldsymbol{u}$ and $\boldsymbol{v}$ it follows that there is a constant $M > 0$ such that

$$|a(\boldsymbol{u},\boldsymbol{v})| \le M\|\boldsymbol{u}\|_{H^1}\,\|\boldsymbol{v}\|_{H^1}.$$

The proof of $V$-ellipticity follows very closely the arguments in Chapter 8, Example 32; indeed, from (8.66) and Korn's inequality (8.67) we see that the bilinear form is $V$-elliptic provided that the part $\Gamma_1$ of the boundary on which $\boldsymbol{u} = \boldsymbol{0}$ is not empty. Thus this problem has a unique solution.

8. Consider next the problem of an elastic plate whose boundary is rigidly clamped; the problem is thus one of finding $w$ that satisfies

$$\nabla^4 w = f \quad \text{in } \Omega,$$

$$\left.\begin{aligned} u &= 0 \\ \partial u/\partial\nu &= 0 \end{aligned}\right\} \quad \text{on } \Gamma.$$

Both boundary conditions are essential, and so the space of admissible functions is $V = H_0^2(\Omega)$, whereas the bilinear form $a(\cdot,\cdot)$ and linear functional $\ell$ are as in (9.24).

To show that $a$ is continuous, consider the first term in the bilinear form. We have, using the Cauchy–Schwarz inequality and the definition of the $H^2$–norm,

$$\begin{aligned}
\left|\int_\Omega \frac{\partial^2 w}{\partial x^2}\frac{\partial^2 v}{\partial x^2}\,dx\right| &= \left(\frac{\partial^2 w}{\partial x^2},\frac{\partial^2 v}{\partial x^2}\right)_{L^2} \\
&\le \left\|\frac{\partial^2 w}{\partial x^2}\right\|_{L^2}\left\|\frac{\partial^2 v}{\partial x^2}\right\|_{L^2} \le \|w\|_{H^2}\,\|v\|_{H^2}\,.
\end{aligned}$$

The remaining two terms in the integrand may be bounded similarly, and it is then trivial to show that $|a(w,v)| \le 4\|w\|_{H^2}\|v\|_{H^2}$.

To show that $a$ is $V$-elliptic we simply note that

$$a(v,v) \geq \sum_{|\alpha|=2} \int_\Omega (D^\alpha v)^2 \, dx,$$

and the desired bound is obtained by applying (7.36). Continuity of the linear functional is easy to prove, so we find that the problem of a plate that is clamped all around its boundary has a unique solution.

9. An example of a problem that does not have a unique solution is the BVP

$$\begin{aligned} -\nabla^2 u &= f \quad &&\text{in } \Omega, \\ \partial u/\partial\nu &= 0 \quad &&\text{on } \Gamma. \end{aligned} \tag{9.33}$$

The corresponding VBVP is: find $u \in H^1(\Omega)$ such that

$$\int_\Omega \nabla u \cdot \nabla v \, dx = \int_\Omega fv \, dx \tag{9.34}$$

for all $v \in H^1(\Omega)$. Thus (9.34) is similar in many respects to Example 5, one exception being that the space of admissible functions is $H^1(\Omega)$ and *not* $H_0^1(\Omega)$. As before, $\ell$ and $a$ are continuous; but $a$ is *not* $H^1$-*elliptic*: indeed, the inequality

$$a(v,v) \geq \alpha\|v\|^2$$

ceases to hold for any function $v$ that is constant (for which case $a(v,v)=0$). Hence, although the lack of $H^1$-ellipticity does not necessarily mean that a unique solution does not exist (Theorem 1 gives *sufficient* conditions for the existence of a solution; if these conditions are not satisfied, it does not imply nonexistence or nonuniqueness), we are unable to guarantee the existence of a unique solution.

Now the problem (9.33) has in fact been treated previously, in Chapter 8, Example 31; indeed, recall that in order for a unique solution to exist, it was necessary and sufficient that the data $f$ and the solution $u$ satisfy

$$\int_\Omega f \, dx = 0 \quad \text{and} \quad \int_\Omega u \, dx = 0. \tag{9.35}$$

The reason why we cannot prove existence of a unique solution to the corresponding VBVP (9.34) is essentially that the conditions (9.35) are not satisfied in the statement of (9.34) as it stands. First of all, the space $H^1(\Omega)$ in which $u$ is sought is too large; for uniqueness we must restrict attention to the subspace of $H^1(\Omega)$ consisting of elements orthogonal to constants, that is, elements satisfying $(9.35)_2$.

Second, the compatibility condition $(9.35)_1$ is a *necessary* condition for existence of a solution, since we get from (9.34), setting $v = c$, a constant,

$$0 = \int_\Omega fc\, dx \quad \text{or} \quad \int_\Omega f\, dx = 0.$$

We show that $(9.35)_1$ is in fact also a *sufficient* condition for existence, as in Section 8.5.

THEOREM 2. *Let $V$ be a closed subspace of $H^m(\Omega)$, let $a$ be a continuous bilinear form on $V \times V$, and $\ell$ a continuous linear functional on $V$. Let $P$ be a closed subspace of $V$ such that*

$$a(u + p, v + \bar{p}) = a(u, v) \quad \textit{for all } u, v \in V \quad \textit{and} \quad p, \bar{p} \in P. \tag{9.36}$$

*Also, denote by $Q$ the subspace of $V$ consisting of functions orthogonal to $P$ in the $L^2$-norm; that is,*

$$Q = \left\{ v \in V : \int_\Omega vp\, dx = 0 \quad \textit{for all } p \in P \right\},$$

*and assume that $a(\cdot, \cdot)$ is $Q$-elliptic: there is a constant $\alpha > 0$ such that*

$$a(q, q) \geq \alpha \|q\|_Q^2 \quad \textit{for } q \in Q,$$

*the norm on $Q$ being the same as that on $V$. Then*

*(i) there exists a unique solution to the problem of finding $u \in Q$ such that*

$$a(u, v) = \langle \ell, v \rangle \quad \textit{for all} \quad v \in V \tag{9.37}$$

*if and only if the compatibility condition*

$$\langle \ell, p \rangle = 0 \quad \textit{for} \quad p \in P \tag{9.38}$$

*holds;*

*(ii) (continuous dependence on the data) the solution $u$ satisfies*

$$\|u\|_Q \leq \alpha^{-1} \|\ell\|_{Q'}.$$

REMARK. Note that Theorem 1 is a special case for which $P = \{0\}$, so that $Q = V$. Also, observe that for Example 9 we have $V = H^1(\Omega), P = P_0$, the set of constant functions, and $Q$ consists of functions satisfying $(9.35)_2$, whereas (9.38) reduces to $(9.35)_1$.

PROOF. First we show the necessity of (9.38). Assuming that (9.37) holds, we have from (9.36),

$$a(u, v + p) = a(u, v) = \langle \ell, v + p \rangle$$

so that, subtracting this from (9.37) and using the linearity of $\ell$, (9.38) follows.

Conversely, assume that (9.38) holds; we want to show that (9.37) has exactly one solution. This we do by showing first that (9.37) is equivalent to the problem of finding $u \in Q$ such that

$$a(u, q) = \langle \ell, q \rangle \quad \text{for all} \quad q \in Q. \tag{9.39}$$

Once this has been done, the proof follows that in Theorem 1, since $a(\cdot, \cdot)$ is continuous and $Q$-elliptic and, furthermore, it is readily shown that $Q$ is a closed subspace and therefore complete in the $H^m$-norm.

Now since $P$ is closed in $V$, which is in turn closed in $H^m(\Omega)$ and hence also in $L^2(\Omega)$, we have by Theorem 8 of Chapter 4 that

$$V = P \oplus Q.$$

Hence we can write any $v \in V$ uniquely in the form $v = p + q$ for $p \in P$ and $q \in Q$. Since (9.38) is assumed to hold we have, from (9.36) and (9.38),

$$a(u, p) = 0 \quad \text{and} \quad \langle \ell, p \rangle = 0$$

so that, if $u \in Q$ is a solution of (9.39), then

$$a(u, q) + a(u, p) = \langle \ell, q \rangle + \langle \ell, p \rangle$$

or, with $p + q = v$,

$$a(u, v) = \langle \ell, v \rangle \quad \text{for all } v \in V.$$

Hence a solution of (9.39) is also a solution of (9.37). Conversely, (9.37) holds a fortiori for all $v \in Q$ since $Q$ is in $V$. Hence a solution of (9.37) is also a solution of (9.39), so that problems (9.37) and (9.39) are equivalent.

The remainder of the proof of Part (i), as well as the proof of Part (ii), now follow in much the same way as that of Theorem 1. □

**Example**

10. We return to (9.33), but consider instead the nonhomogeneous boundary condition

$$\frac{\partial u}{\partial \nu} = g \quad \text{in } \Gamma.$$

The VBVP is now: find $u \in H^1(\Omega)$ such that

$$\int_\Omega \nabla u \cdot \nabla v \, dx = \int_\Omega fv \, dx + \int_\Gamma gv \, ds \quad \text{for all} \quad v \in H^1(\Omega). \qquad (9.40)$$

Here $V = H^1(\Omega)$. Of course, (9.40) does not have a unique solution as it stands; we note that

$$a(u+p, v+\bar{p}) = \int_\Omega \nabla(u+p) \cdot \nabla(v+\bar{p}) \, dx = a(u,v) \quad \text{for all } p, \bar{p} \in P_0,$$

where $P_0$ is the space of constant functions. Thus $P = P_0$ and the solution must be sought in the space

$$Q = \left\{ q \in H^1(\Omega) : \int_\Omega q \, dx = 0 \right\}. \qquad (9.41)$$

We know that $a(\cdot,\cdot)$ and $\langle \ell, \cdot \rangle$ are continuous; to show that $a$ is $Q$-elliptic we use the Poincaré inequality (7.19) with (9.41) to show that

$$\int_\Omega \nabla q \cdot \nabla q \, dx \geq \alpha \|q\|_{H^1}^2 = \alpha \|q\|_Q^2.$$

Hence there exists a *unique* solution $u$ in $Q$ to (9.40) if and only if

$$\langle \ell, p \rangle = 0 \quad \Rightarrow \quad \int_\Omega f \, dx + \int_\Gamma g \, ds = 0.$$

Furthermore, we have

$$\|u\|_Q = \|u\|_{H^1} \leq \alpha^{-1} \|\ell\|_{Q'}. \qquad (9.42)$$

Since $\langle \ell, v \rangle$ is given by the right-hand side of (9.40) we can in fact express (9.42) in the alternative form (9.29).

11. In Example 7, if the boundary condition is

$$\boldsymbol{t} = \boldsymbol{0} \quad \text{on } \Gamma,$$

then $\Gamma_1 = \varnothing$ and this is the same situation as that encountered in Chapter 8, Example 33. Returning to Theorem 2, that holds for vector-valued functions with minor modifications, we first note that $V = [H^1(\Omega)]^3$, and the space of functions $P$ coincides with the set of rigid body displacements:

$$P = \{\boldsymbol{v} : \ \boldsymbol{v}(\boldsymbol{x}) = \boldsymbol{a} + \boldsymbol{b} \times \boldsymbol{x}, \quad \boldsymbol{a}, \, \boldsymbol{b} \in \mathbb{R}^3\}.$$

Thus $Q$, the space that is $L^2$-orthogonal to $P$, is given by

$$Q = \left\{ \boldsymbol{u} \in V : \int_\Omega \boldsymbol{u} \cdot (\boldsymbol{a} + \boldsymbol{b} \times \boldsymbol{x}) \, dx = 0 \right\}$$

or, since $\boldsymbol{a}$ and $\boldsymbol{b}$ are arbitrary,

$$Q = \left\{ \boldsymbol{u} \in V : \int_\Omega \boldsymbol{u}\, dx = \boldsymbol{0}, \quad \int_\Omega \boldsymbol{u} \times \boldsymbol{x}\, dx = \boldsymbol{0} \right\}.$$

Korn's inequality (8.67) holds on $Q$, and thus the bilinear form is $Q$-elliptic. According to Theorem 2, a unique solution exists in $Q$ provided that the compatibility condition (9.38) is satisfied: this is precisely the pair of conditions

$$\int_\Omega \boldsymbol{Q}\, dx = \boldsymbol{0} \quad \text{and} \quad \int_\Omega \boldsymbol{Q} \times \boldsymbol{x}\, dx = \boldsymbol{0}$$

encountered in Chapter 8, Example 33.

**Regularity of solutions.** A few words are in order regarding the regularity or degree of smoothness of the solution to a VBVP. Recall that we discussed this issue in Section 8.5 in the context of the conventional or classical formulation, and we showed there that the solution $u$ belongs to $H^s(\Omega)$ (or rather, a subspace $\tilde{H}^s(\Omega)$ of $H^s(\Omega)$) if the data $f$ are in $H^{s-2m}(\Omega)$. Here $s \geq 2m$.

In the context of variational boundary value problems we also showed earlier that, when the linear functional $\ell$ is of the form

$$\langle \ell, v \rangle = \int_\Omega f v\, dx \tag{9.43}$$

with $f \in L^2(\Omega)$, then the problem of finding $u \in V$ that satisfies

$$a(u, v) = \langle \ell, v \rangle, \quad v \in V \tag{9.44}$$

is equivalent to the problem of $u \in V$ that satisfies

$$Au = f \quad \text{in} \quad L^2(\Omega) \tag{9.45}$$

for appropriate $A$. Thus, for a differential operator of order $2m$ we conclude that $u$ is "almost" in $H^{2m}(\Omega)$, in the sense that $u$ lies in the domain of an operator of order $2m$. If $A$ contains *all* derivatives of order $2m$, then $u$ in fact belongs to $H^{2m}(\Omega)$.

The preceding characterization of the smoothness of solutions of VBVPs can be strengthened considerably with the aid of the existence and uniqueness results given earlier in this section, and the theory of Section 8.5. Indeed, suppose that (9.43) through (9.45) hold, and that $a$ and $\ell$ satisfy the requirements for existence and uniqueness; since the solution $u$ is unique, we may conclude from (9.45) and from Theorem 1 of Chapter 8 that $u$ in fact belongs to $H^{2m}(\Omega)$, or rather the subspace of $H^{2m}(\Omega)$ consisting of functions that satisfy all of the homogeneous boundary conditions. Going one step further, if $f$ is even more smooth so that it lies in $H^s(\Omega)$, say, with $s \geq 2m$, then we are assured that $u$ is in $H^s(\Omega)$. We summarize these observations in the following theorem.

THEOREM 3. *Let $\Omega$ be a smooth domain and let $u \in V$ be a solution of the VBVP*

$$a(u, v) = \int_\Omega fv \, dx, \quad v \in V,$$

*where $V \subset H^m(\Omega)$. If $f \in H^{s-2m}(\Omega)$ with $s \geq 2m$, then $u \in H^s(\Omega)$ and the estimate*

$$\|u\|_{H^s} \leq C\|f\|_{H^{s-2m}}$$

*holds.*

## 9.4 Minimization of functionals

It should now be clear that an elliptic boundary value problem of the form

$$\begin{aligned} Au &= f && \text{in } \Omega \\ B_j u &= g_j && \text{on } \Gamma \ (j = 0, \ldots, m-1) \end{aligned} \tag{9.46}$$

can be posed in the alternative form of a variational boundary value problem

$$u \in V, \quad a(u, v) = \langle \ell, v \rangle, \quad v \in V. \tag{9.47}$$

The formulation (9.47) has been shown also to hold certain advantages over (9.46).

It turns out that if the bilinear form in (9.47) is *symmetric*, that is, if

$$a(v, u) = a(u, v) \quad \text{for all } u, \ v \in V$$

– and this is has been the case for the examples treated here – then a third formulation is possible: the variational boundary value problem is equivalent to a *minimization problem*, in which a function $u$ in $V$ is sought that renders the value of a functional $J : V \to \mathbb{R}$ a minimum; that is,

$$J(u) \leq J(v) \quad \text{for all } v \in V, \tag{9.48}$$

where $J$ is defined by

$$J(v) = \tfrac{1}{2}a(v, v) - \langle \ell, v \rangle. \tag{9.49}$$

The formulation (9.48) is often taken as a starting point, particularly in physical problems for which $J(v)$ represents the energy of a system, and the solution of the problem is that which renders the energy a minimum. For example, returning to the membrane problem (9.1), the quantity $\frac{1}{2}a(v,v)$ represents the *strain energy* of the membrane, that is, the energy that the membrane possesses by virtue of its having sustained a displacement $v$. The quantity $-\int_\Omega fv\,dx$ represents the *potential energy* of the loading. Adding these two together, we obtain

$$J(v) = \tfrac{1}{2}\int_\Omega |\nabla v|^2 dx - \int_\Omega fv\,dx,$$

the *total potential energy* of the system, whose minimum characterizes the solution; this minimum is sought in $H_0^1(\Omega)$.

Identical remarks apply to the elasticity problem; the total potential energy is again given by

$$J(\boldsymbol{v}) = \tfrac{1}{2}a(\boldsymbol{v},\boldsymbol{v}) - \langle \ell, \boldsymbol{v}\rangle$$

in which $a$ and $\ell$ are defined in (9.25). Of course, if the elastic body undergoes a rigid body displacement, then $a(\boldsymbol{v},\boldsymbol{v}) = 0$, as would be expected; the body has undergone no deformation, and therefore possesses no strain energy.

The aim of this section, then, is to investigate the consequences of formulating problems such as (9.48), (9.49), and to show how these relate to VBVPs. Before embarking on this task, it is instructive to review the situation for functions of a single variable, wherein many of the ideas for general functionals are present.

Consider a $C^1$ function $f : \mathbb{R} \to \mathbb{R}$, and suppose that we wish to locate the minimum value of $f(x)$, as well as the point $x_0$ at which the minimum occurs. That is, we wish to find $x_0$ such that

$$f(x_0) \le f(x) \quad \text{for all} \quad x \in \mathbb{R}.$$

From elementary calculus we know that a necessary condition for $f$ to have an extremum (that is, a minimum, a maximum, or an inflection point) at $x_0$ is that

$$f'(x_0) = 0.$$

The point $x_0$ is a minimum if the function $f$ is *convex*, that is, if a straight line drawn between any two points on the curve of $f(x)$ lies on or above the graph of the function. Mathematically, $f$ is convex if, for $0 < \theta < 1$,

$$f(y + \theta(x-y)) \le f(y) + \theta(f(x) - f(y))$$

or

$$f(\theta x + (1-\theta)y) \le \theta f(x) + (1-\theta)f(y).$$

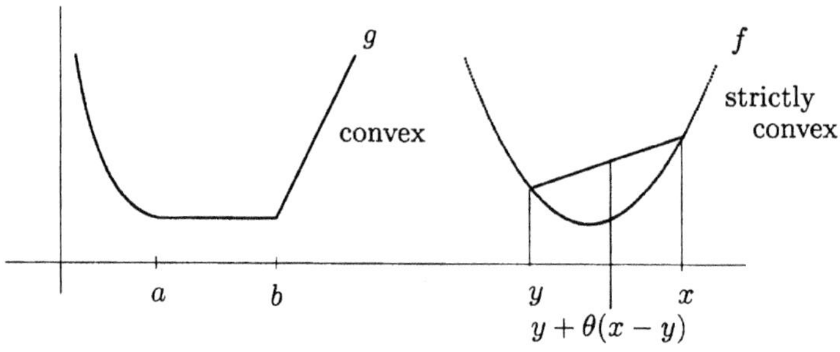


FIGURE 9.4. Examples of a convex and a strictly convex function

The minimum may not be unique. For example, consider the function $g$ shown in Figure 9.4; the minimum value of $g(x)$ occurs at all points between $a$ and $b$. But for a *strictly convex* function, that is, one for which

$$f(\theta x + (1-\theta)y) < \theta f(x) + (1-\theta) f(y), \quad 0 < \theta < 1, \quad x \neq y,$$

the minimum is unique; this is the case for the function $f$ in Figure 9.4. Remarkably, all of these ideas extend in a very simple way to functionals defined on arbitrary spaces. In order to see how this is done we first introduce the required generalizations of convex functions and their derivatives.

**Convex functionals.** Let $J : V \to \mathbb{R}$ be a functional defined on a vector space $V$. Then $J$ is said to be *convex* if

$$J(\theta u + (1-\theta)v) \leq \theta J(u) + (1-\theta) J(v),$$

and *strictly convex* if

$$J(\theta u + (1-\theta)v) < \theta J(u) + (1-\theta) J(v),$$

for all $u, v \in V$ with $u \neq v$, and for $0 < \theta < 1$.

**Gateaux derivative.** A functional $J : V \to \mathbb{R}$ on a normed space $V$ is said to be Gateaux-differentiable, or simply differentiable, at $u \in V$ if there exists an operator $DJ : V \to V'$ defined by

$$\langle DJ(u), v\rangle = \lim_{\theta \to 0} \frac{[J(u+\theta v) - J(u)]}{\theta} \tag{9.50}$$

for all $v \in V$. Equivalently, $DJ$ is defined by

$$\langle DJ(u), v\rangle = \frac{d}{d\theta}[J(u+\theta v)]\bigg|_{\theta=0} \tag{9.51}$$

(see Exercise 9.12). The operator $DJ$ is called the *gradient of* $J$ and $DJ(u) : V \to \mathbb{R}$ is the (Gateaux) derivative of $J$ at $u$. Observe from (9.50) or (9.51) that $DJ$ maps $V$ to its dual space $V'$, so that $DJ(u)$ is required to be *a bounded linear functional* on $V$.

The Gateaux derivative does not always exist; it may be verified, for example, that if $J$ is defined by $J : \mathbb{R}^2 \to \mathbb{R}$,

$$J(\boldsymbol{x}) = \begin{cases} x_1^2(1 + 1/x_2), & x_2 \neq 0, \\ 0, & x_2 = 0, \end{cases}$$

then

$$\lim_{\theta \to 0} \theta^{-1}[J(\boldsymbol{x} + \theta \boldsymbol{y}) - J(\boldsymbol{x})] = y_1^2/y_2,$$

which is not linear in $\boldsymbol{y}$.

### Examples

12. If $V$ is an interval in $\mathbb{R}$, then we see that $DJ$ reduces to the conventional derivative: $DJ(x) = dJ/dx$. Furthermore, if $V \subset \mathbb{R}^n$, then according to (9.50) or (9.51) we have (noting that $(\mathbb{R}^n)' = \mathbb{R}^n$)

$$DJ : \mathbb{R}^n \to \mathbb{R}^n, \quad \langle DJ(\boldsymbol{x}), \boldsymbol{y} \rangle = \sum_{i=1}^n \frac{\partial J}{\partial x_i} y_i = \nabla J \cdot \boldsymbol{y};$$

that is, the Gateaux derivative is the directional derivative (see Exercise 9.13).

13. Let $J : H^1(\Omega) \to \mathbb{R}$ be defined by

$$J(v) = \tfrac{1}{2} \int_\Omega \nabla v \cdot \nabla v \, dx - \int_\Omega fv \, dv.$$

Then $J$ is convex: indeed,

$$J(\theta u + (1 - \theta)v) = \tfrac{1}{2}\theta^2 \int_\Omega |\nabla u|^2 \, dx + \tfrac{1}{2}(1 - \theta)^2 \int_\Omega |\nabla v|^2 \, dx$$
$$+\theta(1 - \theta) \int_\Omega \nabla u \cdot \nabla v \, dx - \theta \int_\Omega fu \, dx - (1 - \theta) \int_\Omega fv \, dx.$$

Now $\int_\Omega (\nabla u - \nabla v) \cdot (\nabla u - \nabla v) \, dx \geq 0$ for $v \neq u$ (equality occurring for the case in which $u$ and $v$ are constant functions), so that

$$2 \int_\Omega \nabla u \cdot \nabla v \, dx \leq \int_\Omega (|\nabla u|^2 + |\nabla v|^2) \, dx.$$

Hence

$$\begin{aligned}
&J(\theta u + (1-\theta)v) \leq \tfrac{1}{2}\theta^2 \int_\Omega |\nabla u|^2 \, dx + \tfrac{1}{2}(1-\theta)^2 \int_\Omega |\nabla v|^2 \, dx \\
&+\tfrac{1}{2}\theta(1-\theta) \int_\Omega (|\nabla u|^2 + |\nabla v|^2) \, dx - \theta \int_\Omega fu \, dx - (1-\theta) \int_\Omega fv \, dx \\
&= \theta J(u) + (1-\theta) J(v).
\end{aligned}$$

We also observe that $J$ is *strictly convex* on $H_0^1(\Omega)$, since the only constant function in $H_0^1(\Omega)$ is $u = 0$.

To find the derivative of $J$ we use

$$\begin{aligned}
&\langle DJ(u), v \rangle \\
&= \frac{d}{d\theta} \left[ \int_\Omega [\tfrac{1}{2}(|\nabla u|^2 + 2\theta \nabla u \cdot \nabla v + \theta^2 |\nabla v|^2) - f(u + \theta v)] \, dx \right]_{\theta=0} \\
&= \int_\Omega (\nabla u \cdot \nabla v + \theta |\nabla v|^2 - fv) \, dx \bigg|_{\theta=0}
\end{aligned}$$

or

$$\langle DJ(u), v \rangle = \int_\Omega (\nabla u \cdot \nabla v - fv) \, dx.$$

Note that $DJ$ is an operator from $H^1(\Omega)$ to $[H^1(\Omega)]'$, and so $DJ(u)$ is a bounded linear functional on $H^1(\Omega)$.

We are now in a position to demonstrate the relationship between minimization problems and VBVPs, and start with the following fundamental result.

THEOREM 4. *Let $J$ be a convex differentiable functional defined on a subspace $V$ of a normed space $X$. An element $u \in V$ is a solution of the minimization problem*

$$J(u) \leq J(v) \quad \textit{for all } v \in V \tag{9.52}$$

*if and only if $u$ is a solution of the VBVP of finding $u \in V$ that satisfies*

$$\langle DJ(u), v \rangle = 0 \quad \textit{for all } v \in V. \tag{9.53}$$

PROOF. We show first that (9.52) implies (9.53). Assume that (9.52) holds; then, replacing $v$ by $u + \theta v$ for any $u$, $v \in V$ and $\theta \in (0, 1)$, we have

$$J(u + \theta v) - J(u) \geq 0.$$

Dividing by $\theta$ and allowing $\theta$ to go to zero, we obtain

$$DJ(u)v \geq 0. \tag{9.54}$$

But $v$ is arbitrary, so (9.54) holds if we replace $v$ by $-v$. Using the linearity of $DJ(u)$ we get $\langle DJ(u), v\rangle \leq 0$, and so $\langle DJ(u), v\rangle = 0$.

To show that (9.53) implies (9.52), we start with

$$\begin{aligned} J(\theta v + (1-\theta)u) = J(u + \theta(v-u)) &\leq \theta J(v) + (1-\theta)J(u) \\ &= J(u) + \theta(J(v) - J(u)) \end{aligned}$$

by the convexity of $J$. Hence

$$J(v) - J(u) \geq \frac{J(u + \theta(v-u)) - J(u)}{\theta}$$

so that, as $\theta \to 0$,

$$J(v) - J(u) \geq \langle DJ(u), v - u\rangle = 0;$$

hence (9.53) implies (9.52). □

**Examples**

14. Suppose that $X = H^1(\Omega)$, $V = H_0^1(\Omega)$, and

$$J(v) = \tfrac{1}{2}\int_\Omega \nabla v \cdot \nabla v \, dx - \int_\Omega fv \, dx. \tag{9.55}$$

We found $DJ(u)$ in the previous example, so it follows that the problem of finding $u \in H_0^1(\Omega)$ that minimizes (9.55) is equivalent to the problem of finding $u \in H_0^1(\Omega)$ that satisfies

$$\langle DJ(u), v\rangle = 0 \quad \text{or} \quad \int_\Omega \nabla u \cdot \nabla u \, dx = \int_\Omega fv \, dx \quad \text{for all } v \in H_0^1(\Omega). \tag{9.56}$$

We recognize (9.56) as a VBVP.

15. The preceding example is just a special case of the general minimization problem that involves *quadratic functionals* of the form

$$J : V \to \mathbb{R}, \quad J(v) = \tfrac{1}{2}a(v,v) - \langle \ell, v\rangle \tag{9.57}$$

in which $a(\cdot\,,\cdot)$ is a symmetric bilinear form on $V$ and $\ell$ is a linear functional on $V$. Here $V$ will generally be a subspace of a Sobolev space $H^m(\Omega)$, or perhaps of $[H^m(\Omega)]^n$ in the case of a problem such as that of elasticity. When $J$ takes the form (9.57) then we have

$$\begin{aligned}
&\langle DJ(u), v\rangle \\
&= \lim_{\theta\to 0} \theta^{-1}\left[\tfrac{1}{2}a(u+\theta v, u+\theta v) - \langle \ell, u+\theta v\rangle - \tfrac{1}{2}a(u,u) + \langle \ell, u\rangle\right] \\
&= \lim_{\theta\to 0} \theta^{-1}\left[\theta(a(u,v) - \langle \ell, v\rangle) + \tfrac{1}{2}\theta^2 a(v,v)\right] \\
&= a(u,v) - \langle \ell, v\rangle
\end{aligned}$$

using the bilinearity and symmetry of $a(\cdot\,,\cdot)$ and the linearity of $\ell$. Hence the problem of minimizing (9.57) is equivalent to the VBVP of finding $u \in V$ satisfying

$$a(u,v) = \langle \ell, v\rangle \quad \text{for all } v \in V, \tag{9.58}$$

*assuming that* $J(\cdot)$ *is convex.* That is not usually a problem; if $a(\cdot\,,\cdot)$ is $V$-elliptic, for example, then it is strictly convex (see Exercise 9.14).

To summarize, then, any VBVP of the form (9.58) in which $a(\cdot\,,\cdot)$ is $V$-elliptic is equivalent to the problem of minimizing the functional (9.57) and vice versa. This equivalence, as a matter of interest, explains the reason for the terminology "variational" in the expression "variational boundary value problem". The classical *calculus of variations* is concerned with the problem of minimizing functionals of a general nature and the expression $\langle DJ(u), v\rangle$ is known in that theory as the *first variation* of $J$. A necessary condition for a minimum is that the *first variation vanish*, that is, $\langle DJ(u), v\rangle = 0$, and this is what we call a variational BVP. It is important to note, though, that problems of the form (9.58) are referred to as VBVPs *even if* $a(\cdot\,,\cdot)$ *is not symmetric*, in which case there is *no* corresponding minimization problem.

We close this section with a theorem that gives conditions for the *existence* and *uniqueness* of solutions to minimization problems involving functionals of the form (9.57). Of course, existence and uniqueness could be discussed in terms of the equivalent VBVP (9.58), using the theory of Section 9.3. But for completeness we discuss problem (9.57) on its own, and show in fact that the requirements for well-posedness of Theorems 1 and 5 coincide.

THEOREM 5. *Let $J : V \to \mathbb{R}$ be the functional given by (9.57), in which $V$ is a closed subspace of a Hilbert space $H$. Assume that $a(\cdot\,,\cdot)$ is bilinear, symmetric, continuous, and $V$-elliptic, and that $\ell$ is bounded and linear. Then the problem of finding $u \in V$ that minimizes $J(v)$ over all $v \in V$ has one and only one solution.*

PROOF. We start by observing that $a(\cdot\,,\cdot)$ defines an *inner product* on $V$; indeed, if we write $a(u,v) \equiv (u,v)_a$, then

$$(u,v)_a = (v,u)_a, \quad (\alpha u + \beta v, w)_a = \alpha(u,w)_a + \beta(v,w)_a,$$

and the positive-definiteness of $(\cdot,\cdot)_a$ follows from the continuity and $V$-ellipticity of $a$, in that

$$M\|u\|_H^2 \geq (u,u)_a \geq \alpha\|u\|_H^2. \tag{9.59}$$

Thus $(u,u)_a \geq 0$ and $(u,u)_a = 0$ if and only if $u = 0$. Furthermore, the norm $\|u\|_a \equiv (u,u)_a$ is *equivalent* to the standard norm on $V$, as (9.59) indicates, so it follows that the space $V$ with the inner product $(\cdot,\cdot)_a$ is a Hilbert space.

We now apply the the Riesz Representation Theorem using $(\cdot,\cdot)_a$: corresponding to the functional $\ell$ there exists $\tilde{\ell} \in V$ such that $\langle \ell, v\rangle = (\tilde{\ell}, v)_a$. Hence (9.57) reads

$$J(v) = \tfrac{1}{2}\|v - \tilde{\ell}\|_a^2 - \tfrac{1}{2}\|\tilde{\ell}\|_a^2, \tag{9.60}$$

where $\|\cdot\|_a$ is the norm generated by $(\cdot,\cdot)_a$. From (9.60) it is clear that the problem amounts to one of finding $u \in V$ such that

$$\|u - \tilde{\ell}\|_a \leq \|\tilde{\ell} - v\|_a \quad \text{for all } v \in V.$$

By Theorem 6 of Chapter 4 such an element exists and is unique. Indeed since $\tilde{\ell} \in V$ we have $u = \tilde{\ell}$. □

We remark in conclusion that Theorem 5 is equivalent to Theorem 1 when the bilinear form is *symmetric*, but that Theorem 1 alone is of use if $a(\cdot,\cdot)$ is nonsymmetric.

## 9.5 Bibliographical remarks

Good accounts and further examples of the theory covered in Sections 9.1 to 9.3 may be found in Dautray and Lions ([13], Chapter VII) and in Rektorys [41]. The discussion of non-$V$-elliptic problems that includes Theorem 2 is adapted from the treatment of this topic by Nečas [34] and Rektorys [41].

The discussion in Section 9.4 of the minimization of convex functionals has focused only on those aspects pertinent to our main goals. The subject is huge, and in itself contains many interesting applications of functional analysis. For more details the texts by Glowinski [16] and Zeidler [55] are good sources.

We have avoided discussion of problems such as (9.52) when $V$ is a *convex subset* but not a subspace. For example, consider the problem of finding a function that satisfies

$$\left.\begin{array}{l} -\nabla^2 u - f \geq 0, \quad u \geq g \\ (u-g)(-\nabla^2 u - f) = 0 \end{array}\right\} \text{in } \Omega$$
$$u = 0 \text{ on } \Gamma.$$

This corresponds to the problem of finding the shape of a membrane stretched over an obstacle, as shown in Figure 9.5. This set of equations describes the fact that the membrane has to be above or on the obstacle ($u - g \geq 0$), that the force acting on the membrane is either zero, when there is no contact, or positive, at those points at which there is contact ($-\nabla^2 u - f \geq 0$). The second equation indicates that these quantities cannot both be positive; one either has contact, in which case $u - g = 0$ and the net force is positive, or the membrane lies above the obstacle, in which case the net force is necessarily zero.

It can be shown (Exercise 9.17) that the corresponding VBVP is the *variational inequality:* find $u \in K$ such that

$$\int_\Omega \nabla u \cdot \nabla(v - u)\, dx - \int_\Omega f(v - u)\, dx \geq 0 \quad \text{for all } v \in K, \tag{9.61}$$

where $K$ is the *convex subset*

$$K = \{v \in H_0^1(\Omega) : v \geq g \text{ a.e. in } \Omega\} \tag{9.62}$$

(note that $K$ is not a subspace), and that the corresponding minimization problem is: find $u \in K$ such that

$$J(u) \leq J(v) = \tfrac{1}{2}\int_\Omega \nabla v \cdot \nabla v\, dx - \int_\Omega fv\, dx$$

for all $v \in K$. For a detailed account of variational inequalities see, for example, Baiocchi and Capelo [4] and Glowinski [16]. The book by Duvaut and Lions [15] is devoted to a thorough study of variational inequalities that arise in mechanics and physics.

## 9.6 Exercises

**Formulation of variational boundary value problems**

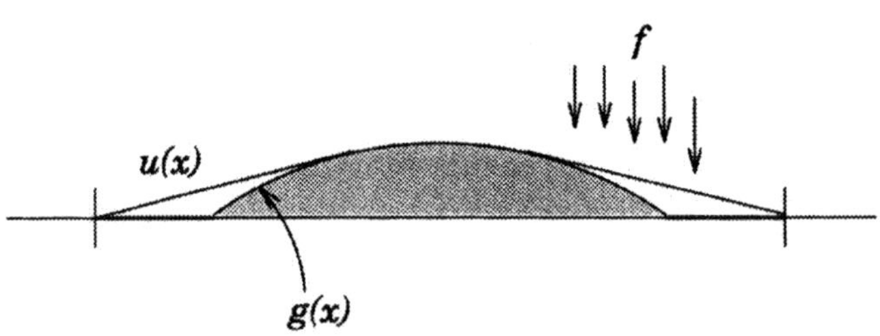


FIGURE 9.5. A membrane stretched over an obstacle

9.1. Formulate the VBVP corresponding to

$$\begin{aligned}
[k(x)u''(x)]'' - [d(x)u'(x)]' + c(x)u(x) &= f(x) \text{ in } (0,1), \\
u(0) = 0, &\quad u'(0) = 0 \\
(ku'')(1) = \alpha, &\quad [-(ku'')' + du'](1) = \beta.
\end{aligned}$$

9.2. Find the VBVP corresponding to

$$\begin{aligned}
-\nabla^2 u &= f \quad \text{in } \Omega \subset \mathbb{R}^2, \\
\frac{\partial u}{\partial \tau} &= g \quad \text{on } \Gamma,
\end{aligned}$$

in which $\partial u/\partial \tau \equiv \nabla u \cdot \boldsymbol{\tau}$ is the oblique directional derivative in the direction of the unit vector $\boldsymbol{\tau}$, which is not generally tangential to the boundary $\Gamma$.

9.3. The VBVP for the plate problem may be derived in a manner that facilitates imposition of the natural boundary conditions, in the following way.

(a) Equations (8.15) imply that $\sum_{\alpha,\beta=1}^{2} \partial^2 M_{\alpha\beta}/\partial x_\alpha \partial x_\beta = -q$. Multiply this equation by an arbitrary function $v$ and use Green's theorem to obtain the identity

$$\begin{aligned}
&\int_\Omega v \sum_{\alpha,\beta=1}^{2} \frac{\partial^2 M_{\alpha\beta}}{\partial x_\alpha \partial x_\beta}\, dx \\
&= -\int_\Gamma \sum_{\alpha,\beta=1}^{2} [v S_\beta \nu_\beta + M_{\alpha\beta}\nu_\alpha \partial v/\partial x_\beta]\, ds \\
&+ \int_\Omega \sum_{\alpha,\beta=1}^{2} M_{\alpha\beta} \frac{\partial^2 v_{\alpha\beta}}{\partial x_\alpha \partial x_\beta}\, dx.
\end{aligned}$$

(b) Assuming that the same conditions as in Example 2 hold, derive the VBVP simply by imposing the natural boundary condition $M_{11} = 0$ or $M_{22} = 0$ on $\Gamma$, and by defining $V$ as in the example. Show that the bilinear form becomes

$$\begin{aligned}
a(w,v) = (1-\nu)\int_\Omega \left\{ \frac{\partial^2 w}{\partial x^2}\frac{\partial^2 v}{\partial x^2} + 2\frac{\partial^2 w}{\partial x \partial y}\frac{\partial^2 v}{\partial x \partial y} + \frac{\partial^2 w}{\partial y^2}\frac{\partial^2 v}{\partial y^2} \right\} dx \\
+ \int_\Omega \nu(\nabla^2 w)(\nabla^2 v)\, dx.
\end{aligned}$$

How would you reconcile this expression with that given in (9.24)?

**Existence, uniqueness, and regularity of solutions**

9.4. Verify that the VBVP in Exercise 9.1 has a unique solution if the functions $k$, $d$, and $c$ are all strictly positive, with $k \in C^2[0,1]$, $c \in C[0,1]$, and $d \in C^1[0,1]$.

9.5. Consider the BVP for nonhomogeneous, anisotropic heat conduction with a temperature-dependent heat source (Chapter 8), viz.

$$\begin{aligned} -\sum_{i,j} \frac{\partial}{\partial x_i}\left(k_{ij}\frac{\partial u}{\partial x_j}\right) + bu &= f \quad \text{in } \Omega, \\ u &= 0 \quad \text{on } \Gamma; \end{aligned}$$

the coefficients $k_{ij}$ of the thermal conductivity matrix are such that the operator is strongly elliptic. Derive the corresponding VBVP and show that the bilinear form is $V$-elliptic provided that $b(\boldsymbol{x}) \geq 0$. Show also that the bilinear form is continuous provided that $|k_{ij}(\boldsymbol{x})| \leq K$.

9.6. For a plate occupying a domain $\Omega$ with arbitrary, nonrectangular boundary $\Gamma$, it can be shown (see, for example, Rektorys [41], Chapter 23) that the moment acting on the boundary is given by $M_n = \boldsymbol{n}^T\boldsymbol{M}\boldsymbol{n} = \sum_{\alpha,\beta=1}^{2} M_{\alpha\beta}n_\alpha n_\beta$. In this exercise the unit normal is denoted by $\boldsymbol{n}$, to avoid confusion with Poisson's ratio $\nu$.

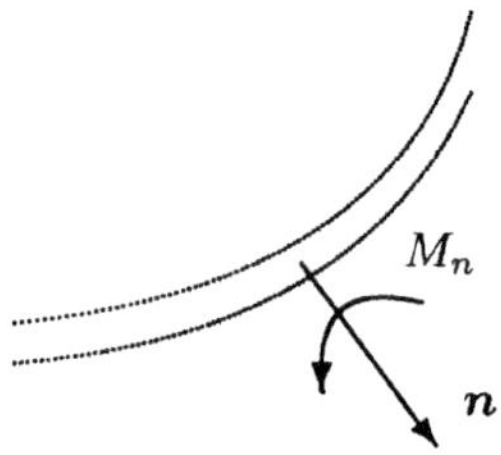


Consider the problem of a plate that is simply supported on its boundary, so that the displacement and moment are zero along the boundary. Show that the moment boundary condition becomes

$$\nu\nabla^2 w + (1-\nu)\partial^2 w/\partial n^2 = 0 \quad \text{on } \Gamma.$$

If the second boundary condition is $w = 0$, show that the corresponding bilinear form (after an appropriate definition of the space $V$) is unchanged from that in Exercise 9.3. Show also that $a(\cdot,\cdot)$ is $V$-elliptic provided that $\nu$ lies in the range $0 \leq \nu < 1$. [Use the inequality (7.33).]

9.7. Show that the bilinear form associated with the BVP

$$\begin{gathered} -(pu')' + ru = f \qquad \text{in } \Omega = (0,1), \\ u(0) = 0, \quad u'(1) + u(1) = 0, \end{gathered}$$

is $V$-elliptic and continuous; here $p$ and $r$ satisfy the usual conditions for the left-hand side to be a Sturm–Liouville operator.

9.8. Investigate the well-posedness of the problem of an elastic beam that has the set of boundary conditions

(a) as in Example 4;

(b) $u''(0) = g_0, \quad u''(1) = g_1,$
$u'''(0) = h_0, \quad u'''(1) = h_1.$

9.9. Show that the bilinear form in Example 7 is $V$-elliptic provided that the Lamé constants satisfy the conditions given in Exercise 8.8.

9.10. Derive the identity

$$\sum_{i,j,k,l=1}^{n} C_{ijkl}\epsilon_{ij}(\boldsymbol{u})\epsilon_{kl}(\boldsymbol{v}) = \lambda(\operatorname{div}\boldsymbol{u})(\operatorname{div}\boldsymbol{v}) + 2\mu\epsilon(\boldsymbol{u})\cdot\epsilon(\boldsymbol{v})$$

used in Example 7.

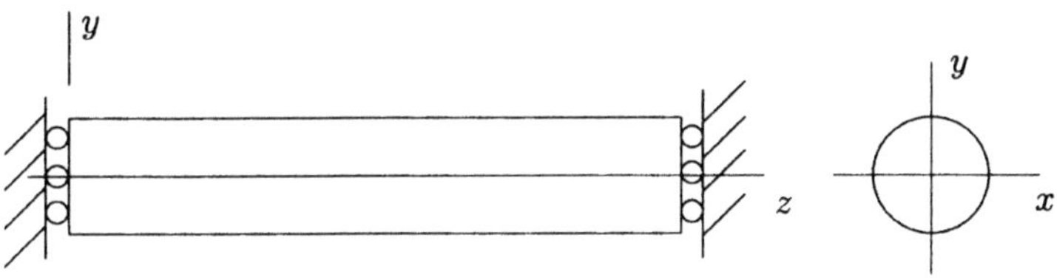


9.11. An elastic cylinder is subjected to a body force $\boldsymbol{f}$ in its domain $\Omega = S \times (0, L)$, where $S = \{(r,\theta) : \ 0 \le \theta < 2\pi, \ r < R\}$. The curved boundary $r = R$ is free of applied forces, and on its two ends the cylinder is restrained from axial displacement on the boundary, and a system of (ideally) frictionless bearings results in there being no tangential force there. Use Theorem 2 to investigate the conditions under which this problem has a unique solution.

**Minimization of functionals**

9.12. Show that an equivalent definition of the Gateaux derivative is

$$\langle DJ(u), v\rangle = \frac{d}{d\theta}\left[J(u+\theta v)\right]_{\theta=0}.$$

9.13. Consider the functional $J : \mathbb{R}^n \to \mathbb{R}$; show that

$$\langle DJ(\boldsymbol{x}), \boldsymbol{y}\rangle = \sum_{i=1}^{n} \frac{\partial J}{\partial x_i} y_i.$$

9.14. If $a : V \times V \to \mathbb{R}$ is a $V$-elliptic, symmetric bilinear form and $\ell : V \to \mathbb{R}$ is a linear functional, show that

$$J(v) = \tfrac{1}{2}a(v,v) - \langle \ell, v \rangle$$

is strictly convex.

9.15. Show that $J(v) = \frac{1}{2}a(v,v) - \langle \ell, v \rangle$ is convex if $a$ is positive, that is, if $a(v,v) \geq 0$ for all $v \in V$. Hence prove the converse of Theorem 3: if $u$ satisfies $a(u,v) = \langle \ell, v \rangle$ for all $v \in V$, then $u$ minimizes $J$.

9.16. Formulate the minimization problem corresponding to the VBVP of Exercise 9.1.

9.17. Consider, in the context of Theorem 5, the situation in which $V$ is a closed and convex *subset* of $H$. Verify that the theorem still holds, but that the condition (9.58) for a minimum is replaced by the *variational inequality*

$$u \in V \quad \text{and} \quad a(u, v-u) \geq \langle \ell, v-u \rangle \quad \text{for all } v \in V.$$

The obstacle problem (9.61), (9.62) is a special case of this abstract problem.

# 10
# Approximate methods of solution

In the two preceding chapters we have devoted considerable attention to various aspects of boundary value problems. The stage has now been reached where we can quite justifiably ask: how does one actually obtain solutions? The answer is rather disappointing, unfortunately; except for problems involving very simple PDEs and geometries, it is quite impossible, using existing methods, to obtain exact solutions to most BVPs in either the conventional or variational formulations.

This state of affairs naturally leads to the question of whether it is possible to obtain *approximate* solutions. Here matters are far more encouraging, in that there are available many good methods for finding approximate solutions. Some, such as the finite difference method, are based on the classical formulation whereas others, such as the Galerkin method, take as their starting point the variational formulation.

The methods that make use of variational formulations have enjoyed a great upsurge in popularity in the past three decades, particularly since the establishment of the finite element method, which is probably the best known special case of the Galerkin method. In the sections that follow we show how approximate solutions to VBVPs or, equivalently, to the corresponding minimization problems, can be obtained. The emphasis is on the Galerkin method and the finite element method, although we also give some indication of other related methods in Section 10.3.

## 10.1 The Galerkin method

The basic idea behind the Galerkin method is an extremely simple one. Consider the VBVP of finding $u \in V$ that satisfies

$$a(u, v) = \langle \ell, v \rangle \quad \text{for all } v \in V, \tag{10.1}$$

where $V$ is a subspace of a Hilbert space $H$. We assume for convenience that all spaces are defined over the real numbers. The difficulty in trying to solve (10.1) lies with the fact that $V$ is a very large space (infinite-dimensional, in the language of Chapter 6), with the result that it is not possible to set up a practical method for finding the solution. But suppose that, instead of posing the problem in $V$, we pick a few linearly independent functions $\phi_1, \phi_2, \ldots, \phi_N$ in $V$ and define the space $V^h$ to be the finite-dimensional subspace of $V$ spanned by the functions $\phi_i$. That is,

$$V^h \subset V, \quad \text{span}\,\{\phi_i\}_{i=1}^N = V^h. \tag{10.2}$$

The index $h$ is a parameter that lies between 0 and 1, and whose magnitude gives some indication of how close $V^h$ is to $V$; $h$ is related to the dimension of $V^h$, and as the number $N$ of basis functions chosen gets larger, $h$ gets smaller (for example, we could set $h = 1/N$). In the limit, as $N \to \infty$, $h \to 0$ and we would like to choose $\{\phi_i\}$ in such a way that $V^h$ will approach $V$, in a manner made precise later.

Having defined the space $V^h$, problem (10.1) is now posed in $V^h$ instead of in $V$. That is, we try to find a function $u_h \in V^h$ that satisfies

$$a(u_h, v_h) = \langle \ell, v_h \rangle \qquad \text{for all}\, v_h \in V^h. \tag{10.3}$$

This is the essence of the Galerkin method. In order to solve for $u_h$, we simply note that both $u_h$ and $v_h$ must be linear combinations of the basis functions of $V^h$, so that

$$u_h = \sum_{i=1}^N c_i \phi_i \quad \text{and} \quad v_h = \sum_{j=1}^N d_j \phi_j. \tag{10.4}$$

Of course, since $v_h$ is arbitrary, so are the coefficients $d_j$. Substitution of (10.4) in (10.3) and use of the fact that $a$ is bilinear and $\ell$ is linear, lead to the equation

$$\sum_{i=1}^N \sum_{j=1}^N a(\phi_i, \phi_j) c_i d_j = \sum_{j=1}^N \langle \ell, \phi_j \rangle d_j$$

or, more concisely,

$$\sum_{j=1}^N d_j \left( \sum_{i=1}^N K_{ij} c_i - F_j \right) = 0 \tag{10.5}$$

in which

$$K_{ij} = a(\phi_i, \phi_j) \quad \text{and} \quad F_j = \langle \ell, \phi_j \rangle. \tag{10.6}$$

Note that $K_{ij}$ and $F_j$ can be evaluated in practice since the $\phi_i$ are known functions and the forms of $a$ and $\ell$ are also known.

Since the coefficients $d_j$ are arbitrary, it follows that (10.5) only holds if the term in brackets is zero. The problem is thus reduced to one of solving the set of *simultaneous linear equations*

$$\sum_{i=1}^{N} K_{ij} c_i = F_j, \quad j = 1, \ldots, N \tag{10.7}$$

or, more compactly,

$$\boldsymbol{K}^t \mathbf{c} = \boldsymbol{F} \tag{10.8}$$

in which $\boldsymbol{K}$ and $\boldsymbol{F}$ are, respectively, the matrix and vector with entries $K_{ij}$ and $F_i$. Once these equations are solved, the approximate solution $u_h$ can be found from the first of equations (10.4).

**Examples**

1. Consider the BVP

$$-\frac{d^2u}{dx^2} = \sin\frac{\pi x}{2} \quad \text{in } \Omega = (0,1), \quad u(0) = u'(1) = 0.$$

The corresponding VBVP is: find $u \in V$ such that

$$\int_0^1 u'v' \, dx = \int_0^1 (\sin \pi x/2) v \, dx \quad \text{for all } v \in V,$$

where $V = \{v \in H^1(0,1) : u(0) = 0\}$ (note that $u'(1) = 0$ is a natural boundary condition). Now define $V^h$ to be the subspace of $V$ spanned by the $N$ monomials

$$\phi_i(x) = x^i, \quad i = 1, 2, \ldots, N.$$

Then

$$K_{ij} = a(\phi_i, \phi_j) = \int_0^1 \phi_i' \phi_j' \, dx$$

and

$$F_i = \langle \ell, \phi_i \rangle = \int_0^1 (\sin \pi x/2) \phi_i \, dx.$$

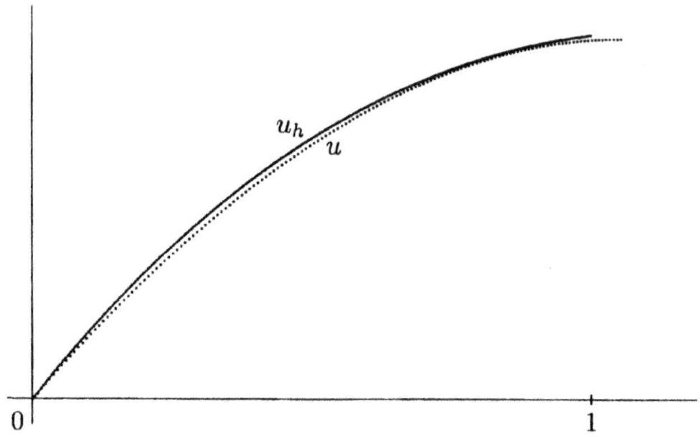


FIGURE 10.1. Exact and approximate solutions to the problem in Example 1

Suppose that we take $N = 2$; then we obtain the set of simultaneous equations

$$\boldsymbol{K}\boldsymbol{c} = \boldsymbol{F} \iff \begin{aligned} c_1 + c_2 &= 0.405 \\ c_1 + \tfrac{4}{3}c_2 &= 0.295 \end{aligned}$$

($\boldsymbol{K}$ is symmetric; that is, $\boldsymbol{K}^t = \boldsymbol{K}$) and the solution to these equations is

$$c_1 = 0.738, \quad c_2 = -0.33.$$

The approximate solution is thus

$$\begin{aligned} u_h(x) &= c_1\phi_1(x) + c_2\phi_2(x) \\ &= 0.738x - 0.33x^2. \end{aligned}$$

This problem can be solved in closed form, and the exact solution is

$$u(x) = (2/\pi)^2 \sin(\pi x/2),$$

which is compared with the approximate solution in Figure 10.1. We see that even the crude approximation in a two-dimensional subspace produces in this case a solution that compares very favorably with the exact solution.

2. Consider the VBVP of finding $u \in H_0^1(\Omega)$ that satisfies

$$\int_\Omega \nabla u \cdot \nabla v \, dx = \int_\Omega fv \, dx \quad \forall v \in H_0^1(\Omega),$$

where $f(x, y) = xy$. [The corresponding BVP is

$$\begin{aligned} -\nabla^2 u &= xy \quad \text{in } \Omega, \\ u &= 0 \quad \text{on } \Gamma.] \end{aligned}$$

Here $\Omega$ is the unit square $(0,1) \times (0,1)$ in $\mathbb{R}^2$. We now choose as a basis for $V^h$ the set of functions

$$\phi_1 = \sin \pi x \sin \pi y, \qquad \phi_2 = \sin \pi x \sin 2\pi y,$$
$$\phi_3 = \sin 2\pi x \sin \pi y, \qquad \phi_4 = \sin 2\pi x \sin 2\pi y,$$

which of course all belong to $H_0^1(\Omega)$. The next step is to evaluate

$$K_{ij} = a(\phi_i, \phi_j) \text{ and } F_i = (f, \phi_i)_{L^2},$$

which is straightforward if we make use of the identity

$$\begin{aligned} \int_0^1 \sin n\pi x \sin m\pi x \, dx &= \int_0^1 \cos n\pi x \cos m\pi x \, dx \\ &= \begin{cases} 0 & \text{if } n \neq m, \\ \frac{1}{2} & \text{if } n = m. \end{cases} \end{aligned}$$

Then

$$K_{ij} = \int_0^1 \int_0^1 \left( \frac{\partial \phi_i}{\partial x} \frac{\partial \phi_j}{\partial x} + \frac{\partial \phi_i}{\partial y} \frac{\partial \phi_j}{\partial y} \right) dx \, dy,$$

and because of the orthogonality of the trigonometric functions the only nonzero terms of $K_{ij}$ are

$$\begin{aligned} K_{ii} &= \int_0^1 \int_0^1 \left( \frac{\partial \phi_i}{\partial x} \right)^2 + \left( \frac{\partial \phi_i}{\partial y} \right)^2 dxdy \\ &= \pi^2 \int_0^1 \int_0^1 n^2 \cos^2 n\pi x \sin^2 m\pi y \, dxdy \\ &\quad + \pi^2 \int_0^1 \int_0^1 m^2 \sin^2 n\pi x \cos^2 m\pi y \, dxdy, \end{aligned}$$

where $n$ and $m$ take the values:

| $i$ | 1 | 2 | 3 | 4 |
|---|---|---|---|---|
| $n$ | 1 | 1 | 2 | 2 |
| $m$ | 1 | 2 | 1 | 2. |

After carrying out the integration we obtain

$$K_{ii} = \frac{\pi^2}{4}(n^2 + m^2) \qquad \text{or} \qquad \boldsymbol{K} = \frac{\pi^2}{4} \begin{bmatrix} 2 & 0 & 0 & 0 \\ 0 & 5 & 0 & 0 \\ 0 & 0 & 5 & 0 \\ 0 & 0 & 0 & 8 \end{bmatrix}.$$

Similarly,

$$\begin{aligned} F_i &= \int_0^1 \int_0^1 xy \, \phi_i \, dx \, dy \\ &= \int_0^1 \int_0^1 xy \sin n\pi x \sin m\pi y \, dx \, dy \\ &= \frac{1}{\pi^2}(1, -2, -2, 4). \end{aligned}$$

Hence the solution is

$$c = \frac{4}{\pi^4}\left(\tfrac{1}{2}, -\tfrac{2}{5}, -\tfrac{2}{5}, \tfrac{1}{2}\right)$$

and so

$$\begin{aligned} u_h(x,y) &= \frac{4}{\pi^4}\left[\tfrac{1}{2}(\sin\pi x\,\sin\pi y + \sin 2\pi x\,\sin 2\pi y)\right. \\ &\quad \left. - \tfrac{2}{5}(\sin\pi x\,\sin 2\pi y + \sin 2\pi x \sin\pi y)\right]. \end{aligned}$$

We recall from Section 5.5 that the bilinear form $a(\cdot,\cdot)$ defines an *inner product* on $V$ if $a$ is symmetric and $V$-elliptic; indeed, the properties of linearity and symmetry are obvious, whereas the property of positive-definiteness comes from the $V$-ellipticity of $a$:

$$a(v,v) \geq \alpha\|v\|_V^2 > 0 \qquad \text{for all nonzero } v. \tag{10.9}$$

Furthermore, we have seen in the proof of Theorem 5, Chapter 9, that if $a$ is also continuous, then the norm $\|v\|_a \equiv a(v,v)$ generated by this inner product is equivalent to the standard norm on $V$, so that if $V$ is complete with respect to the standard norm, it is also complete with respect to the norm $\|\cdot\|_a$. As before, this inner product is denoted by $(\cdot,\cdot)_a$ and referred to as the *energy inner product* (the rationale behind this terminology has been discussed in Section 9.4), and the corresponding norm is called the *energy norm.*

Now if the set of basis functions $\{\phi_i\}_{i=1}^N$ is chosen in such a way that they are *orthogonal with respect to the energy inner product,* then the system of equations (10.7) simplifies considerably, since

$$K_{ij} = a(\phi_i,\phi_j) = (\phi_i,\phi_j)_a = 0 \text{ if } i \neq j,$$

and so

$$K_{ii}c_i = F_i, \quad \text{or} \quad c_i = F_i/K_{ii}.$$

This is in fact the case in Example 2.

However, a word of warning is appropriate. Although for the preceding example it was quite simple to find a basis that was orthogonal with respect to $(\cdot,\cdot)_a$, in general this is quite difficult. One could of course choose any non-orthogonal basis and use the Gram–Schmidt procedure of Section 6.2 to orthogonalize or even orthonormalize, but for all except the most trivial problems this is a laborious procedure, and little is to be gained from it.

The problem of constructing a basis $\{\phi_i\}_{i=1}^N$ in such a way that $V^h$ approaches $V$ as $N \to \infty$ can be rather awkward. Remember that although orthonormal bases for spaces such as $L^2$ are well known, at least for spaces of functions on the real line or on simple two- and three-dimensional domains (see, for example, Section 6.4), when using the Galerkin method we are required to find bases for spaces $V$ that are subspaces of Sobolev spaces $H^m(\Omega)$, and that are defined on domains $\Omega$ which may be quite irregular in shape. A very simple and elegant method for constructing such bases is provided by the finite element method. This is the topic of discussion in the next two chapters.

**The Rayleigh–Ritz method.** The Rayleigh-Ritz method is very closely linked to the Galerkin method. It takes as its starting point the minimization problem (9.52) and, as with the Galerkin method, proceeds to pose this problem on a finite-dimensional subspace. That is, problem (9.52) is replaced by the problem of finding $u_h \in V^h$ such that

$$J(u_h) \le J(v_h) \quad \text{for all}\, v_h \in V^h,$$

where $V^h$ is a finite-dimensional subspace of $V$. If $\{\phi_k\}_{k=1}^N$ is a basis for $V^h$, then substitution of $v_h = \sum_{k=1}^N c_k\phi_k$ in the expression for $J$ yields the function

$$\hat{J}(c_k) \equiv J\left(\sum_{k=1}^{N} c_k\phi_k\right),$$

which is a function of the $N$ variables $c_1, \ldots, c_N$. In order to minimize $J(v_h)$, therefore, we require that

$$\frac{\partial \hat{J}}{\partial c_k} = 0, \quad k = 1, \ldots, N,$$

and this yields a set of $N$ simultaneous algebraic equations in the $N$ unknowns $c_1, \ldots, c_N$. Solution of these equations then gives the components $c_k$ of $u_h$. In particular, if $J$ is given by (9.57), then

$$\hat{J}(c_k) = \tfrac{1}{2}\sum_{i,j=1}^{N} K_{ij}c_ic_j - \sum_{j=1}^{N} F_jc_j,$$

where $K_{ij}$ and $F_j$ are defined by (10.6), and minimization with respect to the $c_k$ yields the set of linear equations

$$\sum_{j=1}^{N} K_{ij}c_j = F_i, \quad i = 1, \ldots, N$$

which is precisely (10.7). Here, though, $\boldsymbol{K}$ is always symmetric.

## 10.2 Properties of Galerkin approximations

In Section 10.1 we introduced the Galerkin method and illustrated how the method is used in practice. It is not very satisfactory, however, simply to leave things at that; we ought to know, first of all, whether the Galerkin method always works and, if so, how significantly the approximate solution differs from the exact solution. Also, we would like to be confident that as the number of functions $\phi_i$ in the basis of $V^h$ is increased, $V^h$ approaches in some sense the space $V$ and $u_h$ approaches the exact solution $u$. This last consideration is of course one of *convergence* of the approximate solution as $h \to 0$ (or as $N \to \infty$).

**Existence and uniqueness.** The question of existence and uniqueness of a solution is easily resolved, in view of the results in Section 9.3. Since $V^h$ is a finite-dimensional subspace of a Hilbert space, it is necessarily complete. Assuming then that $a$ is a $V$-elliptic continuous bilinear form on $V \times V$ and $\ell$ is a continuous function on $V$, the same obviously holds true when their domains are restricted to $V^h$. Hence, since it has already been shown that a unique solution to (10.1) exists, the same will hold true for the approximate problem (10.3). This information is recorded in the following theorem.

THEOREM 1. *Let $V^h$ be a finite-dimensional subspace of a Hilbert space $V$, $a : V^h \times V^h \to \mathbb{R}$ a continuous, $V$-elliptic bilinear form, and $\ell : V^h \to \mathbb{R}$ a bounded linear functional. Then there exists a unique function $u_h \in V^h$ that satisfies*

$$a(u_h, v_h) = \langle \ell, v_h \rangle \quad \text{for all } v_h \in V^h.$$

*Futhermore, if $\ell$ is of the form*

$$\langle \ell, v_h \rangle = \int_\Omega f v_h \, dx$$

*with $f \in L^2(\Omega)$, then*

$$\|u_h\|_V \leq \frac{1}{\alpha} \|f\|_{L^2},$$

*where $\alpha$ is the constant in (10.9).*

**Errors in Galerkin approximations.** Having established the conditions under which an approximate solution can be found, we proceed now to characterize the *error* $e$, which is defined to be the difference between the exact and approximate solutions:

$$e = u - u_h.$$

For this purpose we return to (10.1). Since $V^h$ is a subspace of $V$ it is in order to choose $v$ to be a member of $V^h$; denoting this member by $v_h$ we have

$$a(u, v_h) = \langle \ell, v_h \rangle \quad \text{for all } v_h \in V^h. \tag{10.10}$$

Furthermore, the Galerkin approximation $u_h$ satisfies (10.3). Subtracting (10.3) from (10.10) and making use of the bilinearity of $a$, it is found that

$$a(u_h, v_h) = a(u, v_h) \tag{10.11}$$

or

$$a(u_h - u, v_h) = 0;$$

that is,

$$a(e, v_h) = 0. \tag{10.12}$$

This seemingly innocuous result has a useful geometrical interpretation in the event that $a$ is *symmetric*; for then the inner product $(\cdot, \cdot)_a = a(\cdot, \cdot)$ is available, and the theory of Sections 5.3 and 6.4 on orthogonal projections comes into play. Indeed, according to Exercise 6.21, if $\{\phi_k\}_{k=1}^N$ is an orthonormal basis of $V^h$ with respect to $(\cdot, \cdot)_a$, then the orthogonal projection onto $V^h$ is defined by

$$Pv = \sum_{k=1}^{N} (v, \phi_k)_a \phi_k. \tag{10.13}$$

But if we set $v_h = \phi_k$ in equation (10.11) we find that

$$(u, \phi_k)_a = (u_h, \phi_k)_a$$

so that

$$\begin{aligned} Pu &= \sum_{k=1}^{N} (u_h, \phi_k)_a \phi_k \\ &= Pu_h = u_h. \end{aligned} \tag{10.14}$$

*Hence the orthogonal projection of the solution $u$ onto $V^h$, with respect to the inner product $(\cdot, \cdot)_a$, is the approximate solution $u_h$.* Clearly, then, the error $e = u - u_h = u - Pu$ belongs to $N(P)$, that is,

$$(e, u_h)_a = 0,$$

which confirms (10.12). In other words, relative to the inner product $(\cdot, \cdot)_a$ *the error is orthogonal to the subspace $V^h$.*

The geometrical analogy may be carried a step further. It would appear that the distance $\|u - v_h\|$, when measured using the norm $\|\cdot\|_a$, is a *minimum* when $v_h = u_h$. That this is indeed so is borne out by the following. We have

$$\begin{aligned} a(u - v_h, u - v_h) &= a(u - u_h + u_h - v_h, u - u_h + u_h - v_h) \\ &= a(e + (u_h - v_h), e + (u_h - v_h)) \\ &= a(e, e) + 2a(e, u_h - v_h) + a(u_h - v_h, u_h - v_h), \end{aligned}$$

using the bilinearity of $a$. Now the second term on the right-hand side is zero, since $e$ is orthogonal to all members of $V^h$ with respect to the inner product $(\cdot,\cdot)_a$. Thus

$$\|u - v_h\|_a^2 = \|e\|_a^2 + \|u_h - v_h\|_a^2$$

and for fixed $u$ and $u_h$ (and hence for fixed $e$) we conclude that $\|u - v_h\|_a$ is smallest when $v_h = u_h$,; that is,

$$\|u - u_h\|_a \leq \|u - v_h\|_a \quad \text{for all } v_h \in V^h. \tag{10.15}$$

In other words, the function in $V^h$ that is closest to $u$ is the Galerkin approximation. In this sense *the Galerkin approximation is the best approximation to $u$ in $V^h$*. Of course we could have deduced (10.15) directly from Theorem 7 of Chapter 4 and (10.14).

**Convergence of Galerkin approximations.** As mentioned earlier, each value of the parameter $h$ defines a subspace $V^h$ of $V$: the smaller $h$ is, the larger the dimension of $V^h$ will be. We could use for $h$ the reciprocal of the number of basis functions that span $V^h$, although in Chapter 11 we give $h$ a more geometrical meaning, in the context of the finite element method. In any case, $h$ lies in (0,1) and equation (10.3) can be regarded as representing a *family* of Galerkin approximations, each value of $h$ having associated with it a problem (10.3). Corresponding to this family of problems we have also a family of solutions $u_h$, and once again a particular solution $u_h$ is associated with each value of $h$. Of course, if we define $h$ by $h = 1/N$, where $N = \dim V^h$, then $h$ cannot take on all values in (0,1) but only those of the form $1/N$ for integer $N$.

With these ideas at our disposal, it is quite simple to give a definition of the convergence of a family of Galerkin approximations; we say that the family of solutions $u_h$ *converges to the exact solution $u$* if

$$\lim_{h \to 0} \|u_h - u\|_V = 0. \tag{10.16}$$

The task of proving convergence, once a basis or a family of bases has been identified, is made easier by a deceptively simple result, which has far-reaching implications.

LEMMA 1 (CÉA'S LEMMA). *Let $V$ be a closed subspace of a Hilbert space, and let $a$ and $\ell$ be, respectively, a continuous, $V$-elliptic bilinear form and a bounded linear functional on $V$. Then there exists a constant $C$, independent of $h$, such that*

$$\|u - u_h\|_V \leq C \inf_{v_h \in V^h} \|u - v_h\|_V. \tag{10.17}$$

*Consequently, a sufficient condition for the Galerkin approximation $u_h$ to converge to the solution $u$ of problem (10.1) is that there exists a family of $\{V^h\}$ of subspaces with the property that*

$$\inf_{v_h \in V^h} \|u - v_h\|_V \to 0 \quad as \quad h \to 0. \tag{10.18}$$

PROOF. Since the bilinear form $a$ is $V$-elliptic by assumption, we have (denoting the norm on $V$ simply by $\|\cdot\|$, for convenience)

$$\begin{aligned} \alpha\|u - u_h\|^2 &\leq a(u - u_h, u - u_h) \\ &= a(u - u_h, u - v_h - u_h + v_h) \\ &= a(u - u_h, u - v_h) - a(e, u_h - v_h). \end{aligned}$$

The last term on the right-hand side is zero by (10.12) so, using the continuity of $a$,

$$\begin{aligned} \alpha\|u - u_h\|^2 &\leq a(u - u_h, u - v_h) \\ &\leq M\|u - u_h\| \, \|u - v_h\|. \end{aligned}$$

The inequality (10.18) now follows, with $C = M/\alpha$. □

Céa's Lemma effectively transforms the problem of estimating the error $u - u_h$ to one of estimating the distance of $u$ from the subspace $V^h$. That is, we can gain an idea of the quality of the approximation $u_h$ by estimating how far off $u$ is from $V^h$. Since $\inf_{v_h \in V^h} \|u - v_h\| \leq \|u - \tilde{v}_h\|$ for any particular $\tilde{v}_h \in V^h$, it also follows that one may obtain a suitable estimate by choosing $\tilde{v}$ in an appropriate and convenient way. It turns out that the most convenient choice is to make use of the *interpolate* of $u$; this is a function $\tilde{u}_h$ in $V^h$ whose value coincides with that of $u$ at $N$ points $\boldsymbol{x}_1, \boldsymbol{x}_2, \ldots, \boldsymbol{x}_N$ in $\Omega$. Since $\tilde{u}_h$ has the representation

$$\tilde{u}_h = \sum_{k=1}^{N} \tilde{c}_k \phi_k, \tag{10.19}$$

where $\phi_k$ is any basis of $V^h$, we can determine the coefficients $\tilde{c}_k$ from the fact that

$$u(\boldsymbol{x}_k) = \tilde{u}_h(\boldsymbol{x}_k), \qquad k = 1, \ldots, N, \tag{10.20}$$

for a given function $u$. That is, we solve for $\tilde{c}_j$ the $N$ simultaneous equations

$$\sum_{j=1}^{N} \tilde{c}_j \phi_j(\boldsymbol{x}_k) = u(\boldsymbol{x}_k), \qquad k = 1, \ldots, N.$$

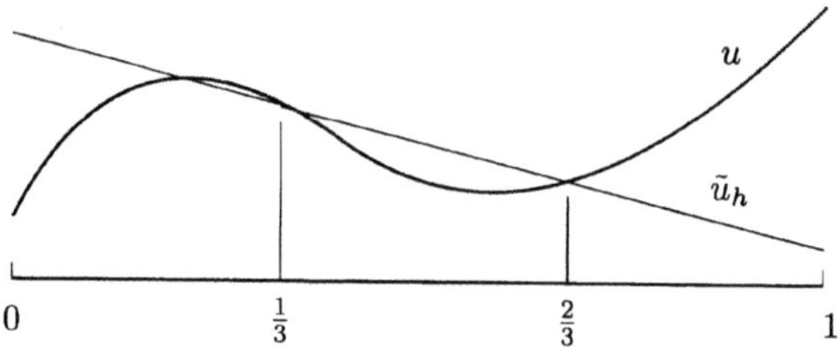


FIGURE 10.2. A function and its interpolate in the space spanned by (10.22)

We observe that the operator $P : V \to V$ defined by

$$Pu = \tilde{u}_h \tag{10.21}$$

is a *projection operator.*

**Example**

3. Choose $V = H^1(0,1)$ and $V^h = P_1(0,1)$, the space of polynomials of degree not greater than 1, which is spanned by the functions

$$\phi_1(x) = x, \qquad \phi_2(x) = 1 - x. \tag{10.22}$$

Suppose that we require the interpolate $\tilde{u}$ to be equal to $u$ at the points $x_1 = \frac{1}{3}$ and $x_2 = \frac{2}{3}$ (Figure 10.2); then (10.20) and (10.22) give

$$\begin{aligned} \tilde{c}_1 \cdot \tfrac{1}{3} + \tilde{c}_2 \cdot \tfrac{2}{3} &= u(\tfrac{1}{3}) \\ \tilde{c}_1 \cdot \tfrac{2}{3} + \tilde{c}_2 \cdot \tfrac{1}{3} &= u(\tfrac{2}{3}). \end{aligned}$$

Solving, we have

$$\tilde{c}_1 = 2u(\tfrac{2}{3}) - u(\tfrac{1}{3}), \qquad \tilde{c}_2 = 2u(\tfrac{1}{3}) - u(\tfrac{2}{3}).$$

With the choice (10.19) for $v_h$, (10.17) yields

$$\|u - u_h\|_V \le C \inf_{v \in V^h} \|u - v_h\|_V \le C\|u - \tilde{u}_h\|_V, \tag{10.23}$$

and so the problem of convergence reduces to one of finding out whether $\tilde{u}_h \to u$ as $h \to 0$, and if so at what rate this occurs. We have thus reduced the problem of *convergence of Galerkin approximations* to one of *convergence of interpolates.*

In the case of the finite element method, which is distinguished by the fact that the basis functions are piecewise polynomials, we show that the distance between $u$ and its interpolate $\tilde{u}_h$ satisfies an inequality of the form

$$\|u - \tilde{u}_h\|_V \le ch^\beta, \tag{10.24}$$

where the constant $c$ is independent of $h$, and $\beta$ is positive. Then (10.23) immediately implies that

$$\|u - u_h\|_V \le \frac{cM}{\alpha} h^\beta. \tag{10.25}$$

Hence, as the approximation is progressively improved so that $N$ gets larger and $h$ gets smaller, we can expect $u_h$ to converge to $u$ at a rate that is determined by the magnitude of $\beta$. This is expressed in the form

$$\|u - u_h\|_V = 0(h^\beta), \tag{10.26}$$

and we say that *the convergence of $u_h$ to $u$ is of order $\beta$*. Clearly the aim is to have $\beta$ as large as possible.

A result such as (10.24) is called an *interpolation error estimate*, for obvious reasons, whereas (10.25) or (10.26) is called a *Galerkin (asymptotic) error estimate*. Our means of examining the convergence of Galerkin approximations is always via estimates of the form (10.25).

We conclude this section by noting that the constant in the error estimate (10.25) can be improved in the event that $a$ is symmetric. Indeed, returning to (10.15), this inequality implies that

$$\|u - u_h\|_a \le \inf_{v_h \in V^h} \|u - v_h\|_a;$$

using the $V$-ellipticity and continuity of the bilinear form, we find that this reduces to

$$\|u - u_h\|_V \le \sqrt{\frac{M}{\alpha}} \inf_{v_h \in V^h} \|u - v_h\|_V,$$

which represents an improvement over (10.17) since $M \ge \alpha$.

## 10.3 Other methods of approximation

The Galerkin–Rayleigh–Ritz approach is very widely used, particularly in the context of finite element methods. But a number of other approximate methods also exist, and although these are perhaps not as ubiquitous as the Galerkin method, they are nevertheless worth knowing about as they make interesting, viable, and in some cases superior alternative approaches. We discuss the Petrov–Galerkin method, the method of weighted residuals, the method of least squares, collocation methods, and $H^{-1}$-methods.

We start by returning to the BVP

$$\begin{aligned} Au &= f \qquad \text{in } \Omega \\ B_0 u = \cdots &= B_{m-1} u = 0 \quad \text{on } \Gamma \end{aligned} \tag{10.27}$$

in which, as before, $A$ is an elliptic operator of order $2m$. For convenience attention is confined to problems with homogeneous boundary conditions. As in Chapter 9, we multiply both sides of (10.27) by a function $v$ and integrate to obtain

$$(Au, v) = (f, v) \tag{10.28}$$

in which $(\cdot, \cdot)$ represents the $L^2$-inner product. Now in Chapter 9 Green's theorem was used to shift half the derivatives in $Au$ over to $v$, and in so doing to arrive at the VBVP

$$a(u, v) = (f, v) \tag{10.29}$$

for all $v \in V$ ($V$ being a subspace of $H^m(\Omega)$). The other methods discussed here all rely on the observation that there are other ways besides (10.29) of formulating VBVPs. At one extreme we could consider (10.28) as it stands and seek

$$u \in U, \qquad (Au, v) = (f, v) \qquad \text{for all } v \in L^2(\Omega) \tag{10.30}$$

in which, for example,

$$U = \{u \in H^{2m} : \ B_0 u = \cdots = B_{m-1} u = 0\}. \tag{10.31}$$

In this case no derivatives of $u$ will have been shifted over to $v$.

A second alternative is to pose the VBVP in the form (10.29), but to seek the approximate solution $u_h$ in a space $U^h$ which is distinct from that in which the admissible functions $v_h$ are sought; this is known as the *Petrov–Galerkin method*, to distinguish it from the standard Galerkin method which has hitherto been the main focus of attention.

Continuing with the process of considering alternative formulations, at the other extreme we could shift *all* derivatives over to $v$ by repeated application of Green's theorem, and then pose the problem of finding $u \in L^2(\Omega)$ such that

$$(u, A^* v) = (f, v) \qquad \text{for all } v \in V, \tag{10.32}$$

where $A^*$ is the formal adjoint of $A$ and $V = H_0^{2m}(\Omega)$, for example.

All of these new approaches are characterized by the feature that the solution is sought in a particular space, and the equation is required to be satisfied for all functions $v$ belonging to a space that is generally distinct from the solution space. The Petrov–Galerkin method is based on (10.29), and we find that weighted residual, least squares and collocation methods are based on (10.30) whereas $H^{-1}$-methods are based on (10.32).

**Example**

4. Consider the problem

$$\begin{aligned} -u'' + u &= f \qquad \text{in}\, \Omega = (0,1), \\ u(0) &= u(1) = 0. \end{aligned} \tag{10.33}$$

Equation (10.30) reads, for this problem,

$$\int_0^1 (-u'' + u)v \; dx = \int_0^1 fv \; dx,$$

where $U = H^2(0,1) \cap H_0^1(0,1)$ and $V = L^2(0,1)$. Integrating by parts once we obtain

$$\int_0^1 (u'v' + uv) \; dx = \int_0^1 fv \; dx$$

with $U = V = H_0^1(0,1)$. Finally, by integrating by parts once more we may pose the problem of finding $u \in L^2(0,1)$ such that

$$\int_0^1 (-uv'' + uv) \; dx = \int_0^1 fv \; dx$$

for all $v \in H_0^2(0,1)$.

We now proceed to discuss the group of approximate methods based on the formulation (10.30). Given finite-dimensional subspaces $U^h \subset U$ and $V^h \subset V$, where $U$ is given by (10.31) and $V = L^2(\Omega)$, consider the problem of finding a function $u_h \in U^h$ such that

$$(Au_h - f, v_h) = 0 \quad \text{or} \quad (r(u_h), v_h) = 0 \tag{10.34}$$

for all $v_h \in V^h$. The expression $r(u_h) \equiv Au_h - f$ is called the *residual*; if $u_h$ is the exact solution, then of course the residual vanishes.

**The method of weighted residuals.** In this method $U^h$ and $V^h$ are chosen such that dim $U^h = \dim V^h = N$, say, with bases

$$\{\phi_k\}_{k=1}^N \quad \text{for } U^h \text{ and } \{\psi_k\}_{k=1}^N \quad \text{for } V^h. \tag{10.35}$$

Then

$$u_h = \sum_{k=1}^N c_k \phi_k \quad \text{and} \quad v_h = \sum_{k=1}^N b_k \psi_k, \tag{10.36}$$

where the coefficients $b_k$ are arbitrary in view of the arbitrariness of $v_h$. Substitution of (10.36) in (10.34) yields a set of $N$ simultaneous equations

$$\sum_{k=1}^N M_{kl} c_k = F_l, \qquad l = 1, \ldots, N,$$

in which

$$M_{kl} = (A\phi_k, \psi_l) \quad \text{and} \quad F_l = (f, \psi_l),$$

$(\cdot,\cdot)$ denoting the $L^2$-inner product.

**The method of least squares.** This method entails finding a function $u_h \in U^h$ that *minimizes* the residual, the magnitude of the residual being measured in the $L^2$-norm (hence the name of the method). That is, we define a functional

$$J : U^h \to \mathbb{R}, \qquad J(v_h) = \|r(v_h)\|_{L^2}^2 = \int_\Omega (Av_h - f)^2 \, dx$$

and we seek $u_h \in U^h$ such that

$$J(u_h) \le J(v_h) \quad \text{for all } v_h \in U^h.$$

From Section 9.4 this is equivalent to the problem of finding $u_h \in U^h$ such that

$$\langle DJ(u_h), v_h \rangle = 0$$

or

$$\int_\Omega (Au_h - f)Av_h \, dx = 0 \qquad \text{for all } v_h \in U^h. \tag{10.37}$$

Comparison with (10.34) indicates that the method of least squares is equivalent to the method of weighted residuals if in the latter we choose the space $V^h$ to be the span of the functions $\{A\phi_k\}_{k=1}^N$, where $\{\phi_k\}_{k=1}^N$ is the basis for $U^h$.

**Collocation methods.** The idea behind these methods is to force the residual to vanish at a finite number of points in the domain $\Omega$. That is, if $\dim U^h = N$, then we seek $u_h \in U^h$ such that

$$Au_h - f = 0 \quad \text{at} \quad \boldsymbol{x} = \boldsymbol{x}_k, \quad k = 1, 2, \ldots, N.$$

The collocation method is also a variant of the method of weighted residuals, as can be appreciated from the following considerations: corresponding to each $v_h$ in $V^h$ we can define a linear functional $\ell_h$ such that

$$\langle \ell_h, Au_h - f \rangle = (Au_h - f, v_h).$$

Now suppose for definiteness that $u_h$ and $f$ are smooth enough for the residual $r(u_h) = Au_h - f$ to belong to $H_0^1(\Omega)$; that is,

$$Au_h - f \in H \subset H_0^1(\Omega),$$

where $H$ is assumed to be dense in $H_0^1(\Omega)$. Then $\ell_h$ is a linear functional on $H$, and $\ell_h$ lies in the dual space $H'$ of $H$ and we must find $u_h \in U^h$ such that

$$\langle \ell_h, r(u_h)\rangle = 0 \quad \text{for all } \ell_h \in H'. \tag{10.38}$$

The collocation method arises from choosing the functionals $\ell_h$ to be the *N Dirac delta functionals* $\delta_{\boldsymbol{x}_j}, j = 1, \ldots, N$, defined by

$$\delta_{\boldsymbol{x}_j} : H \to \mathbb{R}, \qquad \langle \delta_{\boldsymbol{x}_j}, w\rangle = w(\boldsymbol{x}_j).$$

**The Petrov–Galerkin method.** This method takes as a point of departure the formulation (10.29); but whereas $U = V$, the subspaces $U^h$ and $V^h$ are distinct. Suppose that these two finite-dimensional subspaces have bases $\{\phi_i\}_{i=1}^N$ and $\{\psi_i\}_{i=1}^N$; then, following the sequence of manipulations that lead to (10.8), we once again arrive at the equation

$$\boldsymbol{K}^t\boldsymbol{c} = \boldsymbol{F},$$

but this time the matrix $\boldsymbol{K}$ and vector $\boldsymbol{F}$ are defined by

$$K_{ij} = a(\phi_i, \psi_j) \quad \text{and} \quad F_j = \langle \ell, \psi_j\rangle.$$

The methods of weighted residuals, least squares, and collocation all possess advantages that make them, in principle at least, viable alternatives to the Galerkin method. In particular, in all three cases a greater degree of smoothness is expected of the approximate solution: if $A$ is an operator of order $2m$, then both weighted residuals and least squares require that $U^h \subset H^{2m}(\Omega)$, whereas in the case of the collocation method, the assumption that $Au_h - f \in H_0^1(\Omega)$ will require that $u_h \in H^3(\Omega)$.

The rationale behind opting for the the Petrov–Galerkin procedure rather than the standard Galerkin method is perhaps less clear, since the advantage of greater smoothness is not present. However, there are various situations in which the Petrov–Galerkin method provides approximations of far superior quality. This is particularly true of *convection-diffusion problems*, of the form

$$a\frac{\partial u}{\partial x} - k\frac{\partial^2 u}{\partial x^2} = f,$$

in which the standard heat or diffusion equation is supplemented by a term involving the first derivative, and which accounts for convective transport. When $a$ is much bigger than $k$, so that convective effects dominate, the Galerkin method gives solutions that are oscillatory, and bear little relation to the exact solution. The Petrov–Galerkin method, on the other hand, is one way of overcoming this deficiency, by a judicious choice of $U^h$ and $V^h$, the latter often being chosen to be of the form $V^h = U^h \oplus W^h$ in which $W^h$ is a space of functions that, when added to $U^h$, serve to overcome the drawbacks of the standard approach.

**$H^{-1}$-methods.** Unlike the methods discussed previously, the $H^{-1}$-method is based on the formulation (10.32) in which all derivatives are transferred by means of successive applications of Green's theorem. Working in finite-dimensional subspaces, we now seek $u_h \in U^h$ such that

$$(u_h, A^* v_h) = (f, v_h) \quad \text{for all } v_h \in V^h, \tag{10.39}$$

where $U^h \subset L^2(\Omega)$ and $V^h \subset H_0^{2m}(\Omega)$. When using this approach we are free to choose $U^h$ in such a way that some or all of its members are discontinuous, while at the same time the functions making up $V^h$ have to belong to $H_0^{2m}(\Omega)$.

Returning to the earlier example (10.33), the $H^{-1}$-method entails finding $u_h \in U^h \subset L^2(0,1)$ such that

$$\int_0^1 (-u_h v_h'' + u_h v_h)\, dx = \int_0^1 f v_h\, dx$$

for all $v_h \in V^h \subset H_0^2(0,1)$. So, for example, we could choose as a basis for the $U^h$ a set of piecewise-constant functions whereas for $V^h$ we need a basis of functions that are at least continuously differentiable and which, together with their first derivatives, vanish on the boundary.

In Exercise 10.10 the methods introduced here are applied to a simple problem.

## 10.4 Bibliographical remarks

There is a large body of literature on the Galerkin and Rayleigh–Ritz methods, with texts ranging from those that deal mainly with computational aspects to those that also consider questions of convergence. The texts by Dhatt and Touzot [14] and Hughes [22] are good examples of the former, and Johnson [23] and Strang and Fix [51] cover both aspects. The books by Becker, Carey and Oden [5] and Oden and Carey [37] all give full coverage of the Galerkin method, although they emphasize the use of the method as part of the finite element method. Further details on the alternative methods discussed in Section 10.3 may be found in Carey and Oden [10] and in Rektorys [41]. The Petrov–Galerkin method has been the subject of extensive research in the context of the finite element method, and is discussed in the text by Johnson [23]. An interesting application of the method to the finite element analysis of eastic beams is presented in the paper by Loula, Hughes, and Franca [32].

## 10.5 Exercises

### The Galerkin method

10.1. The BVP $(xu')' = x$ in $\Omega = (1,2)$, $u(1) = u(2) = 0$, has the exact solution $u(x) = \frac{1}{4}x^2 - (3\ln x/4\ln 2) - \frac{1}{4}$. Use the Galerkin method to find an approximate solution $u_h$ in the subspace of $H_0^1(1,2)$ spanned by $\phi_1(x) = (x-1)(x-2)$ and $\phi_2(x) = x(x-1)(x-2)$. Compare the exact and approximate solutions by (a) sketching graphs of $u$ and $u_h$; (b) evaluating the errors $\|e\|_{L^\infty}$, $\|e\|_{L^2}$, and $\|e\|_{H^1}$, where $e = u - u_h$.

10.2. Use the Galerkin method with basis functions $\phi_1(x,y) = (-x^2 + x)(-y^2+y)$ and $\phi_2(x,y) = (x^3 - \frac{3}{2}x^2 + \frac{1}{2}x)(y^3 - \frac{3}{2}y^2 + \frac{1}{2}y)$ to solve the BVP

$$\begin{aligned} -\nabla^2 u &= xy \quad \text{on } \Omega = (0,1)\times(0,1), \\ u &= 0 \quad \text{on } \Gamma. \end{aligned}$$

10.3. Let $J : H \to \mathbb{R}$ be a functional on a Hilbert space $H$ defined by

$$J(v) = \tfrac{1}{2}a(v,v) - \langle \ell, v\rangle, \qquad v \in H,$$

where $a(\cdot\,,\cdot)$ is continuous and $H$-elliptic and $\ell(\cdot)$ is continuous, and let $\{u_n\}$ be a sequence in $H$ such that

$$\lim_{n\to\infty} J(u_n) = J(u) \le J(v).$$

Then $u_n$ is called a *minimizing sequence.* The aim of this exercise is to show how such a minimizing sequence can be generated, and also to show that $u_n \to u$ in the norm $\|\cdot\|_a$.

Let $\{\phi_k\}_{k=1}^\infty$ be an orthonormal basis for $H$ with respect to the inner product $(\cdot\,,\cdot)_a = a(\cdot\,,\cdot)$, and let $H^{(n)} = \operatorname{span}\{\phi_1,\ldots,\phi_n\}$. Let $u_n$ be the minimizer of $J$ in the space $H^{(n)}$. Show that $u_n = \sum_{k=1}^n \langle \ell, \phi_k\rangle\phi_k$ and that $\langle \ell,\phi_k\rangle = (u,\phi_k)_a$, where $u$ is the minimizer of $J$ in $H$. Hence deduce that $u_n$ is the $n$th partial sum of the Fourier series expansion for $u$, and conclude that $\|u_n - u\|_a \to 0$ and $\|u_n - u\|_H \to 0$.

10.4. Use the Rayleigh–Ritz method with basis function $\phi_1(x,y) = (x^2 - \alpha^2)(y^2-\beta^2)$ to find an approximate solution to the problem of minimizing the functional

$$J:\ H_0^2(\Omega) \to \mathbb{R}, \quad J(v) = \frac{D}{2}\int_{-\alpha}^{\alpha}\int_{-\beta}^{\beta}\left[(\nabla^2 v)^2 - \frac{2qv}{D}\right]\,dxdy$$

corresponding to the problem of deflection of an elastic plate occupying the domain $\Omega = (-\alpha,\alpha)\times(-\beta,\beta)$. The corresponding classical problem is

$$\begin{aligned} \nabla^4 u &= q \ \text{ in } \Omega, \\ u = \partial u/\partial\nu &= 0 \ \text{ on } \Gamma, \end{aligned}$$

and the exact solution satisfies $u(0,0) = 0.0202q\alpha^4/D$ at the origin, if $\alpha = \beta$. Compare this with the approximate solution.

## Properties of Galerkin approximations

10.5. Given the VBVP

$$a(u,v) = \langle \ell, v \rangle, \qquad v \in V$$

in which $a$ is symmetric and $V$-elliptic, show that the Galerkin approximation $u_h$ satisfies

$$\|u - u_h\|_a^2 = \|u\|_a^2 - \|u_h\|_a^2;$$

that is, the error in the energy norm equals the error in the energy, and therefore $\|u_h\|_a \le \|u\|_a$.

10.6. If $u$ minimizes the functional $J : V \to \mathbb{R}$ given by

$$J(v) = \tfrac{1}{2}a(v,v) - \langle \ell, v \rangle,$$

show that $J(u) = -\frac{1}{2}a(u,u)$.

10.7. Verify that the operator $P$ defined by (10.21) is a projection.

10.8. Let $V^h$ be the subspace of $H^1(-1,1)$ spanned by the three functions

$$\phi_1(x) = \tfrac{1}{2}x(x-1), \qquad \phi_2(x) = 1 - x^2, \qquad \phi_3(x) = \tfrac{1}{2}x(x+1),$$

and define the interpolate $\tilde{u}_h$ of $u \in H^1(-1,1)$ by

$$P_h u = \tilde{u}_h = \sum_{k=1}^{3} a_k \phi_k(x),$$

where $a_k = u(x_k)$ and $x_1 = -1, x_2 = 0, x_3 = +1$. Find $\tilde{u}_h(x)$ if $u(x) = \sin(\pi/2)(x - \frac{1}{2})$, and evaluate $\|u - u_h\|_{H^1}$.

## Other methods of approximation

10.9. Use the Green's formula

$$(Au, v)_{L^2} = (u, A^* v) + G(u,v)$$

(see Section 8.4) to show that the least-squares problem (10.36) is equivalent to the higher-order problem

$$A^* A u_h = A^* f \text{ in } \Omega,$$

plus appropriate boundary conditions. If $A = -\nabla^2$ (the Laplacian), show that this is equivalent to the problem

$$\nabla^4 u_h = \nabla^2 f,$$

where $\nabla^4$ is the biharmonic operator.

10.10. Consider the problem

$$\begin{aligned} -u'' + u &= \sin x \qquad \text{in } \Omega = (0,1) \\ u(0) &= u(1) = 0. \end{aligned}$$

(a) Reformulate this problem in a manner suitable for solution using the method of weighted residuals. Then find an approximate solution in a two-dimensional subspace spanned by polynomials of appropriate order (that is, make suitable choices for $U^h$ and $V^h$ in (10.35)).

(b) Find approximate solutions using the method of least squares and the method of collocation.

(c) Apply integration by parts successively to transform the problem into one suitable for solution using $H^{-1}$ methods. Again choose suitable subspaces of $U^h$ and $V^h$, and find an approximate solution.

# Part III

# The Finite Element Method

# 11
# The finite element method

In practical situations the determination of suitable basis functions for use in the Galerkin method can be extremely difficult, especially in cases for which the domain $\Omega$ does not have a simple shape. The finite element method overcomes this difficulty by providing a systematic means for generating basis functions on domains of fairly arbitrary shape. What makes the method especially attractive is the fact that these basis functions are piecewise polynomials that are nonzero only on a relatively small part of $\Omega$, so that computations may be carried out in a modular fashion, which is well suited to computer-based approaches. As we show a little later, the family of spaces $V^h$ $(h \in (0,1))$ defined by the finite element procedure possesses the property that $V^h$ approaches $V$ as $h$ approaches zero, in an appropriate sense. This is, of course, an indispensable property for convergence of the Galerkin method.

In Section 11.1 we outline the steps that lead to the construction of finite element bases for *second-order BVPs* defined on *domains in* $\mathbb{R}$ *and* $\mathbb{R}^2$. This is the simplest and most commonly encountered class of problems, and most of the features of finite element approximations can be demonstrated in this context. The generalization to higher-order BVPs and to problems in $\mathbb{R}^n$ for $n \geq 3$ follows in a natural way, and is covered in detail in all texts dealing with the finite element method.

Then in Sections 11.2 and 11.3 we describe, for problems in $\mathbb{R}$ and $\mathbb{R}^2$, respectively, some commonly used basis functions. We also show by means of examples how these are used to solve actual boundary value problems.

Section 11.4 is devoted to the construction of bases that are relevant to fourth-order problems; for this case it is necessary to impose additional continuity requirements, which in turn lead to bases that are quite different from those used in second-order problems.

The elements discussed in Section 11.2 to 11.4 are all polygonal (in two dimensions). Domains with curved boundaries can, however, be accommodated by resorting to the use of isoparametric elements; exactly how this is done is the subject of Section 11.5.

The final section of this chapter deals with an issue that is central to all finite element computations, viz. numerical integration. For general problems, involving inhomogeneous media, say, or for domains that are approximated using isoparametric elements, the integrals that arise cannot be evaluated exactly. For this reason it makes sense to resort to methods that allow the integrals to be evaluated approximately, although with a degree of accuracy that can be estimated and therefore improved upon if desired.

## 11.1 The finite element method for second-order problems

Suppose that we have a VBVP of the form: find $u \in V$ such that

$$a(u, v) = \langle \ell, v \rangle \quad \text{for all } v \in V. \tag{11.1}$$

For a second-order BVP the space of admissible functions $V$ consists of all those functions in $H^1(\Omega)$ that satisfy the essential boundary conditions.

A Galerkin approximation $u_h$ to the solution of (11.1) may be sought by constructing a finite-dimensional subspace $V^h$ of $V$, which is spanned by a finite number of basis functions $N_i$, and we then pose the problem of finding $u_h \in V^h$ that satisfies

$$a(u_h, v_h) = \langle \ell, v_h \rangle \quad \text{for all } v_h \in V^h. \tag{11.2}$$

If $\{N_i\}_{i=1}^G$ is a basis for $V^h$, then we have seen in Chapter 10 that expansion of $u_h$ and $v_h$ in terms of this basis and substitution in (11.2) leads to the set of simultaneous linear equations

$$\sum_i K_{ij} c_i = F_j \tag{11.3}$$

or, in matrix form,

$$\boldsymbol{K}^t \boldsymbol{c} = \boldsymbol{F},$$

where, as before,

$$K_{ij} = a(N_i, N_j) \quad \text{and} \quad F_j = \langle \ell, N_j \rangle. \tag{11.4}$$

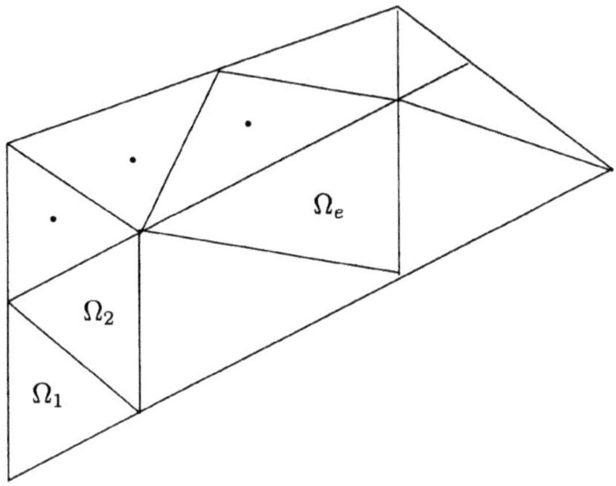


FIGURE 11.1. A polygonal domain in $\mathbb{R}^2$ and its subdivison into finite elements

The aim of this section is to describe a method for constructing those special bases $\{N_i\}_{i=1}^G$ that are associated with the finite element method.

**The finite element mesh.** We start by partitioning the domain $\Omega$ into a finite number $E$ of subdomains $\Omega_1, \Omega_2, \ldots, \Omega_E$, called *finite elements.* These elements are nonoverlapping and cover $\Omega$, in the sense that

$$\Omega_e \cap \Omega_f = \varnothing \quad \text{for} \quad e \neq f, \qquad \bigcup_{e=1}^{E} \overline{\Omega}_e = \overline{\Omega}.$$

To avoid complicating matters unnecessarily, we assume that the domain $\Omega$ is *polygonal* if it is a subset of $\mathbb{R}^2$. That is, the boundary $\Gamma$ of $\Omega$ is made up of straight segments. Under these conditions, it is easy to see that the entire domain can be covered exactly by polygonal elements (Figure 11.1). One more condition is imposed on the subdivision of $\Omega$: it is required that every side of the boundary of an element in $\mathbb{R}^2$ be *either* part of the boundary $\Gamma$, *or* a side of another element. This condition rules out a situation such as that shown in Figure 11.2, in which $AB$ is a side of $\Omega_2$ but not of $\Omega_1$.

**Nodal points.** We next identify certain points called *nodes* or *nodal points* in the subdivided domain; these points play a key role in the finite element method, as will soon become evident. Nodes are allocated at least at the vertices of elements, as shown in Figure 11.3(a), but in order to improve the approximation, further nodes may be introduced, for example, at the midpoints of the sides of elements as shown in Figure 11.3(b). In any case there is a total of $G$ nodes, say, which are numbered $1, 2, \ldots, G$ and which have position vectors $\boldsymbol{x}_1, \boldsymbol{x}_2, \ldots, \boldsymbol{x}_G$. The set of elements and nodes that

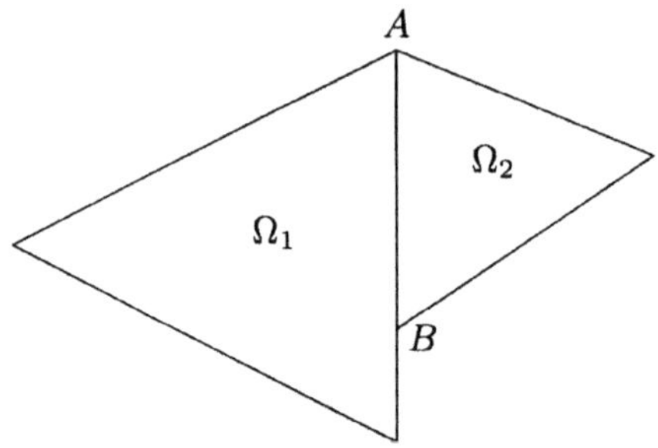


FIGURE 11.2. An inadmissible subdivison

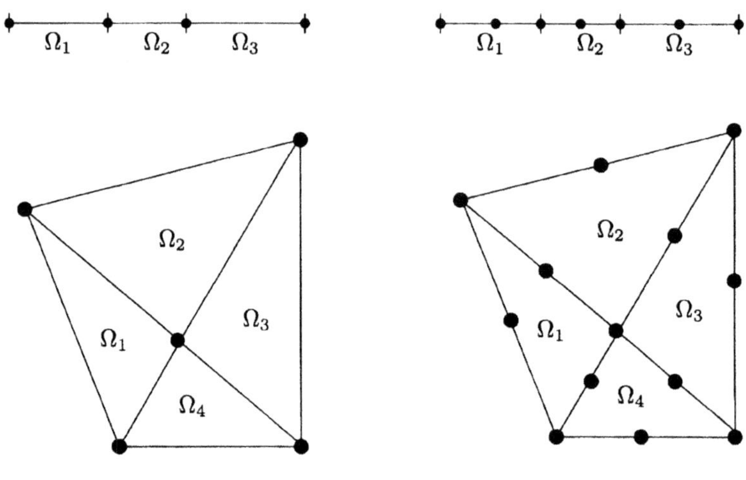


(a) (b)

FIGURE 11.3. Finite element meshes comprising elements and nodal points

make up the domain $\Omega$ is called a *finite element mesh.*

**Basis functions $N_i$.** We are now ready to describe how the finite element basis functions are formed. In carrying out this procedure it must be borne in mind that the basis functions define a subspace of $V$, so that they must be functions in $H^1(\Omega)$ (for second-order problems) that satisfy the essential boundary conditions. The question of boundary conditions is left aside for now, and we proceed to construct a set of basis functions with the following properties.

(i) The functions $N_i$ are *bounded and continuous*, that is,

$$N_i \in C(\overline{\Omega}); \tag{11.5}$$

(ii) there is a total of $G$ basis functions, that is, one for each node, and each function $N_i$ is nonzero only on those elements that are connected to node $i$:

$$N_i(\boldsymbol{x})|_{\Omega_e} \equiv 0 \quad \text{if } \boldsymbol{\xi} \notin \bar{\Omega}_e; \tag{11.6}$$

(iii) $N_i$ is equal to 1 at node $i$, and equal to zero at the other nodes:

$$N_i(\boldsymbol{x}_j) = \begin{cases} 1 & \text{if } i = j, \\ 0 & \text{otherwise}; \end{cases} \tag{11.7}$$

(iv) the *restriction* $N_i^{(e)}$ of $N_i$ to $\Omega_e$ is a polynomial:

$$N_i|_{\Omega_e} \equiv N_i^{(e)}, \quad N_i^{(e)} \in P_k(\Omega_e) \ \text{ for some } k \geq 1, \tag{11.8}$$

where $P_k(\Omega_e)$ is the space of polynomials of degree at most $k$ on $\Omega_e$.

From (iii) and (iv) it is clear that the function $N_i^{(e)}$ defined on element $\Omega_e$ will have the property that

$$N_i^{(e)}(\boldsymbol{x}_j) = \begin{cases} 1 & \text{if } i = j, \\ 0 & \text{otherwise}, \end{cases} \tag{11.9}$$

$i$ and $j$ running over all nodes in $\Omega_e$. We call $N_i^{(e)}$ a *local basis function.* These ideas are illustrated in Figure 11.4. It is not difficult to show that Conditions (i) and (iv) ensure that the functions $N_i$ belong to $H^1(\Omega)$, as required. We are thus going to set up basis functions that are *piecewise polynomials*, and that have *small supports*, in that they are nonzero only in a "small" region. It should be clear that we may regard a typical basis function $N_i$ as having been built up by patching together the local basis functions $N_i^{(e)}$ associated with node $i$, as shown in Figure 11.5. To distinguish the basis functions $N_i$ from the local basis functions $N_i^{(e)}$, we refer to the former as *global basis functions.*

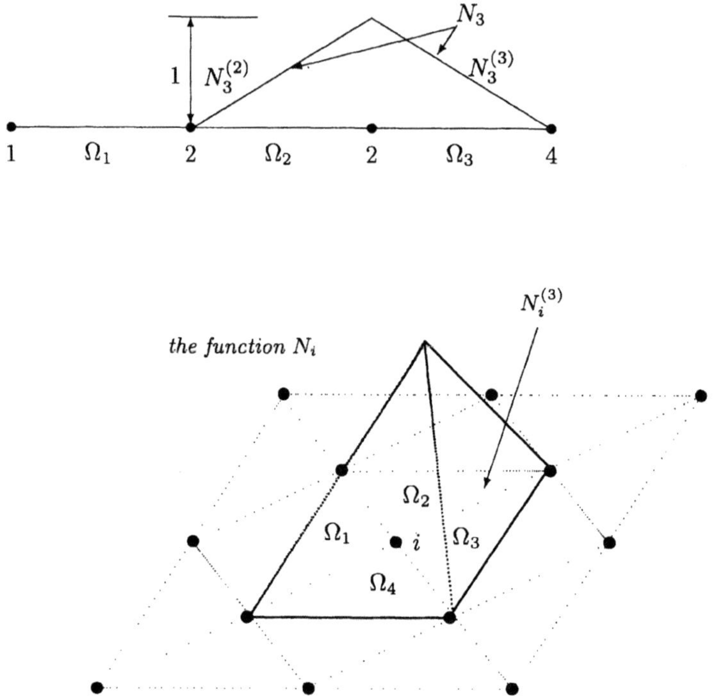


FIGURE 11.4. Local and global basis functions

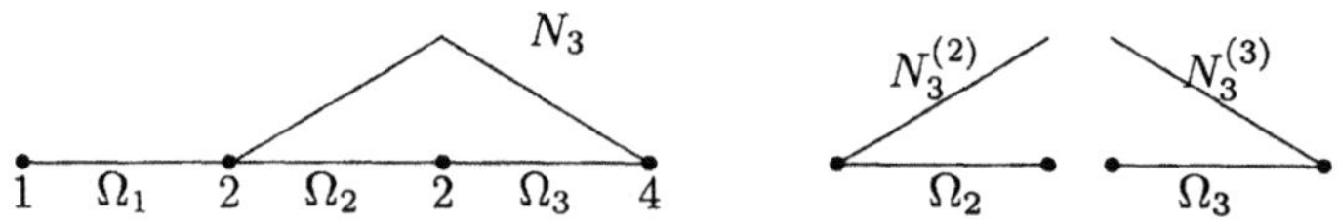


FIGURE 11.5. A basis function $N_i$ formed by patching together local basis functions $N_i^{(e)}$

Specific examples are given in the following section, but in the meantime we observe that if we write

$$v_h(\boldsymbol{x}) = \sum_{i=1}^{G} b_i N_i(\boldsymbol{x}), \tag{11.10}$$

then as a consequence of (11.7),

$$v_h(\boldsymbol{x}_j) = \sum_{i=1}^{G} b_i N_i(\boldsymbol{x}_j) = b_j; \tag{11.11}$$

that is, the coefficient $b_j$ is simply *the value of $v_h$ at node $j$.*

We denote by $X^h$ the space spanned by the basis functions $\{N_i\}_{i=1}^G$. Note that nothing has as yet been done about the essential boundary conditions, which are required to form part of the definition of $V^h$. This is now addressed by simply setting

$$\begin{aligned} V^h &= \{v_h \in X^h : \ v_h \text{ satisfies essential BCs}\} \\ &= \text{span}\,\{N_i : \ N_i \text{ satisfies essential BCs}\}. \end{aligned} \tag{11.12}$$

In the same way we denote by $X_e$ the space spanned by the functions $N_i^{(e)}$. That is,

$$X_e = \text{span}\,\{N_i^{(e)}\} = \{v_h|_{\Omega_e},\ v_h \in X^h\},$$

that is, $X_e$ is the space consisting of the restriction of all functions in $X^h$ to $\Omega_e$. In view of Condition (iv) we see that $X_e$ consists of all polynomials up to a given degree.

**The approximate solution.** We go straight to (11.3) and (11.4) and note that the bilinear form $a(N_i, N_j)$ will be an integral over $\Omega$, and that this integral can in turn be written as a sum of integrals over $\Omega_e$. Denoting the integrand by $\mathcal{F}(N_i, N_j)$, we thus have

$$\begin{aligned} a(N_i, N_j) &= \int_\Omega \mathcal{F}(N_i, N_j)\, dx \\ &= \sum_{e=1}^{E} \int_{\Omega_e} \mathcal{F}(N_i, N_j)\, dx \\ &= \sum_{e=1}^{E} \int_{\Omega_e} \mathcal{F}(N_i^{(e)}, N_j^{(e)})\, dx \\ &\equiv \sum_{e=1}^{E} a^{(e)}(N_i^{(e)}, N_j^{(e)}), \end{aligned}$$

in which (11.8) has been introduced in the penultimate line. Likewise, if the integrand appearing in $\langle \ell, N_j \rangle$ is denoted by $\mathcal{G}(N_j)$, then

$$\begin{aligned}
\langle \ell, N_j \rangle &= \int_\Omega \mathcal{G}(N_j)\, dx \\
&= \sum_{e=1}^{E} \int_{\Omega_e} \mathcal{G}(N_j)\, dx \\
&= \sum_{e=1}^{E} \int_{\Omega_e} \mathcal{G}(N_j^{(e)})\, dx \\
&\equiv \sum_{e=1}^{E} \langle \ell^{(e)}, N_j^{(e)} \rangle.
\end{aligned}$$

In other words, the matrix $\boldsymbol{K}$ and vector $\boldsymbol{F}$ in (11.4) can now be expressed in the form

$$\boldsymbol{K} = \sum_{e=1}^{E} \boldsymbol{K}^{(e)} \quad \text{and} \quad \boldsymbol{F} = \sum_{e=1}^{E} \boldsymbol{F}^{(e)}, \tag{11.13}$$

in which

$$K_{ij}^{(e)} = a^{(e)}(N_i^{(e)}, N_j^{(e)}) \quad \text{and} \quad F_j^{(e)} = \langle \ell^{(e)}, N_j^{(e)} \rangle. \tag{11.14}$$

So the actual evaluation of $K_{ij}$ and $F_j$ reduces to the evaluation of a number of matrices $K_{ij}^{(e)}$ and vectors $F_j^{(e)}$ for each element, and then summing these contributions over all elements. The Condition (ii) (equation (11.6)) results in an additional simplifying feature: since $N_i = 0$ for all elements that do not have node $i$ as a node, clearly $K_{ij}^{(e)} = 0$ *if nodes $i$ and $j$ do not belong to* $\Omega_e$. If follows that a judicious numbering of nodes will result in the matrix $\boldsymbol{K}$ having a *banded* structure in which all nonzero entries are clustered around the main diagonal. From a computational viewpoint this represents a distinct advantage.

The matrix $\boldsymbol{K}$ is known as the *stiffness matrix* and $\boldsymbol{F}$ is known as the *load vector*; this terminology derives from the early development of the finite element method in the context of structural mechanics, and tends to be used whatever the actual physical context. Likewise, the matrix $\boldsymbol{K}^{(e)}$ and vector $\boldsymbol{F}^{(e)}$ are referred to, respectively, as the *element stiffness matrix* and *element load vector.*

In the next section we give details of a systematic procedure for setting up local basis functions $N_i^{(e)}$, and hence also the basis function $N_i$, for the case of one-dimensional problems.

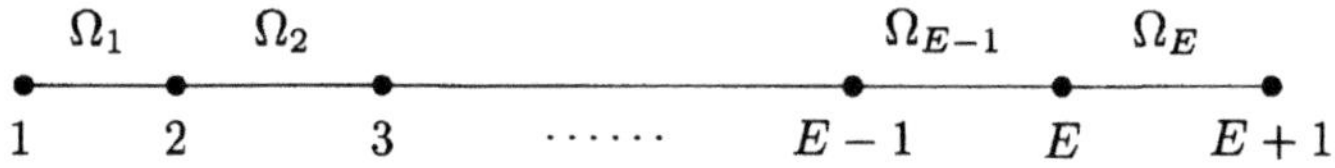


FIGURE 11.6. The domain $(a, b)$ and its subdivision into elements

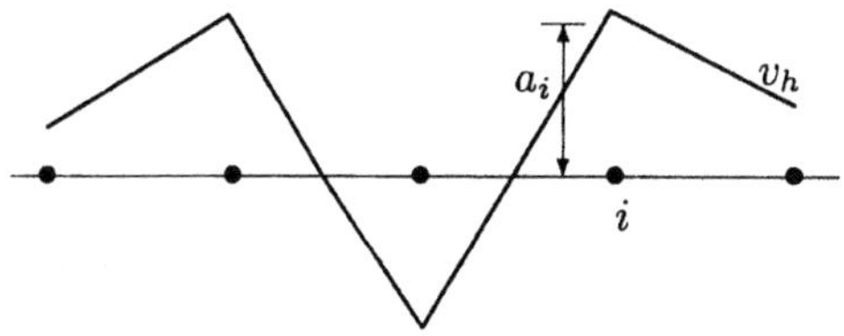


FIGURE 11.7. A typical member of $X^h$

## 11.2 One-dimensional problems

Consider a second-order problem defined on a subset $\Omega = (a, b)$ of the real line. The domain is divided into elements $\Omega_1, \Omega_2, \ldots, \Omega_E$, each element $\Omega_e$ being a segment of length $h_e$, say. Suppose now that we would like $X^h$ to be the space of piecewise polynomials of degree one, that is, piecewise straight lines. Then it follows that $X_e$ is the same as $P_1(\Omega_e)$. Furthermore, since every straight line is of the form $f(x) = a + bx$, a knowledge of the value of $f$ at two points in $\Omega_e$ suffices to determine $f$ uniquely on $\Omega_e$. With this in mind we define nodal points at the ends of all elements, so that each element has *two* nodal points. As shown in Figure 11.6, this simple arrangement allows the nodes to be numbered sequentially in such a way that $\Omega_e$ will be connected to nodes $e$ and $e + 1$.

The space $X_e$ has dimension two, so the next step is to define two basis functions $N_i^{(e)}$ and $N_{i+1}^{(e)}$ that satisfy (11.9). The only functions that fit all requirements are

$$N_i^{(e)}(x) = \frac{x_{i+1} - x}{h_e}, \quad N_i^{(e)}(x) = \frac{x - x_i}{h_e}. \tag{11.15}$$

By patching together the functions in the manner outlined in the previous section we see that each basis function $N_i(x)$ is a piecewise linear "hat" function made up of the local basis functions associated with node $i$. Hence every function $v_h$ in $X^h$ is a *piecewise linear function* of the form

$$v_h(x) = \sum_{i=1}^{G} d_i N_i(x)$$

in which $d_i$ is the value of $v_h$ at node $i$ (Figure 11.7).

**The reference element.** Instead of defining local basis functions for each

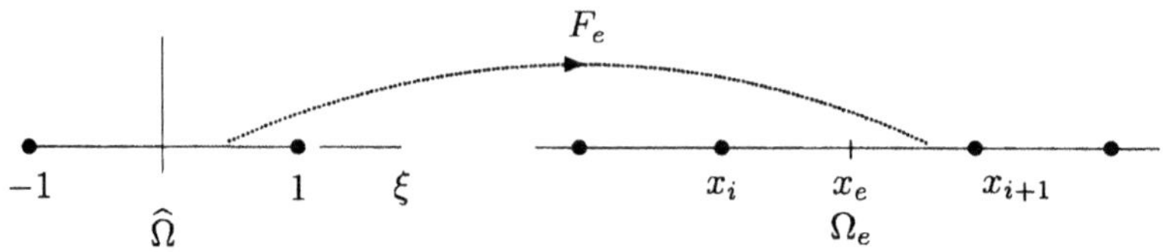


FIGURE 11.8. Reference element $\widehat{\Omega}$ and the image $\Omega_e$ under the affine map $F_e$

element as in (11.15), matters can be simplified considerably by setting up a *reference element* $\widehat{\Omega}$, say, which is isolated from the actual finite element mesh and which is referred to its own coordinate system $\xi$. The reference element extends from $\xi = -1$ to $\xi = +1$ and has the same system of nodal points as the elements $\Omega_e$ in the actual mesh (two nodes in this case). This situation is shown in Figure 11.8. Each element $\Omega_e$ can now be thought of as having been generated by an invertible map $F_e$ from $\widehat{\Omega}$ to $\Omega_e$, of the form

$$x = F_e(\xi) = \frac{h_e}{2}\xi + x_e \quad \text{or} \quad \xi = F_e^{-1}(x) = \frac{2}{h_e}(x - x_e), \tag{11.16}$$

where $x_e$ is the coordinate of the *center* of $\Omega_e$ ($x_e = \frac{1}{2}(x_i + x_{i+1})$ in Figure 11.8) and $h_e$ is the length of $\Omega_e$. Thus as $\xi$ goes from $-1$ to $+1$, $x$ goes from $x_i$ to $x_{i+1}$. In particular, nodes 1 and 2 on the reference element are mapped, respectively, to nodes $i$ and $i+1$ connected to element $\Omega_e$.

The map $F_e$ is an example of an *affine map*; a map or function $f$ defined on the real line is said to be affine if it is of the form $f(x) = a + bx$, where $a$ and $b$ are constants. Such a map transforms an interval most generally by stretching and translating it, the constant term accounting for the translation (Figure 11.8).

One advantage of introducing a reference element in this way is that we can define, once and for all, local basis functions $\widehat{N}_1$ and $\widehat{N}_2$ on $\widehat{\Omega}$ that have the requisite properties, and having done this simply use (11.16) to map $\widehat{N}_1$ and $\widehat{N}_2$ to $N_i^{(e)}$ and $N_{i+1}^{(e)}$, respectively, for each element, by defining $N_i^{(e)}$ and $N_{i+1}^{(e)}$ to be functions on $\Omega_e$ satisfying

$$N_i^{(e)} \circ F_e = \widehat{N}_1 \quad \text{and} \quad N_{i+1}^{(e)} \circ F_e = \widehat{N}_2, \tag{11.17}$$

or

$$N_i^{(e)}(x) = \widehat{N}_1(\xi), \quad N_{i+1}^{(e)}(x) = \widehat{N}_2(\xi), \tag{11.18}$$

in which $x$ and $\xi$ are related through (11.16) (Figure 11.9).

For a piecewise linear basis we thus define

$$\begin{aligned} \widehat{N}_1(\xi) &= \tfrac{1}{2}(1 - \xi), \\ \widehat{N}_2(\xi) &= \tfrac{1}{2}(1 + \xi), \end{aligned} \tag{11.19}$$

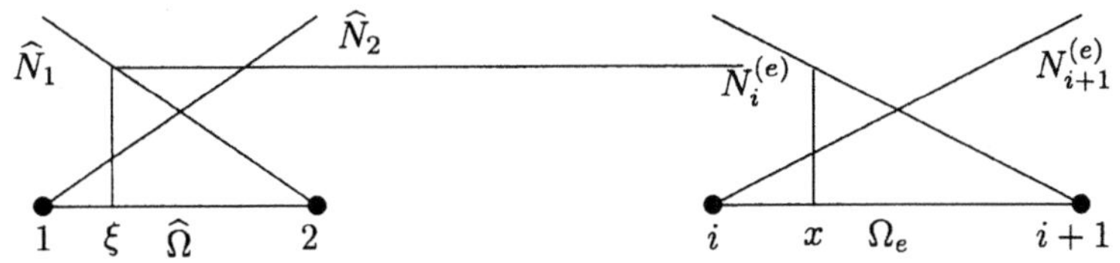


FIGURE 11.9. Local basis functions defined on the reference element $\widehat{\Omega}$ and on the element $\Omega_e$

and from (11.18) and (11.16) we then recover (11.15) since, for example,

$$N_i^{(e)}(x) = \widehat{N}_1(\xi) = \tfrac{1}{2}\left(1 - \frac{2}{h_e}(x - x_e)\right) = \frac{1}{h_e}(x_{i+1} - x),$$

as is readily verified. In future local basis functions are always defined on a reference element, with the assumption that the actual local basis functions can be recovered by means of a relation such as (11.17). Later we show that the process of evaluating the matrices and vectors defined in (11.13) and (11.14) is also rendered more straightforward by carrying out such integrations on the reference element.

The procedure is readily extended to higher-order approximations. For example, suppose that we wish to construct a space of *piecewise quadratic* functions, with the restrictions to each element being a quadratic function (that is, a member of $P_2(\Omega_e)$). Every function $f \in X_e$ is thus of the form $f(x) = a + bx + cx^2$ and so specification of $f$ at *three* points in $\Omega_e$ determines the function uniquely in $\Omega_e$. Hence we place nodes at the ends as well as at the *midpoints* of elements, so that an arbitrary element will have associated with it nodes $i$, $i+1$, and $i+2$.

Next, a reference element $\widehat{\Omega}$ is set up with nodes at the ends and at the center, and we define quadratic basis functions

$$\begin{aligned} \widehat{N}_1(\xi) &= \tfrac{1}{2}\xi(\xi - 1), \\ \widehat{N}_2(\xi) &= 1 - \xi^2, \\ \widehat{N}_3(\xi) &= \tfrac{1}{2}\xi(\xi + 1), \end{aligned} \tag{11.20}$$

as shown in Figure 11.10.

Naturally the functions in (11.20) satisfy (11.9), as do the local basis functions $N_i^{(e)}$, $N_{i+1}^{(e)}$, $N_{i+2}^{(e)}$ which are generated using

$$N_i^{(e)}(x) = \widehat{N}_1(\xi), \quad N_{i+1}^{(e)}(x) = \widehat{N}_2(\xi), \quad N_{i+2}^{(e)}(x) = \widehat{N}_3(\xi).$$

A few typical piecewise quadratic basis functions $N_i$ that result from patching together the quadratic local basis functions are also shown in Figure 11.10.

Polynomial bases of the form (11.19) and (11.20) are often referred to as *Lagrange* bases or families, because of their close association with Lagrange

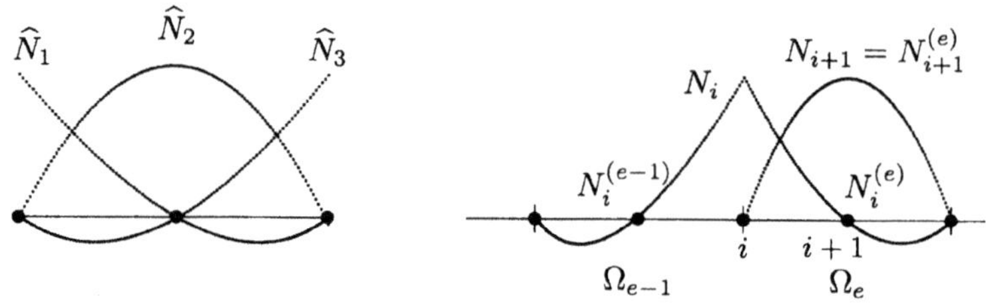


FIGURE 11.10. Quadratic local basis functions and piecewise quadratic global functions

interpolation. Generally, if the interval $(-1,1)$ has a total of $K+1$ equally spaced nodes, then the Lagrange polynomial corresponding to node $I$ is given by

$$\widehat{N}_I(\xi) = \frac{(\xi-\xi_1)\cdots(\xi-\xi_{I-1})(\xi-\xi_{I+1})\cdots(\xi-\xi_{K+1})}{(\xi_I-\xi_1)\cdots(\xi_I-\xi_{I-1})(\xi_I-\xi_{I+1})\cdots(\xi_I-\xi_{K+1})};$$

that the requirement (11.9) is indeed met can be deduced by inspection. This general formula allows piecewise cubic and higher-order approximations to be generated in a systematic manner.

We now show how an approximate solution to a two-point BVP may be found, using piecewise linear basis functions.

**Example**

1. Consider the BVP

$$\begin{aligned} &-u'' + u = \sin \pi x, \quad x \in \Omega = (0,1), \\ &u(0) = u(1) = 0. \end{aligned} \tag{11.21}$$

The corresponding VBVP is: find $u \in V = H_0^1(0,1)$ such that

$$\int_0^1 (u'v' + uv)\, dx = \int_0^1 v \sin \pi dx \quad \text{for all } v \in H_0^1(0,1),$$

and the Galerkin approximation is: find $u_h \in V^h$ such that

$$\int_0^1 (u_h' v_h' + u_h v_h)\, dx = \int_0^1 v_h \sin \pi \, dx \quad \text{for all } v_h \in V^h,$$

where $V^h = \{v_h \in X^h : \; v_h(0) = v_h(1) = 0\}$.

Suppose that we use three elements for this problem. Since $X^h$ is to be spanned by *piecewise linear* basis functions, nodes are required at the ends of elements only, and the basis functions $N_i$ are formed by

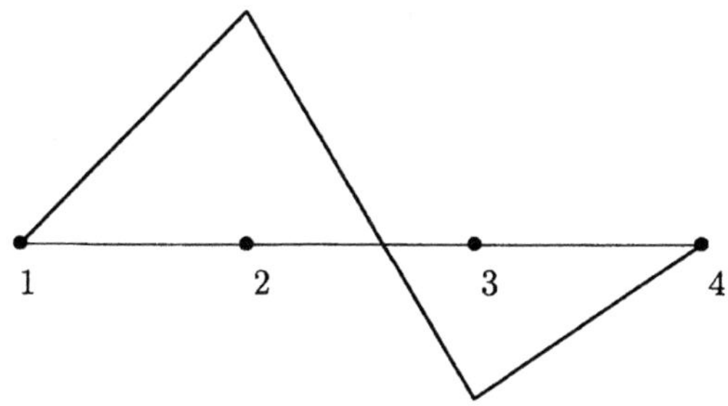


FIGURE 11.11. A typical member of $V^h$

patching together the functions defined in (11.15). These functions span $X^h$.

Next, we construct $V^h$ by requiring that (see (11.12))

$$V^h = \text{span}\,\{N_i \in X^h : N_i(0) = N_i(1) = 0\} = \text{span}\,\{N_2, N_3\},$$

since $N_1$ and $N_4$ do not satisfy $(11.21)_2$. Hence every member of $V^h$ is a linear combination of $N_2$ and $N_3$ (Figure 11.11). From (11.13) and (11.14) we have, for $i$ and $j = 2, 3$,

$$\begin{aligned} a(N_i, N_j) &= \int_0^1 (N_i' N_j' + N_i N_j)\, dx \\ &= \underbrace{\int_0^{1/3} \left[ N_i^{(1)\prime} N_j^{(1)\prime} + N_i^{(1)} N_j^{(1)} \right] dx}_{K_{ij}^{(1)}} \\ &\quad + \underbrace{\int_{1/3}^{2/3} \left[ N_i^{(2)\prime} N_j^{(2)\prime} + N_i^{(2)} N_j^{(2)} \right] dx}_{K_{ij}^{(2)}} \\ &\quad + \underbrace{\int_{2/3}^{1} \left[ N_i^{(3)\prime} N_j^{(3)\prime} + N_i^{(3)} N_j^{(3)} \right] dx}_{K_{ij}^{(3)}} \end{aligned}$$

using the fact that the restriction of $N_i$ to element $\Omega_e$ is $N_i^{(e)}$. Of course, $N_i^{(e)} = 0$ if node $i$ is not a node of $\Omega_e$.

Now recall that $K_{ij}^{(e)} = 0$ if either node $i$ or node $j$ does not belong to $\Omega_e$. Hence the only nonzero contributions that have to be calculated are $K_{22}^{(1)}, K_{22}^{(2)}, K_{23}^{(2)}, K_{33}^{(2)}, K_{33}^{(3)}$ (note that $K_{ij}$ is symmetric).

The computational work is facilitated by evaluating integrals on the reference element; thus

$$\begin{aligned} K_{22}^{(1)} &= \int_{\Omega_1} \left( \frac{dN_2^{(1)}}{dx}\frac{dN_2^{(1)}}{dx} + N_2^{(1)}N_2^{(1)} \right) dx \\ &= \int_{\widehat{\Omega}} \left( \frac{d\widehat{N}_2}{d\xi}\frac{d\widehat{N}_2}{d\xi}\left(\frac{2}{h_1}\right)^2 + \widehat{N}_2\widehat{N}_2 \right) \frac{h_1}{2}\, d\xi \qquad (11.22) \end{aligned}$$

using the fact that

$$\frac{dN_2^{(1)}}{dx} = \frac{d\widehat{N}_2}{d\xi}\frac{d\xi}{dx} \quad \text{and} \quad d\xi = \frac{2}{h_1}dx$$

(here $h_e = \frac{1}{3}$ for all elements). We now substitute for $\widehat{N}_2$ from (11.19) and integrate between $\xi = -1$ and $\xi = +1$ to get $K_{22}^{(1)} = 28/9$. After transforming to $\widehat{\Omega}$, it is found that $K_{33}^{(2)}$ is exactly the same as (11.22); $K_{22}^{(2)}$ and $K_{33}^{(3)}$ differ only in that $\widehat{N}_2$ is replaced by $\widehat{N}_1$ in (11.22), but since $(\widehat{N}_1)^2$ and $(\widehat{N}_2)^2$ have the same integrals, as do $(\widehat{N}_2')^2$ and $(\widehat{N}_2')^2$, we arrive at the same answer. Hence

$$K_{22}^{(2)} = K_{33}^{(2)} = K_{33}^{(3)} = \frac{28}{9}.$$

The term $K_{23}^{(2)}$ is obtained from

$$\begin{aligned} K_{23}^{2} &= \int_{1/3}^{2/3} \left[ N_2^{(2)'}N_3^{(2)'} + N_2^{(2)}N_3^{(2)} \right] dx \\ &= \int_{\widehat{\Omega}} \left( \left(\widehat{N}_1'\widehat{N}_2'\right)\left(\frac{2}{h_2}\right)^2 + \widehat{N}_1\widehat{N}_2 \right) \frac{h_2}{2}\, d\xi = -\frac{53}{18}. \end{aligned}$$

Collecting all terms, then, we have

$$K_{ij} = \sum_{e=1}^{3} K_{ij}^{(e)} \implies \begin{cases} K_{22} &= \frac{28}{9} + \frac{28}{9} = \frac{56}{9} = K_{33}, \\ K_{23} &= -\frac{53}{18}. \end{cases}$$

Next we have to evaluate

$$F_i = \int_0^1 \sin(\pi x) N_i\, dx. \qquad (11.23)$$

Now, although evaluation of this integral over each element is a simple enough matter for this problem, it can be quite tedious in general, and in the case of more complicated data, impossible to carry out exactly. We observe, however, that it is a *piecewise linear* approximation to the exact solution that is being sought, so it would make

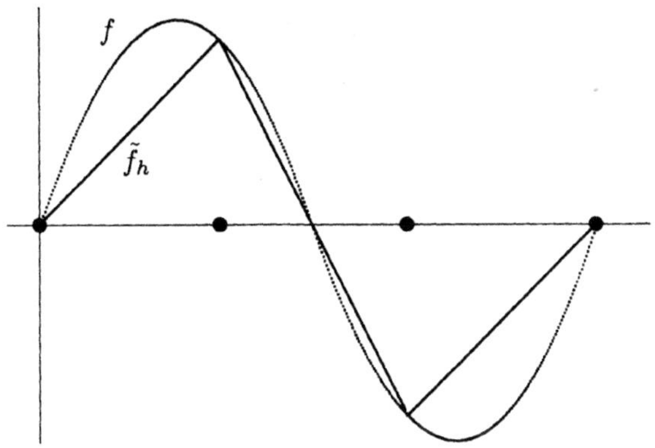


FIGURE 11.12. The interpolate of the function $f(x) = \sin(\pi x)$

sense to replace $f(x) = \sin \pi x$ by its *linear interpolate* $\tilde{f}_h$ in $X^h$. This would enable evaluation of the terms $F_i$ to be carried out very easily while preserving the piecewise linear nature of the approximation. The interpolate of $f$ is, we recall, a function of the form

$$\tilde{f}_h(x) = \sum_{i=1}^{4} f_i N_i(x), \quad f_i = f(x_i)$$

with the property that $\tilde{f}_h$ is linear between nodes and equal to $f$ at the nodes (Figure 11.12).

When $f(x) = \sin \pi x$ we have $f_1 = f_4 = 0$ and $f_2 = f_3 = \sqrt{3}/2$. Equation (11.23) is thus replaced by

$$\tilde{F}_i \equiv \int_0^1 \left( \sum_{j=1}^{4} f_j N_j \right) N_i \, dx = \sum_{e=1}^{3} \int_{\Omega_e} \left( \sum_{j=1}^{4} f_j N_j^{(e)} \right) N_i^{(e)} \, dx$$

$$= \underbrace{\int_0^{1/3} N_2^{(1)} N_i^{(1)} \, dx}_{\tilde{F}_i^{(1)}} + \underbrace{\int_{1/3}^{2/3} (N_2^{(2)} + N_3^{(2)}) N_i^{(2)} \, dx}_{\tilde{F}_i^{(2)}}$$

$$+ \underbrace{\int_{2/3}^{1} \frac{\sqrt{3}}{2} N_3^{(3)} N_i^{(3)} dx}_{\tilde{F}_i^{(3)}} .$$

Transforming each of these integrals to integrals on $\widehat{\Omega}$ as before, and evaluating, we find that

$$\tilde{F}_1^{(1)} = \tilde{F}_4^{(3)} = \sqrt{3}/36, \quad \tilde{F}_2^{(1)} = \tilde{F}_2^{(3)} = \sqrt{3}/18, \quad \tilde{F}_2^{(2)} = \tilde{F}_3^{(2)} = \sqrt{3}/12,$$

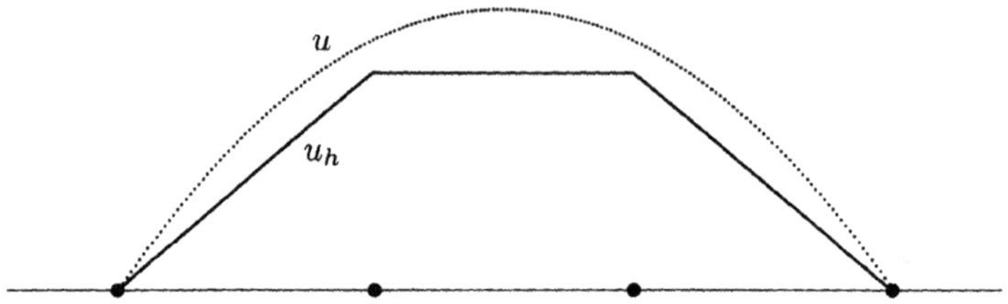


FIGURE 11.13. The exact and approximate solutions to Example 1

the other components of $\tilde{F}_i^{(e)}$ being zero. Thus

$$\begin{aligned} \tilde{F}_i &= \frac{\sqrt{3}}{18}[\underbrace{(\tfrac{1}{2},\,1,\,0,\,0)}_{\tilde{F}_i^{(1)}} + \underbrace{(0,\,\tfrac{3}{2},\,\tfrac{3}{2},\,0)}_{\tilde{F}_i^{(2)}} + \underbrace{(0,\,0,\,1,\,\tfrac{1}{2})}_{\tilde{F}_i^{(3)}}] \\ &= \frac{\sqrt{3}}{36}(1,\,5,\,5,\,1). \end{aligned}$$

Finally, then, we have to solve

$$\begin{aligned} K_{22}c_2 + K_{23}c_3 &= \tilde{F}_2, \\ K_{32}c_2 + K_{33}c_3 &= \tilde{F}_3, \end{aligned}$$

which gives $c_2 = c_3 = 0.0734$. Hence the approximate solution is

$$u_h(x) = 0.0734(N_2(x) + N_3(x)).$$

We compare this with the *exact* solution $u(x) = (1 + \pi^2)^{-1} \sin \pi x$ in Figure 11.13 and see that the finite element solution is a fair approximation of the exact solution, given the very small subspace $V^h$ that has been used. Furthermore, the approximate solution, like the exact, is symmetric about $x = \frac{1}{2}$. The approximate solution could of course be improved in two ways: by subdividing the domain into a larger number of elements, and by using a higher-order element, such as the quadratic element. Of course, either of these refinements will result in a greater amount of computational work, which must be taken into consideration. It is clearly of interest to know beforehand by how much an approximate solution will improve as a result of either of the two refinements mentioned, so that one may decide whether the refinement is worth the effort. This is a question that is addressed in the following chapter. In the next section we apply the ideas developed here to second-order problems defined on domains $\Omega$ in $\mathbb{R}^2$.

## 11.3 Two-dimensional problems

We have already given a rough indication in Section 11.1 of how a finite element mesh is constructed for problems defined on domains in $\mathbb{R}^2$. The subject is taken up again here, and the ideas in Section 11.2 are generalized to two-dimensional second-order problems.

Recall from Section 11.1 that the domain $\Omega \subset \mathbb{R}^2$ is assumed to be *polygonal*, so that the boundary $\Gamma$ is made up of straight segments. Next, $\Omega$ is partitioned into elements $\Omega_1, \Omega_2, \ldots, \Omega_E$, and it is required that every side of an element be either part of the boundary $\Gamma$, or the side of another element. We proceed now to look at a few commonly used elements.

**Triangular elements.** Triangles are the simplest polygonal shapes in $\mathbb{R}^2$, so it is not surprising that triangular elements possess features that make for very simple means of approximation.

Suppose that we want $X^h$ to be the space of piecewise polynomials of degree 1, so that we require $X_e$ to be the space $P_1(\Omega_e)$ (see the corresponding discussion at the beginning of Section 11.2). The most general function in $P_1(\Omega_e)$ is of the form $f(x,y) = a + bx + cy$, so that if the values of the function are known at *three* points, then it is uniquely determined (in other words $a, b$, and $c$ can be found). *Three* nodal points are thus required, and these are placed at the *vertices* of the triangle. This positioning of nodes ensures that if any of the sides $ij, jk$, or $ki$ of $\Omega_e$ is shared with an adjacent element $\Omega_f$, say, the *piecewise linear function formed by patching together the functions defined on* $\Omega_e$ and $\Omega_f$ *will be continuous across the interface of these elements.* Rather than having to deal with an element that has nodes numbered in some arbitrary way, the development of the theory is made considerably more convenient by adopting a *local* numbering system when evaluating the element stiffness matrix and load vector, since the numbering system is then identical to that on the reference element. Once these have been evaluated, the components can then simply be placed in the correct rows and columns of the global matrix and vector by recalling the global node numbers of the element. Thus consider again the triangle shown in Figure 11.14; we assign local node numbers 1 through 3 to the three nodes in the manner shown, and record the association

$$\begin{aligned} i &\leftrightarrow 1 \\ j &\leftrightarrow 2 \\ k &\leftrightarrow 3 \end{aligned}$$

between global and local node numbers. The coordinates of each node can likewise be expressed in global or local form, and we may write $\boldsymbol{x}_I^{(e)}$ ($I = 1, 2, 3$) for $\boldsymbol{x}_i$, $\boldsymbol{x}_j$, $\boldsymbol{x}_k$. Finally, the local basis functions may also be numbered locally, in the form $N_I^{(e)}$. Wherever it is necessary to make the distinction, local quantities are always indexed by upper case letters.

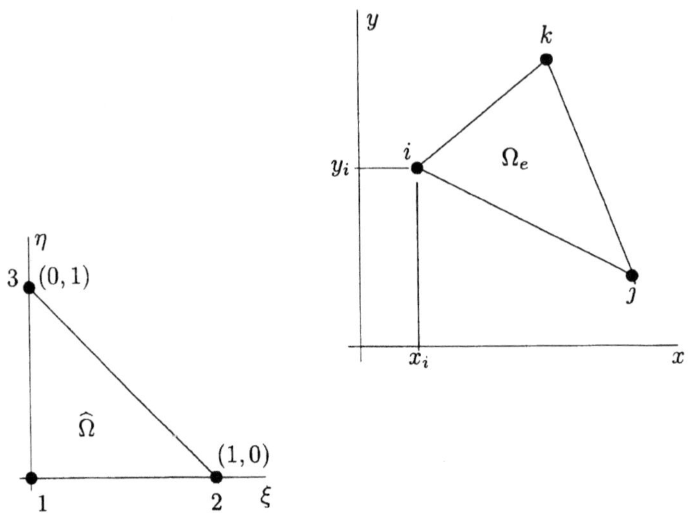


FIGURE 11.14. A triangular element and the corresponding reference element

The element stiffness matrix $\boldsymbol{K}^{(e)}$ here is a $3 \times 3$ matrix, and the manner in which its contribution to $\boldsymbol{K}$ is made (according to (11.13)) is as shown in Figure 11.15. The process whereby $\boldsymbol{K}^{(e)}$ and $\boldsymbol{F}^{(e)}$ are computed for each element, and then added to the global matrix, is known as *assembly*. The reference triangular element $\widehat{\Omega}$ is shown in Figure 11.14. Now it is required, as in the one-dimensional case, to construct a map $F_e$ that will transform $\widehat{\Omega}$ to the element $\Omega_e$, and the most general map that takes triangles to triangles is the *affine* map, which is of the form

$$f(\boldsymbol{\xi}) = \boldsymbol{T}\boldsymbol{\xi} + \boldsymbol{b}, \tag{11.24}$$

in which $\boldsymbol{T}$ is a constant matrix and $\boldsymbol{b}$ a constant vector.

The reference element is a right-angled isosceles triangle, and it is not difficult to verify that the transformation

$$x = x_1^{(e)}(1-\xi-\eta) + x_2^{(e)}\xi + x_3^{(e)}\eta$$

$$y = y_1^{(e)}(1-\xi-\eta) + y_2^{(e)}\xi + y_3^{(e)}\eta$$

or

$$\boldsymbol{x} = F_e(\boldsymbol{\xi}) \equiv (1-\xi-\eta)\boldsymbol{x}_1^{(e)} + \xi\boldsymbol{x}_2^{(e)} + \eta\boldsymbol{x}_3^{(e)}$$

maps the nodal points $1, 2, 3$ of $\widehat{\Omega}$ to local nodal points $1, 2, 3$ of $\Omega_e$, and indeed maps each point $\boldsymbol{\xi} \in \widehat{\Omega}$ to a point $\boldsymbol{x} \in \Omega_e$. The matrix $\boldsymbol{T}$ and vector $\boldsymbol{b}$ in (11.24) are thus seen to be given by

$$\boldsymbol{T} = \begin{pmatrix} x_2^{(e)} - x_1^{(e)} & x_3^{(e)} - x_1^{(e)} \\ y_2^{(e)} - y_1^{(e)} & y_3^{(e)} - y_1^{(e)} \end{pmatrix} \quad \text{and} \quad \boldsymbol{b} = \begin{pmatrix} x_1^{(e)} \\ y_1^{(e)} \end{pmatrix},$$

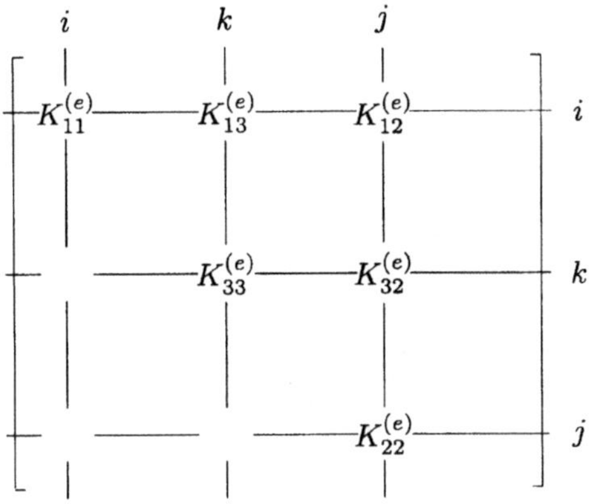


FIGURE 11.15. Assembly of the global stiffness matrix

whereas the inverse of this transformation is

$$\begin{aligned}
\xi &= \frac{1}{2A_e}\left[(y_3^{(e)} - y_1^{(e)})(x - x_1^{(e)}) - (x_3^{(e)} - x_1^{(e)})(y - y_1^{(e)})\right], \\
\eta &= \frac{1}{2A_e}\left[-(y_2^{(e)} - y_1^{(e)})(x - x_1^{(e)}) + (x_2^{(e)} - x_1^{(e)})(y - y_1^{(e)})\right],
\end{aligned}$$

where $A_e$ is the area of $\Omega_e$.

The next step is to define local basis functions $\widehat{N}_1, \widehat{N}_2, \widehat{N}_3$ on $\widehat{\Omega}$, and then to obtain the basis functions on $\Omega_e$ from (11.17), or

$$N_I^{(e)}(\boldsymbol{x}) = \widehat{N}_I(\boldsymbol{\xi}), \quad I = 1, 2, 3.$$

The local basis functions on $\widehat{\Omega}$ must satisfy (11.9), and are

$$\begin{aligned}
\widehat{N}_1(\boldsymbol{\xi}) &= 1 - \xi - \eta, \\
\widehat{N}_2(\boldsymbol{\xi}) &= \xi, \\
\widehat{N}_3(\boldsymbol{\xi}) &= \eta;
\end{aligned}$$

the function $\widehat{N}_1$ is shown in Figure 11.16, whereas Figure 11.4 shows the images of $\widehat{N}_A$ on the elements attached to node $i$, and the basis function $N_i(\boldsymbol{x})$ that results from patching together all local basis functions associated with node $i$. Clearly the condition (11.9) is satisfied. The basis function $N_i$ formed by patching together all the local functions $N_i^{(e)}$ associated with node $i$ is the two-dimensional counterpart of the "hat" function in one dimension, and is pyramidal in shape. Naturally $N_i$ is piecewise linear, and is nonzero only on those elements that have node $i$ as a node.

*Piecewise quadratic* triangular elements are obtained by adding a further three nodes to an element, at the midpoints of the sides, as in Figure 11.17. The most general function in $P_2(\widehat{\Omega})$ has the form $f(\xi, \eta) =$

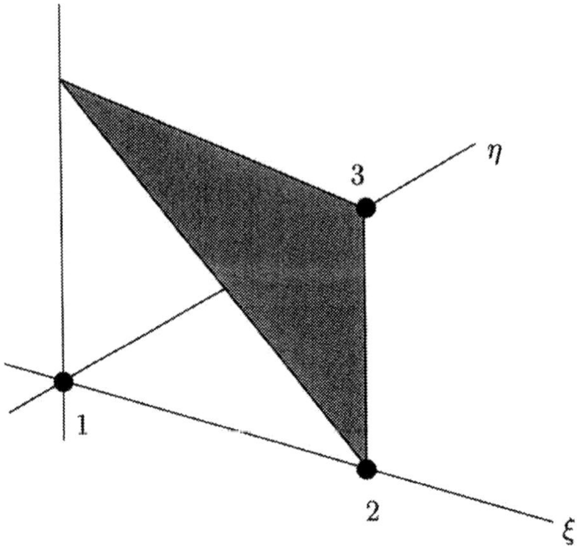


FIGURE 11.16. The local basis function $\widehat{N}_1$ on the reference triangle $\widehat{\Omega}$

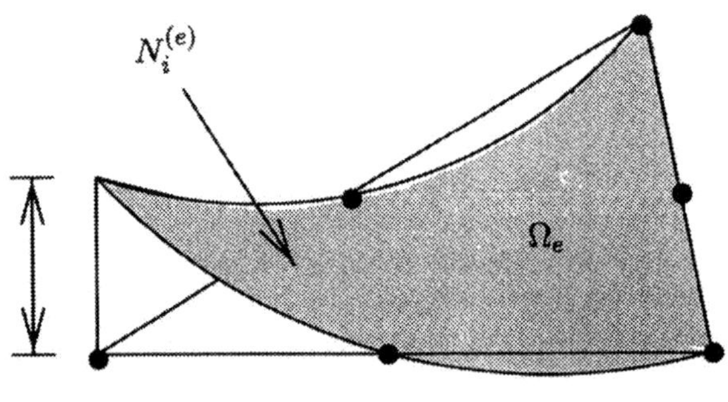


FIGURE 11.17. A triangular element with quadratic local basis functions

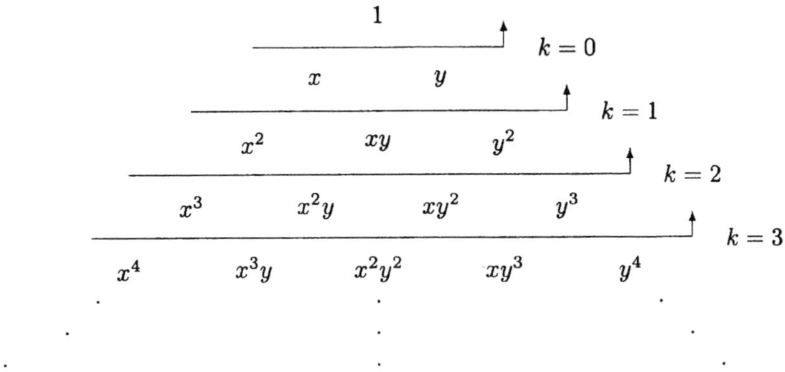


FIGURE 11.18. The Pascal triangle

$a_1 + a_2\xi + a_3\eta + a_4\xi^2 + a_5\xi\eta + a_6\eta^2$, and is thus uniquely determined on $\widehat{\Omega}$ (and hence on $\Omega_e$) by its values at the six nodes. Furthermore, the basis functions $N_i$ formed by patching together all local basis functions $N_i^{(e)}$ are continuous (see Exercise 11.6). Some of the local basis functions are shown in Figure 11.17.

**The Pascal triangle.** The task of constructing bases of ever-increasing orders on elements in $\mathbb{R}^2$ is greatly facilitated by making use of the Pascal triangle; this is an arrangement in triangular form of the terms in a polynomial, the $k$th row containing all the terms of the form $x^p y^q$ such that $p + q = k$ (Figure 11.18). Thus the terms in the polynomial of degree $k$ can be identified by inspection, as can their number, and therefore also the number of nodal points required.

**Rectangular elements.** We turn now to a second category of finite elements, namely, those that are rectangular or, more generally, quadrilateral in shape. If we are to adhere to the policy of having nodal points at least at the vertices of elements, then clearly the simplest rectangular element will be one with *four* nodes, one node at each corner (Figure 11.19). The question now arises: what kind of space of polynomials $X_e$ can be defined on $\Omega_e$ so that any function in $X_e$ is uniquely determined by its values at the four vertices? Functions in $P_1(\Omega_e)$ are completely determined by *three* nodal values, so they will not do. On the other hand, quadratic functions require *six* nodal values, which is more than we have at our disposal. The solution to the problem is to examine the first few terms of the polynomial

$$f(x, y) = a_1 + a_2 x + a_3 y + a_4 xy + a_5 x^2 + a_6 y^2 + \dots$$

and to resolve that four terms be retained. The constant and linear terms are obviously required, and it remains to decide which additional term to retain, in order to arrive at a total of four terms. It is inadvisable to keep the

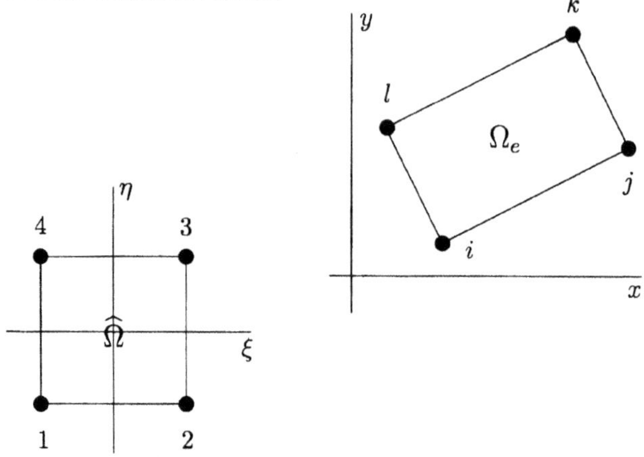


FIGURE 11.19. A rectangular element and corresponding reference element

terms involving $x^2$ or $y^2$, since this would result in a lopsided approximation in which a quadratic term appears for only one of the coordinates. However, there is no objection to retaining the term involving $xy$: this ensures that the coordinates $x$ and $y$ are equally represented, for then the approximation is of the form

$$f(x, y) = a_1 + a_2x + a_3y + a_4xy.$$

We call $f(x, y)$ a *bilinear polynomial*; in general, the space of polynomials containing terms of degree $\leq k$ *in each of the variables* is denoted by $Q_k(\Omega)$, so that $f(x, y)$ is a member of $Q_1(\Omega)$. Note that the inclusions $P_k(\Omega) \subset Q_k(\Omega) \subset P_{2k}(\Omega)$ hold for $\Omega \subset \mathbb{R}^2$. The situation now is that $X_e = Q_1(\Omega_e)$, so that $X^h$ consists of *piecewise bilinear polynomials*.

As before, we set up a reference element $\hat{\Omega}$ which this time is the square $(-1, 1) \times (-1, 1)$, shown in Figure 11.19. The reference element is mapped onto an arbitrary rectangular element $\Omega_e$ by the *affine* transformation

$$\boldsymbol{x} = F_e\boldsymbol{\xi} \equiv \boldsymbol{T\xi} + \boldsymbol{b} \quad \text{or} \quad \begin{pmatrix} x \\ y \end{pmatrix} = \begin{pmatrix} T_{11} & T_{12} \\ T_{21} & T_{22} \end{pmatrix} \begin{pmatrix} \xi \\ \eta \end{pmatrix} + \begin{pmatrix} b_1 \\ b_2 \end{pmatrix} \tag{11.25}$$

in which the matrix $\boldsymbol{T}$ is given by

$$\boldsymbol{T} = \tfrac{1}{2} \begin{pmatrix} x_2^{(e)} - x_1^{(e)} & y_2^{(e)} - y_1^{(e)} \\ x_4^{(e)} - x_1^{(e)} & y_4^{(e)} - y_1^{(e)} \end{pmatrix} \tag{11.26}$$

and $\boldsymbol{b}$ is the position vector of the centroid of the rectangle $\Omega_e$. Since affine maps take straight lines to straight lines, it is worth noting that the most general such map would transform the reference element in Figure 11.19 to a *parallelogram*, so that parallelograms could be as easily accommodated.

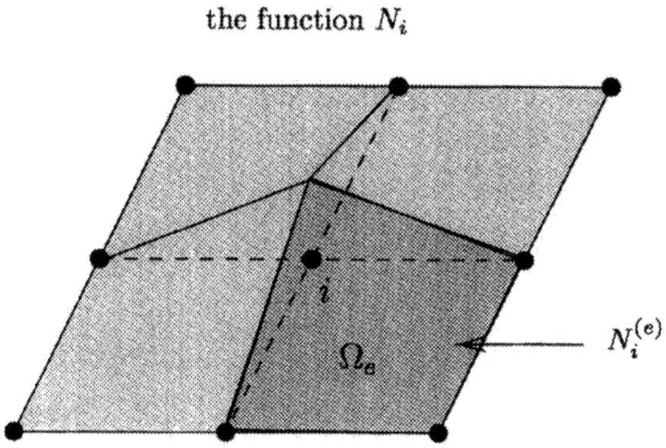


FIGURE 11.20. A piecewise bilinear global basis function

Next, we set up bilinear local basis functions on $\Omega$ that satisfy (11.9); just as the reference element $(-1,1)\times(-1,1)$ may be regarded as the Cartesian product of the one-dimensional reference element, in the same way local basis functions satisfying (11.9) may be generated from products of the functions (11.19). Thus we obtain

$$\widehat{N}_I(\boldsymbol{\xi}) = \tfrac{1}{4}(1+\xi_I\xi)(1+\eta_I\eta) \quad (I = 1,\ldots,4), \tag{11.27}$$

where $(\xi_i\ \eta_i)$ are the coordinates of node $i$ on the reference element; in full this reads

$$\begin{aligned}
\widehat{N}_1(\boldsymbol{\xi}) &= \tfrac{1}{4}(1-\xi)(1-\eta),\\
\widehat{N}_2(\boldsymbol{\xi}) &= \tfrac{1}{4}(1+\xi)(1-\eta),\\
\widehat{N}_3(\boldsymbol{\xi}) &= \tfrac{1}{4}(1+\xi)(1+\eta),\\
\widehat{N}_4(\boldsymbol{\xi}) &= \tfrac{1}{4}(1-\xi)(1+\eta).
\end{aligned}$$

Then the functions $N_I^{(e)}$ are obtained by setting $N_I^{(e)}(\boldsymbol{x}) = \widehat{N}_I(\boldsymbol{\xi})$, with $\boldsymbol{x}$ and $\boldsymbol{\xi}$ being related through (11.25). As in the case of triangular elements, the positioning of the nodes and the choice of local basis functions ensures that the basis functions $N_i$ will be continuous across element boundaries, as shown in Figure 11.20. Higher-order approximations on rectangular elements may be generated by once again appealing to Pascal's triangle. Figure 11.21 shows the triangle, on which are marked the four terms that give the bilinear approximation. By extending the diamond-shaped pattern associated with the bilinear approximation, we arrive at a *biquadratic* approximation that contains nine terms, so that nine nodes are required, as shown in Figure 11.22. The local basis $\{N_I^{(e)}\}_{I=1}^{9}$ on $\Omega_e$ spans $Q_2$; these functions are again most conveniently found by constructing basis functions on $\widehat{\Omega}$, and then using the relationship $\widehat{N}_I(\boldsymbol{\xi}) = N_I^{(e)}(\boldsymbol{x})$. The nine functions on $\widehat{\Omega}$ may be generated from products of the one-dimensional quadratic

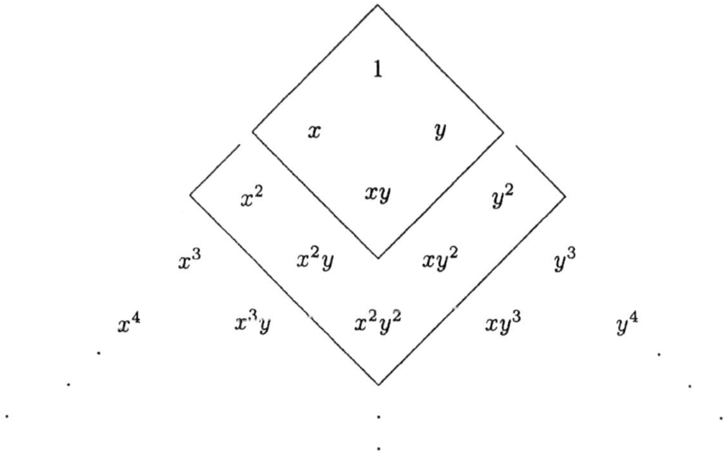


FIGURE 11.21. Pascal's triangle as a tool for generating bases in $Q_k$

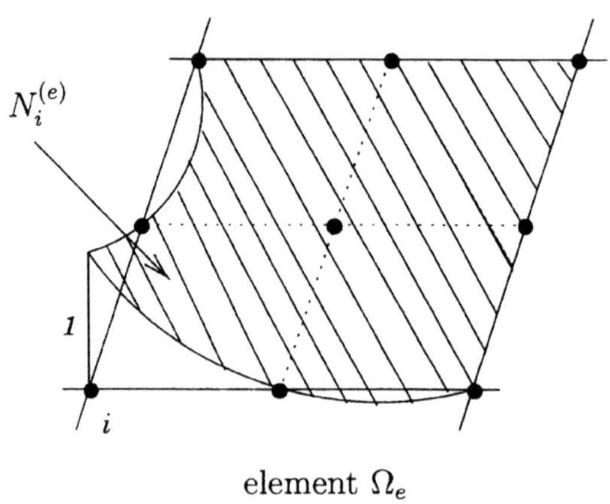


FIGURE 11.22. A biquadratic local basis function

functions, thus ensuring that (11.9) is satisfied. Indeed, suppose that we denote by $\widehat{n}_1$, $\widehat{n}_2$, $\widehat{n}_3$ the three functions given in (11.20); then the nine basis functions for the nine-noded element follow from

$$\begin{aligned} N_1(\xi,\eta) &= \widehat{n}_1(\xi)\widehat{n}_1(\eta), \\ N_2(\xi,\eta) &= \widehat{n}_3(\xi)\widehat{n}_1(\eta), \\ &\vdots \\ N_9(\xi,\eta) &= \widehat{n}_2(\xi)\widehat{n}_2(\eta). \end{aligned}$$

The basis functions $N_i$ formed by patching together the function $N_i^{(e)}$ associated with global node $i$ are piecewise biquadratic polynomials that are continuous across interelement boundaries.

This concludes the discussion on elements for second-order problems in two dimensions. We now work through a simple example involving rectangular elements.

**Example**

2. Consider the problem

$$\begin{aligned} -\nabla^2 u &= 2-(x^2+y^2) \quad \text{in } \Omega=(0,1)\times(0,1), \\ u &= 0 \text{ on } \Gamma_1, \\ \partial u/\partial n &= 0 \text{ on } \Gamma_2, \end{aligned}$$

where $\Gamma_1$ and $\Gamma_2$ are the parts of the boundary $\Gamma$ shown in Figure 11.23.

The corresponding VBVP is: find $u \in V$ such that

$$\int_\Omega \nabla u \cdot \nabla v \, dxdy = \int_\Omega [2-(x^2+y^2)]v \, dxdy \quad \text{for all } v \in V,$$

where $V = \left\{v \in H^1(\Omega) : v = 0 \text{ on } \Gamma_1\right\}$, and the approximate problem is: find $u_h \in V^h \subset V$ such that

$$\int_\Omega \nabla u_h \cdot \nabla v_h \, dxdy = \int_\Omega [2-(x^2+y^2)]v_h \, dxdy \quad \text{for all } v_h \in V^h.$$

We divide the domain into four square elements and choose for $X^h$ the space of piecewise *bilinear* functions, so that nodes are required at the corners of elements only (see Figure 11.23).

Next we construct $v_h$. From (11.12) it is required that

$$V^h = \text{span}\{N_i \in X^h : \; N_i(\boldsymbol{x}) = 0 \;\text{ on } \Gamma_1\} = \text{span}\{N_1, N_2, N_3, N_4\},$$

the functions $N_i$ being piecewise bilinear functions; the restriction of $N_i$ to $\Omega_e$ is of course $N_i^{(e)}$.

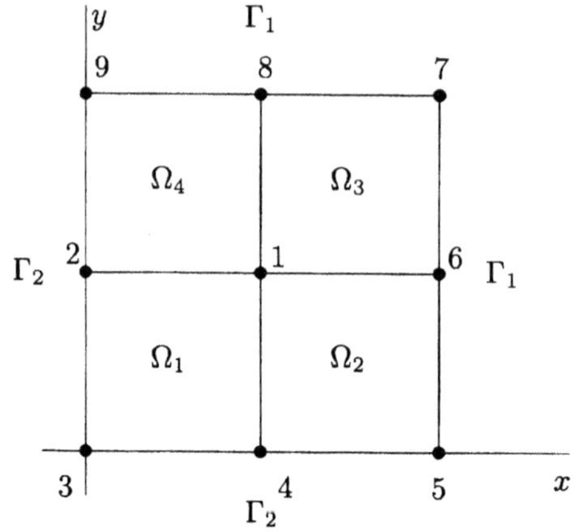


FIGURE 11.23. The domain and finite element mesh for Example 2

Now we require

$$\begin{aligned} K_{ij} &= \iint_{\Omega} \nabla N_i \cdot \nabla N_j \; dxdy \\ &= \sum_{e=1}^{4} \underbrace{\iint_{\Omega_e} \nabla N_i^{(e)} \cdot \nabla N_j^{(e)} \; dxdy}_{K_{ij}^{(e)}} . \end{aligned}$$

Since all elements have the same geometry, the amount of computational work can be reduced considerably by observing that many of the integrals have the same value. Indeed, if nodes $i$ and $j$ both belong to $\Omega_e$, then an moment's thought will convince us that

$$K_{ij}^{(e)} = \iint_{\Omega_e} \nabla N_i^{(e)} \cdot \nabla N_j^{(e)} \; dxdy = \begin{cases} K_{11}^{(1)} & I = J, \\ K_{12}^{(1)} \;\text{ if } & I, J \text{ are adjacent}, \\ K_{13}^{(1)} & I, J \text{ otherwise}, \end{cases} \tag{11.28}$$

so in fact only three integrals have to be evaluated.

The four nodes associated with element 1 are numbered 1 through 4, so that in this case the local and global node numbers coincide. We have

$$K_{11}^{(1)} = \iint_{\Omega_1} \nabla N_1^{(1)} \cdot \nabla N_1^{(1)} \, dxdy$$

$$= \iint_{\Omega_1} \left[ \left( \frac{\partial N_1^{(1)}}{\partial x} \right)^2 + \left( \frac{\partial N_1^{(1)}}{\partial y} \right)^2 \right] dxdy$$

$$= \int_{\widehat{\Omega}} \left[ \left( \frac{\partial \widehat{N}_1}{\partial \xi} \frac{\partial \xi}{\partial x} + \frac{\partial \widehat{N}_1}{\partial \eta} \frac{\partial \eta}{\partial x} \right)^2 + \left( \frac{\partial \widehat{N}_1}{\partial \xi} \frac{\partial \xi}{\partial y} + \frac{\partial \widehat{N}_1}{\partial \eta} \frac{\partial \eta}{\partial y} \right)^2 \right] |j| \, d\xi d\eta \tag{11.29}$$

using the chain rule and the rule $dxdy = |j| d\xi d\eta$ for changing variables in area integrals; here $j = \det \boldsymbol{T}$, where $\boldsymbol{T}$ is given by (11.26). For this element,

$$\boldsymbol{T} = \frac{1}{4} \begin{pmatrix} -1 & 0 \\ 0 & -1 \end{pmatrix} \quad \text{and} \quad \boldsymbol{T}^{-1} = \begin{pmatrix} -4 & 0 \\ 0 & -4 \end{pmatrix}.$$

By inverting (11.25) we obtain $\xi$ and $\eta$ in terms of $x$ and $y$, and find that

$$\frac{\partial \xi}{\partial x} = \frac{\partial \eta}{\partial y} = -4, \quad \frac{\partial \xi}{\partial y} = \frac{\partial \eta}{\partial x} = 0.$$

Also, $J = \frac{1}{16}$. With all of this available and with the use of (11.27) we can now evaluate (11.29), to obtain

$$K_{11}^{(1)} = \int_{-1}^{1} \int_{-1}^{1} \left( [\tfrac{1}{16}(\eta - 1)(-4)]^2 + [\tfrac{1}{4}(\xi - 1)(-4)]^2 \right) \frac{1}{16} \, d\xi d\eta = \frac{2}{3}.$$

Similarly, $K_{12}^{(1)} = -\frac{1}{6}, K_{13}^{(1)} = -\frac{1}{3}$. Using (11.28) and (11.29) we get

$$6\boldsymbol{K} = \begin{pmatrix} 4 & -1 & -2 & -1 \\ & 4 & -1 & -2 \\ & \text{sym} & 4 & 1 \\ & & & 4 \end{pmatrix} + \begin{pmatrix} 4 & 0 & 0 & -1 \\ & 0 & 0 & 0 \\ & \text{sym} & 0 & 0 \\ & & & 4 \end{pmatrix}$$

$$+ \begin{pmatrix} 4 & 0 & 0 & 0 \\ & 0 & 0 & 0 \\ & \text{sym} & 0 & 0 \\ & & & 0 \end{pmatrix} + \begin{pmatrix} 4 & -1 & 0 & 0 \\ & 4 & 0 & 0 \\ & \text{sym} & 0 & 0 \\ & & & 0 \end{pmatrix}$$

$$= \begin{pmatrix} 16 & -2 & -2 & -2 \\ & 8 & -1 & -2 \\ & \text{sym} & 4 & -1 \\ & & & 8 \end{pmatrix}.$$

The next task is to evaluate $F_i = \iint_\Omega f N_i \, dxdy$. As in the one-dimensional case, we replace $f(x,y)$ by its *interpolate* $\tilde{f}_h(x,y)$ which is given by

$$\tilde{f}_h(x,y) = \sum_{i=1}^{9} f_i N_i(x,y), \quad f_i = f(x_i, y_i).$$

Then

$$F_j = \iint_\Omega \left( \sum_{i=1}^{9} f_i N_i \right) N_j dxdy.$$

Thus we need integrals of the form $\iint_\Omega f_I N_I^{(e)} N_J^{(e)} dxdy$, and once again a great deal of effort can be avoided by noting that

$$\iint_{\Omega_e} N_I^{(e)} N_J^{(e)} dxdy = \begin{cases} \iint_{\Omega_e} N_1^{(1)} N_1^{(1)} \, dxdy & \text{if } I = J, \\ \iint_{\Omega_e} N_1^{(1)} N_2^{(1)} \, dxdy & I \text{ and } J \text{ adjacent}, \\ \iint_{\Omega_e} N_1^{(1)} N_3^{(1)} \, dxdy & \text{otherwise}, \end{cases}$$

$$= \begin{cases} 1/24 \\ 1/72 \\ 1/144. \end{cases}$$

Hence, taking cognizance of the relationship between local and global node numbers, we have

$$\boldsymbol{F} = \sum_{e=1}^{4} \boldsymbol{F}^{(e)} = \begin{pmatrix} \frac{1}{24}(f_1) + \frac{1}{72}(f_2 + f_4) + \frac{1}{144}(f_3) \\ \frac{1}{24}(f_2) + \frac{1}{72}(f_1 + f_3) + \frac{1}{144}(f_4) \\ \frac{1}{24}(f_3) + \frac{1}{72}(f_2 + f_4) + \frac{1}{144}(f_1) \\ \frac{1}{24}(f_4) + \frac{1}{72}(f_1 + f_3) + \frac{1}{144}(f_2) \end{pmatrix} = \begin{pmatrix} 0.0938 \\ 0.0972 \\ 0.1007 \\ 0.0972 \end{pmatrix}.$$

Finally, we solve $\boldsymbol{Kc} = \boldsymbol{F}$ to obtain

$$\boldsymbol{c} = \begin{pmatrix} 0.1585 \\ 0.2568 \\ 0.4213 \\ 0.2568 \end{pmatrix}.$$

The approximate solution is

$$u_h(\boldsymbol{x}) = 0.1585 N_1(\boldsymbol{x}) + 0.2568(N_2(\boldsymbol{x}) + N_4(\boldsymbol{x})) + 0.4213 N_3(\boldsymbol{x}),$$

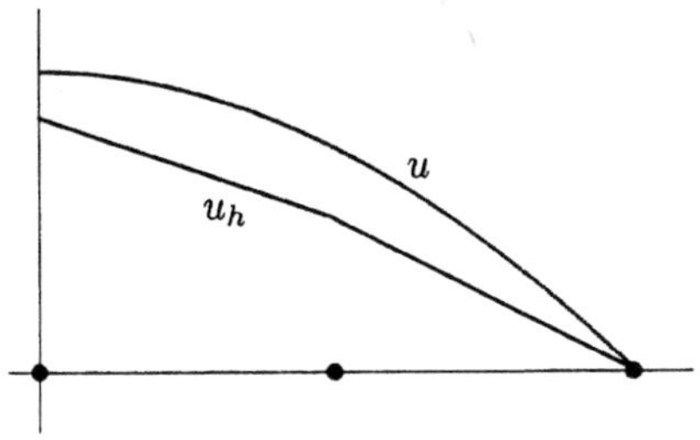


FIGURE 11.24. The exact and approximate solutions to Example 2

and the exact solution is

$$u(x,y) = \tfrac{1}{2}(x^2-1)(y^2-1).$$

These two solutions are compared in Figure 11.24, where we see that the approximation is quite favorable, notwithstanding the relatively crude mesh.

3. Suppose that we wish to find an approximate solution, using finite elements, to a two-dimensional problem in linear elasticity. The variational problem takes the form (9.25), and it is assumed that $\boldsymbol{u} = u_1(x,y)\boldsymbol{e}_1 + u_2(x,y)\boldsymbol{e}_2$, so that the only nonzero components of the strain are, from (8.7), $\epsilon_{11}$, $\epsilon_{12} = \epsilon_{21}$, and $\epsilon_{22}$. It is assumed furthermore that the material is isotropic, so that the bilinear form is given by (see Exercise 9.10)

$$a(\boldsymbol{u}_h, \boldsymbol{v}_h) = \int_\Omega [\lambda(\operatorname{div}\boldsymbol{u}_h)(\operatorname{div}\boldsymbol{v}_h) + 2\mu\boldsymbol{\epsilon}(\boldsymbol{u}_h)\cdot\boldsymbol{\epsilon}(\boldsymbol{v}_h)]\,dx\,dy.$$

The purpose of this example is to give some idea of the changes that are necessary in the event that the principal unknown is vector-valued; we focus on the task of constructing the element stiffness matrix. Now in problems of this nature, it makes a great deal of sense to arrange the nonzero components of the strain matrix in the form of a vector, also denoted by $\boldsymbol{\epsilon}$, and defined by

$$\boldsymbol{\epsilon} = \begin{bmatrix} \epsilon_{11} \\ \epsilon_{22} \\ \sqrt{2}\epsilon_{12} \end{bmatrix}. \tag{11.30}$$

On element $\Omega_e$ the displacement is approximated according to

$$\boldsymbol{u}_h(\boldsymbol{x}) = \sum_{I=1}^{N} \boldsymbol{c}_I N_I(\boldsymbol{x}),$$

using a local numbering system; the vector $\boldsymbol{c}$ represents the value of $\boldsymbol{u}_h$ at node $I$. Superscripts $(e)$ which identify $\boldsymbol{c}_I$ and $N_I$ as quantities associated with $\Omega_e$ are omitted without any danger of ambiguity.

Substitution in (11.30) then yields the representation

$$\epsilon(\boldsymbol{u}_h) = \sum_{I=1}^{N} \boldsymbol{B}_I \boldsymbol{c}_I \equiv \boldsymbol{B}\boldsymbol{c},$$

in which the matrix $\boldsymbol{B}_I$ contains derivatives of $N_I$ with respect to $x$ and $y$, and $\boldsymbol{B}$ and $\boldsymbol{c}$ are the matrix and vector defined by $\boldsymbol{B} = [\boldsymbol{B}_1 \dots \boldsymbol{B}_N]$, $\boldsymbol{c}^t = [\boldsymbol{c}_1^t \dots \boldsymbol{c}_N^t]$. The contribution $\boldsymbol{K}^{(e)}$ of $\Omega_e$ to the stiffness matrix is now found from

$$\int_{\Omega_e} [\lambda(\operatorname{div} \boldsymbol{u}_h)(\operatorname{div} \boldsymbol{v}_h) + 2\mu\epsilon(\boldsymbol{u}_h)\cdot\epsilon(\boldsymbol{v}_h)]\, dx\, dy =$$
$$\boldsymbol{d}^t \underbrace{\left[\int_{\Omega_e} [\lambda \boldsymbol{B}^t \boldsymbol{C}^t \boldsymbol{C} \boldsymbol{B} + 2\mu \boldsymbol{B}^t \boldsymbol{B}]\, dx\, dy\right]}_{\boldsymbol{K}^{(e)}} \boldsymbol{c},$$

where $\boldsymbol{d}$ is the vector of nodal variables corresponding to the arbitrary function $\boldsymbol{v}$.

As before, it makes sense to evaluate this integral on the reference element, and to do so it is necessary to express $\boldsymbol{B}$ as a function of $\xi$ and $\eta$. This is achieved by observing that a typical term in $\boldsymbol{B}_I$ is, for example, $\partial N_I/\partial x$, and we have

$$\frac{\partial N_I}{\partial x} = \frac{\partial \widehat{N}_I}{\partial \xi}\frac{\partial \xi}{\partial x} + \frac{\partial \widehat{N}_I}{\partial \eta}\frac{\partial \eta}{\partial x},$$

which is easily evaluated with the aid of expressions for $\widehat{N}_I$ and the affine maps (11.24) or (11.25). In Section 11.5 it is shown that this transformation can be carried out without explicit inversion of the map between the reference and the actual element. If the matrix thus transformed is denoted as $\widehat{\boldsymbol{B}}$, then we have finally (cf. (11.29))

$$\boldsymbol{K}^{(e)} = \int_{\widehat{\Omega}} [\lambda \widehat{\boldsymbol{B}}^t \boldsymbol{C}^t \boldsymbol{C} \widehat{\boldsymbol{B}} + 2\mu \widehat{\boldsymbol{B}}^t \widehat{\boldsymbol{B}}] |j|\, d\xi\, d\eta.$$

## 11.4 Fourth-order problems and Hermite families of elements

The introduction to the finite element method presented in this chapter has been, up to now, geared towards second-order problems, for the simple reason that these problems lead to the simplest examples of the method; the space in which second-order problems are formulated is a subspace $V$ of $H^1(\Omega)$, and it suffices to construct finite-dimensional spaces $V^h$ that are spanned by continuous functions. In the case of fourth-order problems, however, the situation is less straightforward. The "parent" space for the variational problem is $H^2(\Omega)$, and because not all continuous functions belong to this space (consider, for example, the basis functions constructed in the last two sections), it follows that we have to go a stage further in order to obtain suitable finite element subspaces $V^h$ that satisfy $V^h \subset V \subset H^2(\Omega)$.

Consider, for example, the problem of an elastic beam that is constrained against both displacement and rotation at its two ends. The problem is thus (see also (8.20))

$$\begin{aligned} &\frac{d^4w}{dx^4} = \frac{f}{EI} \quad \text{on } (0, L), \\ &w(0) = w'(0) = 0, \\ &w(L) = w'(L) = 0. \end{aligned} \tag{11.31}$$

The space of admissible functions is $V = H_0^2(\Omega)$, and the VBVP is: find $w \in V$ such that

$$\int_0^L w''v''\,dx = \int_0^L (f/EI)v\,dx \quad \text{for all } v \in V.$$

It is clear then that the space $V^h$ must comprise functions whose second derivatives exist, at least in a weak sense. By analogy with the set of conditions (11.5) through (11.8) for second-order problems, we therefore stipulate that the basis functions of $V^h$ must satisfy the following properties.

(i) The global basis functions comprise two sets, denoted by $N_i$ ($i = 1, \ldots, G$) and $M_j$ ($j = 1, \ldots, K$); these functions are *bounded and continuously differentiable*, that is,

$$N_i,\ M_i \in C^1(\overline{\Omega});$$

(ii) each of the functions $M_i$ and $N_i$ is nonzero only on those elements that are connected to node $i$:

$$\left.\begin{array}{l} M_i(x)|_{\Omega_e} \\ N_i(x)|_{\Omega_e} \end{array}\right\} \equiv 0 \ \text{ if } x \notin \Omega_e;$$

(iii) the basis functions have the properties

$$N_i(x_j) = \frac{dM_i}{dx}(x_j) = \begin{cases} 1 & \text{if } i = j, \\ 0 & \text{if } i \neq j, \end{cases} \qquad M_i(x_j) = \frac{dN_i}{dx}(x_j) = 0;$$

(iv) let $N_i^{(e)}$ and $M_i^{(e)}$ be, respectively, the *restrictions* of $N_i$ and $M_i$ to $\Omega_e$; then $N_i^{(e)}$ and $M_i^{(e)}$ are *polynomials*.

This time it is clear from (iii) and (iv) that the local basis function $N_i^{(e)}$ defined on element $\Omega_e$ will have the properties

$$\begin{aligned} N_i^{(e)}(x_j) &= \begin{cases} 1 & \text{if } i = j, \\ 0 & \text{otherwise,} \end{cases} \\ N_i^{(e)'}(x_j) &= 0 \text{ at all nodal points } x_j, \end{aligned}$$

and the local function $M_i^{(e)}$ will have the properties

$$\begin{aligned} M_i^{(e)}(x_j) &= 0 \text{ at all nodal points } x_j, \\ M_i^{(e)'}(x_j) &= \begin{cases} 1 & \text{if } i = j, \\ 0 & \text{otherwise.} \end{cases} \end{aligned}$$

A basis that satisfies the properties (i) through (iv) is known as a *Hermite family*. The conditions (i) and (iv) ensure that the Hermite basis functions $N_i$ and $M_i$ belong to $H^2(\Omega)$, as required. In contrast to (11.10) and (11.11), if we now write

$$v_h = \sum_{i=1}^{G} b_i N_i + \sum_{j=1}^{K} d_j M_j,$$

then it follows from the construction of the basis functions that

$$v_h(x_j) = \sum_{i=1}^{G} b_i N_i(x_j) + \sum_{i=1}^{K} d_i M_i(x_j) = b_j,$$

whereas

$$v_h'(x_j) = \sum_{i=1}^{G} b_i N_i'(x_j) + \sum_{i=1}^{K} d_i M_i'(x_j) = d_j.$$

So both the value of a function *and its derivative* are interpolated by Hermite basis functions.

The space $X^h$ is now simply defined by

$$X^h = \text{span}\{N_1, \ldots, N_G, M_1, \ldots, M_K\},$$

and $V^h$ by $V^h = V \cap X^h$.

**Example**

4. Returning to a one-dimensional problem such as (11.30) for the elastic beam, suppose we try the simplest mesh, consisting of a set of elements with nodes only at the interelement boundaries (Figure 11.25). The restriction of the local basis functions to element $\Omega_e$ is required to be polynomials whose values *and* slopes at the nodes are uniquely

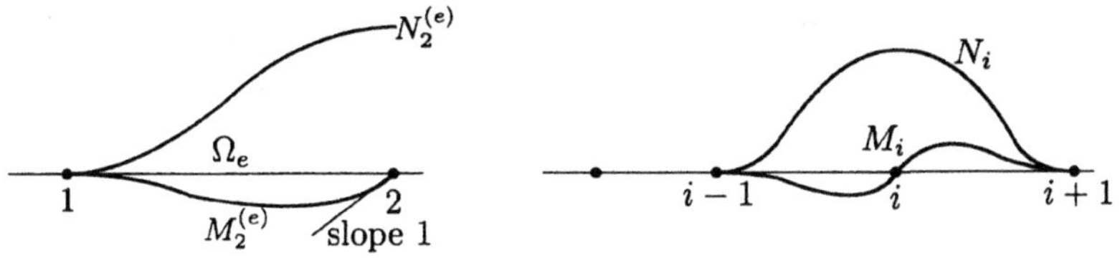


FIGURE 11.25. Local and global Hermite basis functions

determined. Thus an allowance must be made for four coefficients to be determined in each element, and so it follows that the local basis functions are *cubic* polynomials. On the element $\Omega_1 = (x_1, x_2)$, for example, the basis functions are

$$N_1^{(1)}(x) = \frac{-(x-x_2)^2[-h+2(x_1-x)]}{h^3}$$

$$N_2^{(1)}(x) = \frac{(x-x_1)^2[h+2(x_2-x)]}{h^3}$$

$$M_1^{(1)}(x) = \frac{(x-x_1)(x-x_2)^2}{h^2}$$

$$M_2^{(1)}(x) = \frac{(x-x_1)^2(x-x_2)}{h^2}.$$

These are illustrated in Figure 11.25. Global basis functions corresponding to an arbitrary node are also shown in the figure.

The boundary conditions in (11.31) will require that $N_1 = M_1 = N_{E+1} = M_{E+1} = 0$, so that

$$V^h = \text{span}\,\{N_2, \ldots, N_E,\, M_2, \ldots, M_E\}.$$

5. Suppose that it is required to solve the variational problem (9.24) corresponding to deflection of a plate. The space $V^h$ is required to belong to $H^2(\Omega)$, and once again this is achieved by constructing basis functions that are continuously differentiable.

   Consider a triangular element with nodes at the vertices (Figure 11.26), and suppose that we begin by naively extending Example 4, in that the local basis functions on $\Omega_e$ are assumed to be complete cubic functions, that is, members of $P_3(\Omega_e)$. A cubic function of two variables has 10 terms, so it follows that if the function $u$ as well as its first derivatives $u_x \equiv \partial u/\partial x$ and $u_y \equiv \partial u/\partial x$ are going to be the unknown nodal variables, this will account for 9 of the 10 coefficients in the cubic polynomial (3 nodes, and 3 unknowns per node). A fourth node is therefore introduced at the centroid of the element, and only

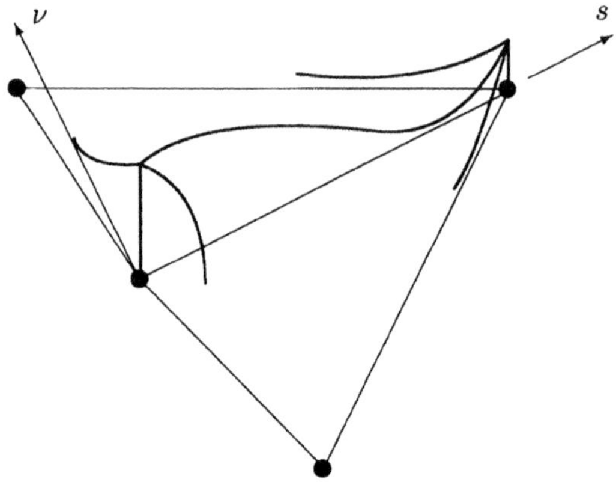


FIGURE 11.26. The restriction of a piecewise cubic function

the value of the function is interpolated at this point. Adopting a local numbering convention, the function $u$ has the representation

$$u = \sum_{I=1}^{3} (b_I N_I^{(e)} + c_I M_I^{(e)} + d_I L_I^{(e)}) + b_4 N_4$$

in which $b_I, c_I$, and $d_I$ are, respectively, the values of $u$, $u_x$ and $u_y$ at node $I$, and $N_I^{(e)}$, $M_I^{(e)}$, and $L_I^{(e)}$ are the corresponding local basis functions. Next we examine whether this local basis will allow a global basis of $C^1$ functions to be constructed. Consider any one of the edges of the triangle, and set up coordinates $(s, n)$ with axes parallel to the tangent and normal to this edge (Figure 11.26). The function $u$ is cubic and, when transformed to the coordinate system $(s, \nu)$, is a cubic function of $s$ along the edge $\nu = 0$. Since $u$ and its first derivatives are specified at the two nodes that constitute the boundary of this edge, it follows that a unique cubic function may be defined along the edge (since $u$ and the tangential derivative $u_s \equiv \partial u/\partial s$ are known at the nodes). Thus, when the function $u$ over the domain $\Omega$ is obtained by patching together its restrictions to the various elements, the resulting function will be *continuous* across adjacent elements.

For this function also to be continuously differentiable across the element boundary, it is necessary that the *normal derivative* $u_\nu \equiv \partial u/\partial \nu$ be continuous there. But $u_\nu$ is a quadratic function of $s$, and is therefore not uniquely determined by its two nodal values. So the global function is *not continuously differentiable* and, as things stand, is not a candidate for a Hermite basis.

The remedy to the problem encountered in Example 5 lies in increasing the degree of polynomial approximation on the element. For this purpose

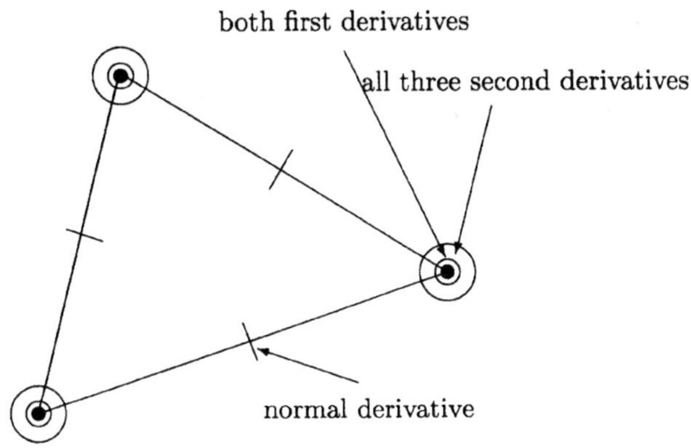


FIGURE 11.27. The degrees of freedom corresponding to a quintic polynomial

the following result is required.

THEOREM 1. *Let $\Omega_e \subset \mathbb{R}^2$ be an arbitrary triangular element with nodes* $1, 2, 3$ *at the vertices and* $4, 5, 6$ *at the midpoints of the sides. Then any complete polynomial function* $p$ *of degree* 5 *is uniquely determined by the nodal values:*

$$\left.\begin{array}{r} p(\boldsymbol{x}_I) \\ p_x(\boldsymbol{x}_I),\ p_y(\boldsymbol{x}_I) \\ p_{xx}(\boldsymbol{x}_I),\ p_{xy}(\boldsymbol{x}_I),\ p_{yy}(\boldsymbol{x}_I) \end{array}\right\} \text{ at the vertices } (I = 1, 2, 3)$$

$p_\nu$ *at the midpoints of the sides.*

Inspection of Pascal's triangle (Figure 11.18) verifies that 21 nodal values are required in order to determine a quintic polynomial uniquely. This element is normally depicted as in Figure 11.27, in which the various degrees of freedom are denoted by different symbols. The proof of this result is left to Exercise 11.11. Bearing in mind the main aim, which is that of constructing a basis which is piecewise polynomial and in $C^1(\Omega)$, and hence in $H^2(\Omega)$, it remains to verify that the function obtained by patching together the quintic interpolation of Theorem 1 will fulfill this purpose. Consider one of the sides which constitutes a boundary between elements: by transformation to the coordinates $(s, \nu)$ we see that the function $u$ is a quintic polynomial $g(s)$, say, along this edge. Now $g$, $g'$, and $g''$ are all determined uniquely at the vertex nodes, and thus the function $g$ is uniquely determined along the edge (since a quintic has six coefficients). Thus we know that $V^h \subset C(\bar{\Omega})$.

Next, consider the restriction to the edge of the normal derivative $\partial u/\partial \nu$, and denote this function by $f(s)$. After transformation to the coordinates $(s, \nu)$ and evaluation at $\nu = 0$, clearly $f$ will be a *quartic* polynomial in $s$. A total of five of its values are uniquely determined: $f$ and $f'$ at the two

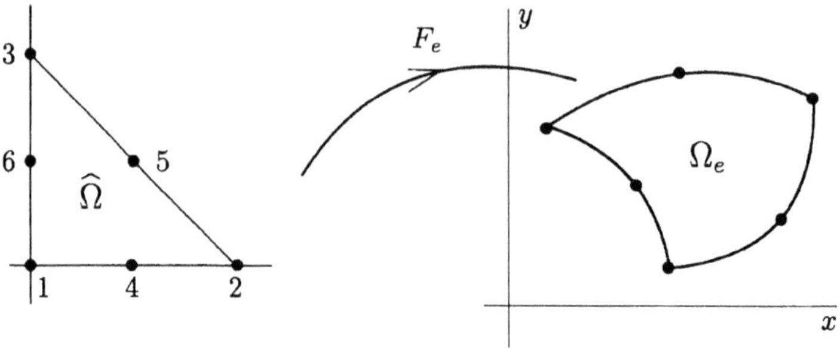


FIGURE 11.28. A curvilinear triangular element

vertex nodes, and $f$ at the midside node. Thus the normal derivative is uniquely determined at the interelement boundary, and so $V^h \subset C^2(\bar{\Omega})$.

## 11.5 Isoparametric elements

For domains in $\mathbb{R}^2$ the finite elements discussed up to now have been restricted to two geometrical types, viz. triangles and rectangles (or, more generally, parallelograms). Such elements are of course adequate for the discretization of domains with polygonal boundaries; but for boundaries of more general shape, and for curved boundaries in particular, it is necessary to extend the ideas of the earlier sections. This section will give an idea of how this extension is carried out, in the context of Lagrangian elements in $\mathbb{R}^2$.

Consider the six-noded reference triangle $\widehat{\Omega}$ which is shown in Figure 11.17, together with the set of quadratic local basis functions. We now generate an element $\Omega_e$ by choosing six nodes $\boldsymbol{x}_I^{(e)}$ $(I = 1, \ldots, 6)$ and by stipulating that $\Omega_e$ be the image of $\widehat{\Omega}$ under the map $F_e : \widehat{\Omega} \to \Omega_e$ defined by

$$\boldsymbol{x} = F_e(\boldsymbol{\xi}) \equiv \sum_{I=1}^{6} \boldsymbol{x}_I^{(e)} \widehat{N}_I(\boldsymbol{\xi}). \tag{11.32}$$

The six nodes $\boldsymbol{x}_I^{(e)}$ of $\Omega_e$ are the images of the six nodes of the reference element, as shown in Figure 11.28, and as may be deduced by using the properties of the local basis functions, and the sides of the reference element are mapped to curves which are described by quadratic polynomials in $\xi$ and $\eta$. In this way we have used the basis functions to generate an element with curved sides: this is known as an *isoparametric element*, and a mesh of elements generated in this way is known as an *isoparametric mesh*. The local basis functions on $\Omega_e$ are generated in the usual way, by using the relationship

$$N_I^{(e)} = \widehat{N}_I \circ F_e \quad \text{or} \quad N_I^{(e)}(\boldsymbol{x}) = \widehat{N}_I(\boldsymbol{\xi}),$$

in which $\boldsymbol{x}$ and $\boldsymbol{\xi}$ are related through (11.32). In this way much of the process developed for affine elements carries over to this more general case. What must be recognized, though, is that the basis functions $N_I^{(e)}$ no longer inherit the polynomial structure of the functions $\widehat{N}_I$, for the simple reason that the map $F_e$ is no longer affine. The manner in which computations are carried out on the reference element is best illustrated through a concrete example.

### Example

6. Consider the VBVP

$$u_h \in V^h, \quad \int_\Omega \nabla u_h \cdot \nabla v_h \, dx = \int_\Omega f v_h \, dx \quad \text{for all } v_h \in V^h,$$

where $V^h \subset V \subset H^1(\Omega)$. The contribution to the stiffness matrix from element $e$ is thus

$$K_{IJ}^{(e)} = \int_{\Omega_e} \nabla N_I^{(e)} \cdot \nabla N_J^{(e)} \, dx,$$

using a local numbering system, in which $I$ and $J$ range over 1 to 6 for a quadratic triangle.

The $2 \times 2$ *Jacobian matrix* $\boldsymbol{J}$ is defined by

$$J_{kl} = \frac{\partial x_k}{\partial \xi_l} \tag{11.33}$$

and is obtained from (11.32). This plays a key role in the evaluation of (11.33) on the reference element, as does its determinant, which is denoted by $j$:

$$j = \det \boldsymbol{J}.$$

It is required that $j(\boldsymbol{\xi}) > 0$ for all $\boldsymbol{\xi} \in \widehat{\Omega}$, in order that the map (11.32) be invertible, and to maintain the orientation of the reference element (for invertibility alone, $j \neq 0$ would suffice). We also observe that for isoparametric elements $j$ is in general a function defined on $\widehat{\Omega}$; for affine maps it is constant.

For computational purposes the integrand of (11.33) is best expressed in matrix form; thus, denoting by $\boldsymbol{B}_I$ the $2 \times 1$ vector consisting of the components of $\nabla N_I^{(e)}$, (11.33) becomes

$$K_{IJ}^{(e)} = \int_{\Omega_e} \boldsymbol{B}_I^t \boldsymbol{B}_J \, dx. \tag{11.34}$$

Now considering that the aim is to evaluate these terms on the reference element, it follows that we have to transform the vectors $\boldsymbol{B}_I$. We have

$$\frac{\partial N_I^{(e)}}{\partial x_j} = \sum_{i=1}^{2} \frac{\partial \widehat{N}_I}{\partial \xi_i} \frac{\partial \xi_i}{\partial x_j}$$

or, in matrix form,

$$\boldsymbol{B}_I = \boldsymbol{J}^{-t} \widehat{\boldsymbol{B}}_I, \tag{11.35}$$

where $\widehat{\boldsymbol{B}}_I^t = [\partial \widehat{N}_I / \partial \xi_1 \; \partial \widehat{N}_I / \partial \xi_2]^t$. This is very convenient, except for one problem: the elements of the inverse Jacobian $\boldsymbol{J}^{-1}$ are given by

$$J_{kl}^{-1}(\boldsymbol{x}) = \frac{\partial \xi_k}{\partial x_l},$$

and evaluation of this matrix would require that (11.32) be inverted, which is a nontrivial task in general.

Fortunately there is a way around this difficulty; indeed, if $\boldsymbol{J}$ is written for convenience in the form $\boldsymbol{J} = \begin{pmatrix} a & b \\ c & d \end{pmatrix}$, then its inverse is given by

$$\boldsymbol{J}^{-1} = \frac{1}{j} \begin{pmatrix} d & -b \\ -c & a \end{pmatrix}.$$

Thus the elements of $\boldsymbol{J}^{-1}$ can be expressed entirely in terms of partial derivatives of $x_k$ with respect to $\xi_l$, which are easily evaluated. This now clears the way for the evaluation of (11.34) since direct substitution of (11.35), together with transformation to the reference element, give

$$\begin{aligned} K_{IJ}^{(e)} &= \int_{\widehat{\Omega}} (j^{-2}) \widehat{\boldsymbol{B}}_I^t \boldsymbol{J}^{-1} \boldsymbol{J}^{-T} \widehat{\boldsymbol{B}}_J \, j \; d\xi \, d\eta \\ &= \int_{\widehat{\Omega}} (j^{-1}) \widehat{\boldsymbol{B}}_I^t \boldsymbol{J}^{-1} \boldsymbol{J}^{-t} \widehat{\boldsymbol{B}}_J \; d\xi \, d\eta. \end{aligned} \tag{11.36}$$

It is important to note that the integrand is no longer a polynomial; rather, due to the presence of the jacobian determinant in the denominator, it is a rational polynomial. In practice integrals such as that appearing in (11.36) are evaluated approximately using numerical integration, a procedure that will be discussed in the next section.

The procedure for generating isoparametric elements from a reference square is unaltered. Whereas there was no point in starting with the three-noded triangle, since an isoparametric map would simply take a triangle to

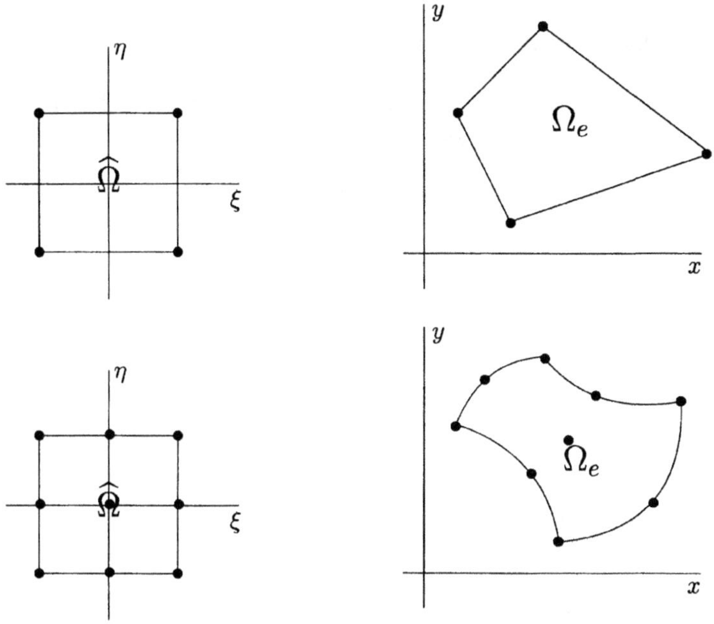


FIGURE 11.29. Isoparametric maps from a reference square element

a triangle, in the case of a square reference element there is every reason to begin with the four-noded element: the basis functions are bilinear, so that the map

$$\boldsymbol{x} = \sum_{I=1}^{4} \boldsymbol{x}_I \widehat{N}_I(\boldsymbol{\xi})$$

with $\widehat{N}_I$ given by (11.27), will give most generally an arbitrary quadrilateral of the kind shown in Figure 11.29. Thus while this element does not have curved sides, the isoparametric concept permits the possibility of working with quadrilaterals whose sides are not parallel.

The next step up is the nine-noded element; here the isoparametric map does lead to an element with curved sides, as shown in Figure 11.29.

The ideas embodied in Example 6 lie at the heart of finite element computations. Indeed, the convenience of this procedure lies not only in the ease with which the element stiffness matrix may be evaluated for arbitrary elements, but also in the fact that the same procedure may be used to carry out these computations also for elements generated by affine maps: (11.32) may be used to generate an element $\Omega_e$ simply by choosing the reference element geometry (triangle or square), the number of nodes and their placement, and the coordinates of the nodal points. The rest of the computations follow as in Example 6.

## 11.6 Numerical integration

A key stage in the implementation of the finite element method is the construction of the stiffness matrix and load vector, and these require that a number of terms be integrated, generally over the reference element. Now while such integrations can be carried out in closed form for simple problems such as the Poisson equation, more complex problems arising, for example, from the modelling of non-homogeneous media, will give integrands which may well not be integrable in closed form, particularly if the coefficients representing the non-homogeneities are anything other than straightforward functions. The existence of such complex integrands will also arise from the use of isoparametric elements, as we have seen in the previous section.

There is thus the need to find an alternative, possibly approximate, way of computing integrals. Two criteria which any such alternative must meet are: (a) its degree of accuracy must be known; and (b) it must be amenable to easy implementation in finite element computer programs. The basis of most numerical integration schemes is the identification of selected points, known as sampling points, at which the value of the function is sampled, and the specification of a set of weights, one for each sampling point.

Suppose that integration is to be carried out over one of the reference elements $\widehat{\Omega}$; then if the sampling points are denoted by $\tilde{\boldsymbol{\xi}}_\ell$ $(\ell = 1, \dots, r)$ and the weights by $w_\ell$ $(\ell = 1, \dots, r)$, a numerical integration formula *of order r* is defined to be a formula of the kind

$$\int_{\widehat{\Omega}} f(\boldsymbol{\xi})\, d\xi \simeq \sum_{\ell=1}^{r} w_\ell f(\tilde{\boldsymbol{\xi}}_\ell). \tag{11.37}$$

The main aim then is to have available a systematic means of choosing the sampling points and weights in such a way as to be able to minimize the error $\left|\int_{\widehat{\Omega}} f(\boldsymbol{\xi})\, d\xi - I_r(f)\right|$ for an integration scheme of given order, where $I_r(f)$ represents the righthand side of (11.37). The choice is usually carried out in such a way that the integration scheme is exact for polynomials of a given degree.

**One-dimensional problems.** For integration of functions over the interval $(-1, 1)$, *Gauss quadrature* is a popular option. The Gauss quadrature rule may be defined for any order, though schemes up to those of order three are most common in finite element calculations. Sampling points and weights for the rules of orders 1, 2 and 3 are as follows:

| Order | $\xi_\ell$ | $w_\ell$ |
|---|---|---|
| 1 | $0$ | $2$ |
| 2 | $-1/\sqrt{3}$<br>$1/\sqrt{3}$ | $1$<br>$1$ |
| 3 | $-\sqrt{3/5}$<br>$0$<br>$\sqrt{3/5}$ | $5/9$<br>$8/9$<br>$5/9$ |

The weights and sampling points corresponding to Gauss quadrature are chosen in an optimal fashion, so as to integrate exactly polynomials of as high a degree as possible. Thus a polynomial of order $2r$ is integrated exactly by a Gauss rule of order $r+1$. Alternatively, a rule of order $r$ integrates exactly a polynomial of order $2r-1$.

**Example**

7. We show in this example how the sampling points and weights for the scheme of order 2 may be obtained. Suppose that an arbitrary cubic function $f(\xi) = a_0 + a_1\xi + a_2\xi^2 + a_3\xi^3$ is to be integrated exactly over the interval $(-1, 1)$, using a scheme of order 2. Now

$$\int_{-1}^{1} f(\xi)\, d\xi = 2a_0 + \tfrac{2}{3}a_2,$$

so it is required to find sampling points $\tilde{\xi}_1$ and $\tilde{\xi}_2$, and weights $w_1$ and $w_2$, such that

$$w_1 f(\tilde{\xi}_1) + w_2 f(\tilde{\xi}_2) = 2a_0 + \tfrac{2}{3}a_2. \tag{11.38}$$

Suppose that we simplify matters by assuming that the sampling points are located symmetrically about the origin, and the two weights are equal, so that $\tilde{\xi}_2 = -\tilde{\xi}_1$ and $w_2 = w_1$; then we obtain

$$2w_1(a_0 + a_2\tilde{\xi}_1^2) = 2a_0 + \tfrac{2}{3}a_2.$$

This must hold for all values of $a_0$ and $a_1$, and so it follows that $w_1 = 1$ and $\tilde{\xi}_1 = 1/\sqrt{3}$.

Gauss quadrature is closely related to properties of the Legendre polynomials, which were introduced in Chapter 6. Indeed, it can be shown that the sampling points and weights corresponding to an integration rule of order $r$ are given by

$$\tilde{\xi}_\ell = \ell\text{th zero of the Legendre polynomial } P_r,$$

$$w_\ell = \frac{2}{[P_r'(\tilde{\xi}_\ell)]^2(1-\tilde{\xi}_\ell)^2}, \qquad \ell = 1, \dots, r.$$

**Integration over a square reference element.** The extension of the Gauss quadrature rule to the reference element $(-1,1)\times(-1,1)$ is straightforward, and relies simply on the application of the one-dimensional rule in each coordinate direction: thus

$$\begin{aligned}
\int_{\widehat{\Omega}} f(\xi,\eta)\,d\xi d\eta &= \int_{-1}^{1}\int_{-1}^{1} f(\xi,\eta)\,d\xi d\eta \\
&\simeq \int_{-1}^{1}\sum_{\ell=1}^{r} w_\ell f(\tilde{\xi}_\ell,\eta)\,d\eta \\
&\simeq \sum_{\ell=1}^{r} w_\ell \sum_{m=1}^{r} w_m f(\tilde{\xi}_\ell,\tilde{\eta}_m) \\
&= \sum_{\ell,m=1}^{r} w_\ell w_m f(\tilde{\xi}_\ell,\tilde{\eta}_m).
\end{aligned}$$

**Integration over triangles.** An integration rule of order 1 may be defined on a triangle $\Omega_e$ by

$$\int_{\Omega_e} f(x,y)\,dx\,dy \simeq A_e f(\widehat{x},\widehat{y}) \tag{11.39}$$

in which $(\widehat{x},\widehat{y})$ are the coordinates of the *centroid* of the triangle (Figure 11.30). Likewise, a rule of order 3 may be defined by

$$\int_{\Omega_e} f(x,y)\,dx\,dy \simeq \tfrac{1}{3}A_e \sum_{\ell=1}^{3} f(\tilde{x}_\ell,\tilde{y}_\ell), \tag{11.40}$$

where $(\tilde{x}_\ell,\tilde{y}_\ell)$ $(\ell = 1,2,3)$ are the coordinates of the midpoints of the sides (Figure 11.30). It is not too difficult to show (see Exercise 17) that the rule of order 1 is exact for polynomials of degree 1, while the rule of order 3 is exact for polynomials of order 2.

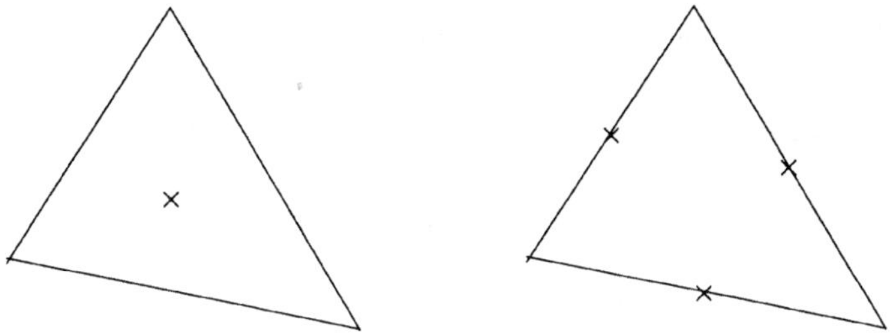


FIGURE 11.30. Integration rules on the reference triangle

## 11.7 Bibliographical remarks

The basic ideas set out in Sections 11.1 and 11.2 may be found in most books on finite elements, though the style and emphasis vary considerably from one book to another. The subject first gained prominence through its use in the solution of problems in solid and structural mechanics, and the vast majority of texts, though covering most of the essential ideas, are directed at those whose interests lie in mechanics. For some insight into the 'real' applications of the method the books of Zienkiewicz and Taylor [56, 57] are recommended. Burnett [9] also gives examples of a number of physical applications, many of them from outside mechanics. The Finite Element Handbook [24] is an encyclopaedic work which covers just about every aspect of the subject, including a wealth of physical applications.

Computational considerations have not been discussed in detail in this exposition. This is a huge topic in its own right, and a number of textx provide comprehensive coverage of the procedures which are relevant to writing efficient computer programs. The texts [9, 56, 57] are again good sources, as are the works by Hughes [22] and by Dhatt and Touzot [14]. Much valuable theoretical background may also be found in these works.

Other useful elementary texts include those by Becker, Carey and Oden [5], and by Johnson [23]. The latter provides a fairly comprehensive treatment of convection-diffusion problems, for which it is necessary to deviate from the standard Galerkin-based approach.

## 11.8 Exercises

### The finite element method for second order problems

11.1. Assume that the space $X_e$ spanned by local basis functions belongs to $H^1(\Omega_e)$, and that $X_h \subset C(\Omega)$. Show that $X_h \subset H^1(\Omega)$. [Take any $v \in X_h$: apply Green's theorem to $\int_{\Omega_e} (\partial v/\partial x_i)\psi\, dx$, where $\psi \in C_0^\infty(\Omega_e)$; then sum over all elements.]

11.2. The half-bandwidth of a symmetric matrix $\boldsymbol{K}$ is the smallest number $B$ corresponding to which $K_{ij} = 0$ for all $i$ and for all $j$ satisfying $|j - i + 1| > B$. Number the nodes in the finite element mesh shown, in such a way that the resulting stiffness matrix has as small a half-bandwidth as possible.

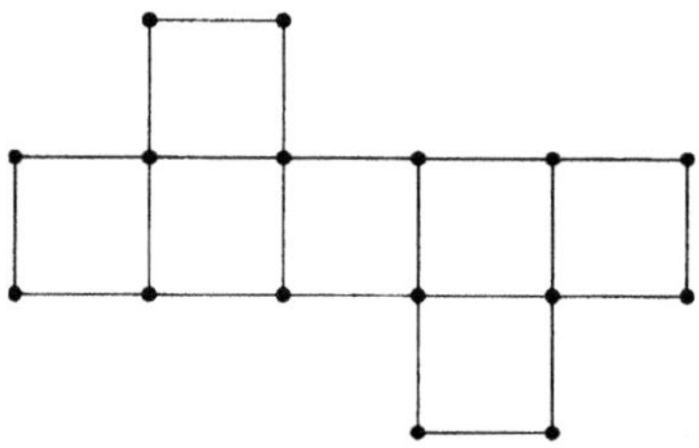

**One-dimensional problems**

11.3. Rework Example 1 using a mesh of two elements and the *quadratic* local basis functions
$\widehat{N}_1(\xi) = \frac{1}{2}\xi(\xi - 1)$, $\widehat{N}_2(\xi) = 1 - \xi^2$, $\widehat{N}_3(\xi) = \frac{1}{2}\xi(\xi + 1)$.

11.4. Let $X_h$ be the space spanned by piecewise linear functions, that is, $X_e = P_1(\Omega_e)$, where $\Omega_e \subset \Omega \subset \mathbb{R}$. Let $f$ be any function defined on $\Omega$, and assume that $f$ can be differentiated as many times as desired. Let $\tilde{f}_h$ be the interpolate of $f$ in $X_h$. The purpose of this Exercise is to show that the *interpolation error* $\tilde{e} = f - \tilde{f}_h$ satisfies the *error bound*

$$||e||_\infty = \max_{0 \le x \le l} |\, f(x) - \tilde{f}_h(x) \,| \le \frac{h^2}{8} \max_{0 \le x \le l} |\, f''(x) \,|$$

where $h$ is the length of an element. Expand $\tilde{e}(x)$ in a Taylor series about any point $\bar{x}$ in $\Omega_e$, that is,

$$\tilde{e}(x) = \tilde{e}(\bar{x}) + \tilde{e}'(\bar{x})(x - \bar{x}) + \tfrac{1}{2}\tilde{e}''(z)(x - \bar{x})^2$$

where $z$ is a point between $x$ and $\bar{x}$. Select $\bar{x}$ to be the point at which $\tilde{e}$ is a maximum; then derive the result

$$|\tilde{e}(\bar{x})| = \tfrac{1}{2}|\tilde{e}''(z)|(x_i - \bar{x})^2$$

where $x_i$ is one of the nodes of $\Omega$. Assuming that $x_i$ is the node nearer to $\bar{x}$, obtain the error estimate.

11.5. Use Exercise 4 to estimate the error $||f - \tilde{f}_h||_\infty$ if $f$ is the function $f(x) = x \sin \pi x$ on the domain $\Omega = (0, 1)$. Compute the actual error using two, three and four elements, and compare with the estimate. Plot a log-log graph of error vs. $h$ and plot the three points corresponding to the three actual errors obtained. Do these points indicate a quadratic rate of convergence?

### Two-dimensional elements

11.6. Show that the basis functions $N_i$ obtained by patching together quadratic local basis functions $N_i^{(e)}$ on triangular elements are continuous.

11.7. It is possible to eliminate the interior node in elements such as the nine-noded quadrilateral, and in so doing to arrive at an element which has nodal points only at the vertices and the midpoints of the sides. Using Pascal's triangle, consider which terms should be contained in such an approximation, and derive the local basis functions. This eight-noded element is known as a *serendipity* element, presumably as a result of its accidental discovery.

11.8. Rework Example 2 using the mesh shown below:

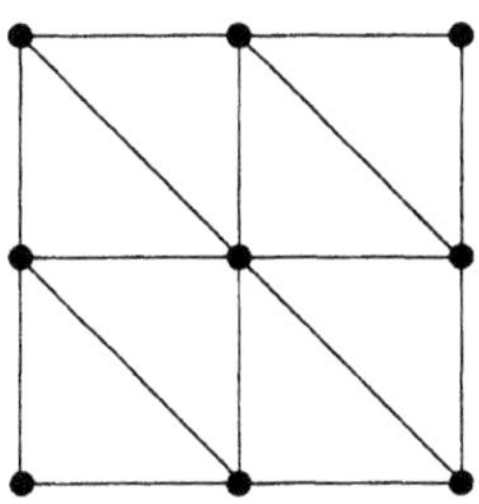

### Fourth-order problems and Hermite families of elements

11.9. Using a mesh of two elements, find an approximate solution to the beam problem (11.31), and compare this with the exact solution

$$w(x) = \frac{fL^4}{EI}\left[\frac{1}{24}\left(\frac{x}{L}\right)^4 - \frac{1}{12}\left(\frac{x}{L}\right)^3 + \frac{1}{24}\left(\frac{x}{L}\right)^2\right].$$

11.10. Prove Theorem 1.

### Isoparametric elements

11.11. Prove that the isoparametric map from the reference element to a parallelogram is necessarily affine.

11.12. Determine the range of values of $d$ for which the quadrilateral element shown below has a jacobian determinant which is everywhere positive.

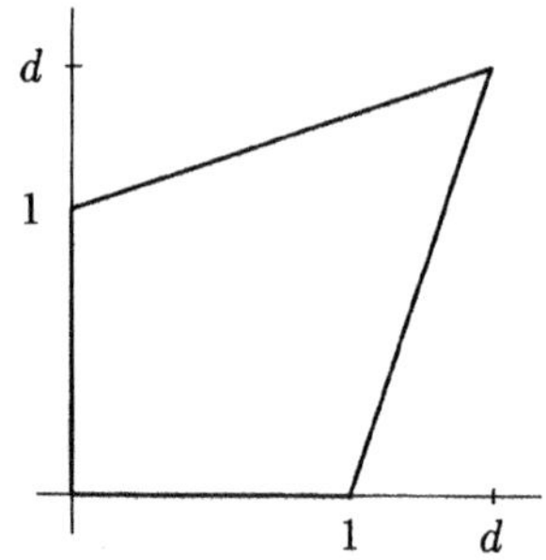


## Numerical integration

11.13. Following the procedure used in Example 7, find the sampling points and weights corresponding to a Gauss quadrature rule of order 3 on the reference triangle.

11.14. Rework Example 2 using the method of Example 6, with $2 \times 2$ Gauss quadrature.

11.15. The purpose of this exercise is to explore the consequences of under-integration, the process whereby the terms in the stiffness matrix are obtained by using an integration scheme of a lower order than that required for exact integration. Consider an element in the form of the reference square $(-1,1) \times (-1,1)$ (that is, $\Omega_e = \widehat{\Omega}$) and suppose that the bilinear form is that corresponding to the Laplacian operator.

(a) The basis functions (11.27) may be expressed in vectorial form as

$$\widehat{\boldsymbol{N}} = \tfrac{1}{4}[\boldsymbol{a} + \boldsymbol{b}\xi + \boldsymbol{c}\eta + \boldsymbol{d}\xi\eta];$$

find the constant vectors $\boldsymbol{a}$, $\boldsymbol{b}$, $\boldsymbol{c}$ and $\boldsymbol{d}$.

(b) The element stiffness matrix is given by (11.33) and the integrand may be expressed in the alternative form $(\nabla \boldsymbol{N})(\nabla \boldsymbol{N})^t$, where $\nabla \boldsymbol{N}$ is the $4 \times 2$ matrix with entries $\partial \widehat{N}_I / \partial \xi_k$. Evaluate the stiffness matrix by integrating exactly, and show that the null space of this matrix is spanned by the single vector $\boldsymbol{a}$.

(c) Evaluate the stiffness matrix again, this time using a one-point integration rule with sampling point $(0,0)$ and weight $w = 4$. Show that the resulting matrix has a *two*-dimensional null space, spanned by $\boldsymbol{a}$ and $\boldsymbol{d}$.

Underintegration has an obvious economical advantage when large problems are required to be solved; but in making use of this procedure, it is necessary remove the additional vector $\boldsymbol{d}$ from the null space, since the desired solution will be polluted by this vector. Highly effective schemes exist for achieving this end.

11.16. Show that the integration rule (11.39) is exact for polynomials of degree 1, while the rule (11.40) is exact for polynomials of order 2.

# 12
# Analysis of the finite element method

Chapter 11 has been devoted to a detailed account of the finite element method, with the focus being on the basic ideas underlying the method, as well as a number of issues that arise in practice. The goal of this chapter is to take developments a step forward, and to provide a mathematical justification for the method. In other words, we return to the problem posed in Chapter 10, in the context of the Galerkin method: given a variational boundary value problem with solution $u$ and approximate solution $u_h$, estimate the error $u - u_h$, and determine the rate of convergence of $u_h$ to $u$ as $h \to 0$. This problem is now addressed in the context of the finite element method.

It was seen in Chapter 10 that the error $\|u - u_h\|_V$ is bounded, up to a multiplicative constant, by the shortest distance from $u$ to the subspace $V^h$ (Theorem 2, Chapter 10). This is Céa's Lemma, and it forms the cornerstone of the analysis of the finite element method; indeed, since this shortest distance is in turn bounded above by the distance $\|u - \tilde{u}_h\|_V$ between $u$ and its *interpolate* $\tilde{u}_h \in V^h$, sharp estimates of the interpolation error will suffice to obtain a knowledge of the finite element approximation error.

The aim of this chapter, therefore, is to obtain such interpolation estimates. The theory is developed in the context of elements that are obtained by affine maps from a reference element, so that the domain $\Omega$ is assumed to have a boundary that is polygonal in $\mathbb{R}^2$, and polyhedral in $\mathbb{R}^3$. Otherwise the theory presented here is quite general in nature.

Section 12.1 is devoted to a discussion of affine families of elements, and of interpolation operators. In Section 12.2 the aim is to derive estimates of

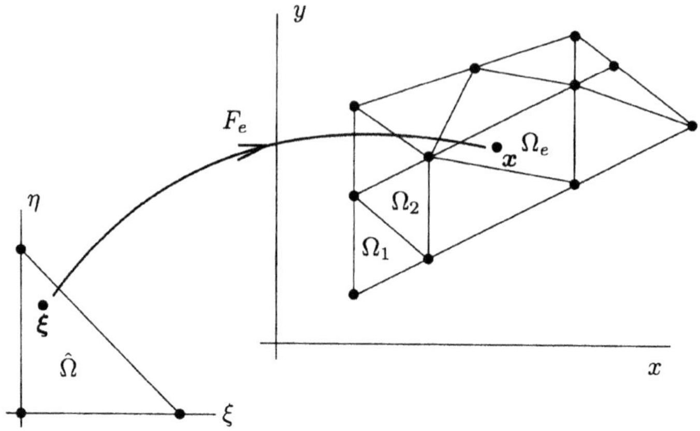


FIGURE 12.1. Generation of a finite element mesh by a family of affine maps

the interpolation error on a single element. This estimate takes the form of a bound on the $H^m$-seminorm of $u - \tilde{u}_h$, in terms of geometrical properties of the element. Then in Section 12.3 error estimates are derived for second-order problems, in appropriate Sobolev norms. The final section of this chapter is devoted to a discussion of the modifications that must be made to the theory in order to accommodate the presence of curved boundaries, and also to incorporate into the estimates the error due to numerical integration.

## 12.1 Affine families of elements

In this section we start to set up the machinery that is vital to a proper development of error estimates for finite element approximations.

**Affine-equivalent elements.** We consider a situation in which a domain $\Omega$ has been partitioned into $E$ finite elements, all elements being of the same geometrical type (for example, all triangles) and having the same degree of approximation (for example, all three-noded triangles). Such a finite element mesh may be generated simply by setting up a single *reference element* $\hat{\Omega}$, say, and by mapping or transforming $\hat{\Omega}$ into each one of the elements $\Omega_e$ in turn (Figure 12.1).
The basic idea has been encountered in Chapter 11, and is very simple. First, define the *reference element* $\hat{\Omega}$, this element being of the same geometrical type as the elements that make up $\Omega$. Next, define an *affine transformation*, that is, a transformation that maps straight lines into straight lines, by

$$F_e : \hat{\Omega} \to \Omega_e \subset \mathbb{R}^n, \quad F_e(\boldsymbol{\xi}) \equiv \boldsymbol{T}_e\boldsymbol{\xi} + \boldsymbol{b}_e = \boldsymbol{x}, \tag{12.1}$$

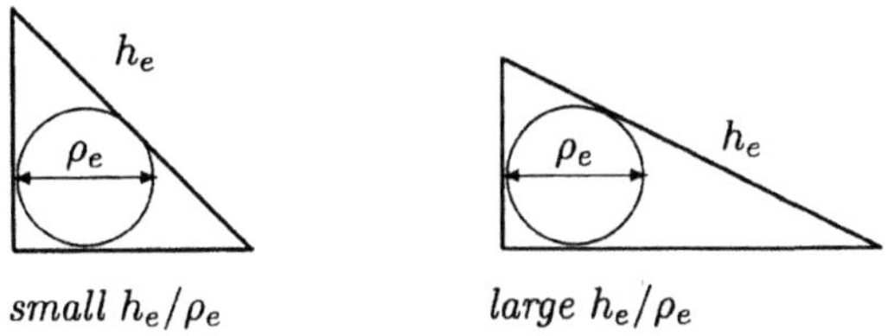


FIGURE 12.2. The constants $h_e$ and $\rho_e$ associated with an element

so that $F_e$ maps each point $\boldsymbol{\xi}$ of $\hat{\Omega}$ to a point $\boldsymbol{x}$ of $\Omega_e$. Here $\boldsymbol{T}_e$ is an invertible $n \times n$ matrix and $\boldsymbol{b}_e$ is a translation vector. We also require of $F_e$ that it maps the nodal point $\boldsymbol{\xi}_I$ of $\hat{\Omega}$ to the (locally numbered) nodal point $\boldsymbol{x}_I^{(e)}$ of $\Omega_e$:

$$F_e(\boldsymbol{\xi}_I) = \boldsymbol{x}_I^{(e)}, \quad I = 1, \ldots, N. \tag{12.2}$$

Once a set of affine transformations has been constructed in this way for each element, we need to focus attention only on the reference element $\hat{\Omega}$ and the family of transformations $F_1, F_2, \ldots, F_E$, since *these provide a complete description of the mesh.*

When two elements $\hat{\Omega}$ and $\Omega_e$ are related to each other by a transformation of the type (12.1), (12.2), they are said to be *affine-equivalent.* Also, a set of finite elements $\Omega_1, \ldots, \Omega_E$ is called an *affine family* if all elements are affine-equivalent to a single reference element $\hat{\Omega}$.

It should be clear from the discussion in Section 11.3 that affine maps of the form (12.1), (12.2) exist in $\mathbb{R}$, and in $\mathbb{R}^2$ from one triangle to another, and as far as quadrilaterals go, most generally from one parallelogram to another. Similar results hold in $\mathbb{R}^3$ for tetrahedra and 3-rectangles or "bricks". We are thus assured that affine maps are always available for the elements with which we are concerned.

The relative size and shape of an arbitrary element $\Omega_e$ are quantified in a natural way by defining the constants

$$h_e = \operatorname{diam}(\Omega_e) = \max\{|\boldsymbol{x} - \boldsymbol{y}|, \ \boldsymbol{x}, \boldsymbol{y} \in \Omega_e\} \tag{12.3}$$

and

$$\rho_e = \sup\{\text{diameters of all spheres contained in } \Omega_e\}. \tag{12.4}$$

When dealing with the reference element $\hat{\Omega}$ we denote the corresponding constants by $\hat{h}$ and $\hat{\rho}$. These quantities are illustrated in Figure 12.2; whereas $h_e$ gives some idea of the "size" of $\Omega_e$, the ratio $h_e/\rho_e$ gives an indication of how "thin" the element is.

We now summarize some useful properties of the affine transformation

(12.1).

LEMMA 1. *Let $F_e : \hat{\Omega} \to \Omega_e$ be the affine map from $\hat{\Omega}$ to $\Omega_e$ defined by (12.1), for $\hat{\Omega}$, $\Omega_e \subset \mathbb{R}^n$. If the matrix norm $\|\boldsymbol{T}_e\|$ is defined by*

$$\|\boldsymbol{T}_e\| = \sup\left\{\frac{\|\boldsymbol{T}_e\boldsymbol{\xi}\|}{\|\boldsymbol{\xi}\|} : \boldsymbol{\xi} \neq 0\right\}$$

*with $\|\boldsymbol{\xi}\| = (\sum_{i=1}^n \xi_i\xi_i)^{1/2}$ for any $\boldsymbol{\xi} \in \mathbb{R}^n$, then*

$$\|\boldsymbol{T}_e\| \leq \frac{h_e}{\hat{\rho}} \text{ and } \|\boldsymbol{T}_e^{-1}\| \leq \frac{\hat{h}}{\rho_e}.$$

PROOF. Let $\boldsymbol{z} = \hat{\rho}\boldsymbol{\xi}/\|\boldsymbol{\xi}\|$; then $\|\boldsymbol{z}\| = \hat{\rho}$ and, for $\boldsymbol{\xi} \neq \boldsymbol{0}$,

$$\|\boldsymbol{T}_e\| = \sup\frac{\|\boldsymbol{T}_e\boldsymbol{\xi}\|}{\|\boldsymbol{\xi}\|} = \sup\left\{\frac{\|(\|\boldsymbol{\xi}\|/\hat{\rho})\boldsymbol{T}_e\hat{\boldsymbol{z}}\|}{\|\boldsymbol{\xi}\|}\right\} = \frac{\|\boldsymbol{T}_e\boldsymbol{z}\|}{\hat{\rho}}.$$

Now pick any two points $\boldsymbol{\xi}$ and $\boldsymbol{\eta}$ in $\hat{\Omega}$ that lie on the sphere of diameter $\hat{\rho}$; then $\|\boldsymbol{\xi} - \boldsymbol{\eta}\| = \hat{\rho}$ and so

$$\begin{aligned}
\|\boldsymbol{T}_e\| &= \hat{\rho}^{-1}\sup\|\boldsymbol{T}_e(\boldsymbol{\xi} - \boldsymbol{\eta})\| \\
&= \hat{\rho}^{-1}\sup\|(\boldsymbol{T}_e\boldsymbol{\xi} + \boldsymbol{b}_e) - (\boldsymbol{T}_e\boldsymbol{\eta} + \boldsymbol{b}_e)\| \\
&= \hat{\rho}^{-1}\sup\|\boldsymbol{x} - \boldsymbol{y}\| \leq h_e/\hat{\rho}.
\end{aligned}$$

The second inequality follows similarly (see Exercise 12.1). □

**Mappings of functions.** Suppose that we are given a continuous function $v$ defined on $\Omega_e$; making use of the affine map (12.1), we can set up an operator $K_e : C(\Omega_e) \to C(\hat{\Omega})$ that maps $v$ to a function $\hat{v}$ in $C(\hat{\Omega})$, the function $\hat{v}$ being defined by

$$K_e v = \hat{v}, \quad \hat{v}(\boldsymbol{\xi}) = v(\boldsymbol{x}), \tag{12.5}$$

where $\boldsymbol{x} = F_e(\boldsymbol{\xi})$ (Figure 12.3). The operator $K_e$ is invertible with inverse $K_e^{-1}$, so that

$$K_e^{-1} : C(\hat{\Omega}) \to C(\Omega_e), \quad K_e^{-1}\hat{v} = v. \tag{12.6}$$

Now suppose that $\{\hat{N}_I\}_{I=1}^M$ is a set of *local basis functions* defined on $\hat{\Omega}$ with the usual property that

$$\hat{N}_I(\boldsymbol{\xi}_J) = \begin{cases} 1 & \text{if } J = I \\ 0 & \text{otherwise,} \end{cases}$$

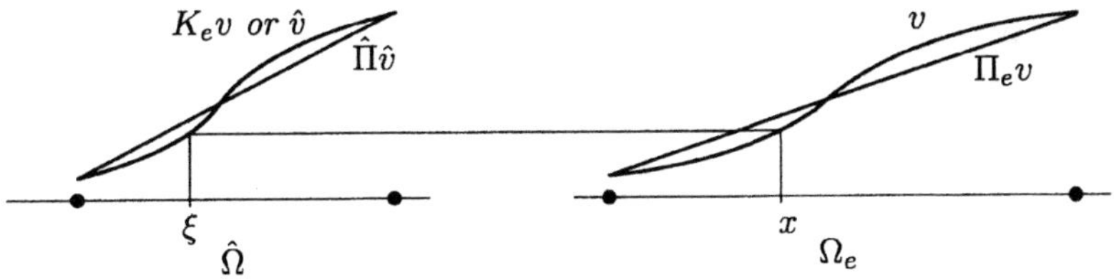


FIGURE 12.3. The map $K_e$

for nodal points $\boldsymbol{\xi}_J$. The function $\hat{N}_I$ is a polynomial of degree $k$, say, that can be mapped to $C(\Omega_e)$ using (12.6):

$$K_e^{-1}\hat{N}_I = N_I^{(e)}.$$

Here $\{N_I^{(e)}\}_{I=1}^M$ is the corresponding set of polynomial *local basis functions* defined on $\Omega_e$; these functions also have the property that $N_I^{(e)}(\boldsymbol{x}_I) = 1$ and $N_I^{(e)}(\boldsymbol{x}_J) = 0$ for $J \neq I$ since (12.5) implies that $\hat{N}_I(\boldsymbol{\xi}_J) = N_I^{(e)}(\boldsymbol{x}_J)$ (we have in fact carried out this transformation for one- and two-dimensional problems in Sections 11.2 and 11.3).

As usual, $\{\hat{N}_I\}$ spans a space $\hat{X}$ (of polynomials, in our case) and so we can construct a *projection operator* $\hat{\Pi}$ that maps any $\hat{v} \in C(\hat{\Omega})$ to its *interpolate* $\hat{v}$ in $\hat{X}$, according to

$$\hat{\Pi} : C(\hat{\Omega}) \to \hat{X}, \quad \hat{\Pi}\hat{v} = \sum_{I=1}^{M} \hat{v}(\boldsymbol{\xi}_I)\hat{N}_I. \tag{12.7}$$

Similarly, we define the projection operator $\Pi_e$ by

$$\Pi_e : C(\Omega_e) \to X_e, \quad \Pi_e v = \sum_{I=1}^{M} v(\boldsymbol{x}_I)N_I^{(e)}, \tag{12.8}$$

where $X_e = \text{span}\,\{N_I^{(e)}\}$ and $\Pi_e v$ is the interpolate of $v$ in $X_e$. We come now to a crucial question about such interpolations: given a function $v$ in $C(\Omega_e)$ and its image $K_e v$ or $\hat{v}$ in $C(\hat{\Omega})$, are $\hat{\Pi}(K_e v)$ and $K_e(\Pi_e v)$ the same functions? That is, if we map $v$ to $\hat{v}$ and then interpolate in $\hat{\Omega}$, is this the same as first interpolating $v$ and then mapping it? A glance at the sketch in Figure 12.3 (for linear interpolations) would seem to indicate that this is plausible; we now prove the assertion.

THEOREM 1. *Let $\hat{\Omega}$ and $\Omega_e$ be affine-equivalent finite elements. Then the interpolation operators $\hat{\Pi}$ and $\Pi_e$ are such that*

$$\hat{\Pi}(K_e v) = K_e(\Pi_e v) \quad \text{or} \quad \hat{\Pi}\hat{v} = \widehat{\Pi_e v}.$$

PROOF. We have

$$\Pi_e v = \sum_{I=1}^{M} v(\boldsymbol{x}_I) N_I^{(e)}$$

by virtue of (12.8). Hence

$$\begin{aligned} K_e(\Pi_e v) &= K_e\left(\sum_{I=1}^{M} \hat{v}(\boldsymbol{\xi}_I) N_I^{(e)}\right) \\ &= \sum_{I=1}^{M} \hat{v}(\boldsymbol{\xi}_I) K_e N_I^{(e)} \quad (K_e \text{ is a linear operator}) \\ &= \sum_{I=1}^{M} \hat{v}(\boldsymbol{\xi}_I) \hat{N}_I \end{aligned}$$

which is precisely $\hat{\Pi}\hat{v}$. □

## 12.2 Local interpolation error estimates

Recall from the discussion of the convergence of Galerkin approximations in Chapter 10 that the error $\|u - u_h\|$, measured in some appropriate norm, can be bounded above by the *interpolation error* $\|u - \tilde{u}_h\|$, where $\tilde{u}_h$ is the interpolate of $u$ in $V^h$. The task of estimating the Galerkin error consequently reduces to one of estimating the interpolation error. We go one step further towards obtaining such an estimate by deriving in this section an estimate of the interpolation error $\|v - \Pi_e v\|$ for functions defined on a *single* finite element $\Omega_e$. Once this estimate has been found, it can be used to obtain an estimate for functions defined over the entire domain $\Omega$.

As before, the finite-dimensional space $X_e$ spanned by local basis functions $N_I^{(e)}$ contains polynomials of degree $\leq k$, for some $k \geq 1$. In other words, either $X_e = P_k(\Omega_e)$ or (as in the case of rectangular elements in $\mathbb{R}^2$) $X_e = Q_l(\Omega_e)$ with $l$ large enough so that $P_k(\Omega_e \subset Q_l(\Omega_e)$. We show eventually that an interpolation error estimate in the $H^m$-norm can be derived for a function $v$ that is smooth enough to be in $H^{k+1}(\Omega_e)$, and so consider the situation in which there are two spaces $H^{k+1}(\Omega_e)$ and $H^m(\Omega_e)$ with $k + 1 \geq m$, and a projection operator $\Pi_e$ that maps members of $H^{k+1}(\Omega_e)$ to $H^m(\Omega_e)$, the images $\Pi_e v$ all lying in $X_e$ (Figure 12.4):

$$\Pi_e : H^{k+1}(\Omega_e) \to H^m(\Omega_e), \quad R(\Pi_e) = X_e. \tag{12.9}$$

The projection operator $\Pi_e$ is defined by (12.8), and since $P_k(\Omega_e) \subset X_e$ by assumption, it has the property that

$$\Pi_e v = v \quad \text{for any} \quad v \in P_k(\Omega_e). \tag{12.10}$$

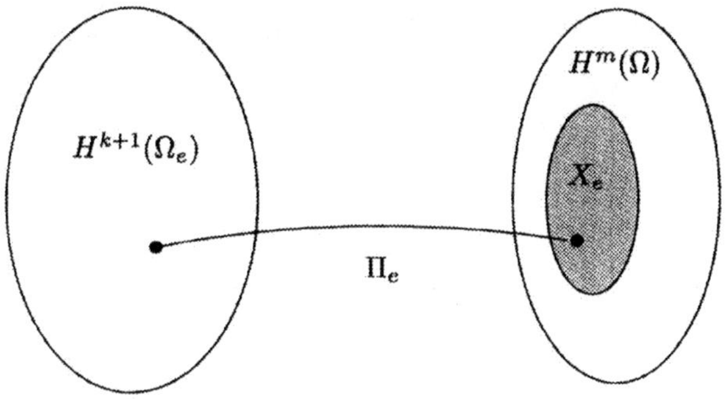


FIGURE 12.4. The action of the operator $\Pi_e$

Similarly,

$$\hat{\Pi}\hat{v} = \hat{v} \quad \text{for any} \quad \hat{v} \in P_k(\hat{\Omega}). \tag{12.11}$$

The main result in this section is: for $v \in H^{k+1}(\Omega_e)$ and $\Pi_e$ satisfying the preceding properties, *the interpolation error in the $H^m$-norm* can be estimated by

$$\|v - \Pi_e v\|_{m,\Omega_e} \leq C h_e^{k+1-m} |v|_{k+1,\Omega_e},$$

where $h_e$ is defined in (12.3) and $|\cdot|_{s,\Omega_e}$ denotes the Sobolev *seminorm:*

$$|v|_{s,\Omega_e}^2 = \sum_{|\alpha|=s} \int_{\Omega_e} [D^\alpha v(\boldsymbol{x})]^2 \, dx$$

(recall also that the Sobolev norm $\|\cdot\|_{s,\Omega_e}$ is given by $\|v\|_{s,\Omega_e}^2 = \sum_{l=1}^{s} |v|_{l,\Omega_e}^2$). Here and subsequently the norm on $H^s(\Omega)$ is denoted by $\|\cdot\|_{s,\Omega}$ rather than the more cumbersome $\|\cdot\|_{H^s(\Omega)}$. We start the development by recording an important result that is required later.

THEOREM 2. *There is a constant $C$, depending only on the geometry of $\Omega$, such that for all $v \in H^{k+1}(\Omega)$,*

$$\inf_{p \in P_k(\Omega)} \|v + p\|_{k+1,\Omega} \leq C |v|_{k+1,\Omega}. \tag{12.12}$$

PROOF. We use the Poincaré inequality (7.19); replacing $u$ by $v + p$ and noting that $D^\alpha p = 0$ for $|\alpha| = k+1$, we have

$$\|v + p\|_{k+1}^2 \leq C \left( |v|_{k+1}^2 + \sum_{|\alpha|<k+1} \left\{ \int_\Omega D^\alpha (v+p) \, dx \right\}^2 \right)$$
$$\text{for } v \in H^{k+1}(\Omega), \quad p \in P_k(\Omega). \tag{12.13}$$

Now construct a polynomial $\bar{p}$ in $P_k(\Omega)$ that has the property that

$$\int_\Omega D^\alpha(v+\bar{p})\,dx = 0 \;\text{ for }\; |\alpha| \le k. \tag{12.14}$$

This can always be done: set $|\alpha| = k$; then $D^\alpha \bar{p}$ equals the coefficient of $\boldsymbol{x}^\alpha$, that can be solved for using (12.14). Having solved for all coefficients of terms of order $k$, set $|\alpha| = k-1$, and use (12.14) to solve for coefficients of terms of order $k-1$. Proceeding in this way, we find $\bar{p}$ for any given $v$.

With $p = \bar{p}$ in (12.13), we have

$$\inf_{p\in P_k(\Omega)} \|v+p\|_{k+1}^2 \le \|v+\bar{p}\|_{k+1}^2 \le C|v|_{k+1}^2,$$

from which (12.12) follows. □

Next we need to know how the seminorms of the functions $v$ and of $\hat{v}$ are related.

THEOREM 3. *Let $\Omega_e$ and $\hat{\Omega}$ be two affine-equivalent open subsets of $\mathbb{R}^n$. Then for any functions $v \in H^s(\Omega_e)$ and $\hat{v} = K_e v \in H^s(\hat{\Omega})$,*

$$|\hat{v}|_{s,\hat{\Omega}} \le \|\boldsymbol{T}_e\|^s |\det \boldsymbol{T}_e|^{-1/2} |v|_{s,\Omega_e} \tag{12.15}$$

*and*

$$|v|_{s,\Omega_e} \le \|\boldsymbol{T}_e^{-1}\|^s |\det \boldsymbol{T}_e|^{1/2} |\hat{v}|_{s,\hat{\Omega}}, \tag{12.16}$$

*where $\boldsymbol{T}_e$ is the matrix occuring in the affine map (12.1).*

PROOF. We prove (12.15); (12.16) is proved in a similar fashion. Now

$$\begin{aligned} |\hat{v}|_{s,\hat{\Omega}}^2 &= \sum_{|\alpha|=s} \int_{\hat{\Omega}} \left(D^\alpha \hat{v}(\boldsymbol{\xi})\right)^2\, d\xi \\ &= \sum_{|\alpha|=s} \int_{\Omega_e} \left(D^\alpha \hat{v}(\boldsymbol{\xi})\right)^2 |\det \boldsymbol{T}_e|^{-1}\, dx \end{aligned} \tag{12.17}$$

(using the result from multivariable calculus that if $\xi_i = f_i(x_j)$, then $d\xi \equiv d\xi_1 d\xi_2 \dots d\xi_n = |\det(\partial f_i/\partial x_j)| dx_1 dx_2 \cdots dx_n$).

By an application of the chain rule we have (see Exercise 12.6), for fixed $\boldsymbol{x}$ and $\boldsymbol{\xi}$,

$$|D^\alpha \hat{v}(\boldsymbol{\xi})| \le \|\boldsymbol{T}_e\|^s |D^\alpha v(\boldsymbol{x})|, \quad |\alpha| = s \tag{12.18}$$

(since $\boldsymbol{\xi}$ and $\boldsymbol{x}$ are fixed, $D^\alpha \hat{v}(\boldsymbol{\xi})$ and $D^\alpha v(\boldsymbol{x})$ are simply real numbers). Hence (12.17) becomes

$$|\hat{v}|_{s,\hat{\Omega}}^2 \le \sum_{|\alpha|=s} \int_{\Omega_e} \left(D^\alpha v(\boldsymbol{\xi})\right)^2 \|\boldsymbol{T}_e\|^{2s} (\det \boldsymbol{T}_e)^{-1}\, dx$$

from which (12.15) follows, since $\|\boldsymbol{T}_e\|$ and $\det \boldsymbol{T}_e$ are constant. □

We come now to the interpolation error estimate for the seminorm $|v - \Pi_e v|_{m,\Omega_e}$.

THEOREM 4. *Let $k$ and $m$ be nonnegative integers such that*

$$H^{k+1}(\hat{\Omega}) \subset C(\hat{\Omega}), \quad H^{k+1}(\hat{\Omega}) \subset H^m(\hat{\Omega}),$$

*and*

$$P_k(\hat{\Omega}) \subset \hat{X} \subset H^m(\hat{\Omega}).$$

*Let $\Pi_e$ and $\hat{\Pi}$ be the operators defined in* (12.7) *and* (12.8). *Then for any affine equivalent element $\Omega_e$ and for all functions $v \in H^{k+1}(\Omega_e)$,*

$$|v - \Pi_e v|_{m,\Omega_e} \le \hat{C} \frac{h_e^{k+1}}{\rho_e^m} |v|_{k+1,\Omega_e}, \tag{12.19}$$

*where $h_e$ and $\rho_e$ are defined in (12.3) and (12.4), and $\hat{C}$ is a constant depending on $\hat{\Omega}$ and $\hat{\Pi}$.*

PROOF. We have, for all $\hat{v} \in H^{k+1}(\hat{\Omega})$ and all $\hat{p} \in P_k(\hat{\Omega})$, and using (12.11),

$$\begin{aligned} |\hat{v} - \hat{\Pi}\hat{v}|_{m,\hat{\Omega}} &\le \|\hat{v} - \hat{\Pi}\hat{v}\|_{m,\hat{\Omega}} = \|\hat{v} - \hat{\Pi}\hat{v} + \hat{p} - \hat{\Pi}\hat{p}\|_{m,\hat{\Omega}} \\ &\le \|I(\hat{v} + \hat{p}) - \hat{\Pi}(\hat{v} + \hat{p})\|_{m,\hat{\Omega}} \\ &\le \|I(\hat{v} + \hat{p})\|_{m,\hat{\Omega}} + \|\hat{\Pi}(\hat{v} + \hat{p})\|_{m,\hat{\Omega}} \\ &\le \underbrace{(\|I\| + \|\hat{\Pi}\|)}_{\hat{C}} \|\hat{v} + \hat{p}\|_{k+1,\hat{\Omega}}. \end{aligned}$$

The last line follows from the fact that $I$ and $\hat{\Pi}$ are bounded operators from $H^{k+1}(\hat{\Omega})$ to $H^m(\hat{\Omega})$. The use of Theorem 2 now yields

$$|\hat{v} - \hat{\Pi}\hat{v}|_{m,\hat{\Omega}} \le \hat{C} \inf_{\hat{p} \in P_k(\hat{\Omega})} \|\hat{v} + \hat{p}\|_{k+1,\hat{\Omega}} \le C\hat{C}|\hat{v}|_{k+1,\hat{\Omega}}. \tag{12.20}$$

From Theorem 1 we have $\hat{\Pi}(K_e v) = K_e(\Pi_e v)$, so that

$$\hat{v} - \hat{\Pi}\hat{v} = K_e v - \hat{\Pi}(K_e v) = K_e(v - \Pi_e v); \tag{12.21}$$

consequently, using (12.16) (replace $v$ by $v - \Pi_e v$ and set $s = m$) and (12.21) we obtain

$$\begin{aligned} |v - \Pi_e v|_{m,\Omega_e} &\le \|\boldsymbol{T}_e^{-1}\|^m |\det \boldsymbol{T}_e|^{1/2} |K_e(v - \Pi_e v)|_{m,\hat{\Omega}} \\ &= \|\boldsymbol{T}_e^{-1}\|^m |\det \boldsymbol{T}_e|^{1/2} |\hat{v} - \hat{\Pi}\hat{v}|_{m,\hat{\Omega}}; \end{aligned} \tag{12.22}$$

furthermore, from (12.15) with $s = k+1$,

$$|\hat{v}|_{k+1,\hat{\Omega}} \leq \|\boldsymbol{T}_e\|^{k+1} |\det \boldsymbol{T}_e|^{-1/2} |v|_{k+1,\Omega_e}. \tag{12.23}$$

Finally, substituting (12.20) in (12.22), then (12.23) in that result we obtain

$$|v - \Pi_e v|_{m,\hat{\Omega}} \leq \hat{C} \|\boldsymbol{T}_e^{-1}\|^m \|\boldsymbol{T}_e\|^{k+1} |v|_{k+1,\Omega_e}$$

which, with the use of Lemma 1, leads to (12.19). □

REMARKS.

1. Since we wish to evaluate $|v - \Pi_e v_{m,\Omega_e}|$, it follows that both $v$ and $\Pi_e v$ must be in $H^m(\Omega_e)$ for this term to make sense. Equivalently, $\hat{v}$ and $\hat{\Pi}\hat{v}$ must be in $H^m(\hat{\Omega})$. This accounts for the inclusions $H^{k+1}(\hat{\Omega}) \subset H^m(\hat{\Omega})$ and $\hat{X} \subset H^m(\hat{\Omega})$. Note that $v \in H^{k+1}(\Omega_e)$ implies $\hat{v} \in H^{k+1}(\hat{\Omega})$. The inclusion $H^{k+1}(\hat{\Omega}) \subset H^m(\hat{\Omega})$ of course holds if $m \leq k+1$.

2. In evaluating the interpolant $\Pi_e v$ of $v$, it is necessary to know the nodal values of $v$. This in turn requires that $v$ be continuous, so that we must have $v \in H^{k+1}(\Omega_e) \subset C(\Omega_e)$ or equivalently, $\hat{v} \in H^{k+1}(\hat{\Omega}) \subset C(\hat{\Omega})$. By the Sobolev Embedding Theorem, this inclusion holds if $k+1 > n/2$ for a problem in $\mathbb{R}^n$.

The two parameters $h_e$ and $\rho_e$ appearing in (12.19) may be reduced to one if attention is restricted to finite elements for which the ratio $h_e/\rho_e$ is bounded above, so that elements are not allowed to become too "flat". For this purpose we introduce the notion of a *regular* family of finite elements. A family $\{\Omega_1, \ldots, \Omega_E\}$ of finite elements is said to be regular if

(i) there exists a constant $\sigma$ such that $h_e/\rho_e \leq \sigma$ for all elements;

(ii) the diameters $h_e$ approach zero.

In the case of regular families the error estimate of Theorem 4 can be expressed in terms of a *norm*; this is recorded in the following.

COROLLARY TO THEOREM 4. *Let the conditions of Theorem 4 hold, and let $\{\Omega_1, \ldots, \Omega_E\}$ be a regular family of finite elements. Then there is a constant $C$ such that, for any element $\Omega_e$ in the family, and all functions $v \in H^{k+1}(\Omega_e)$,*

$$\|v - \Pi_e v\|_{m,\Omega_e} \leq C h_e^{k+1-m} |v|_{k+1,\Omega_e}. \tag{12.24}$$

It is not difficult (see Exercise 12.5) to deduce this result, and in particular to show that it depends on Property (ii) of regular families of finite elements.

**Examples**

1. Let $\Omega_e$ be the three-noded triangle in $\mathbb{R}^2$. The space $X_e$ spanned by the local interpolation functions is $P_1(\Omega_e)$, so that $k = 1$. Assuming that $v$ is smooth enough to belong to $H^2(\Omega_e)$, (12.24) gives

$$\|v - \Pi_e v\|_{m,\Omega_e} \leq Ch^{2-m}|v|_{2,\Omega_e}. \tag{12.25}$$

   We confirm that the conditions of Theorem 3 hold: $H^{k+1}(\hat{\Omega}) = H^2(\hat{\Omega}) \subset C(\hat{\Omega})$ by the Sobolev Embedding Theorem. Second, the estimate (12.25) holds for all $m$ such that $m \leq k+1$; that is, $0 \leq m \leq 2$.

2. For problems such as those arising in linear elasticity, for which the unknown variable is vector-valued, the set of results culminating in (12.25) carries over virtually unchanged. We return to Chapter 9, Example 3, for which case $V = \{\boldsymbol{v} : \boldsymbol{v} \in [H^1(\Omega)]^2, \ \boldsymbol{v} = \boldsymbol{0} \text{ on } \Gamma_1\}$. This problem is posed on a domain in $\mathbb{R}^2$, so suppose that we make use of four-noded rectangular elements, generated by a family of affine maps (11.27) from the reference square.

   Now the basis functions (11.27) corresponding to this element are bilinear; thus the restriction to $\Omega_e$ of any function $v_h \in V^h$ will belong to $Q_1(\Omega_e)$, and since $P_1 \subset Q_1 \subset P_2$ it follows that the value of $k$ appropriate to this problem is $k = 1$. If the $H^m$-norm for vector-valued functions is defined on $\Omega_e$ according to

$$\|\boldsymbol{v}\|^2_{m,\Omega_e} = \|v_1\|^2_{m,\Omega_e} + \|v_2\|^2_{m,\Omega_e},$$

   then for all functions $\boldsymbol{v} \in [H^2(\Omega_e)]^2$ there exists a constant $C$ such that

$$\|\boldsymbol{v} - \Pi_e \boldsymbol{v}\|_{m,\Omega_e} \leq Ch_e^{2-m}|v|_{2,\Omega_e}.$$

## 12.3 Error estimates for second-order problems

Having established properties of finite element interpolations over individual elements, we turn now to the question of interpolation of a function defined on the entire domain $\Omega$. Specifically, we have a function $v \in C(\Omega)$, and we construct its interpolant $\tilde{v}_h$ in the finite element space $X_h$ according to

$$\tilde{v}_h(\boldsymbol{x}) = \sum_{i=1}^{G} v(\boldsymbol{x}_i)N_i(\boldsymbol{x}),$$

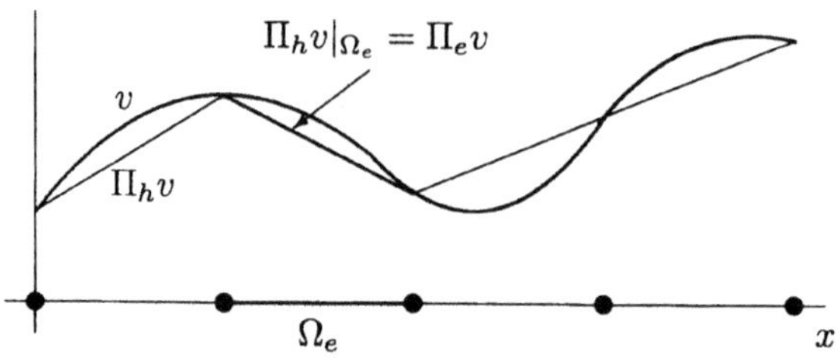


FIGURE 12.5. Global interpolation of a function

where $N_i$ are the global basis functions that span $X_h$. As in Section 12.1, we define a *projection operator* $\Pi_h$ that maps $v$ to its interpolant $\tilde{v}_h$ or $\Pi_h v$ :

$$\Pi_h : C(\Omega) \to X_h, \quad \Pi_h v = \sum_{i=1}^{N} v(\boldsymbol{x}_i) N_i. \tag{12.26}$$

From the way in which the functions $N_i$ are constructed from local basis functions $N_i^e$, it should be clear that the restriction of $\Pi_h v$ to any element $\Omega_e$ is in fact $\Pi_e v$ (Figure 12.5):

$$\Pi_h v|_{\Omega_e} = \Pi_e v, \quad \Pi_h v|_{\Omega_e} = \Pi_e v.$$

Since we are primarily interested in this section in obtaining error estimates for second-order problems we must estimate $\|u - v_h\|_{1,\Omega}$ for any $v_h \in V^h$, in accordance with Céa's Lemma. We choose for convenience $v_h = \Pi_h u$, and so seek an estimate of the *interpolation error* $\|u - \Pi_h u\|_{1,\Omega}$ (recall that $m = 1$ for second-order problems). In the same way as $\|u - \Pi_e u\|_{m,\Omega_e}$ is estimated in terms of the parameter $h_e$, a suitable parameter is required for the global estimate. For this purpose, suppose that we are dealing with a *regular* family of finite elements, and set

$$h = \max_{1 \le e \le E} (h_e). \tag{12.27}$$

The constant $h$ is called the *mesh parameter*, and is a measure of how refined the mesh is: the smaller $h$ is, the larger the number of elements for a given domain $\Omega$. Hence, if it is possible to obtain an interpolation error estimate of the form

$$\|u - \Pi_h u\|_{1,\Omega} \le C h^{\beta} |u|_{k+1,\Omega},$$

then we are assured of convergence as $h \to 0$, provided that $\beta > 0$.

The mesh parameter provides a natural way of quantifying the dimension of the spaces $X_h$ or $V^h$ that occur in Galerkin approximations. Recall from Chapter 10 that we discussed the notion of a *family* of problems, parametrized by a real parameter $h$. The idea is that for each value of $h$ the approximate solution is sought in a finite-dimensional space $V^h$, with the hope that the error $\|u - u_h\|$ approaches zero as $h \to 0$. At the time $h$ was thought of as being, for example, $1/(\dim V^h)$. In the context of the finite element method, though, the mesh parameter gives a measure of how fine the subdivision of $\Omega$ is: the smaller $h$ is, the finer the subdivision. Furthermore, the smaller $h$ is, the larger the number of elements and nodal points will be, and hence the larger the dimension of $V^h$ will be. Furthermore, there is now a clear sense in which $V^h$ can be said to approach $V$ as $h \to 0$: it is required that $\|v - \Pi_h v\| \to 0$ as $h \to 0$. Consequently we may use $h$, as defined in (12.27), as a measure of the size of the subspace $V^h$ relative to $V$.

The following global *interpolation* error estimate establishes the precise sense in which $V^h \to V$.

THEOREM 5. *Assume that all the conditions of Theorem 4 and its corollary hold. Then there exists a constant c independent of h such that, for any* $v \in H^{k+1}(\Omega)$,

$$\|v - \Pi_h v\|_{m,\Omega} \le c h^{k+1-m} |v|_{k+1,\Omega} \quad \textit{for } m = 0 \textit{ or } m = 1. \tag{12.28}$$

PROOF. When $m = 1$, then $\hat{X} \subset H^1(\hat{\Omega})$ and $X_h \subset C(\overline{\Omega})$ imply that $X_h \subset H^1(\Omega)$ (see Exercise 12.7). Hence $\Pi_h u \in H^1(\Omega)$ with $\Pi_h u|_{\Omega_e} = \Pi_e u$ and we thus have, applying the Corollary to Theorem 4 with $m = 0$ or 1,

$$\begin{aligned}
\|u - \Pi_h u\|_{m,\Omega} &= \left( \sum_{e=1}^{E} \|u - \Pi_e u\|^2_{m,\Omega_e} \right)^{1/2} \\
&\le \left( \sum_{e=1}^{E} C^2 h_e^{2(k+1-m)} |u|^2_{k+1,\Omega_e} \right)^{1/2} \\
&\le C h^{k+1-m} \left( \sum_{e=1}^{E} |u|^2_{k+1,\Omega_e} \right)^{1/2} \\
&= C h^{k+1-m} |u|_{k+1,\Omega}.
\end{aligned}$$

This proves the theorem. □

Finally, we come to the error estimate for second-order problems.

THEOREM 6. *Consider the* VBVP *of finding* $u \in V$ *such that*

$$a(u, v) = \langle \ell, v \rangle \quad \text{for all} \quad v \in V \subset H^1(\Omega), \tag{12.29}$$

*where $a(\cdot,\cdot)$ is continuous and $V$-elliptic and $\langle \ell, \cdot\rangle$ is continuous on $V$. If $u_h$ is the finite element approximation of the solution in $V^h$, then there exists a constant $C$ independent of $h$ such that*

$$\|u - u_h\|_{1,\Omega} \leq Ch^k |u|_{k+1,\Omega}.$$

PROOF. From Theorem 2 of Chapter 10, with $v_h = \Pi_h u$ and (12.28) with $m = 1$ we obtain

$$\|u - u_h\|_{1,\Omega} \leq (M/\alpha)\|u - \Pi_h u\|_{1,\Omega} \leq Ch^k |u|_{k+1,\Omega}$$

with $C = cM/\alpha$. □

It may happen in practice that the solution $u$ is not smooth enough to belong to $H^{k+1}(\Omega)$. For example, if we know from the theory of elliptic BVPs that $u$ is in $H^2(\Omega)$, then the use of quadratic six-noded triangles for a problem in $\mathbb{R}^2$ means that $k = 2$ or $k + 1 = 3$, and the seminorm $|v|_{3,\Omega}$ in (12.28) does not necessarily make sense. We overcome this problem by going back to Section 12.2, and by noting that the entire theory developed there still holds if we replace $k + 1$ by $r$, and hence also $k$ by $r - 1$, where $r \leq k + 1$ is any positive integer. Specifically, we do this in Theorems 2 and 4, and in the Corollary to Theorem 4. Of course, $r$ must be such that $H^r(\hat{\Omega}) \subset C(\hat{\Omega})$ (that is, $r > n/2$ and $r \geq m$). The estimate (12.24) then reads, for $v \in H^r(\Omega_e)$,

$$\|v - \Pi_e v\|_{m,\Omega_e} \leq Ch_e^{\mu} |v|_{r,\Omega_e},$$

where $\mu = k + 1 - m$ if $r \geq k + 1$ (since in this case $v \in H^{k+1}(\Omega_e)$ also) and $\mu = r - m$ if $r < k + 1$. Coming to the global estimate (12.28), we may alter this accordingly so that, for $v \in H^r(\Omega)$,

$$\|u - u_h\|_{1,\Omega} \leq Ch^{\alpha} |u|_{r,\Omega}, \tag{12.30}$$

where $\alpha = \min(k, r - 1)$.

We make one more improvement to the error estimate (12.30). As it stands, it involves the *unknown* quantity $|u|_{r,\Omega}$ on the right-hand side. This dependence on $u$ is easily removed, however, if we know that the solution depends continuously on the data. The theory of Chapter 8 leads to the result that if the original PDE is of the form $Au = f$ with $f \in H^s(\Omega)$ and with $\Omega$ having a *smooth* boundary, then the solution $u$ lies in $H^{s+2}(\Omega)$ and

$$\|u\|_{s+2} \leq C_1 \|f\|_s, \tag{12.31}$$

for some constant $C_1 > 0$. The finite element theory developed here is applicable only to polygonal domains (in $\mathbb{R}^2$), but if it is known that the estimate holds even for such a case, then we may set $r = s + 2$, and since

$$|u|_r = |u|_{s+2} \leq \|u\|_{s+2} \leq C_1 \|f\|_s,$$

the dependence on $|u|_r$ in (12.30) may be removed.

COROLLARY TO THEOREM 6. *Let the conditions for Theorem 6 hold, and let the data $f$ be given in $H^s(\Omega)$, $s \geq 0$. Furthermore, assume that* (12.31) *holds. Then a constant $C$ exists such that, as $h \to 0$,*

$$\|u - u_h\|_{1,\Omega} \leq ch^{\beta}\|f\|_s, \tag{12.32}$$

*where $\beta = \min(k, s+1)$.*

According to the theorem and its corollary, since the order of convergence $\beta$ is governed by the smaller of $k$ and $s+1$, when $s \leq k-1$ then convergence is governed by the smoothness of $f$. For example, if $f$ is only in $L^2(\Omega) = H^0(\Omega)$, then it suffices to use elements that contain only polynomials of degree $\leq 1$ (such as two-noded elements in $\mathbb{R}$, three-noded triangles, and four-noded rectangles in $\mathbb{R}^2$).

For problems posed on domains in $\mathbb{R}$ the issue of the smoothness of the boundary does not arise, and so the estimate (12.32) holds in such cases.

**Example**

4. Consider the problem

$$\begin{aligned} -\nabla^2 u &= f \text{ in } \Omega \subset \mathbb{R}^n, \\ u &= 0 \text{ on } \Gamma. \end{aligned}$$

The corresponding VBVP is: find $u \in H_0^1(\Omega)$ such that

$$\int_\Omega \nabla u \cdot \nabla u \, dx = \int_\Omega fv \, dx \quad \text{for all } v \in H_0^1(\Omega),$$

and this problem has a unique solution. Similarly, the VBVP corresponding to the approximate solution is: find $u_h \in V^h$ such that

$$\int_\Omega \nabla_h \cdot \nabla v_h \, dx = \int_\Omega fv_h \, dx \quad \text{for all } v_h \in V^h,$$

and this problem also has a unique solution. Here $V^h$ consists of those piecewise polynomial functions in $X_h$ that satisfy the boundary condition, so that $V^h \subset H_0^1(\Omega)$. If $f \in H^s(\Omega)$, then the error is estimated by

$$\|u - u_h\|_{1,\Omega} \leq ch^{\beta}\|f\|_{s,\Omega},$$

where $\beta = \min(k, s+1)$. Thus if linear ($k = 1$) elements are used, the error is of order $h$ since $s+1$ will not be less than 1.

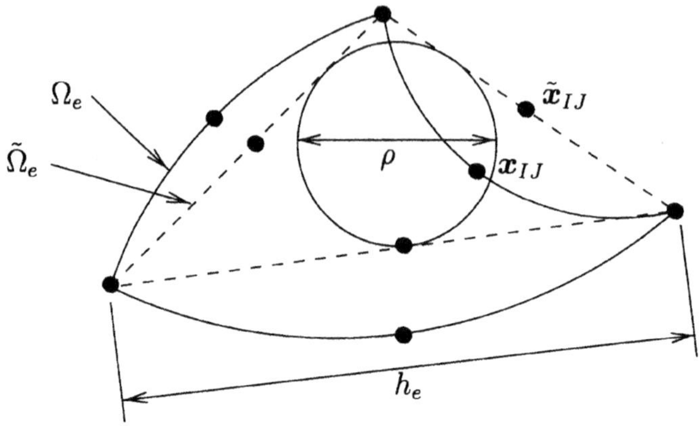


FIGURE 12.6. The triangle generated by a quadratic isoparametric map

## 12.4 Isoparametric families and numerical integration

The theory that culminates in the error estimate (12.28) is based entirely on the assumption that finite element meshes are generated by affine maps from a reference element. The theory therefore does not take into account deviations in the form of isoparametric maps, nor indeed does it account for errors induced by numerical integration. In this section we give some indication of how these deviations are accommodated in the error estimates.

**Isoparametric maps.** There are various complications that arise when dealing with this more general family of elements: in particular, the Jacobian matrix $\boldsymbol{J}$ defined in (11.33) is no longer constant. The theory appropriate to isoparametric maps is outlined for the special case of the six-noded triangle, shown in Figure 12.6. This element is of course obtained by the map

$$\boldsymbol{x} = F_e(\boldsymbol{\xi}) = \sum_{I=1}^{6} \boldsymbol{x}_I \hat{N}_I(\boldsymbol{\xi}), \tag{12.33}$$

in which the functions $\hat{N}_I$ are quadratic. Also shown in the figure is the element $\tilde{\Omega}_e$ generated by the *affine* map $\tilde{F}_e$ from the reference triangle. The definitions (12.3) and (12.4) of the quantities $h_e$ and $\rho_e$ are retained, but these refer to the *affine* element $\tilde{\Omega}_e$, as shown in Figure 12.6. Then under these conditions a family of isoparametric elements is said to be *regular* if

1. there exists a constant $\sigma$ such that

$$\frac{h_e}{\rho_e} \leq \sigma \quad \text{for } e = 1, \ldots, E;$$

2. the quantities $h_e$ approach zero;

3. if $\boldsymbol{x}_{IJ}$ and $\tilde{\boldsymbol{x}}_{IJ}$ are, respectively, the coordinates of the midpoint nodes of $\Omega_e$ and $\tilde{\Omega}_e$, then

$$\|\boldsymbol{x}_{IJ} - \tilde{\boldsymbol{x}}_{IJ}\| = O(h_e^2) \quad \text{for } 1 \leq I < J \leq 3. \tag{12.34}$$

Thus a comparison with the definition given in Section 12.2 shows that a family of regular isoparametric elements has to satisfy the criteria that are set for affine families, but in addition $\Omega_e$ is required to be not very different from $\tilde{\Omega}_e$, in the sense of (12.34). Under these conditions it is possible to prove the following analogue of Theorem 4 and its corollary.

THEOREM 7. *For any regular family of isoparametric elements generated by the map (12.33) corresponding to the six-noded triangle, and for any function $v \in H^3(\Omega_e)$, there exists a constant $C$ such that*

$$\|v - \Pi_e v\|_{m,\Omega_e} \leq C h_e^{3-m}(|v|_{2,\Omega_e} + |v|_{3,\Omega_e}), \tag{12.35}$$

*for integers $m \leq 3$.*

Thus the estimate (12.35) differs from (12.24) (with $k = 3$ there) only in that the term $|v|_{2,\Omega_e}$ also appears on the right-hand side.

One of the reasons for using isoparametric families is that these permit the construction of domains with curved boundaries. It is often the case, though, that the actual curved boundary $\Gamma$ of the domain $\Omega$ cannot be represented exactly using isoparametric elements. When attempting to arrive at an error estimate of the kind (12.28) for second-order problems, therefore, the theory must take account of the fact that the domain $\Omega_h$ which is defined by the finite element mesh may be distinct from $\Omega$. Such a situation is of course also true in the case of affine families, which would at best represent a polygonal approximation to a domain with a curved boundary.

Let $\Omega_h$ be the domain represented by a regular family of isoparametric elements, and let $\Omega$ be the actual domain (Figure 12.7). The space $V^h$ in which approximate solutions are sought is now defined as a subspace of $H^1(\Omega_h)$ (for second-order problems). Then for a second-order problem and a regular isoparametric mesh comprising six-noded triangles, Theorem 7 may be used to derive the counterpart to (12.28); that is,

$$\|v - \Pi_h v\|_{m,\Omega_h} \leq C h^{3-m} \|v\|_{3,\Omega_h}; \tag{12.36}$$

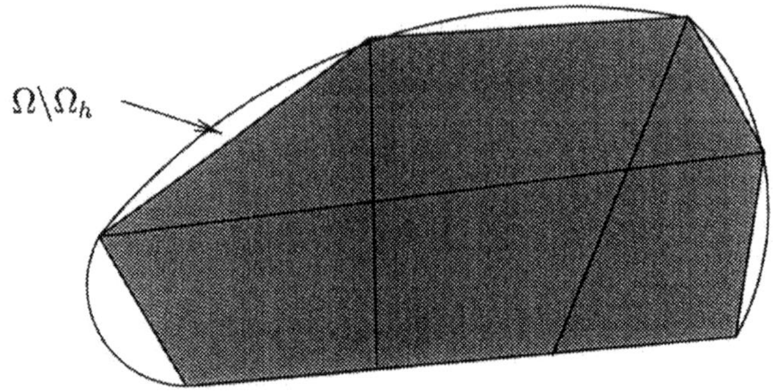


FIGURE 12.7. The domain $\Omega$ and its approximation $\Omega_h$

note that the norms are defined here on the domain $\Omega_h$. In going from (12.35) to (12.36) we also make use of the elementary fact that $|v|_{2,\Omega_e} + |v|_{3,\Omega_e} \le c\|v\|_{3,\Omega_e}$.

**Numerical integration.** We consider now the modifications that have to be made to the standard theory, in the event that numerical integration procedures of the kind discussed in Section 11.6 are used. Take, for example, the problem of finding $u \in V = H_0^1(\Omega)$ that satisfies

$$a(u,v) = \langle \ell, v\rangle \quad \text{for all } v \in V, \tag{12.37}$$

where

$$a(u,v) = \int_\Omega \boldsymbol{k}\nabla u \cdot \nabla v \, dx,$$

$\boldsymbol{k}$ being a matrix of functions; thus the integrand reads, when expanded,

$$\sum_{i,j=1}^{n} k_{ij}\frac{\partial u}{\partial x_i}\frac{\partial v}{\partial x_j}.$$

The matrix $\boldsymbol{k}$ is assumed to be symmetric and the coefficients $k_{ij}$ are such that the bilinear form is continuous and $V$-elliptic (and hence also $V^h$-elliptic): in particular we assume that $k_{ij} \in C(\overline{\Omega})$, and that a constant $k_0 > 0$ exists such that

$$\boldsymbol{ka}\cdot\boldsymbol{a} \ge k_0|\boldsymbol{a}|^2 \tag{12.38}$$

for any vector $\boldsymbol{a}$. The linear functional is assumed to be given by

$$\langle \ell, v\rangle = \int_\Omega fv \, dx.$$

The discrete problem entails finding $u_h \in V^h$ such that

$$a(u_h, v_h) = \langle \ell, v_h\rangle \quad \text{for all } v_h \in V^h. \tag{12.39}$$

Now if numerical integration is used to evaluate the integrals, the discrete problem that is solved is not in fact (12.39), but rather the problem

$$a_h(u_h, v_h) = \langle \ell_h, v_h \rangle \quad \text{for all } v_h \in V^h,$$

in which the bilinear form $a_h(u_h, v_h)$ and linear functional $\langle \ell_h, v_h \rangle$ are obtained by integrating numerically over each element and summing over all elements. For an integration rule of order $r$, therefore,

$$\begin{aligned} a_h(u, v) &= \textstyle\sum_{e=1}^{E} \sum_{l=1}^{r} \sum w_l \boldsymbol{k}(\boldsymbol{\xi}_l) \nabla u(\boldsymbol{\xi}_l) \cdot \nabla v_h(\boldsymbol{\xi}_l), \\ \langle l_h, v \rangle &= \textstyle\sum_{e=1}^{E} \sum_{l=1}^{r} w_l f(\boldsymbol{\xi}_l) v_h(\boldsymbol{\xi}_l). \end{aligned} \tag{12.40}$$

Since $a \neq a_h$ and $\ell \neq \ell_h$, the theory leading to Theorem 6 needs to be modified in order to arrive at an error estimate. In particular, Céa's Lemma (Theorem 10.2) does not hold any longer, and must be replaced by a suitable extension; this is provided by the following result.

THEOREM 8 (STRANG'S LEMMA). *Suppose that the bilinear form $a_h(\cdot,\cdot)$ is uniformly $V^h$-elliptic, in the sense that a constant $\alpha$, independent of $h$, exists such that*

$$a_h(v_h, v_h) \geq \alpha \|v_h\|_V^2.$$

*Then there exists a constant $C$ independent of $h$ such that*

$$\begin{aligned} \|u - u_h\|_V \leq & \left( \inf_{v_h \in V^h} \left[ \|u - v_h\|_V + \sup_{w_h \in V^h} \frac{|a(v_h, w_h) - a_h(v_h, w_h)|}{\|w_h\|_V} \right] + \right. \\ & \left. \sup_{w_h \in V^h} \frac{|\langle \ell, w_h \rangle - \langle \ell_h, w_h \rangle|}{\|w_h\|_V} \right). \end{aligned} \tag{12.41}$$

The proof of this theorem is discussed in Exercise 12.10. We see that it reduces to Céa's Lemma in the event that integration is exact, since in that case $a_h = a$ and $\ell_h = \ell$.

There are thus two additional tasks that need to be carried out in order to arrive at an error estimate: the two new terms on the right-hand side of (12.41) have to be estimated, and it is necessary also to establish conditions under which the approximate bilinear form $a_h$ is $V^h$-elliptic. The former is usually achieved by deriving *consistency error estimates* of the form

$$\begin{aligned} \sup_{w_h \in V^h} \frac{|a(\Pi_h u, w_h) - a_h(\Pi_h u, w_h)|}{\|w_h\|_V} &\leq C_1 h^k, \\ \sup_{w_h \in V^h} \frac{|\langle \ell, w_h \rangle - \langle \ell_h, w_h \rangle|}{\|w_h\|_V} &\leq C_2 h^k, \end{aligned} \tag{12.42}$$

in which $\Pi_h$ denotes the interpolation operator defined in (12.26) and $C_1$ and $C_2$ are constants that depend, respectively, on $\boldsymbol{k}$ and $u$, and on $f$. These estimates would then permit the necessary extension of Theorem 5. The theory leading to the desired estimates is rather complex, and the details are omitted.

We examine the issue of $V^h$-ellipticity, for the special case of an affine family generated by the three-noded reference triangle, and with the use of one-point integration on this triangle; recall that such a rule is exact for polynomials of degree one.

THEOREM 9. *Suppose that integration on the triangle $\Omega_e$ is carried out using the rule*

$$\int_{\Omega_e} f(\boldsymbol{x})\, dxdy \simeq I_1(f) \equiv A_e f(\hat{\boldsymbol{x}}),$$

*where $A_e$ is the area of $\Omega_e$ and $\hat{\boldsymbol{x}}$ the location of its centroid. If $X_e = P_1(\Omega_e)$, where $X_e$ is the space spanned by the local basis functions, then there exists a constant $\alpha$, independent of $h$, such that*

$$a_h(v_h, v_h) \geq \alpha \|v_h\|_V^2.$$

PROOF. For $v_h|_{\Omega_e} \in P_1(\Omega_e)$, the vector $\nabla v_h$ is constant on $\Omega_e$; therefore, using (12.38),

$$\begin{aligned} \int_{\Omega_e} \boldsymbol{k}\nabla v_h \cdot \nabla v_h \, dx\, dy &\simeq I_1(\boldsymbol{k}\nabla v_h \cdot \nabla v_h) \\ &= A_e \boldsymbol{k}(\hat{\boldsymbol{x}}) \nabla v_h(\hat{\boldsymbol{x}}) \cdot \nabla v_h(\hat{\boldsymbol{x}}) \\ &\geq A_e k_0 |\nabla v_h|^2(\hat{\boldsymbol{x}}) \\ &= k_0 |v_h|_{1,\Omega_e}^2. \end{aligned}$$

The desired result then follows from summing over all elements, and then using the Poincaré–Friedrichs inequality (7.34). □

Theorem 8, together with the consistency error estimates, gives the following result.

THEOREM 10. *Assume that the conditions of Theorem 8 hold. Then if the solution $u \in H_0^1(\Omega)$ of the problem (12.37) belongs to $H^2(\Omega)$, and if the data satisfy (12.37) and*

$$k_{ji} = k_{ij}, \quad k_{ij} \in C(\overline{\Omega}), \quad f \in H^2(\Omega),$$

*then there exists a constant $C$ dependent on $u$, $\boldsymbol{k}$, and $f$ but independent of $h$,*

$$\|u - u_h\|_{1,\Omega} \leq Ch.$$

## 12.5 Bibliographical remarks

This chapter draws heavily on the work by Ciarlet [11], which may be consulted for further details of the topics presented here, and for extensions of the theory. The texts by Brenner and Scott [8], Oden and Reddy [38], Oden and Carey [37], Raviart and Thomas [39], and Strang and Fix [51] are also very useful sources for the mathematical theory of finite elements, as is the Finite Element Handbook [24]. Johnson [23] provides a useful expository account of finite element methods for convection-diffusion problems, and for hyperbolic problems generally.

The interpolation theory for isoparametric elements is discussed by Ciarlet [11], mainly in the context of the six-noded triangle. The original, and more general, treatment is by Ciarlet and Raviart [12]. Likewise, the developments leading to error estimates when numerical integration is used are treated in detail in [11] and in [51].

## 12.6 Exercises

### Affine families of elements

12.1. Complete the proof of Lemma 1 by showing that $||\boldsymbol{T}_e^{-1}|| \leq \hat{h}/\rho_e$.

### Local interpolation error estimates

12.2. Show that $I : H^{k+1}(\hat{\Omega}) \to H^m(\hat{\Omega})$ and $\hat{\Pi} : H^{k+1}(\hat{\Omega}) \to H^m(\hat{\Omega})$ are bounded operators, where $\hat{\Pi}$ is the projection operator defined in Section 12.1. [Theorem 2 of Chapter 7 is useful when dealing with $\hat{\Pi}$.]

12.3. Consider a *regular* family of triangular finite elements in $\mathbb{R}^2$, that is, one for which

$$h_e/\rho_e \leq \sigma,$$

for some $\sigma > 0$. Show that this condition is satisfied if the smallest angle $\theta_e$ in an element is bounded below by some constant; that is,

$$\theta_e \geq \theta_0 \qquad \text{for some } \theta_0 > 0.$$

This is known as *Zlámal's condition*; it ensures that elements are not too severely distorted.

12.4. Complete the following table.

| | | | |
|---|---|---|---|
| Largest $k$ for which $P_k(\Omega_e) \subset X_e$ | $k = 1$ | ? | ? |
| $\|\|u - u_h\|\|_{m,\Omega_e}$ | $O(h_e^{2-m})$ $0 \leq m \leq 2$ | ? | ? |
| Regularity | $H^2(\Omega_e)$ | ? | ? |
| | 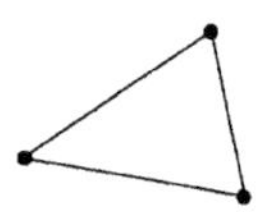 | 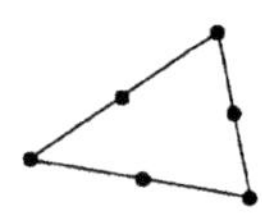 | 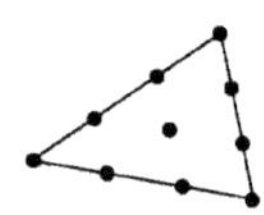 |
| |  | 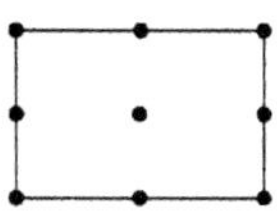 | 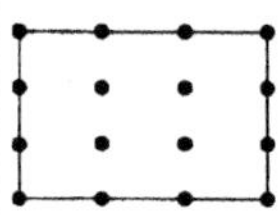 |

12.5. Derive the estimate

$$||v - \Pi_e v||_{m,\Omega} \leq C h_e^{k+1-m} |v|_{k+1,\Omega_e},$$

and explain where in the derivation the condition $h_e \to 0$ is required. Also show that the constant $C$ is proportional to $\sigma^a$ in Exercise 12.4, for some positive number $a$, and explain how this affects the error estimate.

12.6. The purpose of this exercise is to derive the relation (12.18) for functions defined on domains in $\mathbb{R}^2$. We start by defining the *Fréchet derivative* $\mathcal{D}v$ of a function $v$ to be the linear map

$$\mathcal{D}v : \mathbb{R}^2 \to \mathbb{R}, \qquad \mathcal{D}v(\boldsymbol{a}) = \sum_{i=1}^{2} \frac{\partial v}{\partial x_i} a_i.$$

The second Fréchet derivative is defined to be the bilinear map

$$\mathcal{D}^2 v : \mathbb{R}^2 \times \mathbb{R}^2 \to \mathbb{R}, \qquad \mathcal{D}^2 v(\boldsymbol{a}, \boldsymbol{b}) = \sum_{i,j=1}^{2} \frac{\partial^2 v}{\partial x_i \partial x_j} a_i b_j$$

and higher derivatives are defined similarly. Generally $\mathcal{D}^k v$ is an operator from $\mathbb{R}^2 \times \cdots \times \mathbb{R}^2$ ($k$ times) to $\mathbb{R}$, and it is linear in each slot. The space of all $k$th Fréchet derivatives is a normed space with norm

$$||\mathcal{D}^k v|| = \sup |\mathcal{D}^k v(\boldsymbol{a}^{(1)}, \ldots, \boldsymbol{a}^{(k)})|, \qquad ||\boldsymbol{a}^{(k)})|| \leq 1. \tag{i}$$

Clearly then, we have

$$|D^\alpha v(\boldsymbol{x})| \leq ||\mathcal{D}^k v|| \text{ for } |\alpha| = k; \tag{ii}$$

for example, if $\alpha = (1, 1)$, then

$$\begin{aligned} |D^\alpha v(\boldsymbol{x})| &= |\partial^2 v/\partial x \partial y| = |\mathcal{D}^2 v(\boldsymbol{e}_1, \boldsymbol{e}_2)| \\ &\leq \sup\{|\mathcal{D}^2 v(\boldsymbol{a}, \boldsymbol{b})| : ||\boldsymbol{a}|| \leq 1, ||\boldsymbol{b}|| \leq 1\}. \end{aligned}$$

To derive (12.18), show that

$$\mathcal{D}^k \hat{v}(\boldsymbol{a}^{(1)}, \ldots, \boldsymbol{a}^{(k)}) = \mathcal{D}^k v(\boldsymbol{T}_e \boldsymbol{a}^{(1)}, \ldots, \boldsymbol{T}_e \boldsymbol{a}^{(k)})$$

and then use (i) and (ii). [Carry out the calculation first for $k = 1$ and $k = 2$; then proceed to the general case.]

**Error estimates for second-order problems**

12.7. Show that if $\hat{X} \subset H^1(\hat{\Omega})$ and $X_h \subset C(\overline{\Omega})$, then $X_h \subset H^1(\Omega)$.

12.8. The theory of Sections 12.1 and 12.2 does not enable us to obtain optimal error estimates in the $L^2$-norm for second-order problems, mainly because of the central rôle played by the inequality $\|u - u_h\|_V \leq C\|u - v_h\|_V$. It is possible, however, to obtain $L^2$ estimates using what is known as the *Aubin–Nitsche method.* The method is outlined in this exercise, for the problem (12.29).

Consider the *auxiliary* VBVP of finding $w \in V \subset H^1(\Omega)$ such that

$$a(w, v) = (u - u_h, v)_{L^2} \text{ for all } v \in V,$$

and let $\tilde{w}_h$ be the interpolate of $w$ in $V_h$. Show that

$$a(w - \tilde{w}_h, u - u_h) = \|u - u_h\|^2_{L^2}$$

and use the continuity of $a$ and the results of Sections 12.1 and 12.2 to obtain the estimate

$$\|u - u_h\|^2_{L^2} \leq Ch^{\beta+\gamma} \|u\|_{H^r} \|w\|_{H^p},$$

where $\beta = \min(k, r-1)$ and $\gamma = \min(k, p-1)$. Finally, use Theorem 1 of Chapter 8 (remember that $Aw = u - u_h$) to obtain the estimate

$$\|u - u_h\|_{L^2} \leq Ch^\nu \|u\|_{H^r},$$

where $\nu = \min(2k, k+1, r)$.

12.9. Use the result of Exercise 12.7 to obtain $L^2$-estimates of the error for the problem

$$\begin{aligned} \nabla^2 u &= f \text{ in } \Omega \subset \mathbb{R}^2, \\ u &= 0 \text{ on } \Gamma, \end{aligned}$$

assuming that $f \in L^2(\Omega)$, and using linear or bilinear elements.

12.10. Suppose that we have to solve a fourth-order BVP defined on $\Omega = (0,1)$, and assume that we intend using the Hermite basis functions described in Section 11.4. Verify that the theory developed in Sections 12.1 to 12.3 remains essentially unchanged except that, for example, it is necessary to specify that $H^{k+1}(\hat{\Omega}) \subset C^1(\hat{\Omega})$ in Theorem 4. Derive an estimate of the error in finite element approximations of the problem

$$\begin{aligned} &\frac{d^4u}{dx^4} + ku = f \text{ in } (0,1), \\ &u(0) = u(1) = 0, \\ &u'(0) = u'(1) = 0, \end{aligned}$$

assuming that $f \in L^2(0,1)$, and using the cubic Hermite functions in Example 4 of Chapter 11.

### Isoparametric elements and numerical integration

12.11. The purpose of this exercise is to derive the estimate (12.41) in Theorem 8. Use the $V^h$-ellipticity of $a_h$ to obtain the inequality

$$\begin{aligned} \alpha\|u_h - v_h\|_V^2 \leq\ & a(u - v_h, u_h - v_h) + a(v_h, u_h - v_h) \\ & - a_h(v_h, u_h - v_h) + \langle \ell_h, u_h - v_h\rangle - \langle \ell, u_h - v_h\rangle; \end{aligned}$$

then use the continuity of $a$ and the triangle inequality to derive (12.41).

12.12. Show that the bilinear form $a_h(\cdot,\cdot)$ defined by (12.40) is $V^h$-elliptic, if $\Omega_e$ is the six-noded quadratic element and the integration rule used is the three-point rule on the triangle.

# References


[1] Adams R.A., *Sobolev Spaces.* Academic Press (New York) 1975.

[2] Apostol T.M., *Mathematical Analysis: A Modern Approach to Advanced Calculus.* Addison-Wesley (Reading, Mass.) 1957.

[3] Babuška I. and Aziz A.K., Survey Lectures on the Mathematical Foundations of the Finite Element Method, in *The Mathematical Foundations of the Finite Element Method with Applications to Partial Differential Equations* (ed. A.K. Aziz). Academic Press (New York) 1972.

[4] Baiocchi C. and Capelo A., *Variational and Quasi-Variational Inequalities.* Wiley (New York) 1984.

[5] Becker E.B., Carey G.F. and Oden J.T., *Finite Elements, Volume 1: An Introduction.* Prentice-Hall (Englewood Cliffs, N.J.) 1981.

[6] Binmore K.G., *Mathematical Analysis: A Straightforward Approach.* Cambridge University Press (Cambridge) 1977.

[7] Binmore K.G., *The Foundations of Analysis: A Straightforward Introduction. Book 2: Topological Ideas.* Cambridge University Press (Cambridge) 1981.

[8] Brenner S. and Scott L.R., *The Mathematical Theory of Finite Element Methods.* Springer-Verlag (New York) 1994.

[9] Burnett D.S., *Finite Element Analysis.* Addison-Wesley (Reading, Mass.) 1987.

[10] Carey G.F. and Oden J.T., *Finite Elements, Vol. 2: A Second Course.* Prentice-Hall (Englewood Cliffs, N.J.) 1983.

[11] Ciarlet P.G., *The Finite Element Method for Elliptic Problems.* North-Holland (Amsterdam) 1978.

[12] Ciarlet P.G and Raviart P.-A., Interpolation theory over curved elements with applications to finite element methods. *Computer Methods in Applied Mechanics and Engineering* **1** (1972) 217–249.

[13] Dautray R. and Lions, J.-L., *Mathematical Analysis and Numerical Methods for Science and Technology, Vol. 2: Functional and Variational Methods.* Springer-Verlag (Berlin) 1988.

[14] Dhatt G. and Touzot G., *The Finite Element Method Displayed.* Wiley (New York) 1984.

[15] Duvaut G. and Lions J.L., *Inequalities in Mechanics and Physics.* Springer-Verlag (Berlin) 1976.

[16] Glowinski R., *Numerical Methods for Nonlinear Variational Problems.* Springer-Verlag (Berlin) 1984.

[17] Grisvard P., *Elliptic Problems in Nonsmooth Domains.* Pitman (London) 1985.

[18] Halmos P., *Finite Dimensional Vector Spaces.* Van Nostrand Reinhold (New York) 1958.

[19] Hewitt E. and Stromberg K.R., *Real and Abstract Analysis: A Modern Treatment of the Theory of Functions of a Real Variable.* Springer-Verlag (New York) 1965.

[20] Hoffman K. and Künze R., *Linear Algebra.* Addison-Wesley (Reading, Mass.) 1973.

[21] Horgan C.O., Korn's inequalities and their applications in continuum mechanics, *SIAM Review* **37** (1995) 491–511.

[22] Hughes T.J.R., *The Finite Element Method: Linear Static and Dynamic Analysis.* Prentice-Hall (Englewood Cliffs, N.J.) 1987.

[23] Johnson C., *Numerical Solutions of Partial Differential Equations by the Finite Element Method.* Cambridge University Press (Cambridge) 1987.

[24] Kardestuncer H. (ed.), *Finite Element Handbook.* McGraw-Hill (New York) 1987.

[25] Kolmogorov A.N. and Fomin S.V., *Elements of the Theory of Functions and Functional Analysis. Volume 1: Metric and Normed Spaces.* Graylock Press (Rochester, N.Y.) 1957.

[26] Kolmogorov A.N. and Fomin S.V., *Elements of the Theory of Functions and Functional Analysis. Volume 2: Measure, Lebesgue Integrals and Hilbert Space.* Academic Press (New York) 1961.

[27] Kreyszig E., *Introductory Functional Analysis with Applications.* Wiley (New York) 1978.

[28] Lang S., *Introduction to Linear Algebra.* 2nd edition. Springer-Verlag (New York) 1986.

[29] Lang S., *Undergraduate Analysis.* Springer-Verlag (New York) 1983.

[30] Lions J.L. and Magenes E., *Non-Homogeneous Boundary-Value Problems and Applications, Volume 1.* Springer-Verlag (New York) 1972.

[31] Lipschutz S., *Set Theory and Related Topics.* Schaum Outline Series. McGraw-Hill (New York) 1964.

[32] Loula A.F.D., Hughes T.J.R. and Franca L.P., Petrov-Galerkin formulations of the Timoshenko beam problem. *Computer Methods in Applied Mechanics and Engineering* **63** (1987) 115–132.

[33] Naylor A.W. and Sell G.R., *Linear Operator Theory in Engineering and Science.* Springer-Verlag (Berlin) 1982.

[34] Nečas J., *Les Méthodes Directes en Théorie des Equations Elliptiques.* Masson (Paris) 1967.

[35] Noble B., *Applied Linear Algebra.* Prentice-Hall (Englewood Cliffs, N.J.) 1969.

[36] Oden J.T., *Applied Functional Analysis: An Introductory Treatment for Students of Mechanics and Engineering Science.* Prentice-Hall (Englewood Cliffs, N.J.) 1979.

[37] Oden J.T. and Carey G.F., *Finite Elements, Volume 4: Mathematical Aspects.* Prentice-Hall (Englewood Cliffs, N.J.) 1982.

[38] Oden J.T. and Reddy J.N., *An Introduction to the Mathematical Theory of Finite Elements.* Wiley (New York) 1976.

[39] Raviart P.-A. and Thomas J.M., *Introduction a l'Analyse Numérique des Équations aux Dérivées Partielles.* Masson (Paris) 1983.

[40] Reed M. and Simon B., *Methods of Modern Mathematical Physics I: Functional Analysis.* Academic Press (New York) 1980.

[41] Rektorys K., *Variational Methods in Mathematics, Science and Engineering.* 2nd edition. D. Reidel (Dordrecht) 1980.

[42] Roman P., *Some Modern Mathematics for Physicists and Other Outsiders, Volume 1: Algebra, Topology and Measure Theory.* Pergamon (Oxford) 1975.

[43] Roman P., *Some Modern Mathematics for Physicists and Other Outsiders, Volume 2: Functional Analysis with Applications.* Pergamon (Oxford) 1975.

[44] Royden H.L., *Real Analysis.* 3rd edition. Collier-Macmillan (London) (1988).

[45] Rudin W., *Real and Complex Analysis.* 2nd edition. McGraw-Hill (New York) 1974.

[46] Schwartz L., *Théorie des Distributions.* Hermann (Paris) 1950.

[47] Schwartz L., *Mathematics for the Physical Sciences.* Hermann (Paris) 1966.

[48] Showalter R.E., *Hilbert Space Methods for Partial Differential Equations.* Pitman (Boston) 1977.

[49] Smirnov V.I., *A Course of Higher Mathematics, Volume 5: Integration and Functional Analysis.* Pergamon (Oxford) 1964.

[50] Strang G., *Linear Algebra and its Applications.* Academic Press (New York) 1976.

[51] Strang G. and Fix G.J., *An Analysis of the Finite Element Method.* Prentice-Hall (Englewood Cliffs, N.J.) 1973.

[52] Zauderer E., *Partial Differential Equations of Applied Mathematics.* 2nd edition. Wiley (New York) 1989.

[53] Zeidler E., *Nonlinear Functional Analysis and Its Applications. Volume IIA: Linear Monotone Operators.* Springer-Verlag (Berlin) 1990.

[54] Zeidler E., *Applied Functional Analysis: Applications of Mathematical Physics.* Springer-Verlag (Berlin) 1995.

[55] Zeidler E., *Applied Functional Analysis: Main Principles and Their Applications.* Springer-Verlag (Berlin) 1995.

[56] Zienkiewicz O.C. and Taylor R.L., *The Finite Element Method. Volume 1: Basic Formulation and Linear Problems.* McGraw-Hill (London) 1989.

[57] Zienkiewicz O.C. and Taylor R.L., *The Finite Element Method. Volume 2: Solid and Fluid Mechanics, Dynamics and Nonlinearity.* McGraw-Hill (London) 1991.

# Solutions to Exercises

## Chapter 1

1.1. $A = \{-2, 3\}$, $B = \{-3, -2, -1, 0, 1, 2, 3\}$. $A \cup B = B$; $A \cap B = A$; $A \cap \mathbb{Z}^+ = 3$, $A - \mathbb{Z}^+ = \{-2\}$.

1.2. $A \cup C = \{1, 2, 9\}$ so $B \times (A \cup C) = \{(7,1), (7,2), (7,9), (8,1), (8,2), (8,9)\}$; $A \cap C = \{1\}$ so $(A \cap C) \times B = \{(1,7), (1,8)\}$.

1.3. Let $x \in A \cap (B \cup C)$. Then $x \in A$ and $x \in B$ or $C$; i.e., $x \in A$ and $x \in B$, or $x \in A$ and $x \in C$. Hence $x \in (A \cap B) \cup (A \cap C)$. The second identity is proved in a similar way.

1.4. $n(A \cup B \cup C) = n(A) + n(B) + n(C) - n(A \cap B \cap C) - n(A \cap B - C) - n(B \cap C - A) - n(C \cap A - B)$.

1.5. $\mathcal{P}(A) = \{A, \varnothing, \{1\}, \{2\}, \{3\}, \{1,2\}, \{2,3\}, \{1,3\}\}$;
$\mathcal{P}(B) = \{B, \varnothing, \{\{1,2\}\}, \{3\}\}$.

1.6. Consider the table

| | | | | |
|---|---|---|---|---|
| 1/1 | 1/2 | 1/3 | 1/4 | 1/5 ... |
| 2/1 | 2/2 | 2/3 | 2/4 | 2/5 ... |
| 3/1 | 3/2 | 3/3 | 3/4 | 3/5 ... |
| 4/1 | ... | | | |

The rationals can be listed by writing down the numbers in the preceding table in the order shown, omitting those already listed (e.g., omit $2/2 = 1$). This then gives a listing of all rationals whose numerator and denominator add up to 2, then 3, and so on. In this way all positive rationals are covered. Multiply by -1 to get negative rationals.

1.7. (i) $[a, b]$; (ii) $\mathbb{R}$; (iii) $[0, 1]$.

1.8. (i) Not open: $A = \{\pm 1/n\pi,\ n = 1, \dots\}$ and for every $x \in A$ there is a nhd $N(x)$ such that $N(x) - \{x\} \notin A$. Not closed: $0 \notin A$ is a point of accumulation. (ii) Neither open nor closed. (iii) Open, not closed since $\{\pm 1/n\pi\}$ are points of accumulation, but are not in $A$.

1.9. Assume $I$ is closed. Let $x \in I'$; since $x \notin I$, the distance from $x$ to $I$ is finite. Hence we can set up a neighborhood of radius $\epsilon < d$ about $x$ that lies entirely in $I'$. Hence $I'$ is open. Conversely, assume $I'$ is open. We always have $I \subset \bar{I}$, so we want to show that $\bar{I} \subset I$. Let $x \in \bar{I}$ and assume $x \notin I$. Then $x$ is in $I'$. Since $I'$ is open, there is a neighborhood $N$ of $x$ with $N \cap I = \varnothing$, which is a contradiction. Thus $x \in I$ and so $\bar{I} \in I$.

1.10. Points of accumulation: $A = \{z : x^2 - y^2 = 1\}$. $A$ is open.

1.11. (i) $-1, 1/2, -1/6, 1/24, \dots$; (ii) $1, 0, 1, 0, 1 \dots$; (iii) $-3, 6/7, 9/13, 12/19, \dots$

1.12. (a) Converges to $-3/2$; (b) not convergent; (c) converges to 1.

1.13. $|(3n + 2)/(n - 1) - 3| = |5/(n - 1)| < 0.001$. Assume $n > 1$, so that $5 < 0.001(n - 1) \Rightarrow n > 5001$. Take $n = 5001$.

1.14. Suppose $u_n$ is monotone increasing, with sup $= m$. For any $\epsilon > 0$ there exists $N$ such that $|u_n - m| < \epsilon$ for $n > N$, so $u_n \to m$. The same reasoning applies if $u_n$ is monotone decreasing.

1.15. (i) $\max A = 1 = \sup A$, $\min A$ is undefined, $\inf A = 0$. (ii) $\max A$, $\min A$ undefined; $\sup A = 1, \inf A = -1$. (iii) $\min A$, $\max A$ do not exist, $\inf A = -\infty$; $\sup A = c$. (iv) $|z^2 + 1| \le |z^2| + 1 = |z|^2 + 1 \le 2$. Maximum achieved at $z = \pm 1$. Minimum is achieved at $z = \pm i$.

1.16. $y = \inf A \Rightarrow a \le y \le x$ for all $x \in A$ and lower bounds $a$. Thus $-x \le -y \le -a$ so that $-y$ is the least upper bound of $-A$.

1.17. Take $A = (-1, 0)$ and $B = (-2, 0)$; then $a = b = 0$. But $C = (0, 2)$ so that $\sup C = 2 \ne ab$.

1.18. (i) Let $p = \sup I$; then $x \le p$ for any $x \in I$. Let $J = \{\alpha x : x \in I\}$; since $\alpha > 0$, $\alpha x \le \alpha p$. Hence $J$ is bounded above by $\alpha p$. Let the supremum of $J$ be $q$ (we must prove that $q = \alpha p$). Since $\alpha p$ is an upper bound for $J$ and $q$ is the least upper bound, $q \le \alpha p$. But for any $y \in J$ we have $y \leq q \Rightarrow \alpha^{-1}y \leq \alpha^{-1}q$. But $I = \{\alpha^{-1}y : y \in J\}$, hence $\alpha^{-1}q$ is an upper bound for $I$. Thus $p \leq \alpha^{-1}q$ or $\alpha p \leq q$. Since $q \leq \alpha p$ also, we have $\alpha p = q$.

1.19. (i) Closed. (ii) Open. Set of limit points is $\Omega \cup \{\boldsymbol{x} : x^2 + y^2 + z^2 = a^2, z > 0\} \cup \{\boldsymbol{x} : x^2 + y^2 < a^2, z = 0\}$.

1.20. (a) $\sqrt{2}$; ($b$) $2a$.

1.21. (a) Not an equivalence relation, but a partial ordering; (b) equivalence relation, not a partial ordering; (c) neither an equivalence relation nor a partial ordering; (d) not an equivalence relation, but a partial ordering.

1.22. $\{(2,2), (3,3), (4,4), (5,5), (6,6), (2,5), (5,2), (3,6), (6,3)\}$.

1.23. Take $c \in A_a \cap A_b$. Then $c \sim a$ implies that $a \sim c$. Also, $c \sim b$. Thus $a \sim b$ by transitivity, and $b \sim a$ by reflexivity; hence $a \in A_b$ and $b \in A_a$. Take any $x \in A_a : \ x \sim a$; hence $x \sim b$, so $x \in A_b$. Thus $A_a \subset A_b$. Similarly show that $A_b \subset A_a$.

1.24. $A_{\boldsymbol{a}}$ is the set of ordered pairs of integers lying on the "diamond" $\{z : \ |x| + |y| = \text{const}\}$ on which $\boldsymbol{a}$ is located.

**Chapter 2**

2.1. (a) not continuous at $x = \pm 1$; (b) continuous on $(-\infty, 0]$.

2.2. (a) Suppose that $|x-y| < \delta$. Then $|p(x) - p(y)| = |a_1(x-y) + a_2(x^2 - y^2) + \cdots + a_k(x^k - y^k)| \leq |x-y|[|a_1| + |a_2|\,|x+y| + \cdots + |a_k|\,|x^{k-1} + x^{k-2}y + \cdots + y^{k-1}|] < \delta C$ since term in square brackets is bounded above. Set $\delta = \epsilon/C$.
(b) For $0 \leq y \leq x$ we have $\sqrt{y} \leq \sqrt{x} \Rightarrow 2y \leq 2\sqrt{xy} \Rightarrow x - 2\sqrt{xy} + y \leq x - y$ or $(\sqrt{x} - \sqrt{y})^2 \leq x - y$. If $|x-y| < \delta$, then $|u(x) - u(y)| < \delta^{1/2}$. For given $\epsilon$ set $\delta = \epsilon^2$.

2.3. (b) $|f(x) - f(y)| = |x^{-1} - y^{-1}| = |y - x|/|xy|$. But $x > a$, $y > a$, so $xy > x^2$ or $1/xy < 1/a^2$. Hence $|f(x) - f(y)| < a^{-2}|x - y|$.

2.4. $|f(\boldsymbol{x}) - f(\bar{\boldsymbol{x}})| = |(x^2 + 2y) - (\bar{x}^2 + 2\bar{y})| = |(x^2 - \bar{x}^2) + 2(y - \bar{y})| \leq |x^2 - \bar{x}^2| + 2|y - \bar{y}|$. Suppose that $|\boldsymbol{x} - \bar{\boldsymbol{x}}| < \delta$; i.e., $(x-\bar{x})^2 + (y-\bar{y})^2 < \delta^2$. Then$|x^2 - \bar{x}^2| = |x - \bar{x}||x + \bar{x}| < \delta \cdot C$. Also, $|y - \bar{y}| < \delta$. Hence $|f(\boldsymbol{x}) - f(\bar{\boldsymbol{x}})| < (C+2)\delta$. Set $\delta = \epsilon/(C+2)$.

2.5. Set $f(\cdot) = d(\cdot, E)$. Then $|f(\boldsymbol{x}) - f(\boldsymbol{y})| = |\inf_{\boldsymbol{z} \in A} |\boldsymbol{x} - \boldsymbol{z}| - \inf_{\boldsymbol{z} \in A} |\boldsymbol{y} - \boldsymbol{z}|| \leq ||\boldsymbol{x} - \boldsymbol{y}| + \inf |\boldsymbol{y} - \boldsymbol{z}| - \inf |\boldsymbol{y} - \boldsymbol{z}|| = |\boldsymbol{x} - \boldsymbol{y}|$. Given $\epsilon > 0$, choose $\delta = \epsilon$.

2.6. $|f(x_0) - f(x)| < \epsilon$ whenever $|x_0 - x| < \delta$, i.e., for $x \in (x_0 - \delta, x_0 + \delta)$. Pick any such $x$: either $0 < f(x_0) - f(x) < \epsilon$ in which case $f(x) > f(x_0) - \epsilon$ or $0 < f(x) - f(x_0) < \epsilon$ in which case $f(x) < f(x_0) + \epsilon$. For the first case choose $\epsilon$ smaller than $f(x_0)$ so that $f(x)$ is positive. For the second case $f(x) > f(x_0) > 0$.

2.7. Assume that $f(a) < 0, f(b) > 0$. Since $f(a) < 0$, there is an interval $[a, c]$ in which $f(x) < 0$. Let the l.u.b. of such points $c$ be $\bar{c}$; then $f(\bar{c}) \leq 0$. We cannot have $f(\bar{c}) < 0$ since we would then be able to find an interval about $\bar{c}$ for which $f(x) < 0$, which would imply that $\bar{c}$ is not a l.u.b. Hence $f(\bar{c}) = 0$. A similar argument applies if $f(a) > 0$ and $f(b) < 0$.

2.8. (a) $u \in C(-1, 1)$; (b) $u \in C^\infty([0, \pi] \times [0, 1])$; (c) $u \in C^1[0, 1]$.

2.9. $|u(\boldsymbol{x}) - u(\boldsymbol{y})| = \big||\boldsymbol{x}| - |\boldsymbol{y}|\big| \leq |\boldsymbol{x} - \boldsymbol{y}|$ since $|\boldsymbol{x}| = |\boldsymbol{x} - \boldsymbol{y} + \boldsymbol{y}| \leq |\boldsymbol{x} - \boldsymbol{y}| + |\boldsymbol{y}|$.

2.10. Choose $\delta = \epsilon / L$ in the definition of continuity.

2.11. $I = I_Q \cup I'$, where $I_Q$ and $I'$ are the subsets of rationals and irrationals. $\mu(I') = \mu(I) - \mu(I_Q) = \mu(I)$.

2.12. Let $M$ be an arbitrary measurable set in $\mathbb{R}$. If $1 \in M,\ 0 \notin M$, then $\chi_E^{-1}(M) = E$; $1 \notin M,\ 0 \in M \Rightarrow \chi_E^{-1}(M) = E'$; $1 \notin M,\ 0 \notin M \Rightarrow \chi_E^{-1}(M) = \varnothing$; $1 \in M,\ 0 \in M \Rightarrow \chi_E^{-1}(M) = \operatorname{dom} \chi_E$. Thus $\chi_E^{-1}(M)$ is a measurable set. Conversely, if $E$ is not measurable, then $\chi_E$ cannot be measurable.

2.13. Put $\delta_n = 2^{-n}$. For each $n$ and for every $x$ there is an integer $k_n$ such that $k_n\delta_n \leq x < (k_n + 1)\delta_n$. Set $\phi_n(x) = k_n(x)\delta_n$ if $0 \leq x < n$, $\phi_n(x) = 0$ for $n \leq x$. Then $x - \delta_n < \phi_n(x) \leq x$ if $0 \leq x \leq n$; furthermore $0 \leq \phi_1 \leq \phi_2 \leq \ldots \leq x$ and $\phi_n(x) \to x$ as $n \to \infty$, for $x \in [0, \infty]$. Set $s_n = \phi_n \circ f$.

2.14. First calculate $\int_{\mathbb{R}} s_k(x)\, dx = \sum_{k=0}^{2^n - 1} (k/2^n)(1/2^n) = (1/2^{2n}) \sum k$. Then use the formula $\sum_{k=0}^{m-1} k = m(m-1)/2$.

2.15. $f^+(x) = \begin{cases} 1, & 0 \leq x \leq 1 \\ 0 & \text{otherwise}, \end{cases} \quad f^-(x) = \begin{cases} 1, & -1 \leq x < 0 \\ 0 & \text{otherwise}. \end{cases}$

$\int_{\mathbb{R}} f^+\, dx = \int_{\mathbb{R}} f^-\, dx = 1$, so $\int_{\mathbb{R}} f\, dx = 0$.

$\int_{\mathbb{R}} g^+\, dx = +\infty$, $\int_{\mathbb{R}} g^-\, dx = 1$, so $\int_{\mathbb{R}} g\, dx = +\infty$.

2.16. Use the fact that $|f| = f^+ + f^-$, and that integrability of $f$ implies that of $f^+$ and $f^-$. For the converse use $f = f^+ - f^-$. Show that $-\int f^+ - \int f^- \leq \int f^+ - \int f^- \leq \int f^+ + f^-$.

2.17. (a) $ap > -1$; (b) $ap < -1$.

2.18. All real $a$ except $a = -\frac{1}{2}, -\frac{3}{2}$.

2.19. Consider $0 < \int_\Omega |u(\boldsymbol{x}) - \alpha v(\boldsymbol{x})|^2 \, d\boldsymbol{x}$ for any $\alpha \in \mathbb{R}$. Expand and then choose $\alpha = \int_\Omega uv \, dx / \int_\Omega |v|^2 \, dx$.

**Chapter 3**

3.1. (a) Vector space; (b) not a vector space; (c) not a vector space; (d) vector space; (e) not a vector space.

3.2. (a) Subspace; (b) not a subspace: $\mathbf{0} \notin V$.

3.3. (a) Subspace; (b) not a subspace; (c) subspace; (d) subspace.

3.4. Suppose that $U = V \oplus W$, and let $u = v_1 + w_1 = v_2 + w_2$ for $v_1, v_2 \in V$ and $w_1, w_2 \in W$. Then $v_1 - v_2 = w_1 - w_2$. But $v_1 - v_2 \in V$ and $w_1 - w_2 \in W$, so that $v_1 - v_2 = w_1 - w_2 = 0$, or $v_1 = v_2, w_1 = w_2$. Conversely, suppose that $u = v + w$ for $v \in V$, $w \in W$ with $v$ and $w$ uniquely defined. If $V \cap W \neq \{0\}$, then there exists $z \in V \cap W$ with $z \neq 0$. Hence we can write $u = (v + z) + (w - z)$ so that the decomposition of $u$ is not unique, a contradiction.

3.5. For any $u \in C[0,1], u(x) = v(x) + w(x)$, where $v(x) = \frac{1}{2}(u(x) + u(-x))$ and $w(x) = \frac{1}{2}(u(x) - u(-x))$. Thus $v \in V$ and $w \in W$. Also, $V \cap W = \{v : \ v \text{ is even and odd}\} = \{0\}$.

3.6. $\alpha\beta \leq \text{area}A + \text{area}B$, hence $\alpha\beta \leq \alpha^p/p + \beta^q/q$ since $A = \int_0^\alpha x^{p-1} \, dx = \alpha^p/p$, etc. The proof now follows easily from the hints given.

3.7. $(u, w) = (v, w) \Rightarrow (u - v, w) = 0$ for all $w$. Set $w = u - v$.

3.8. $(u, v)_0 = 0$; $(u, v)_1 = (u, v)_0 + (u', v')_0 \neq 0$.

3.9. $\|u\| = \|u - v + v\| \leq \|u - v\| + \|v\|$. Repeat with $u$.

3.10. $\|u + v\|^2 + \|u - v\|^2 = (u + v, u + v) + (u - v, u - v)$. Expand and rearrange.

3.11. If $v = \alpha u$, then $\|u + v\| = (u + \alpha u, u + \alpha u)^{1/2} = (1 + \alpha)\|u\|$. But $\|u\| + \|v\| = (1+\alpha)\|u\|$. Conversely, assume that $\|u+v\| = \|u\| + \|v\|$. Then $\|u + v\|^2 = \|u\|^2 + \|v\|^2 + 2(u, v) = (\|u\| + \|v\|)^2 = \|u\|^2 + \|v\|^2 + 2\|u\|\|v\|$. Hence $\|u\|\,\|v\| = (u, v)$ or $(\hat{u}, \hat{v}) = 1$, where $\hat{u} = u/\|u\|$, $\hat{v} = v/\|v\|$. Suppose $\hat{v} \neq \hat{u}$; then $\hat{v} = \hat{u} + w$, and $1 = (\hat{u}, \hat{u} + w) = 1 + (\hat{u}, w) \Rightarrow (\hat{u}, w) = 0$. Also, $\|\hat{v}\|^2 = 1 = 1 + \|w\|^2 + 2(\hat{u}, w)$; i.e., $\|w\| = 0 \Rightarrow w = 0$. Hence $\hat{v} = \hat{u}$ or $v = \alpha u$ for some $\alpha$.

3.12. Assume that $\|x - y\|\,\|y - z\| = \|x - z\|$. Square and rearrange to get $(\hat{a}, \hat{b}) = 1$, where $\hat{a} = a/\|a\|$, $a = x - y$, $b = y - z$. Thus $\hat{b} = \hat{a}$ which gives $y = \alpha x + (1 - \alpha)z$, where $\alpha = \|y - z\|/(\|x - y\| + \|y - z\|)$. The converse is straightforward.

3.13. Verify that $(\cdot\,,\cdot)$ defined by $(u,v)=\int_a^b u'v'dx$ is an inner product on $X$.

3.14. No.

3.15. Expand the right-hand side and simplify.

3.16. $\|\alpha u+(1-\alpha)v\|\le\alpha\|u\|+(1-\alpha)\|v\|\le 1$.

3.17. $\|u\|^2+2\alpha(u,v)+\alpha^2\|v\|^2=\|u\|^2-2\alpha(u,v)+\alpha^2\|v\|^2$. The result follows from this.

3.18. (i) $(\sqrt{5}-1)/2$; (ii) 1.

3.20.

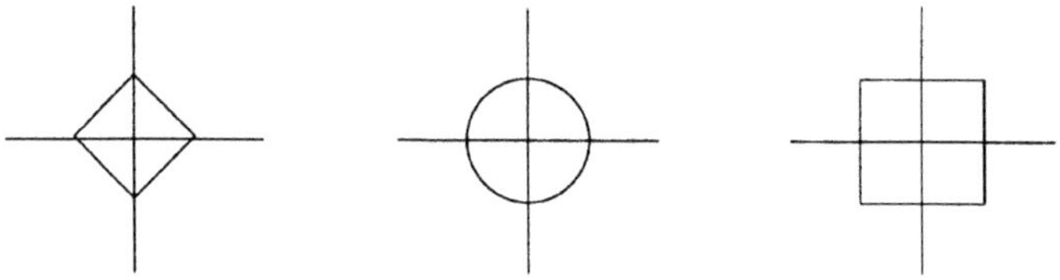

$p=1$ $p=2$ $p=\infty$

3.21. $\|\boldsymbol{x}\|_2^2=x^2+y^2=(|x|+|y|)^2-2|x|\,|y|\le(|x|+|y|)^2=\|\boldsymbol{x}\|_1^2$. $\|\boldsymbol{x}\|_1^2=x^2+y^2+2|xy|\le 2(x^2+y^2)=\|\boldsymbol{x}\|_2^2$.

3.22. $\int|u^r v^r|\;dx\le[\int|u|^{r(p/r)}]^{r/p}[\int|v|^{r(q/r)}]^{r/q}$. Take $r$th roots of both sides.

3.23. Follow the argument of Example 20.

## Chapter 4

4.1. 2.

4.2. $|(u_n,v_n)-(u,v)|=|(u_n-u,v_n-v)+(u,v_n-v)+(v,u_n-u)|\le\|u_n-u\|\,\|v_n-v\|+\|u\|\,\|v_n-v\|+\|v\|\,\|u_n-u\|\to 0$ as $n\to\infty$. Set $v_n=v$ (i.e. the sequence $v,v,\ldots$) to get $(u_n,v)\to(u,v)$. Finally, $|(u_n,v)-(u,v)|\le|(u_n-u,v)|\le\|u_n-u\|\,\|v\|$, hence $(u_n,v)\to(u,v)$. Set $v_n=u_n$ to get the final result.

4.3. $\|u-w\|=\|u-u_n+u_n-w\|\le\|u-u_n\|+\|u_n-w\|<\epsilon+\alpha$. The inequality follows from the arbitrariness of $\alpha$.

4.4. (a) $(-1,1]$; (b) $(-\infty,\infty)$.

4.5. (a) $u_n(x)\to 0$ pointwise. But $\|u_n-u\|_{L^2}^2=\int_{1/n}^{2/n}n^2\,dx=n\to\infty$ as $n\to\infty$; (b) $u_n(x)\to$ pointwise since $u_n(x)=n^{3/2}x/\exp(n^2x^2)=$

$n^{3/2}x/[1+n^2x^2+\frac{1}{2}n^4x^4+\ldots] \to 0$ as $n \to \infty$. But $\|u_n - u\|_{L^2}^2 = \int_{-n}^{n} y^2 \exp(-2y^2)\, dy$ (setting $y = nx$) $= -\frac{1}{2}([y\exp(-2y^2)]_{-n}^{n} - \int_{-n}^{n} \exp(-2y^2)dy = -\frac{1}{2}(0+\sqrt{(\pi/2)})$ as $n \to \infty$.

4.6. $\sup|u_n(x)| = 1/2$at $x = 1/n$. Thus in $[0,1], u_n(x) \to 0$ pointwise but $\|u_n - u\|_\infty = 1/2$, so convergence is not uniform. But convergence is uniform in $(a,1]$ $(a>0)$ : $\sup|u_n(x)| = na/(1+n^2a^2)$ at $x = a$ for $n > 1/a$ (check this by sketching $u_n(x)$) and $\sup|u_n(x)| \to 0$ as $n \to \infty$.

4.7. $\sup|u_n(x) - u(x)| < \epsilon$ for $n > N$. Hence $\int_a^b |u_n(x) - u(x)|^p\, dx \le (\sup|u_n(x) - u(x)|)^p \cdot (b-a) < (b-a)\epsilon^p$.

4.8. $\|u\| = 0$ does not imply that $u = 0$; $|||\cdot|||$ is also not a norm.

4.9. $\|u_n - u_m\|_{L^2}^2 = \frac{n}{n+2} - \frac{2mn}{mn+m+n} + \frac{m}{m+2} = 2\frac{(m-n)^2}{(m+2)(n+2)(mn+m+n)}$. Numerator $(m-n)^2 \le (m+n)^2$. Now show that $\|u_n - u_m\|_{L^2}^2 \to 0$ as $n, m \to \infty$.

4.10. $\|u_n - u_m\|_{L^1} = \int_0^1 |x^n - x^m|\, dx = \frac{1}{n+1} - \frac{1}{m+1}$ (taking $m > n$) $= \frac{m-n}{(n+1)(m+1)} \le \frac{m}{(n+1)(m+1)} \to 0$ as $n, m \to \infty$. Hence $\{u_n\}$ is a Cauchy sequence.

4.11. $\{u_n\}$ is Cauchy, so $\sup|u_n(x) - u_m(x)| < \epsilon$ for $m, n > N$. For any $x_0$, $|u_n(x_0) - u_m(x_0)| < \epsilon$, so $\{u_n(x_0)\}$ is a Cauchy sequence of real numbers. $\mathbb{R}$ is complete, so $u_n(x_0) \to u(x_0)$, say, which defines a function $u(x)$. The rest of the proof follows easily from the hints given.

4.12. Let $\{\boldsymbol{x}^k\}$ be a Cauchy sequence in $\mathbb{R}^n$ : $\|\boldsymbol{x}_k - \boldsymbol{x}_l\| < \epsilon$ for $k, l > N$; i.e., $\sum_i |x_{ki} - x_{li}|^p < \epsilon^p$. Hence $|x_{ki} - x_{li}|^p < \epsilon^p$ for each $i$. But $\mathbb{R}$ is complete so $x_{ki} \to x_i$, say. Hence $\boldsymbol{x} \to \boldsymbol{x}$ in $\mathbb{R}^n$.

4.13. Assume $\{u_n\}$ convergent: $\|u_n - u\| < \epsilon$ for $n > N$. Also $\|u_m - u\| < \epsilon'$ for $m > N'$. Hence $\|u_n - u_m\| = \|(u_n - u) + (u_m - u)\| \le \|u_n - u\| + \|u_m - u\| < \epsilon + \epsilon'$ for $n, m > N$ (assume $N > N'$).

4.14. $\|u_n - u_m\|^2 = \int_{1/2}^{1/2+1/n} [n(x-\frac{1}{2}) - m(x-\frac{1}{2})]^2\, dx + \int_{1/2+1/n}^{1/2+1/m} [1 - m(x-\frac{1}{2})]^2\, dx$. Show that this $\to 0$ as $m, n \to \infty$, so that $\{u_n\}$ is Cauchy. Also, $\|u_n - u\|^2 = \int_{1/2}^{1/2+1/n} [n(x-\frac{1}{2}) - 1]^2\, dx \to 0$ as $n \to \infty$. So $u_n \to u$ in $L^2$.

4.15. Take $v_n \in Y$ with $v_n \to v$. It is required to show that $v \in Y$. From Exercises 3.9 and 3.22, $|\,\|v_n\|_{L^1} - \|v\|_{L^1}| \le \|v_n - v\|_{L^1} \le c\|v_n - v\|_{L^2}$. Thus $|1 - \|v\|_{L^1}| < \epsilon$ so that $v \in Y$.

4.16. Let $v(x) \in C[-1,1]$ be defined by $v(x) = \begin{cases} -1, & -1 \le x < -\epsilon, \\ 1/\epsilon, & -\epsilon \le x \le \epsilon, \\ +1, & \epsilon < x \le 1. \end{cases}$
We have $\|u - v\|_{L^2}^2 = \int_{-\epsilon}^{0}(-1-\epsilon^{-1})^2\,dx + \int_0^{\epsilon}(1-\epsilon^{-1})^2\,dx = \epsilon^3/3 - \epsilon^2 + \epsilon$. Hence $v$ can be made arbitrarily close to $u$ by choosing $\epsilon$ small enough.

4.17. $\|u - v\|_\infty = \sup|1 - v(x)|$, where $|v(x)| < 1$ and $v(0) = 0$. Hence $\|u - v\|_\infty = 1$; neighborhoods of $u$ of radius less than 1 do not contain members of $V$, so $u$ is not a point of accumulation.

4.18. $v \in \bar{B}(u_0, r) \Rightarrow \|u_0 - v\|_\infty \le r$; i.e., $\sup|\sin 2\pi x - \cos 2\pi r| \le r$. $\sup|u_0 - v| = \sqrt{2}$ (at $x = 3/8$) so we require $r \ge 3/8$.

4.19. Cf. solution to Exercise 1.9.

4.20. Assume that $Y$ is complete, and let $v$ be a point of accumulation of $Y$. Then each open ball $B(v, 1/n), n = 1, 2, \ldots$, contains a point $v_n$, say, in $Y$. The sequence $\{u_n\}$ is convergent, hence Cauchy, in $Y$. Since $Y$ is complete, $v \in V$. Hence $Y$ contains all its points of accumulation, and is closed. Conversely, assume that $Y$ is closed, and let $\{v_n\}$ be a Cauchy sequence in $Y$. Then $\{v_n\}$ is a Cauchy sequence in $X$, and so converges to $v$ in $X$. From Theorem 3 of Chapter 4, $v$ is in $Y$ also, so $Y$ is complete.

4.21. $W$ dense in $X \Rightarrow$ for any $v \in X$ there is a $w \in W$ such that $\|w - v\| < \epsilon$. Similarly, for any $u \in Y$ there is $v \in X$ such that $\|v - u\| < \epsilon$. Hence $\|u - w\| \le \|u - v\| + \|v - w\| < 2\epsilon$, so $W$ is dense in $Y$.

4.22. Take $f \in L^p$. For given $\epsilon > 0$ choose a bounded function $g$ in $L^p$, where $g$ has compact support, for example, $|g| \le M$ in $[a, b]$ and $g = 0$ otherwise. Select $g$ so that $\|f - g\|_p < \epsilon$. Bounded functions with compact support are dense in $L^p$, so we can find $\{h_n\}$ in $C_0$ such that $g = \lim h_n$ a.e. Assume that $|h_n| \le M$ in $[a, b]$ and 0 otherwise. Then $|g - h_n|^p \le (2M)^p$ on $[a, b]$ and $\|g - h_n\|_p \to 0$ from the Dominated Convergence Theorem. Choose $n$ so that $\|g - h_n\| \le \epsilon$ and use the Minkowski inequality.

4.23. Suppose that there are two points $v_0, v_0'$ such that $\|u - v_0\| = d$. Then $w = (v_0 + v_0')/2$ is in $M$ hence, using the parallelogram law, it can be shown that $d^2 \le \|u_0 - w\|^2 < \frac{1}{2}\|u - v_0\|^2 + \frac{1}{2}\|u - v'\|^2 = d^2$, a contradiction.

4.24. Consider $\{u_n\} \subset Y^\perp$ with limit $u_0$ in $X$. We must show that $u_0 \in Y^\perp$ also. By definition $(u_n, v) = 0$ for any $v \in Y$; thus $0 = \lim_{n\to\infty}(u_n, v) = (\lim_{n\to\infty} u_n, v) = (u_0, v) = 0 \Rightarrow u_0 \in V^\perp$.

4.25. Theorem 7(b), which requires completeness of $H$, is used in Lemma 1.

4.26. Let $u \in X$ and $w \in Y^{\perp}$. Then $u \in Y$ also, so $(u, w) = 0$. $u$ is arbitrary; hence $w \in X^{\perp} \Rightarrow Y^{\perp} \subset X^{\perp}$.

## Chapter 5

5.1. (i) $R(M)$ = points on the upper unit semicircle, $N(M) = \varnothing$; (ii) $R(K) = [0, \infty), N(K) = \{0\}$; (iii) $R(f) = (0, \infty), N(f) = \varnothing$.

5.2. $N(S) = \{0\}$; $N(T) = \{\alpha(-8, 4, 1)\}$.

5.3. (i) One-to-one, not surjective; (ii) one-to-one, surjective ($T$ is a reflection about a line at 45° through the origin).

5.5. (i) $ST(\boldsymbol{x}) = S(x, -y) = (-2y, x)$; $TS(\boldsymbol{x}) = T(2y, x) = (2y, -x)$; (ii) $ST(x) = S(\sin x) = \sin^2 x - 1$, $TS(x) = T(x^2 - 1) = \sin(x^2 - 1)$.

5.6. $S^{-1} : V \to U$ and $T^{-1} : W \to V$ exist. Clearly $TS : U \to W$ is one-to-one onto $W$, so $(TS)^{-1}$ exists. Furthermore, $(TS)u = w \Rightarrow u = (TS)^{-1}w$. But $(TS)u = T(Su) = w$, so $Su = T^{-1}w$ and $u = S^{-1}T^{-1}w$. Hence $(TS)^{-1} = S^{-1}T^{-1}$.

5.7. (i) linear; (ii) linear; (iii) nonlinear.

5.8. $\boldsymbol{Tx} = \begin{pmatrix} -5 & -1 \\ -3 & -5 \end{pmatrix} \boldsymbol{x} + \begin{pmatrix} 4 \\ 5 \end{pmatrix}$, assuming that $\{(0,0),\ (1,0),\ (0,1)\}$ go to $\{(4,5),\ (-1,2),\ (3,0)\}$.

5.9. Let $Tu_1 = v_1, Tu_2 = v_2$. Then $T(\alpha u_1 + \beta u_2) = \alpha v_1 + \beta v_2$ by the linearity of $T$. Hence $T^{-1}(\alpha v_1 + \beta v_2) = \alpha u_1 + \beta u_2$. But $\alpha T^{-1} v_1 = \alpha u_1$, $\alpha T^{-1} v_2 = \alpha u_2 \Rightarrow T^{-1}(\alpha u_1 + \beta u_2) = \alpha T^{-1} v + \beta T^{-1} v_2$.

5.10. No; e.g., $d(\boldsymbol{x}, B) + d(\boldsymbol{y}, B) \neq d(\boldsymbol{x} + \boldsymbol{y}, B)$ in general. Null space is the set $B$.

5.11. For $u \neq 0$, $\|T\| = \sup(\|Tu\|/\|u\|) = \sup \|T(u/\|u\|)\|$ ($T$ is linear) $= \sup \|Tu\|, \|u\| = 1$. To prove the second result, consider $\|Tu\| \leq \|T\|\,\|u\|$. For every $\epsilon > 0$, there is a $u_0$ such that $\|Tu_0\| > (\|T\| - \epsilon)\|u_0\|$. If $\|u\| \leq 1$, then $\|Au\| \leq \|A\|\,\|u\| \leq \|A\| \Rightarrow \sup \|Au\| \leq \|A\|, \|u\| \leq 1$. But if we put $u_1 = u_0/\|u_0\|$, then $\|Au_1\| = \|u_0\|^{-1}\|Au_0\| > \|A\| - \epsilon$, so for $\|u\| \leq 1, \sup \|Au\| \geq \|Au_1\| > \|A\| - \epsilon$ or $\sup \|Au\| \leq \|A\|$.

5.12. $\|\boldsymbol{Ax}\|_\infty = \max_{1 \leq i \leq n} \left|\sum_{j=1}^n A_{ij} x_j\right| \leq \max_{1 \leq i \leq n} \sum_{j=1}^n |A_{ij}|\,|x_j| \leq \max_{1 \leq i \leq n} \sum_{j=1}^n |A_{ij}| \max_{1 \leq j \leq n} |x_j| = \max_{1 \leq i \leq n} \sum_{j=1}^n |A_{ij}|\,\|\boldsymbol{x}\|_\infty$. Hence $\|\boldsymbol{A}\| = \sup(\|\boldsymbol{Ax}\|_\infty / \|\boldsymbol{x}\|_\infty) \leq \max_{i \leq i \leq n} \sum_{j=1}^n |A_{ij}|$. Suppose maximum occurs for $i = k$. Then for $\boldsymbol{x}$ such that $x_j = +1$ if $A_{kj} \geq 0$, $x_j = -1$ if $A_{kj} < 0$ we have $\|\boldsymbol{Ax}\|_\infty / \|\boldsymbol{x}\|_\infty = \sum_{j=1}^n |A_{ij}|$.

5.13. For $\boldsymbol{x} \neq 0$, $(\|\boldsymbol{Ax}\|/\|\boldsymbol{x}\|)^2 = (a+b)^2 - 2ab(x-y)^2/(x^2+y^2)$. Take the supremum (at $y = x$) to find $\|A\|_2$.

5.14. $\|Iu\| = \|u\|$; $I$ is bounded. Consider $u(x) = \sin nx$ : $\|u\|_V = 1$ but $\|Iu\|_W = 1 + n$ which cannot be bounded.

5.15. $\|ST(u)\| = \|S(Tu)\| \leq \|S\|\,\|Tu\| \leq \|S\|\,\|T\|\,\|u\|$.

5.16. Let $\{u_n\} \subset N(T)$ with limit $u$ in $U$. Then $Tu_n = 0$. Thus $0 = \lim_{n\to\infty} Tu_n = T(\lim_{n\to\infty} u_n) = Tu \Rightarrow u \in N(T)$.

5.17. $T$ is one-to-one since, if $Tu_1 = Tu_2 = v$, then $\|Tu_1 - Tu_2\| = 0 \geq K\|u_1 - u_2\|$. $\|T^{-1}v\| = \|u\| \leq K^{-1}\|Tu\| = K^{-1}\|v\|$.

5.18. $u(x) = \int_0^x u'(s)\, dx \leq \sup_{0\leq x\leq 1} |u'(x)| = \|Du\|$. Take sup of both sides.

5.19. $(I-P)(I-P) = I^2 - PI - IP + P^2 = I - P$. *Range* : $R(I-P) = N(P)$, $R(P) = N(I-P)$.

5.20. From Theorem 8, $\|Pu\| \leq \|u\|$. Thus $\|P\| \leq 1$. But for $u \in R(P)$ we have $Pu = u$, so $\|Pu\| = \|u\|$. Hence $\|P\| = 1$.

5.21. Take, for example, the map on $\mathbb{R}^2$ that takes a point $\boldsymbol{x}$ to the point in $B(\boldsymbol{0}, 1)$ closest to $\boldsymbol{x}$. This is a projection, but the map is not homogeneous.

5.22. Let $u \in N(P)$. By definition $(u, v) = 0$ for $v \in R(P)$. Hence $N(P) \subset R(P)^\perp$. Let $u \in R(P)^\perp$. Then $(u, z) = 0$ for $z \in R(P)$. By Theorem 9, $u = v + w$ for $v \in R(P)$, $w \in N(P)$, so $Pu = Pv + Pw = Pv = v$. Also, $0 = (u, z) = (v, z) + (w, z) = (v, z)$; hence $v = 0$. Thus $Pu = 0 \Rightarrow u \in N(P)$.

5.23. $T$ is a projection since $T$ is linear and $T^2u = Tv$ (where $v = u(x)$ if $|x| < 1$ and 0 otherwise) $= v = Tu$. $R(T) = \{u \in L^2(\mathbb{R}) :\ u(x) = 0$ for $|x| \geq 1\}$, $N(T) = \{u \in L^2(\mathbb{R}) :\ u(x) = 0$ for $|x| < 1\}$.

5.24. $v(y) = Pu(y) = \int_{-1}^{1} \exp(i(y - z))u(z)\, dz$; show that $Pv(x) \equiv P^2u(x) = Pu(x)$. $P$ is an orthogonal projection.

5.25. (i) $\boldsymbol{x}$ satisfies $\boldsymbol{Ax} = \boldsymbol{1}$ where $\boldsymbol{1} = (1, \ldots, 1)$; (ii) $\boldsymbol{x}$ satisfies $\boldsymbol{Ax} = \boldsymbol{\alpha} = (1, 0, \ldots, 0)$.

5.26. $u(x) = \frac{2}{e^3-1}(-e^{3-2x} + e^x) - 2x + 2$, $\langle l, f\rangle = \int_0^1 u(x)\, dx$.
$\frac{2}{e^3-1}\left(\frac{3}{2}e - e^3\right) = \int_0^1 g(x)2x\, dx$; so $g$ satisfies $\int_0^1 (gf - u)dx = 0$.

5.28. Let $\{\ell_n\}$ be a Cauchy sequence in $X'$. Then for any $u \in X$, $|\langle \ell_n, u\rangle - \langle \ell_m, u\rangle| \leq \|\ell_n - \ell_m\|\,\|u\| \to 0$ as $m,\ n \to 0$ so $\{\langle \ell_n, u\rangle\}$ is a Cauchy sequence in $\mathbb{R}$, with limit $\langle \ell, u\rangle$, say. Complete the proof by showing that $\ell$ is bounded and linear, and $\ell_n \to \ell$ in $X'$.

5.29. In the use of the projection theorem.

5.30. If there are two elements $u_1$, $u_2$ such that $(u_1, v) = (u_2, v) = \langle \ell, u \rangle$, then $(u_1 - u_2, v) = 0$. Set $v = u_1 - u_2$ : $\|u_1 - u_2\|^2 = 0$ or $u_1 = u_2$. $\|\ell\| = \sup(|\langle \ell, v\rangle|/\|v\|)$ (for $v \neq 0$) $= \sup((u,v)/\|v\|) \leq \sup(\|u\|\,\|v\|/\|v\|) = \|u\|$. Also, $|\langle \ell, u\rangle| = (u,u) = \|u\|^2 \leq \|\ell\|\,\|u\|$ so $\|\ell\| \geq \|u\|$. Hence $\|\ell\| = \|u\|$.

5.31. Take $f = |g|^{q-1} \operatorname{sgn} g$; then $|f|^p = |g|^q$, so $f \in L^p$, and $\|f\|_{L^p} = \|g\|_{L^q}^{q-1}$. Then show that $\langle \ell_g, f\rangle = \|f\|_{L^p}\|g\|_{L^q}$.

5.32. $\ell = 0 \Leftrightarrow \langle \ell, v\rangle = 0$ for all $v \in X$. Given $v \in X$ there exists a sequence $\{v_n\}$ in $Y$ such that $v_n \to v$. Thus $\langle \ell, v\rangle = \langle \ell, \lim_{n\to\infty} v_n\rangle = \lim_{n\to\infty}\langle \ell, v_n\rangle = 0$.

5.33. $|a(u,v)|^2 \leq [\|u'\|\,\|v'\| + \kappa_1\|u\|\,\|v\|]^2 \leq (\kappa_1^2\|u\|^2 + \|u'\|^2)(\|v\|^2 + \|v'\|^2)$, using Cauchy–Schwarz.

5.34. cf Exercise 4.2.

5.35. $|\langle \ell, v\rangle| = |\int_0^1 (-1-4x)v(x)\,dx| = |(-1-4x, v)_{L^2}| \leq \|-1-4x\|_{L^2}\|v\|_{L^2} \leq k\|v\|_{H^1}$. $|a(u,v)| \leq 2\,|\int_0 u'v'\,dx| \leq 2\|u'\|_{L^2}\|v'\|_{L^2} \leq 2\|u\|_{H^1}\|v\|_{H^1}$, hence continuous. $a(v,v) \geq \int_0^1 (v')^2\,dx$. Now $\|v'\|_{L^2}^2 \geq C^{-2}\|v\|_{L^2}^2$ so $(C^{-2}+1)\|v'\|_{L^2}^2 \geq C^{-2}\|v\|_{H^1}^2$. $\int_0^1(-1-4x)v\,dx = \int_0^1 (x+1)u'v'\,dx = [(x+1)u'v]_0^1 - \int_0^1 (u' + (x+1)u'')\,dx \Rightarrow \int_0^1 \{(x+1)u'' + u' - (1 + 4x)\}v\,dx = 0$.

5.36. $|\tilde{a}(u,v)| \leq |a(u,v)| + |(u, \kappa v)_{L^2}| \leq K\|u\|\,\|v\| + K'\|u\|\,\|v\|$ where $K' = \sup|\kappa(x)|$. $\tilde{a}(v,v) = a(v,v) + (v, \kappa v) \geq \alpha\|v\|^2 + \beta(v,v) \geq \alpha\|v\|^2$ where $\beta = \inf \kappa(x)$.

## Chapter 6

6.1. (a) Linearly dependent; (b) linearly independent.

6.2. $\sum_{k=1}^n \alpha_k e^{ikx} = 0 \Rightarrow \sum_{k=1}^n \alpha_k \cos kx = 0$ and $\sum_{k=1}^n \alpha_k \sin kx = 0$ which holds only for all $\alpha_k = 0$. Hence $\{e^{ikx}\}$ is linearly independent.

6.3. If $u, v \in X$, then $(\alpha u + \beta v)'' - 2(\alpha u + \beta v)' + (\alpha u + \beta v) = \alpha(u'' - 2u' + u) + \beta(v'' - 2v' + v) = 0$, hence $\alpha u + \beta v \in X$. $\dim X = 2$. Basis for $X$ is $\{u_1(x) = e^x,\ u_2(x) = xe^x\}$.

6.4. $\dim M = 9$, $\dim K = 4$.

6.5. Let $\dim V = m$ with basis $\{v_1, \ldots, v_m\}$ and $\dim W = n$ with basis $\{w_1, \ldots, w_n\}$. Every $u \in V \oplus W$ is of the form $u = v + w$ for some $v \in V$, $w \in W$. But $v = \sum_i \alpha_i v_i$ and $w = \sum_j \beta_j w_j$ so $u = \sum_i \alpha_i v_i + \sum_j \beta_j w_j$. Hence $B = \{v_1, \ldots, v_m, w_1, \ldots, w_n\}$ spans $V \oplus W$. It remains to show that $B$ is linearly independent.

6.7. $\boldsymbol{\phi}_1 = (1/\sqrt{2})(1,0,1)$, $\boldsymbol{\phi}_2 = (1/\sqrt{2})(1,0,-1)$, $\boldsymbol{\phi}_3 = (0,1,0)$.

6.8. $\phi_0(x) = \sqrt{1/2}$, $\phi_1(x) = \sqrt{3/2}\,x$, $\phi_2(x) = \frac{1}{2}\sqrt{5/2}\,(3x^2-1)$, $\phi_3(x) = \frac{1}{2}\sqrt{7/2}\,(5x^3 - 3x)$.

6.9. $A_{11} = \frac{1}{2}(e^2-1)$, $A_{12} = A_{21} = \frac{1}{2}(1-e^{-2})$, $A_{22} = \frac{1}{6}(1-e^{-6})$. $\det \boldsymbol{A} \neq 0$.

6.10. Consider $I : X_1 \to X_2 : \ \|Iu\|_2 = \|u\|_2 \leq k\|u\|_1$ (show this using Lemma 1; see also Theorem 4). Similarly, $\|u\|_1 \leq K\|u\|_2$ if we consider $I : X_2 \to X_1$.

6.11. $T_{12} = 2$, $T_{23} = 6$, others zero.

6.12. $T_{11} = 2\pi, T_{22} = \cos x$, others zero.

6.13. $(\boldsymbol{b},\boldsymbol{c}) = (\boldsymbol{Ta},\boldsymbol{c}) = (\boldsymbol{a},\boldsymbol{T}^T\boldsymbol{c}) = 0$ if $\boldsymbol{c} \in N(\boldsymbol{T}^T)$. Let $\boldsymbol{d} \in R(\boldsymbol{T})^{\perp}$. Then $(\boldsymbol{d},\boldsymbol{Tu}) = 0 = (\boldsymbol{T}^T\boldsymbol{d},\boldsymbol{u}) \Rightarrow \boldsymbol{d} \in N(\boldsymbol{T}^T)$. Conversely, if $\boldsymbol{d} \in N(\boldsymbol{T}^T)$, then if $\boldsymbol{Tu} = \boldsymbol{v}$ we have $(\boldsymbol{T}^T\boldsymbol{d},\boldsymbol{u}) = 0 = (\boldsymbol{d},\boldsymbol{v}) \Rightarrow \boldsymbol{d} \in R(\boldsymbol{T})^{\perp}$. Hence $N(\boldsymbol{T}^T) = R(\boldsymbol{T})^{\perp} \Rightarrow N(\boldsymbol{T}^T)^{\perp} = R(\boldsymbol{T})$. $N(\boldsymbol{T}^T) = \{(1,1,-1)\}$, $\boldsymbol{b} = (\alpha,\beta,\alpha+\beta)$.

6.14. $(\alpha_2,-\alpha_1,0)$, $(\alpha_3,0,-\alpha_1)$.

6.15. Let $B_1 = \{e_1,\ldots,e_n\}$ and $B_2 = \{f_1,\ldots,f_n\}$ be orthonormal bases of $X$ and $\mathbb{R}^n$, respectively. For any $u \in X$ we have $u = \sum u_i e_i$, $u_i = (u,e_i)$. Define the map $T : X \to \mathbb{R}^n$ by $T(u) = (u_1,\ldots,u_n)$. Then $T$ is an isomorphism (show this) and $\|u\|_X^2 = (u,u) = (\sum u_i e_i, \sum u_j e_j) = \sum u_i^2 = \|Tu\|_{\mathbb{R}^n}^2$.

6.16. $\|\ell\| = \max|\alpha_i|$.

6.17. (i) $u(x) = \sqrt{2\pi}(1/\sqrt{2\pi})$; (ii) $u(x) = \sum_{k=1}^{\infty}(2/k)(1-(-1)^k)\sin kx$.

6.18. $u_0 = -\sqrt{2}/4$, $u_1 = 5\sqrt{3}/6\sqrt{2}$, $u_2 = \sqrt{5}/8\sqrt{2}$.

6.19. $c_k = \frac{1}{2}(u_{2k} - iu_{2k-1})$ for $k = 1,2,\ldots$, $c_k = \frac{1}{2}(u_{2k} + iu_{2k-1})$ for $k = -1,-2,\ldots$, $c_0 = u_0/\sqrt{2}$.

6.20. $0 \leq \|u - \sum_{i=1}^{N}(u,\phi_i)\phi_i\|^2 = \|u\|^2 - \sum_{i=1}^{N}(u,\phi_i)^2$, hence $\sum_{i=1}^{N}(u,\phi_i)^2 \leq \|u\|^2$. Since sum is bounded, we can let $N \to \infty$.

6.21. Use the property $P\phi_k = \phi_k$ to show that $P^2u = Pu$. Clearly $R(P) \subset V$. Conversely, if $v \in V$, show that $Pv = v$ so that $R(P) = V$. Orthogonality: take $v \in R(P)$ and $w \in N(P)$; then $(w,v) = (w,Pu)$. Use this to show that $(w,v) = 0$.

6.22. See Exercise 6.8. $Pu = \sum_{k=0}^{3}(u,\phi_k)\phi_k = \sqrt{2/5}\,\phi_0 + (8/35)\sqrt{5/2}\,\phi_2$.

6.23. (a) Set $u(r,\theta) = R(r)\Theta(\theta)$ to get $(\Theta' \sin\theta)' + \lambda\Theta\sin\theta = 0$. Set $\xi = \cos\theta$ to get Legendre's equation. General solution is $u(r,\theta) = \sum_{n=0}^{\infty}[a_n r^n + b_n r^{-(n+1)}]P_n(\cos\theta)$.
(b) $a_n = (2n+1)/2 \int_0^{\pi} f(\theta)P_n(\cos\theta)\, d\theta$.

6.24. Eigenvalues satisfy $\sqrt{\lambda}\cos(\sqrt{\lambda_k}\ell) + \beta\sin(\sqrt{\lambda_k}\ell) = 0$. $v_k(x) = [(\ell/2) + (1/2\beta)\cos^2(\sqrt{\lambda_k}\ell)]^{-1/2}\sin(\sqrt{\lambda_k}\ell)$. Heat equation: $u(0,t) = 0$, $(\partial u/\partial x + \beta u)(\ell, t) = 0$.

6.25. Use integration by parts and the boundary conditions to show that $(Lu, u) \geq 0$. Nonnegativity of the eigenvalues follows from $0 \leq (Lu, u) \leq \lambda(u, u)$. Since $L^2$ is separable there is at most a countable number of nonzero mutually orthogonal vectors.

6.26. Let the minimizer be $u$, and set $w = u + \epsilon v$; then consider $R(w) = \hat{R}(\epsilon)$ over all $w$ that satisfy $(w, e_1) = (w, e_2) = \ldots = (w, e_{n-1}) = 0$. Set $[d\hat{R}/d\epsilon]_{\epsilon=0} = 0$; expand and differentiate to find that $\lambda = R(u)$ and $u = e_n$.

6.27. $(Ls_n, r_n) = (L\sum_{k=1}^{n} u_k\phi_k, r_n) = (\sum_{k=1}^{n} u_k\lambda_k\phi_k, r_n) = 0$ since $(r_n, \phi_k) = 0$ (Proof of Theorem 6.12).

6.28. Return to (6.34): for symmetry of $L$, (a) $p(x) \to 0$ as $x \to \pm\infty$; (b) $p(-L) = p(L)$.

6.29. (c) Show that $H_n'(x) = 2xH_n(x) - H_{n+1}(x)$. Set $f(x) = \exp(-x^2)$ and show that $f^{(n+1)} + 2xf^{(n)} + 2nf^{(n-1)} = 0$; multiply by $(-1)^{n+1}\exp(x^2)$ to get $H_{n+1} - 2xH_n + 2nH_{n-1} = 0$.

## Chapter 7

7.1. $|\alpha| = 0 \Rightarrow \alpha = (0,0)$, $(\boldsymbol{x}^\alpha/\alpha!)D^\alpha f(\mathbf{0}) = f(\mathbf{0})$. $|\alpha| = 1 \Rightarrow$

$$\frac{x^1y^0}{1!0!}D^{(1,0)}f(\mathbf{0}) + \frac{x^0y^1}{0!1!}D^{(0,1)}f(\mathbf{0}) = x\left.\frac{\partial f}{\partial x}\right|_0 + y\left.\frac{\partial f}{\partial y}\right|_0 \quad \text{etc.}$$

7.2. $\int_{-a}^{a}\delta(x)\phi(x)\, dx \leq e^{-1}\int_{-a}^{a}\delta(x)\, dx$ since $\sup\phi_a(x) = e^{-1}$. If $\delta$ were locally integrable, then $\lim_{a\to 0}\int_{-1}^{a}\delta(x)\, dx = 0$. But left-hand side $= \phi(0) = e^{-1}$.

7.3. $f(x)\phi(x) \in C(\Omega)$. Assume $f \neq 0$, but $\int f\phi\, dx = 0$. In particular, if $f(x_0) \neq 0$, then $f(x) \neq 0$ for all $x \in (x_0 - h,\ x_0 + h)$ for some $h$. Choose arbitrary $\phi$ with compact support inside $(x_0 - h, x_0 + h)$; can always find $\phi$ such that $\int f\phi\, dx \neq 0$, a contradiction.

7.4. Consider $\Omega \subset \mathbb{R}^2$, for example; for $|\alpha| = m$, $\int_\Omega(D^\alpha u)v\, dx = \int_\Omega(\partial^m u/\partial x^k\partial y^{m-k})v\, dx$, where $0 \leq k \leq m$. Use Green's theorem repeatedly.

7.5. $\langle(\mathrm{sgn})',\phi\rangle = -\langle \mathrm{sgn},\phi'\rangle = -\int_{-1}^{0}(-1)\phi'\,dx - \int_0^1(+1)\phi'\,dx = [\phi]_{-1}^{0} - [\phi]_0^1 = 2\phi(0) = 2\langle\delta,\phi\rangle$.

7.6. $\langle(\sin ax\cdot H(x))'',\phi\rangle = \langle \sin ax\cdot H(x),\phi''\rangle = \langle H(x),\phi''\sin ax\rangle$ $= \int_0^1 \phi''\sin ax\,dx = [\phi'\sin ax - a\phi\cos ax]_0^1 - \int_0^1 a^2\phi\sin ax\,dx = a\phi(0) - a^2\langle\sin ax\cdot H(x),\phi\rangle$.

7.7. $\langle f',\phi\rangle = -\langle f,\phi'\rangle = -\int_{-1}^{0} x\phi'(x)\,dx - \int_0^1(x+c)\phi'(x)\,dx = -[x\phi]_{-1}^{0} + \int_{-1}^{0}\phi(x)\,dx - [(x+c)\phi]_0^1 + \int_0^1\phi(x)\,dx = c\phi(0) + \int_{-1}^{1} 1\cdot\phi(x)\,dx = \langle c\delta,\phi\rangle + \langle 1,\phi\rangle$.

7.8. Set $A = \{\boldsymbol{x}:\ -1 < x < 0,\ -1 < y < 0\}$, $B = \{\boldsymbol{x}: 0 < x < 1,\ 0 < y < 1\}$, $C = A\cup B$, with boundaries $\partial A$, $\partial B$. Then

$$\begin{aligned} D^{(1,1)}\langle f,\phi\rangle = \langle f, D^{(1,1)}\phi\rangle &= \int_C xy\frac{\partial^2\phi}{\partial x\partial y}\,dx\,dy \\ &= \int_{\partial A} xy\nu_x\frac{\partial\phi}{\partial y}\,ds + \int_{\partial B} xy\nu_x\frac{\partial\phi}{\partial y}\,ds - \int_C y\frac{\partial\phi}{\partial y}\,dx\,dy \\ &= -\int_C y\frac{\partial\phi}{\partial y}\,dx\,dy = \int_C \phi\,dx\,dy. \end{aligned}$$

7.9. Solution of homogenous equation is $u(x) = e^{-x}$. Now $\langle u'+u,\phi\rangle = -\langle u,\phi'\rangle + \langle u,\phi\rangle = -\langle H, f\phi'\rangle + \langle H, f\phi\rangle$ (using $u = Hf$) $= \langle\delta,\phi\rangle$ after integrating. Left-hand side $= f(0)\phi(0) + \int_0^1(f'+f)\phi\,dx \Rightarrow f(x) = e^{-x}$. Hence $u(x) = (c + H(x))e^{-x}$.

7.10. (a) $u\in H^2(0,3)$; (b) $u\in H^1((0,1)\times(0,2))$.

7.11. $u\perp v$ in $H^1(0,2)$.

7.13. $D^\alpha u\in L^2(\Omega)$ for $|\alpha| = 2$; so $m = 2 > n/2 = 1$.

7.14. Consider $\{u_n\},\{v_n\}\subset C^1(\bar\Omega)$ such that $u_n\to u$ and $v_n\to u$ in the $H^1$-norm with $u$, $v\in H^1(\Omega)$ ($H^1$ is the closure of $C^1$). Then $D^\alpha u_n\to D^\alpha u$, $D^\alpha v_n\to D^\alpha v$ in $L^2$, for $|\alpha|\le 1$. Also, $v_n\to v$ and $u_n\to u$ in $L^2(\Gamma)$. Thus, for example, $(\partial u_n/\partial x_i, v_n)_{L^2(\Omega)} = (u_n, v_n\nu_i)_{L^2(\Gamma)} - (u_n,\partial v_n/\partial x_i)_{L^2(\Omega)}$. Take $\lim_{n\to\infty}$.

7.15. Assume $\Omega\subset\mathbb{R}^2$; then left-hand side is $\int_\Omega\left(\frac{\partial^2 u}{\partial x^2}+\frac{\partial^2 u}{\partial y^2}\right)\left(\frac{\partial^2 v}{\partial x^2}+\frac{\partial^2 v}{\partial y^2}\right)dx$. Now $\int_\Omega\frac{\partial^2 u}{\partial x^2}\frac{\partial^2 v}{\partial x^2}\,dx = \int_\Gamma\left(\frac{\partial^2 u}{\partial x^2}\frac{\partial v}{\partial x} - \frac{\partial^3 u}{\partial x^3}v\right)\nu_x\,ds + \int_\Omega\frac{\partial^4 u}{\partial x^4}v\,dx$. Proceed in this manner; use $\partial/\partial\nu = \nu_1\partial/\partial x + \nu_2\partial/\partial y$.

7.16. Let $\{v_n\}$ be a sequence in $\mathcal{D}(\Omega)$ with limit $v\in H_0^1(\Omega)$. We have $\|v_n\|_{L^2}\le c|v_n|_{H^1}$; $v_n\to v$ in $H^1$ implies that $\|v_n\|_{L^2}\to\|v\|_{L^2}$ and $|v_n|_{H^1}\to|v|_{H^1}$. $|\cdot|_{H^1}$ is positive-definite since $|v|_1 = 0$ implies that $\int|\nabla v|^2\,dx = 0$, so that $v = \text{const} = 0$, given the boundary value of $v$.

7.17. Show that $(u,v) \equiv \int_\Omega \sum_{|\alpha|=m} D^\alpha u D^\alpha v \, dx$ is an inner product. In particular, $(u,u) = 0 \Rightarrow \int_\Omega (D^\alpha u)^2 \, dx = 0$ for $|\alpha| = m$, hence $D^\alpha u = 0$ for $|\alpha| = m$. But $u \in H_0^m(\Omega)$; so $u = 0$.

7.18. $$\|\nabla^2 v\|_{L^2}^2 = \int_\Omega \left[\left(\frac{\partial^2 v}{\partial x^2}\right)^2 + 2\frac{\partial^2 v}{\partial x^2}\frac{\partial^2 v}{\partial y^2} + \left(\frac{\partial^2 v}{\partial y^2}\right)^2\right] dx.$$

But $$\int_\Omega \left(\frac{\partial^2 v}{\partial x\, \partial y}\right)^2 dx = -\int_\Omega \frac{\partial^3 v}{\partial x^2\, \partial y}\frac{\partial v}{\partial x}\, dx = \int_\Omega \frac{\partial^2 v}{\partial y^2}\frac{\partial^2 v}{\partial x^2}\, dx.$$

7.19. Require $\sup |\langle \delta, v\rangle|$ to be defined, i.e., $v$ continuous. Hence $m > n/2$. For example, $\delta : H_0^1(\Omega) \to \mathbb{R}$ is not defined for $\Omega \subset \mathbb{R}^2$.

7.20. $u \in H_0^1(\Omega)^\perp \Rightarrow (u,v)_{H^1} = 0$ for all $v \in H_0^1(\Omega)$; i.e., $0 = \int_\Omega (uv + \sum_{|\alpha|=1} D^\alpha u D^\alpha v)\, dx$. Set $v = \phi \in \mathcal{D}(\Omega) : \; 0 = \int (u - \nabla^2 u)\phi \, dx$ using Green's theorem $\Rightarrow \nabla^2 u = u$. Since $\mathcal{D}(\Omega)$ is dense in $H_0^1$, we can extend this result in the usual way. $u \in H_0^1(\Omega)^\perp$ for $\Omega = (0,1) \Rightarrow u'' - u = 0$. Basis for $H_0^1(\Omega)^\perp$ is $\{e^x, e^{-x}\}$.

7.21. $\int (\ln x)^2 dx = x(\ln x)^2 - 2x \ln x + 2x$. Then use Theorem 9.

## Chapter 8

8.1. (a) Second order, nonlinear, $\Omega =$ upper unit semicircle. (b) Fourth-order, linear, $\Omega =$ triangle with vertices at $(0,0), (1,0), (0,1)$.

8.2. (a) $\frac{d}{dt}\int_{\Omega'} \rho \boldsymbol{v}\, dx = \int_{\Omega'} \boldsymbol{Q}\, dx + \oint_{\Gamma'} \boldsymbol{t}\, ds$. Use Cauchy's law $\boldsymbol{t} = \boldsymbol{\sigma n}$ and the divergence theorem to rewrite the surface integral as $\int_{\Omega'} \operatorname{div} \boldsymbol{\sigma}\, dx$. The left-hand side equals $\int_{\Omega'} \rho \partial^2 \boldsymbol{u}/\partial t^2 \, dx$. Regroup and invoke the arbitrariness of $\Omega'$ to obtain (8.5). (b) $\sigma_{ij} = \lambda(\operatorname{div} \boldsymbol{u}) I_{ij} + 2\mu\epsilon_{ij}(\boldsymbol{u})$. Substitute in (8.5).

8.3. The argument is as in Example 2 of the Introduction: simply replace $f$ by $f - ku$, $ku$ being the force of the foundation.

8.4. (a) $\sum_j \partial\sigma_{\alpha j}/\partial x_j = \sum_\beta \partial\sigma_{\alpha\beta}/\partial x_\beta + \partial\sigma_{\alpha 3}/\partial z$, where $z = x_3$. Integrate with respect to $z$ and use the definitions of $S_\alpha$ and $M_{\alpha\beta}$. (b) Follows as part (a). (c) Differentiate $(8.14)_1$, with respect to $x_\alpha$, sum on $\alpha$, and use $(8.14)_2$ to eliminate $S_\alpha$; this gives $\sum_{\alpha,\beta} \partial^2 M_{\alpha\beta}/\partial x_\alpha \partial x_\beta = -q$. Next, use (8.13). This gives $\sum_{\alpha,\beta} \partial^2 M_{\alpha\beta}/\partial x_\alpha \partial x_\beta = -D[\nu \sum_\alpha \partial^2(\nabla^2 w)/\partial x_\alpha^2 + (1-\nu)\sum_{\alpha,\beta} \partial^4 w/\partial x_\alpha^2 \partial x_\beta^2]$.

8.5. (a) $w' = 0, \; w''' = 0$; (b) $w = 0, \; w'' = 0$.

8.6. Elliptic in $A = \{\boldsymbol{x} : \; x > 1, y > 1\} \cup \{\boldsymbol{x} : \; x < 1, y < 1\}$; strongly elliptic in any open subset of $A$.

8.7. $\sum_{|\alpha|, |\beta|=1} a_{\alpha\beta}\xi^{\alpha+\beta} = -(1+x^2)\xi^2 + 3\eta^2 + 2(1+x^2)\zeta^2 = 0$ at any $x_0$ for any $\xi$ such that $\xi^2 = [3\eta^2 + 2(1+z_0^2)\zeta^2]/(1+x_0^2)$.

8.8. $\sum_{i,j,k,l} C_{ijkl}\xi_i\eta_j\xi_k\eta_l = \mu|\boldsymbol{\xi}|^2|\boldsymbol{\eta}|^2 + (\lambda+\mu)(\boldsymbol{\xi}\cdot\boldsymbol{\eta})^2 = \mu|\boldsymbol{\xi}|^2|\boldsymbol{\eta}^\perp|^2 + (\lambda + 2\mu)(\boldsymbol{\xi}\cdot\boldsymbol{\eta})^2$, where $\boldsymbol{\eta}^\perp$ is the component of $\boldsymbol{\eta}$ orthogonal to $\boldsymbol{\xi}$. The result follows from the independence of $\boldsymbol{\eta}^\perp$ and $(\boldsymbol{\xi}\cdot\boldsymbol{\eta})$. Pointwise stability: $\epsilon \equiv \sum_{i,j,k,l} C_{ijkl}M_{ij}M_{kl} = (3\lambda+2\mu)|\boldsymbol{M}^S|^2+2\mu|\boldsymbol{M}^D|^2$ $[\boldsymbol{M}^S = \frac{1}{3}(\mathrm{tr}\,\boldsymbol{M})\boldsymbol{I}$ and $\boldsymbol{M}^D = \boldsymbol{M}-\boldsymbol{M}^S]$. Show that $|\boldsymbol{M}|^2 = |\boldsymbol{M}^S|^2+|\boldsymbol{M}^D|^2$: then $\epsilon \geq c|\boldsymbol{M}|^2$ iff $3\lambda+2\mu > k_0$ and $\mu > \mu_0$.

8.9. $\dfrac{\partial^2 u}{\partial x_1^2}\nu_1^2 + 2\dfrac{\partial^2 u}{\partial x_1 \partial x_2}\nu_1\nu_2 + \dfrac{\partial^2 u}{\partial x_2^2}\nu_2^2 = g$. Have to check $\sum_{|\alpha|=2} b_\alpha \boldsymbol{a}^\alpha = \nu_1^2 a_1^2 + 2\nu_1\nu_2 a_1 a_2 + \nu_2 a_2 = (\nu_1 a_1 + \nu_2 a_2)^2 = (\nu_1^2+\nu_2^2)^2 \neq 0$ if $\boldsymbol{a} = \boldsymbol{\nu}$.

8.10. Use (8.13) and (8.14). The BC can be rewritten as $\frac{\partial^3 w}{\partial x^3} + \nu\frac{\partial^3 w}{\partial x \partial y^2} = 0$. With $|\alpha| = 3$, $\sum b_\alpha \boldsymbol{\nu}^\alpha = b_{(3,0)}\nu_1^3 + b_{(1,2)}\nu_1\nu_2^2 = \nu_1[\nu_1^2 + \nu\nu_2^2] \neq 0$ along $x = L$, for which $\nu_1 = 1$, $\nu_2 = 0$.

8.11. $|\gamma| + |\beta - 3| \neq 0$.

8.12. $k\boldsymbol{u}\cdot\boldsymbol{\nu} - \boldsymbol{t}\cdot\boldsymbol{\nu} = 0$. $n = 2$; $b_{11} = k$, $b_{22} = -c_{11} = 1$; all other components are zero. So (8.33) is satisfied.

8.13. $\boldsymbol{u}\cdot\boldsymbol{\nu} = 0$, $\boldsymbol{t}\cdot\boldsymbol{s} = \mu\boldsymbol{t}\cdot\boldsymbol{\nu}$ with $\boldsymbol{t} = \boldsymbol{\sigma\nu}$, and $\boldsymbol{\sigma} = \boldsymbol{C\epsilon}(\boldsymbol{v})$.

8.14. $[u'''v - u''v' + u'v'' - uv''']_0^1 = [-B_1uS_1^*v - B_0uS_0^*v + S_1uB_1^*v + S_0uB_0^*v]_0^1$.

8.15. $B_0^* = \partial/\partial\nu$, $S_0 = -S_0^* = 1$.

8.16. Set $\partial v_i/\partial x_j = e_{ij}$; then since $\boldsymbol{\sigma}$ is symmetric, $\sum_{i,j}\sigma_{ij}e_{ij} = \sum_{i,j}\sigma_{ji}e_{ij}$ (i); also, by swapping indices, $\sum_{i,j}\sigma_{ij}e_{ij} = \sum_{i,j}\sigma_{ji}e_{ji}$ (ii); add (i) and (ii) to get desired result. To obtain (8.49), use the fact that $\int_\Omega \sum_{k,l}\sigma_{kl}\epsilon_{kl}(\boldsymbol{u})\,dx = \int_\Omega \sum_{k,l}\sigma_{kl}(\partial u_k/\partial x_l)\,dx = \int_\Gamma \sum_{k,l}\sigma_{kl}\nu_l u_k\,ds - \int_\Omega \sum_{k,l}(\partial\sigma_{kl}/\partial x_l)u_k\,dx$. Set $\boldsymbol{\sigma} = \boldsymbol{C\epsilon}(\boldsymbol{v})$.

8.19. $N(A) = \{u : u(x) = ax + b\} = N(A^*)$. Solution exists if $\int_0^1 f(x)\,dx = \int_0^1 xf(x)\,dx = 0$. Solution is unique if $\int_0^1 u(x)\,dx = \int_0^1 xu(x)\,dx = 0$.

8.20. $N(A) = \{u : u \text{ const.}\}$; unique solution if $\int_\Omega u(\boldsymbol{x})\,dx = 0$. $N(A^*) = \{u : u(\boldsymbol{x}) = \alpha_1 + \alpha_2(x - y)\}$; solution exists if $\int_\Omega f\,dx = \int_\Omega (x - y)f\,dx = 0$. If $\Omega = (-1,1)\times(-1,1)$, then $\int_\Omega f\,dx = 0$ if $f$ is odd in $x$ or $y$; $\int_\Omega (x-y)f(\boldsymbol{x})\,dx = 0$ if $f(x,y) = f(y,x)$.

8.22. (b) From (a), $A : N(A)^\perp \to R(A)$ is bounded. Hence, using the Banach theorem, $A^{-1} : R(A) \to N(A)^\perp$ is linear, bounded $\Rightarrow \|A^{-1}v\| \leq K\|v\|$ for all $v \in R(A)$, so setting $v = Au$ we have $\|u\| \leq K\|Au\|$ for $u \in N(A)^\perp$. If $\{v_n\}$ is a Cauchy sequence in $R(A)$ with limit $v$, then with $u_n = A^{-1}v_n$ we have $\|u_m - u_n\| \leq K\|v_m - v_n\| \to 0$ as $m, n \to \infty$; so $\{u_m\}$ is a Cauchy sequence in $N(A)^\perp$. $N(A)^\perp$ is closed;

so $u_m \to u$ in $N(A)^\perp$. Since $A$ is continuous, $v_n = Au_n \Rightarrow v = Au$. Hence $v \in R(A) \Rightarrow R(A)$ is closed.

8.23. Flexible foundation: $N(A) = \{\mathbf{0}\}$ and a unique solution exists. Coulomb friction: $N(A) = \{c_1\mathbf{e}_1 + c_2\mathbf{e}_2\}$, where $c_1$ and $c_2$ are constants. A unique solution exists if and only if $f_1 = f_2 = 0$. If friction is not limiting, then $N(A) = \{\mathbf{0}\}$.

## Chapter 9

9.1. $V = \{v \in H^2(0,1) : v(0) = 0,\ v'(0) = 0\}$.
$\int_0^l [ku''v'' + du'v' + cu]\,dx = \int_0^1 fv\,dx + \beta v(1) + \alpha v'(1)$.

9.2. Let angle between $\boldsymbol{\tau}$ and $\boldsymbol{\nu}$ be $\beta$. Boundary term in VBVP is $\int_\Gamma v\nabla u \cdot \boldsymbol{\nu}\,ds$, $v \in H^1(\Omega)$. But $\boldsymbol{\tau} = \boldsymbol{\nu}\cos\beta + \boldsymbol{s}\sin\beta$ ($\boldsymbol{s}$ = tangent $= (-\nu_2, \nu_1)$), or $\boldsymbol{\tau} = (\nu_1\cos\beta - \nu_2\sin\beta,\ \nu_2\cos\beta + \nu_1\sin\beta) \Rightarrow \boldsymbol{\nu} = (\tau_1\cos\beta + \tau_2\sin\beta, -\tau_1\sin\beta + \tau_2\cos\beta)$. Boundary term is thus $\int_\Gamma v(g\cos\beta - \nabla u \cdot \boldsymbol{\mu}\sin\beta)\,ds$, where $\boldsymbol{\mu}$ is normal to $\boldsymbol{\tau}$.

9.3. $a(w,v)$ follows by direct substitution of (8.13).

9.4. For continuity of $a$, use the Sobolev Embedding Theorem to obtain $|v(1)| \le C\|v\|_{H^1} \le C\|v\|_{H^2}$, etc.

9.5. $a(v,v) \ge \int_\Omega \mu \sum_i \frac{\partial v}{\partial x_i}\frac{\partial v}{\partial x_i}\,dx$ using strong ellipticity. Complete by using the Poincaré–Friedrichs inequality.

9.6. Use (8.13) to obtain the first part. For the second part return to Exercise 9.3: the remaining boundary term is $\int_\Gamma M_n(w)\partial v/\partial\nu\,ds = 0$. Use the identity $a^2 + 2\nu ab + b^2 \ge (1-\nu)(a^2+b^2)$ and (7.18) to show that $a(v,v) \ge (1-\nu)\int_\Omega \sum_{|\alpha|=2} |D^\alpha u|^2\,dx$. Here $\nu$ is Poisson's ratio. Use the Poincaré inequality (7.18) to get $\int_\Omega (\frac{\partial w}{\partial x})^2 dx \le c\int_\Omega [(\frac{\partial w}{\partial x})^2 + (\frac{\partial^2 w}{\partial x_1 \partial x_2})^2]\,dx$, etc. Then apply the Poincaré–Friedrichs inequality to obtain a similar bound on $\int_\Omega v^2\,dx$. This leads to $a(v,v) \ge C(1-\nu)\|v\|^2_{H^2}$.

9.7. $a(u,v) = \int_0^1 (pu'v' + ruv)\,dx + p(1)u(1)v(1)$. $V$-ellipticity: use Theorem 1, Chapter 7, to get $a(v,v) \ge \alpha\|v\|^2_{H^1}$, $\alpha = \min(p_0, r_0)$. Continuity: $a(u,v) \le \int_0^1 (p_1u'v' + r_1uv)\,dx + p_1u(1)v(1) \le k\int_0^1 (u'v' + uv)\,dx + p_1\|u\|_\infty\|v\|_\infty$ ($k = \max(p_1, r_1)$) $\le k(u,v)_{H^1} + p_1K\|u\|_{H^1}\|v\|_{H^1}$ (Sobolev Embedding Theorem) $\le (k + p_1K)\|u\|_{H^1}\|v\|_{H^1}$.

9.8. (b) VBVP is: $\int_0^1 u''v''\,dx + [h_1v(1) - g_1v'(1) - h_0v(0) + g_0v'(0)] = \int_0^1 fv\,dx$, $v \in H^2(0,1)$; so $P = P_1(0,1)$. Hence $Q = \{v \in H^2(0,1) : \int_0^1 v\,dx = \int_0^1 xv\ \ dx = 0.\}$ $Q$-ellipticity is tricky, but see Rektorys [39], Chapter 35. A unique solution exists if and only if $0 = \langle \ell, p\rangle = \int_0^1 fp\,dx + [g_1p'(1) - h_1p(1) + h_0p(0) - g_0p'(0)]$ for all $p \in P_1(0,1)$.

9.9. See Exercise 8.8: use Korn's inequality.

9.11. $a(\boldsymbol{u}+\boldsymbol{p}, \boldsymbol{v}+\bar{\boldsymbol{p}}) = a(\boldsymbol{u}, \boldsymbol{v})$ for $\boldsymbol{p} = p_r\boldsymbol{e}_r + p_\theta\boldsymbol{e}_\theta$, where $p_r, p_\theta \in P_0(\Omega)$. $Q = \{\boldsymbol{v} \in V : \int_\Omega v_r\, dx = \int_\Omega v_\theta\, dx = 0\}$; a unique solution exists if $\boldsymbol{f}$ satisfies $\int_\Omega f_r\, dx = \int_\Omega f_\theta\, dx = 0$.

9.12. $\langle DJ(u), v\rangle = \lim_{\theta\to 0}\theta^{-1}[J(u+\theta v) - J(u)]$ by definition. Set $f(\theta) = J(u+\theta v)$ for any given $u, v$. Then $\langle DJ(u), v\rangle = \lim_{\theta\to 0}\theta^{-1}[f(\theta) - f(0)] = f'(0) = (d/d\theta)(DJ(u)v)|_{\theta=0}$.

9.13. $\partial J/\partial x_i = \lim_{\theta\to 0}[J(\boldsymbol{x}+\theta(0,\ldots,y_i,\ldots,0)) - J(\boldsymbol{x})]/(\theta y_i)$ ($y_i$ in $i$th slot). Multiply by $y_i$ and sum over $i$ to get result.

9.14. $J(\theta u + (1-\theta)v) = \frac{1}{2}\{\theta^2 a(u,u) + (1-\theta)^2 a(v,v) - 2\theta(1-\theta)a(u,v)\} - \theta\langle \ell, u\rangle - (1-\theta)\langle \ell, v\rangle$. $a(u-v, u-v) > 0$ since $a$ is $V$-elliptic, so $2a(u,v) < a(u,u) + a(v,v)$. Use this to obtain strict convexity of $J$.

9.15. $J(\theta u + (1-\theta)v) = \theta J(u) + (1-\theta)J(v) - \frac{1}{2}\theta(1-\theta)a(u-v, u-v)$. The last term on the right is nonnegative. To show that $u$ is a minimizer: from convexity, $J(v) - J(u) \geq \theta^{-1}[J(u+\theta(v-u)) - J(u)] = \langle DJ(u), v\rangle$ when $\theta \to 0$ (see Example 15).

9.16. $J(v) = \frac{1}{2}\int_0^1 [k(v'')^2 + d(v')^2 + cu^2]\, dx - \int_0^1 fv\, dx - \beta v(1) - \alpha v'(1)$.

9.17. Since $u$ is a minimizer, $J(u) \leq J((1-\theta)u + \theta v)$ for $0 < \theta < 1$ (since $V$ must be convex). Expand and rearrange to get $a(u, v-u) - \langle \ell, v-u\rangle + \frac{1}{2}\theta a(v-u, v-u) \geq 0$. Let $\theta \to 0$.

## Chapter 10

10.3. $u_n$ satisfies $a(u,v) = \langle \ell, v\rangle$ or $(u_n, \phi_k)_a = \langle \ell, \phi_k\rangle$. Also, $u_n = \sum_{k=1}^n (u_n, \phi_k)_a\phi_k = \sum_{k=1}^n \langle \ell, \phi_k\rangle\phi_k$. Now $J(u) = -\frac{1}{2}\|u\|_a^2$ (show this); but $J(u_n) = \frac{1}{2}\|u_n - u\|_a^2 - \frac{1}{2}\|u\|_a^2$; hence $\|u_n - u\|_a \to 0$. The result $\|u_n - u\|_H \to 0$ follows from continuity of $a(\cdot,\cdot)$.

10.5. $\|u - u_h\|_a^2 = a(u-u_h, u-u_h) = \|u\|_a^2 - \|u_h\|_a^2 - 2a(u_h, u-u_h)$. The last term is zero.

10.6. $a(u,v) = \langle \ell, v\rangle$ so $a(u,u) = \langle \ell, u\rangle$; hence $J(u) = -\frac{1}{2}a(u,u)$.

10.8. $\tilde{u}_h(x) = (\sqrt{2}/2)(-\phi_1(x) - \phi_2(x) + \phi_3(x)) = (\sqrt{2}/2)(x^2 + \frac{1}{2}x - 1)$.

10.9. Replace $v$ by $Av_h$ in Green's formula ($G(u,v) = 0$); (10.37) gives $(Av_h, f) = (Av_h, Au_h) \Rightarrow (v_h, A^*f) = (v_h, A^*Au_h)$. If $A = -\nabla^2$, then $A^* = -\nabla^2$.

10.10. (a) $\int_0^1(-u_h'' + u_h - \sin x)v_h\, dx = 0,\ \ v_h \in V^h \subset L^2(0,1),\ u_h \in U^h \subset H^2(0,1) \cap H_0^1(0,1)$.
(b) Least squares: solve $\boldsymbol{M}^T\boldsymbol{a} = \boldsymbol{F}$, where $M_{ij} = \int_0^1(-\phi_i'' + \phi_i)(-\psi_j'' +$

$\psi_j)\,dx$ and $F_j = \int_0^1 (\sin x)(-\psi_j''+\psi_j)\,dx$. Collocation: solve $\sum_{k=1}^N(-\phi_k''(x_i) + \phi_k(x_i))a_k = f(x_i),\ \ i = 1,\ldots,N$.
(c) Solve $\boldsymbol{M}^T\boldsymbol{a} = \boldsymbol{F}$, where $M_{ij} = \int_0^1 \phi_i(-\psi_j'' + \psi_j)\,dx$ and $F_j = \int_0^1 f\psi_j\,dx$.

## Chapter 11

11.1. Must show that a function $v$, say, exists such that $\int_\Omega v_i\phi\,dx = -\int_\Omega v\partial\phi/\partial x_i\,dx$. For each $\Omega_e$, $\int_{\Omega_e}(\partial v/\partial x_i)\phi\,dx = \int_{\Gamma_e} v\nu_i\phi\,ds - \int_{\Omega_e} v\,\partial\phi/\partial x_i\,dx$ since $v|_{\Omega_e} \in H^1(\Omega)$.

11.2. Optimal $B = 5$.

11.4. $\tilde{e}_{\max}$ exists at point $\bar{x}$, where $\tilde{e}' = 0$; then $\tilde{e}(x_i) = 0 = \tilde{e}(\bar{x}) + \frac{1}{2}\tilde{e}''(z)(x_i - \bar{x})^2 \Rightarrow |\tilde{e}(\bar{x})| = \frac{1}{2}|\tilde{e}''(z)|(x_i - \bar{x})^2$. If $i$ is node nearer to $\bar{x}$, then $|x_i - \bar{x}| \leq \frac{1}{2}h$. Hence $|\tilde{e}(\bar{x})| \leq \frac{1}{8}h^2|\tilde{e}''(x)|$. Maximize over all elements to get result.

11.5. $f''(x) = 2\pi\cos\pi x - \pi^2 x\sin\pi x$. $|Max.value| = |2\pi|$ at $x = 0, 1$. Hence $\|\tilde{e}\|_\infty \leq \frac{1}{8}h^2 \cdot 2\pi = \pi h^2/4$. See whether $\log\|\tilde{e}\|_\infty \simeq 2h + \text{const}$.

11.7. Retain 1, $\xi$, $\eta$, $\xi^2$, $\xi\eta$, $\eta^2$, $\xi^2\eta$, $\xi\eta^2$. Then, for example, if node 5 is located at $(\xi,\eta) = (0,-1)$, $\hat{N}_5(\xi,\eta) = \frac{1}{2}(1-\xi^2)(1-\eta)$.

11.10. One needs to solve a system of 21 equations uniquely, for any given right-hand side. Equivalently, show that any polynomial for which $D^\alpha p = 0$ for $|\alpha| \leq 2$ at the vertices, and $p_\nu = 0$ at the midpoints is identically zero. See Ciarlet [11] (Theorem 2.2.11) for full details.

11.11. $\boldsymbol{x} = \sum_{A=1}^4 \boldsymbol{x}_A\hat{N}_A(\xi,\eta)$, where $\hat{N}_A$ are given by (11.27). Substitute and use the geometry of the parallelogram to verify that $\boldsymbol{x} = \boldsymbol{A}\boldsymbol{\xi} + \boldsymbol{b}$ for suitable $\boldsymbol{A}$ and $\boldsymbol{b}$.

11.12. $j = \frac{1}{8}[2d - (\xi+\eta)(1-d)] > 0$ for all $\boldsymbol{\xi} \in \hat{\Omega}$ if $d > \frac{1}{2}$.

11.15. (a) $\boldsymbol{a} = [1\ 1\ 1\ 1]^T$, $\boldsymbol{b} = 2[-1\ 1\ 1\ -1]^T$, $\boldsymbol{c} = 2[1\ 1\ -1\ -1]^T$, $\boldsymbol{d} = [1\ -1\ 1\ -1]^T$. (b) $(\nabla\boldsymbol{N})^T = [-\frac{1}{4}\boldsymbol{b} + \eta\boldsymbol{d}\ \ \frac{1}{4}\boldsymbol{c} + \xi\boldsymbol{d}]$.

11.16. Show by direct integration over reference element; for example, $\int_{\Omega_e}(a + b_1x + b_2y)\,dxdy = A_e(a + \boldsymbol{b}^T\hat{\boldsymbol{x}})$ where $\boldsymbol{b} = [b_1\ b_2]^T$ and $\hat{\boldsymbol{x}} = (1/6)[1\ 1]^T$, for a polynomial of degree one.

## Chapter 12

12.1. $\|\boldsymbol{T}_e^{-1}\| = \sup\|\boldsymbol{T}_e^{-1}\boldsymbol{y}\|/\|\boldsymbol{y}\|$, $\boldsymbol{y} \neq 0$. Set $\boldsymbol{z} = \rho_e\boldsymbol{y}/\|\boldsymbol{y}\|$; then $\|\boldsymbol{T}_e^{-1}\| = \sup\|\rho_e^{-1}\boldsymbol{T}_e^{-1}\boldsymbol{z}\|$. Pick $\boldsymbol{x},\boldsymbol{y}$ in $\Omega_e$ such that $\|\boldsymbol{x} - \boldsymbol{y}\| = \rho_e$ : $\|\boldsymbol{T}_e^{-1}\| = \rho_e^{-1}\sup\|\boldsymbol{T}_e^{-1}(\boldsymbol{x} - \boldsymbol{b} + \boldsymbol{b} - \boldsymbol{y})\| = \rho_e^{-1}\sup\|\boldsymbol{x} - \boldsymbol{y}\| = \hat{h}/\rho_e$.

12.2. $\|Iv\|_{m,\hat{\Omega}} = \|v\|_{m,\hat{\Omega}} \leq \|v\|_{k+1,\Omega}$ since $m \leq k+1$. $\|\hat{\Pi}v\|_{m,\Omega} = \|\sum_i \hat{v}(\hat{\boldsymbol{x}}_i)\hat{\psi}_i\|_{m,\hat{\Omega}} \leq \sum_i |\hat{v}(\hat{\boldsymbol{x}}_i)|\,\|\hat{\psi}_i\|_{m,\hat{\Omega}} \leq C \sup|\hat{v}(\hat{\boldsymbol{x}}_i)|$ ($C$ is independent of $\hat{v}$).

12.3. Let the triangle have angles $\alpha, \beta, \gamma$ with $\theta_e = \alpha \leq \beta \leq \gamma$. Let the sides opposite $\alpha, \beta, \gamma$ be $a, b, c$, respectively. Then $a \leq b \leq c$ and $h_e = c$. The largest circle inscribed in the triangle touches all sides. Draw a sketch and show that $h_e = (\rho_e/2)(\cot\alpha/2 + \cot\beta/2)$. Now $\alpha < \pi/2, \beta < \pi/2$; so $\cot\beta/2 \leq \cot\alpha/2$. Hence $h_e/\rho_e \leq \cot\alpha/2 \leq \sigma$ if we prescribe $\alpha \leq \theta_0$, so that $\sigma = \cot\theta_0/2$.

12.4. $k = 2$ $\quad 0(h_e^{3-m}),\ 0 \leq m \leq 3,\ u \in H^3(\Omega_e)$
$k = 3$ $\quad 0(h_e^{4-m}),\ 0 \leq m \leq 4,\ u \in H^4(\Omega_e)$.

12.5. $\|v - \Pi_e v\|_m^2 = \sum_{l=0}^m |v - \Pi_e v|_1^2 \leq C^2 h^{2(k+1)}[\sigma^0 h_e^0 + \sigma^2 h_e^{-2} + \cdots + \sigma^{2m} h_e^{-2m}]|v|_{k+1}^2 \leq C^2 c h^{2(k+1-m)}[h_e^{2m} + h_e^{2m-2} + \cdots + 1]|v|_{k+1}^2$ (where $c = \max(\sigma^0, \ldots, \sigma^{2m})$). Given $K > 0$ we can always find $\epsilon > 0$ such that the term in square brackets $< 1 + K$ provided $h_e < \epsilon$.

12.6. $\mathcal{D}\hat{v}(a) = \sum_i \frac{\partial \hat{v}}{\partial \hat{x}_i} a_i = \sum_{i,j} \frac{\partial v}{\partial x_j}\frac{\partial x_j}{\partial \hat{x}_i} a_i = \sum_{i,j} \frac{\partial v}{\partial x_j} T_{ji} a_i = \mathcal{D}v(\boldsymbol{Ta})$. Proceed in the same way for higher derivatives. Then for $k = 2$, for example, $|D^\alpha \hat{v}(\hat{\boldsymbol{x}})| \leq \|\mathcal{D}^2 \hat{v}\| = \sup|\mathcal{D}^2\hat{v}(\boldsymbol{a}, \boldsymbol{b})|$ $(\|\boldsymbol{a}\| \leq 1, \|\boldsymbol{b}\| \leq 1)$ $= \sup|\mathcal{D}^2 v(\boldsymbol{Ta}, \boldsymbol{Tb})| = \sup\left|\mathcal{D}^2 v\left(\frac{\boldsymbol{Ta}}{\|\boldsymbol{T}\|}\frac{\boldsymbol{Tb}}{\|\boldsymbol{T}\|}\right)\right| \cdot \|\boldsymbol{T}\|^2$. Use $\|\boldsymbol{Ta}\| \leq \|\boldsymbol{T}\|\,\|\boldsymbol{a}\| \leq \|\boldsymbol{T}\|$.

12.7. Any $v \in X_h$ also belongs to $L^2(\Omega)$, so it is required to find $v_i \in L^2(\Omega)$ such that $\int_\Omega v_i \phi\, dx = -\int_\Omega v\partial/\partial x_i\, dx\ \forall\phi \in \mathcal{D}(\Omega)$. Use Green's theorem applied to the function $\partial w/\partial x_i$, where $w = v|_{\Omega_e}$; then sum over all elements to get $\int_\Omega v_i\phi\, dx = -\int_\Omega u\partial\phi/\partial x_i\, dx = \sum_{e=1}^E \int_{\partial\Omega e} w\phi\nu_i\, ds$; the boundary integrals vanish.

12.8. $a(w, e) = \|e\|_{L^2}^2$, where $e = u - u_h$. Also, $a(\tilde{w}_h, e) = 0$; so $a(w - \tilde{w}_h, e) = \|e\|_{L^2}^2$. Hence $\|e\|_{L^2}^2 \leq K\|w - \tilde{w}_h\|_{1,\Omega}\|e\|_{1,\Omega} \leq KCh^\mu\|w\|_{p,\Omega}h^\beta\|u\|_{r,\Omega}$ for $w \in H^p(\Omega)$, $u \in H^r(\Omega)$, where $\mu = \min(k, p-1)$ and $\beta = \min(k, r-1)$. Since $Aw = e$, we have $w \in H^2(\Omega)$ and $\|w\|_{2m,\Omega} \leq c\|e\|_{L^2}$; so $\|e\|_{L^2} \leq C_1 h^\nu \|u\|_{r,\Omega}$.

12.9. $\|u - u_h\|_{L^2} \leq C_1 h^\nu \|u\|_{r,\Omega}$, where $\nu = \min(2, r)$ for linear or bilinear elements.

12.10. Using the Hermite basis functions and making appropriate changes (e.g., replace $C(\hat{\Omega})$ by $C^1(\hat{\Omega})$), the estimate (12.24) remains valid. The VBVP is: find $u \in H_0^2(0,1)$ such that $\int_0^1 (u''v'' + k(x)uv)\, dx = \int_0^1 fv\, dx$ for all $v \in H_0^2(0,1)$. We obtain an error estimate from $\|u - u_h\|_{2,\Omega} \leq K\|u - \tilde{u}_h\|_{2,\Omega} = K\left(\sum_e \|u - \tilde{u}_h\|_{2,\Omega_e}^2\right)^{1/2} \leq Kh_e^{k-1}$

$(\sum_e |u|^2_{k+1,\Omega_e})^{1/2} = Kh^{k-1}|u|_{k+1,\Omega}$, provided that $P_k(\hat{\Omega}) \subset \hat{X} \subset H^2(\hat{\Omega})$ and $H^{k+1}(\hat{\Omega}) \subset C^1(\hat{\Omega})$ and $u \in H^{k+1}(\Omega)$.

12.12. This follows as in Theorem 9. In particular $\boldsymbol{k}\nabla v \cdot \boldsymbol{v} \geq k_0|\nabla v|^2$ so that $a_h^e(v_h, v_h) \geq k_0 \sum_{\ell=1}^3 w_l |\nabla v_h(\xi_\ell)|^2 = k_0|v_h|^2_{1,\Omega_e}$, since $\nabla v_h \in [P_1(\Omega_e)]^2$ and a rule of order three is exact for quadratic functions.

# Index

9 780387 983073

GPSR Compliance

*The European Union's (EU) General Product Safety Regulation (GPSR) is a set of rules that requires consumer products to be safe and our obligations to ensure this.*

*If you have any concerns about our products, you can contact us on ProductSafety@springernature.com*

In case Publisher is established outside the EU, the EU authorized representative is:

Springer Nature Customer Service Center GmbH
Europaplatz 3
69115 Heidelberg, Germany

**Batch number: 10682207**

Printed by Printforce, the Netherlands